THE VHF/UHF DX BOOK

DIR PUBLISHING LIMITED
BUCKINGHAM
ENGLAND

THE VHF/UHF DX BOOK

FIRST EDITION

PUBLISHED BY DIR PUBLISHING LTD
PO BOX 771, Buckingham MK18 4HH, England

TYPESET BY DIR PUBLISHING LTD

TEXT PRINTED BY THE BATH PRESS
Bath, Avon BA2 3BL

COVER PRINTED BY CLARK COLOUR
Long Marston, Hertfordshire HP23 4QR

IMAGE SETTING BY CHILTERN PRINT
Toddington, Bedfordshire LU5 6ED
& AITCH EM
Aylesbury, Buckinghamshire HP19 3EY

© 1992 DIR PUBLISHING LTD

All rights reserved. Apart from any fair dealing for the purpose of private study, research, criticism or review, as permitted under the Copyright, Designs and Patents Act 1988, no part of this publication may be reproduced, stored in a retrieval system or transmitted in any form or by any means electronic, electrical, chemical, mechanical, optical, photocopying, recording or otherwise, without prior written permission. All enquiries should be addressed to the Publisher.

DIR Publishing Ltd retains sole and exclusive right to produce, or to authorize others to produce, printed-circuit boards or kits for the transverter designs published in this book. Amateur constructors are authorized to build the equipment designs in this book for their own use only.

Whilst every care has been taken in the production of this book, the Publishers cannot be held legally responsible for accidental errors or consequences arising therefrom.

ISBN 0 9520468 0 6

EDITOR
Ian White G3SEK

AUTHORS
Roger Blackwell G4PMK
David Butler G4ASR
Geoff Grayer G3NAQ
Günter Hoch DL6WU
Sam Jewell G4DDK
John Nelson GW4FRX
Dave Powis G4HUP
Dave Robinson G4FRE/WG3I
Ian White G3SEK
John Wilkinson G4HGT

CONTENTS

CHAPTER 1
Introduction
CHAPTER 2
VHF/UHF Propagation
CHAPTER 3
Operating
CHAPTER 4
Assembling your Station
CHAPTER 5
Receivers and Local Oscillators
CHAPTER 6
Transmitters, Power Amplifiers and EMC
CHAPTER 7
Beam Antennas and Feedlines
CHAPTER 8
144MHz
The *Suffolk* 144MHz Transverter	8-2
The W1SL 144MHz Power Amplifier	8-31
Antennas for 144MHz DX	8-37

CHAPTER 9
50MHz & 70MHz
A High-performance 50MHz Transverter	9-2
The *Cray* High-Performance 70MHz Transverter	9-20
Solid-State Power Amplifiers	9-34
Antennas for 50MHz and 70MHz	9-39

CHAPTER 10
432 MHz
A DXer's Transverter for 432MHz	10-2
A Low-noise 432MHz GaAsFET Preamplifier	10-25
The K2RIW 432MHz Power Amplifier	10-27
Antennas for 432MHz DX	10-34

CHAPTER 11
Power Supplies and Control Units
CHAPTER 12
Test Equipment and Station Accessories

CHAPTER 1
INTRODUCTION

INTRODUCTION

by Ian White G3SEK

Contacting stations in far-away places – 'working DX' – is one of amateur radio's greatest challenges, especially if you choose to do it on the VHF and UHF bands.

When the VHF and UHF bands open up for DX, they can produce some truly exotic signals. If you start to get interested, you could soon be hearing the throaty sound of signals reflected back from an aurora, the stunning strength of sporadic-E signals, the startling meteor-bursts of SSB or high-speed Morse... And one day – if you really go for it – you could be hearing your own signals echoing back from the moon.

If you are a newcomer who seldom works as far as the next county, a 200-kilometre contact is genuine DX. But if you're keen you will grow in confidence and competence until you're nonchalantly chatting with Continental stations every evening – and keeping a keen lookout for "some *real* DX"!

Our expectations of the DX potential of the VHF and UHF bands have increased vastly over the past decade. These bands offer many different modes of propagation, and activity has increased so much that almost every DX opening is recognised and exploited. From the UK, amateurs have worked all over western and central Europe on 144MHz and 432MHz – but our horizons don't end there. 50MHz can cover the world at the height of the sunspot cycle and moonbounce has brought the Worked All Continents award within our reach on the higher bands too.

The DX Century Club award on 50MHz – 100 *countries* confirmed – has already been achieved by several UK stations in only a few years, and even by US moonbouncer W5UN on 144MHz. DXCC will surely be achieved on other VHF/UHF bands within this book's lifetime.

It has become fashionable to claim that the advent of ready-made black boxes signalled the end of "real amateur radio", and that radio amateurs have been left behind by rapidly advancing technology. The critics might be right if we merely used our synthesized, computer-controlled transceivers to make easy contacts which the old-timers could have managed with simple home-built equipment. But they're certainly wrong about VHF/UHF DXers. We have used commercial equipment and modern technology as a springboard to do *more* than was ever possible before.

We are not making these vast strides by copying professional practice. Radio amateurs were among the first to appreciate the long-distance potential of the frequencies above 30MHz, and although we now share the VHF and UHF spectrum with many terrestrial and satellite services which span the world, amateur radio still goes its own way. We amateurs are the only people who use VHF and UHF routinely for long-distance communication between skilled operators, so there are no ready-made answers to our special operating and technical requirements. We have to work things out for ourselves.

VHF/UHF DX is one of the growing-points where amateur radio shows that it still has a real future – and that's why we wrote this book.

INTRODUCTION

CHASING THE DX

DXing is primarily an *operating* activity. The objective is to get on the air and contact other people, rather than building equipment for its own sake – or just sitting and thinking about it!

Chapter 2 is about **Propagation**, which is at the heart of all DX-chasing. Unless you understand the different ways by which VHF and UHF signals can travel far beyond the horizon, you run the risk of missing the best DX. In amateur radio there are a lot of myths and misconceptions about propagation so Geoff Grayer G3NAQ has taken a fresh look at the subject. Starting from basic physical principles and using up-to-date scientific information, he explains how the various VHF/UHF DX propagation modes occur and how to make the best use of them.

VHF/UHF DXers often use short-lived and 'unreliable' propagation modes such as sporadic-E, aurora, meteor scatter and moonbounce. Many of these long-distance modes are regarded by professional radio users as merely an occasional nuisance to everyday communication over shorter ranges, so they still have not been fully explored. By exploiting these modes to work DX – and by doing it as often as possible – amateurs can still contribute to the development of radio science.

Chapter 3 is about **Operating** – the actual process of using your station. Even if you could predict every single DX opening, you won't work much unless you're also a good operator. David Butler G4ASR explains that operating technique is something which can be developed and practised, just like any other skill. You always need to match your operating to the needs of the moment, and the propagation modes which produce only weak or fleeting signals require some specialized techniques, which Chapter 3 explains.

Many amateurs enjoy going on the air and talking to people, happy to work whatever DX comes their way. But some are not content with that – they truly earn their reputation as DX-*chasers*. Both types of people are DXers in some degree and no firm line divides them.

Indeed, the same person may be a dedicated DX-chaser on one band whilst remaining a casual DXer on all the others.

Many DXers are working towards specific goals in terms of contacts with particular geographical locations. Contacting a large number of different 2°x1° IARU Locator 'squares' is a very common long-term objective, especially for people who like an open-ended challenge. Others prefer objectives with a definite end-point, *eg* contacting every UK county. The more popular operating goals earn certificates and awards, such as the RSGB Squares and UK Counties awards and the various awards for working more than 100 squares.

Contests are another popular operating activity with VHF/UHF DXers, for a variety of reasons. Some people view contests as an end in themselves and on contest weekends they set up large, well-equipped portable stations on good sites and in rare locator squares. Many home-station operators take the chance to contact these big stations as an easy way of working DX, while others are content just to "give a few points away". From the viewpoint of this book, contests are a practical way of measuring your own operating performance and the DX potential of your station.

THE TECHNICAL CHALLENGES

The aim of this book is to bring together the operating and the technical sides of VHF/UHF DX. Each needs the other – a station assembled with no technical knowledge can never be operated at full efficiency; and a technically excellent station is sterile if it's never used.

You don't have to be put off by the word "technical". And you don't need to worry if you can't understand everything in one gulp – especially if you're starting from scratch, or fresh from the Radio Amateur's Examination. There's no urgency. It may take years to accumulate the practical experience to match the theoretical knowledge you can gain from books, and the two kinds of understanding grow best together. If you want to work more VHF/UHF DX, sooner or later you will naturally find yourself drawn into its technical and

INTRODUCTION

constructional side. And there's no better motivation for a DX-chaser than the thought of missing some DX!

DESIGNING FOR DX

If you want your station to work at its full potential, you have to understand *how* it works. It's very tempting to jump straight into the details of circuitry and hardware, completely overlooking the fact that all the bits and pieces eventually have to work together as a *system*.

System design is what transforms a collection of boxes, cables and antennas into an effective station. Chapter 4 explains the general concepts and encourages you to develop the capabilities of your station to make difficult DX contacts using weak signals. Chapter 5 expands these ideas in the area of receiver design, and explains the techniques of noise and intermodulation analysis which are used to design the transverters in Chapters 8, 9 and 10.

Chapter 6 is about **Transmitters, Power Amplifiers and EMC**. John Nelson GW4FRX reminds us that a poor-quality signal is a problem for everybody else – and he pulls no punches about the poor quality of many present-day transmissions. Well known for his own impeccable signal, John clearly explains the causes of inferior signals and how to remedy them. GW4FRX builds quality into every aspect of his power amplifiers, and Chapter 6 contains a wealth of sound advice on amplifier design and construction. He then explains how to commission and operate a newly-built amplifier without hazards or heart attacks. Finally he tackles the increasingly important question of how to run high power whilst living amicably with your neighbours.

Until recently the highly successful DL6WU family of long Yagis was virtually unknown in the UK – so we're pleased to welcome Günter Hoch DL6WU as the contributor of Chapter 7 on **Antennas**. He begins with a general survey of VHF/UHF antennas, which clearly shows why the Yagi has become a firm favourite. The chapter continues with advice on choosing, designing and installing Yagis of all sizes, and on the equally important topics of matching the antenna to the feedline and making it work properly. There is also a comparison of computer-predicted performance of a wide variety of commercial antennas for 144MHz and 432MHz, contributed by Rainer Bertelsmeier DJ9BV, the Technical Editor of *DUBUS* magazine.

Chapters 4–7 have thus explained how your station ought to perform, and the basic principles of RF equipment design. Chapters 8–12 then show how it's done in practice.

EQUIPMENT DESIGNS

Chapters 8, 9 and 10 present modern equipment designs for working DX on 50MHz, 70MHz, 144MHz and 432MHz. Every item is real, tried and tested.

No previously published transverter designs met our performance standards for modern-day DX operating, so here are four all-new designs. The transverters were developed by Dave Powis G4HUP (50MHz), Dave Robinson G4FRE/WG3I (70MHz) and John Wilkinson G4HGT (432MHz). Sam Jewell G4DDK developed the 144MHz transverter and co-ordinated the whole transverter project.

Because these new transverters follow the system-design principles of Chapters 4 and 5, they are much more 'future-proof' than normal. Although the transverters are already close to optimum for today's operating conditions, the information in Chapters 4 and 5 will help you to analyse the effects of any changes you may wish to make in the future. Each transverter is divided into modules, so you can easily make changes if new devices become available to do the same jobs better. These transverter designs will help to bring your station close to the leading edge – and keep it there.

Chapters 8, 9 and 10 contain all the information you need to build your own transverters and get them working. But that isn't all: each designer has also explained in detail how he developed, constructed and commissioned his own transverter. All four transverters are based on the system design principles outlined in Chapters 4 and 5 so they show a distinct

INTRODUCTION

family relationship. Yet the circuitry, construction and alignment details reveal four people's quite differing approaches to the task of designing a transverter and making it work. By comparing and contrasting their individual accounts, you can learn how design decisions are made – along with a great deal more about the art of practical RF electronics.

As well as the new transverters, Chapters 8, 9 and 10 contain designs for power amplifiers and antennas for the respective bands. Some of these are well known in other countries but have never been published before in the UK. We have also included updated versions of classic 144MHz and 432MHz power amplifiers published many years ago. GW4FRX returns in Chapter 11 to deal with **Power Supplies and Station Control** and gives practical designs and advice on integrating your equipment into a safe and functional transmitting station.

Chapter 12 is devoted to **Test Equipment**, because measurements go hand-in-hand with design to keep your station at peak performance. Roger Blackwell G4PMK takes delight in building equipment which almost equals the performance of professional testgear, at a fraction of the cost. Chapter 12 is his selection of the most useful testgear and accessories for the VHF/UHF DX station.

HOW IT STARTED

I started this project because no existing amateur-radio book seemed to capture the reality of VHF/UHF DX, with its unique blend of advanced technology and personal skill. Even the more specialized VHF/UHF handbooks are too general; they contain enough information to start an interest in DX, but not enough to satisfy it.

In recruiting the other authors, my aim has been to build up a complete picture of VHF/UHF DXing through contributions from specialists in particular areas. Although we've tried to 'write the book' on VHF/UHF DX, what you'll read are not bland accounts in the detached style of a textbook; they are often personal views on controversial topics. You may not always agree with what we say – and we don't even agree completely amongst ourselves – but that's entirely appropriate because VHF/UHF DX is a developing subject area.

Although we don't claim to have covered the whole of VHF/UHF, the information in this book has lots of applications to related areas. In particular we hope that amateur-satellite users will find a great deal of interest. And even local communications such as simplex FM, repeaters and packet radio can benefit from the RF engineering design principles described here.

OVER TO YOU

This book offers you a flying start into VHF/UHF DXing. There's still a lot to learn and a new generation of ideas and techniques to be developed. So now it's over to you...

Good DX!

CHAPTER 2

VHF/UHF PROPAGATION

VHF/UHF PROPAGATION

by Geoff Grayer G3NAQ

Understanding propagation is vital for every VHF/UHF DXer because the VHF and UHF bands display such a wide variety of propagation modes, each of which needs to be understood in order use it to the best advantage. Propagation is also one of the most absorbing interests in amateur radio. Like the British weather, it is always changing and very occasionally it's marvellous! But unlike our weather, long-distance propagation in the VHF/UHF range has only been observed and studied for about fifty years, so we amateurs are still in a position to contribute to observation and understanding.

It was Heinrich Hertz (1857-1895) who first demonstrated that radio waves exist, and that the metre-wavelength radio waves he generated travel in straight lines like light; and even after the discovery of long-distance MF and HF propagation, it was assumed for a long time that VHF propagation was restricted to line-of-sight paths. Fortunately, that was not the whole story...

In this chapter I will explain VHF and UHF propagation using up-to-date ideas and information. Sometimes these may conflict with what you have previously been told, or may read in other publications. To help you do your own thinking and judge for yourself, this chapter begins with the fundamentals. The basic concepts of radio-wave propagation and the various environments which radio waves may encounter are then used to describe in detail the various modes of practical VHF/UHF propagation.

Various specialized terms will be introduced in *italics* in the course of the chapter.

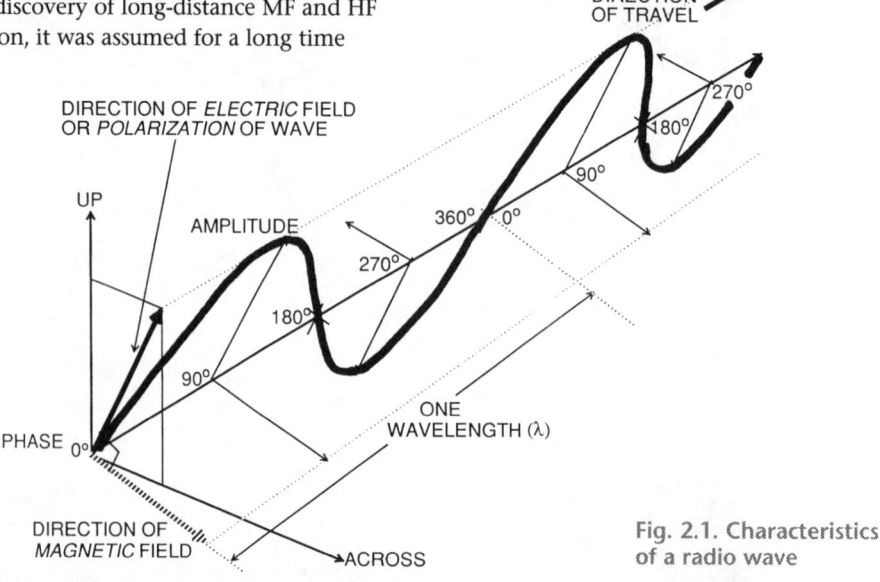

Fig. 2.1. Characteristics of a radio wave

PROPERTIES OF RADIO WAVES

THE ELECTROMAGNETIC SPECTRUM

Radio waves are a class of *electromagnetic (EM) radiation*, along with infra-red radiation, visible light, ultra-violet radiation, X-rays and gamma rays. These various types of EM radiation are distinguished by their wavelength and frequency, which span an enormous range – very-low-frequency (VLF) radio waves are more than 10^{19} times longer in wavelength than gamma rays. They may seem very different because their interactions with matter are so dissimilar, but the propagation of all electromagnetic waves through space can be reduced to two mathematical equations derived by James Clerk Maxwell in the 1870s.

The essence of Maxwell's ideas is as follows. The pioneer investigators of electricity and magnetism found that an electric current passing through a wire affects a magnetic compass needle, so it must be producing a magnetic field. Similarly, the changing magnetic field from a moving magnet makes a current flow through a coil of wire; hence it must be generating a voltage difference. So if an electric charge is made to oscillate back and forth, either along a wire or freely in space, oscillating electric and magnetic fields are **both** produced. These produce magnetic and electric fields respectively, which in turn produce electric and magnetic fields... And so the electromagnetic wave moves off through space – it *propagates*.

THE ANALOGY WITH LIGHT

To illustrate the behaviour of radio waves I will sometimes give examples using light waves, where we can literally see what is happening.

Most large-scale behaviour of radio waves is exactly the same as light except for the difference in scale. VHF radio waves are defined as having wavelengths between 10m and 1m (30–300MHz), and in European amateur parlance UHF is defined as 1m to 30cm (300MHz to 1GHz). In contrast the wavelength of visible light extends from about 350 to 650 nanometres (1nm = 10^{-9}m), several million times shorter. If the earth was reduced in size by the same ratio, it would become a sphere of only a few metres diameter.

Radio waves and light waves also share a property which is unique across the entire EM spectrum: the atmosphere is almost completely transparent to them. That is why both radio and light waves are used for communication.

BASIC PROPERTIES OF ELECTROMAGNETIC WAVES

Below and in Fig. 2.1 opposite we summarize the basic properties of any electromagnetic radiation:

TWO COMPONENTS

An electromagnetic wave is formed from electric and magnetic components. The electric and magnetic field directions (or vectors) are both at right angles to the direction of propagation, and also at right angles to each other.

AMPLITUDE

This is the maximum field-strength attained

VHF/UHF PROPAGATION

during a complete oscillation cycle. Although we could refer to the strength of either the magnetic field (H) or the electric field (E), we usually mean the latter. In space they are related by the formula $E = HZ_0$ which is analogous to the equation $V = IZ$ for electrical circuits. (Remember that an electric field is the result of a potential difference, and a magnetic field results from a flow of electric current). Z_0 is known as the *characteristic impedance of free space*.

PHASE

The phase gives the position of the wave in its oscillation cycle as an angle round a circle. Its value can therefore lie between 0 and 360 degrees (0 to 2π radians). The phase of a wave is changing continuously at any point, and can therefore only be defined relative to itself at another point, or relative to the phase of another wave at the same point.

The electric and magnetic fields of an EM wave are in phase with each other, so Z_0 is a *resistive* impedance whose value is 377Ω.

POLARIZATION

Linearly-polarized waves are defined according to the orientation of the plane containing the *electric* field. Depending how the wave is launched or later modified, this plane may be horizontal, vertical, or any direction in between; in Fig. 2.1 the wave is drawn slant-polarized. By adding waves of different amplitude, polarization and relative phase, circular and elliptical polarizations can be produced. For these, the polarization direction rotates as the wave propagates. Rotation may be right- or left-handed, defined as clockwise or anti-clockwise as viewed along the direction of travel. Note that any receiving antenna will respond only to the component of polarization it has been designed to accept; for example, a horizontally-polarized antenna will only collect the power from the horizontal component of circular, elliptical, or slant-polarized waves.

POWER

The power (W) carried by an electromagnetic wave is shared equally between the electric and magnetic fields and flows along the direction of propagation. Its magnitude is given by:

$$W = E H = \frac{E^2}{Z_0} \text{ watts per m}^2$$

VELOCITY, FREQUENCY, AND WAVELENGTH

All types of electromagnetic waves move with the same velocity in totally empty space, often called *free space*. This velocity is conventionally known as c, and its value is almost exactly 300,000km per second. To bring this down to a more familiar distance scale, EM waves travel about 300mm or one foot in one nanosecond in free space, and for practical purposes in air also.

The frequency f and wavelength λ (lambda) of any wave are related to the *phase velocity* v by:

$$v = \lambda f$$

In free space, v = c, but this is not the case when waves are propagating in the presence of matter. Later in this chapter we will define what is meant by the phase velocity.

INTERACTION OF RADIO WAVES WITH MATTER

We begin this section at the atomic level, where all the fundamental interactions between EM waves and matter take place. Then we move on to show how the same interactions on a larger scale produce directly observable propagation phenomena.

INTERACTIONS AT THE ATOMIC LEVEL

Radio waves normally travel in straight lines. So unless something makes them change direction, your signals will never reach a distant point on the earth's surface; they will shoot straight off into outer space. An electromagnetic wave can only change direction if it encounters some form of matter – atoms, molecules, ions or electrons. In most cases this interaction will change not only the direction of the wave but also its amplitude, phase and polarization. All these effects arise from scattering at the atomic level.

Fig. 2.2. The toroidal radiation pattern produced by an oscillating electron is identical to that of a short dipole antenna

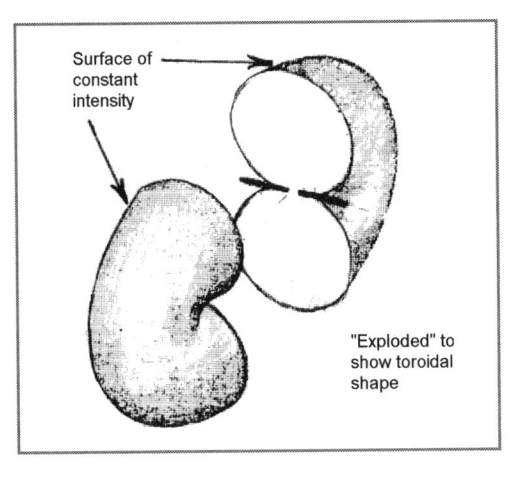

IONIZATION AND ELECTRON OSCILLATORS

All matter is made up of atoms, each consisting of a positively-charged nucleus surrounded by much lighter negatively-charged electrons. Atoms or molecules are normally electrically neutral but electrons can be separated, leaving a positively charged *ion* and one or more free electrons. This *ionization* normally occurs as a result of interactions with sufficiently energetic charged particles or with ultraviolet, X-ray or gamma radiation. Radio waves lack the energy to ionize an atom directly, and hence are classed as *non-ionizing radiation*.

Whether free or bound in atoms, it is the electrons that respond most readily to the field of an incident radio wave because of their smaller mass. The electrons follow the oscillations of the RF field, thus taking up some of its energy. Each electron then re-radiates an independent electromagnetic *wavelet* with a toroidal pattern which is familiar as the radiation pattern of a dipole antenna (Fig. 2.2). The reason why the radiation pattern of a subatomic particle is duplicated by an object so much greater in size is that the electric potential applied to the antenna wire forces many electrons to oscillate together in phase. The contributions from all their wavelets therefore add together to form the same dipole pattern. This is an example of the *coherent addition of waves* – our next topic.

COHERENCE

Coherence is an important concept in the propagation of waves. Waves are defined to be coherent if there is a constant relationship between their phases. There can still be a phase shift between them, in which case their amplitudes will not be completely additive, but if this phase shift remains the same, the waves are coherent.

There is a continuous range of coherence, from *completely coherent*, meaning that the phase relation between waves or wavelets is completely constant (as was the case with the radiating dipole) to *completely incoherent*, a situation in which the phase relationships are both random *and* varying. In this latter case, contributing wavelets add together in a statistical way, so that if *n* equal wavelets were contributing, the result would be $\sqrt{n}$ times a single wavelet, rather than n times as it would be if they were coherent. When the number of contributing wavelets becomes very large, the difference between '$\sqrt{n}$ times' and '*n* times' becomes highly significant, so the degree of coherence is a very important factor in determining signal strengths.

GROUP VELOCITY AND PHASE VELOCITY

Although we conventionally speak of the "velocity of light" as if it were a single quantity, two velocities are required to describe an electromagnetic wave propagating through matter – the *group velocity* and the *phase velocity* – and these are generally different.

If you try to push a heavy object it does not respond immediately, because of its mass or inertia and because of restraining forces such as friction. In the same way, their mass and the forces exerted by other charged particles will prevent the electron oscillators in the propagating medium from immediately following the electric field driving them. Their oscillations will have a phase difference relative to the incident wave, and so too will the wavelets that they re-radiate. To an outside observer, the wave appears to progress through the medium with the same frequency and amplitude, but with a phase that is continually being updated by the process of energy absorption and re-radiation. The effect is to change the velocity and hence the wavelength of the wave compared with free space.

A common misconception is that the velocity of an electromagnetic wave cannot possibly exceed its value in free space. As with most misconceptions, this is partly true. The power and information carried by an EM wave propagates at its *group velocity* which never exceeds the free-space value c. But if you could see the individual oscillations, you would say that the wave appears to propagate at a velocity measured by the distance the wavefront advances in a given time. Since a 'wavefront' is actually a surface joining points of equal phase at any given instant, this *apparent* velocity is called the *phase velocity*. During a single cycle, the wavefront advances by one wavelength; hence the phase velocity is calculated by multiplying the frequency (which remains constant) by the wavelength within the medium.

If the phase delay of the re-radiated wave in the propagating medium (compared with the phase of the incident wave) is between 0° and 180°, it is said to "lag" and the phase velocity appears to decrease. But if the phase delay is in the range 180° to 360°, this is equivalent to a phase **advance** of 180° to 0°, so the phase velocity appears to **increase**. Fig. 2.3 illustrates these effects. Fig. 2.3b shows a delay of $\lambda/6$ (60°) relative to the reference phase (Fig. 2.3a). Fig. 2.3c shows a larger delay of $5\lambda/6$ (300°), which is equivalent to a phase advance of 60°.

The phase velocity in most materials is less than c, but it exceeds c in metals, inside waveguides and – importantly for propagation – in the ionosphere. Only in empty space are the phase and group velocities exactly identical, both being equal to c.

This is a difficult concept, but the distinction between group and phase velocities is very important; failing to grasp it can lead to some bizarre misconceptions! Group velocity is the correct quantity to use when timing radar signals and moon echoes, when making vertical ionospheric soundings or when establishing how long it takes for signals to arrive by an unknown propagation mode; the reason is that all these imply the flow of information or power. But when considering wave properties such as refraction and reflec-

INTERACTION OF RADIO WAVES WITH MATTER

Fig. 2.3. A phase delay greater than 180° ($\lambda/2$) can be equivalent to a phase advance

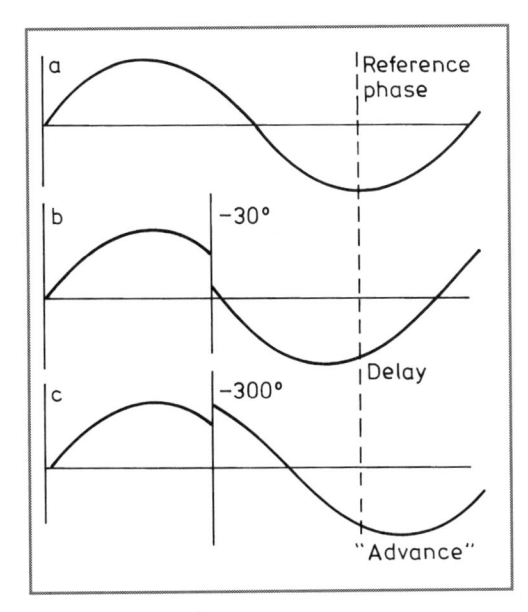

tion – our next topic – you need to think in terms of phase velocity.

LARGE-SCALE EFFECTS

So far we have only considered interactions of EM waves with the electron oscillators on the atomic scale. On the larger scale, these give rise to observable effects such as scattering, refraction, reflection, absorption or diffraction, or any combination of them depending on the object encountered. We will briefly distinguish these, then consider each in more depth.

Refraction, reflection and scattering are closely related. They each cause an EM wave to change direction when it encounters a change in the medium through which it is propagating, and which of these processes is the most important depends on the size, number, density and arrangement of the objects which the wave encounters. I will try to use the terms consistently as follows:

Refraction means a change of direction caused by a change in the velocity of propagation due to the changing properties of the medium. A distinct boundary between two different media will cause a sharp change in direction, while a more gradual change in the properties of the medium will produce more gentle bending of the rays.

Reflection refers to the process where some fraction of the wave is turned back when it encounters a change in velocity at the boundary with a different propagating medium. Only at a distinct boundary are the reflected waves in phase, as from a mirror; reflection from a gradual boundary is negligible. Both refraction and reflection involve the coherent addition of waves.

Scattering is a change of direction associated with objects smaller than or comparable with a wavelength, and generally involves random or partially incoherent phase relationships. An increase in the size or density of these objects is accompanied by an increase in the coherence of the scattered waves, and eventually by a transition from scattering to reflection and refraction.

Two other processes occur when EM waves encounter matter:

Absorption means a reduction of amplitude by conversion to other kinds of energy (usually heat).

Diffraction describes the change of direction of a wave going past a boundary with another medium, but not actually through it.

SCATTERING

When an EM wave encounters an object which is much smaller than a wavelength in size, each electron oscillator within the object will re-radiate a wavelet with its dipole pattern (Fig. 2.2). In situations where these dipoles are randomly oriented, the wave will be scattered *isotropically*, that is equally in all directions. For a larger object, comparable with a wavelength in size, significant phase differences will arise between the re-radiated wavelets from its various parts. Depending on the shape of the object, these will concentrate the resulting wave in directions where the wavelets tend to add in phase, while cancelling the scattered radiation in other directions where the wavelets tend to be out of phase. For larger objects, provided they are not opaque to EM radiation, all the sideways-going radiation tends to cancel, leaving only the forward-going wave. Such a large object can then be

VHF/UHF PROPAGATION

considered a medium of propagation in its own right, and the laws of refraction and reflection described below will apply.

Consider now a collection of small objects. A single small object of a size up to a few wavelengths does not produce a significant scattered signal; it takes many such objects to produce something detectable. If these small objects are spaced at distances large compared with a wavelength and randomly distributed (as naturally occurring objects often will be), their contributions at a distance will be random and the net signal weak. They are said to be *underdense*. However, if the objects are close together compared with a wavelength, their scattered wavelets will reproduce the wavefront of the original wave. They are then behaving like one large object and in this case they are said to be *overdense*. If this overdense collection of objects is extensive enough, it may also be considered as a propagation medium, in which the laws of refraction and reflection will apply. Of course, it is hardly surprising that we have reached the same conclusion as in the previous paragraph, because a large object is merely a very dense collection of small ones, *ie* atoms.

Let us return to the underdense situation. As long as the scattering objects are randomly distributed, their wavelets add statistically and the result is weak (page 2-6). But as soon as some order is introduced, such as arranging the objects in a line, the phase relationship between the scattered waves is no longer random and they will tend to add in certain directions. This is the case with scattering from meteor trails, the aurora and field-aligned ionization.

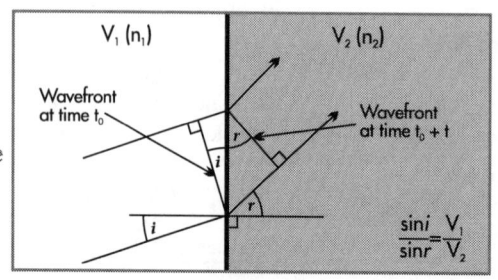

COHERENT AND INCOHERENT SCATTERING

Signals scattered from randomly-distributed objects can be either coherent or incoherent, *ie* with or without a constant phase relationship, as the following two examples demonstrate. A mountain is a random array of scattering surfaces, yet it scatters radio signals coherently because the surfaces are fixed in relation to one another. Although the resulting signal is weak, its strength and phase remain constant and the signal sounds 'normal'.

In contrast, scattering centres in the troposphere or the ionosphere are in random motion relative to one another, so the resulting signal is the sum of incoherent waves. Apart from being weaker, this sum is also continually shifting in phase. Since random phase modulation is equivalent to random frequency modulation, the signal has a superimposed frequency spread. This 'fuzzy' characteristic is typical of incoherently scattered signals received via troposcatter, ionospheric scatter or aurora.

In fact, both the troposphere and the ionosphere can demonstrate the whole range of variation from coherent to incoherent scattering.

REFRACTION

If a wave encounters a boundary where the phase velocity changes, it will undergo a change in direction unless it is perpendicular to the boundary. This is *refraction*, as illustrated in Fig. 2.4.

A measure of the bending power of a medium is called the *refractive index* (usually denoted by n). The refractive index is simply equal to c/v, where c is the velocity in free space and v is the phase velocity in the medium. From the positions of the wavefront just before and just after refraction in Fig. 2.4, a relationship can be derived between the angle of incidence i at one side of the interface, the angle of refraction r at the other side, and the refractive index n:

$$\frac{\sin i}{\sin r} = \frac{v_1}{v_2} = \frac{n_2}{n_1}$$

Fig. 2.4. Refraction at an interface between media with different phase velocities (increasing from left to right)

INTERACTION OF RADIO WAVES WITH MATTER

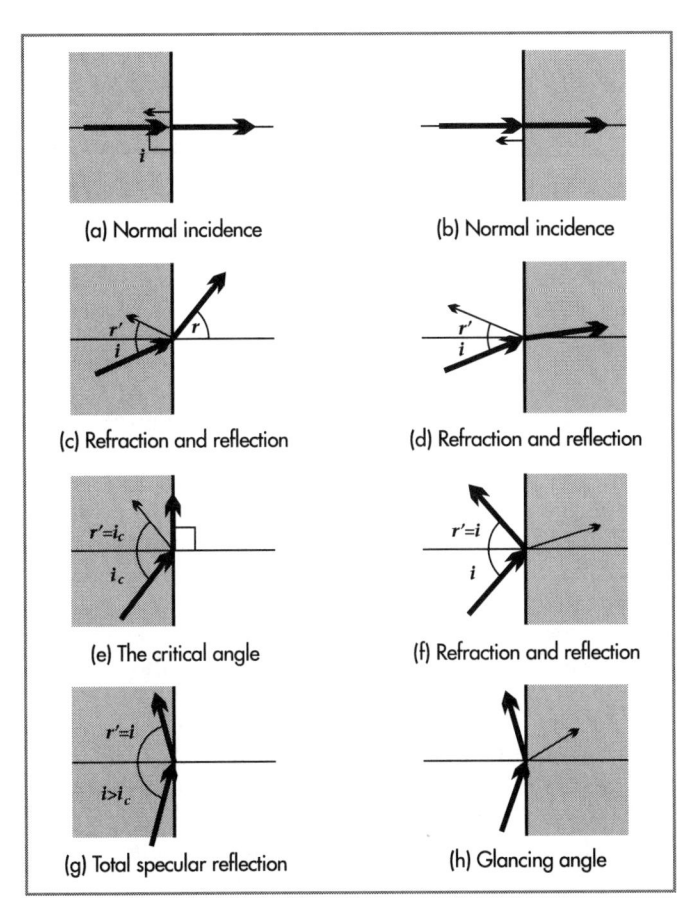

Fig. 2.5. Refraction and reflection at an interface. Left-hand column: wave speeds up. Right-hand column: wave slows down

Fig. 2.6. A radio wave entering a region of increasing ionization density (decreasing refractive index) can be turned back as if it had been reflected. The broken lines join at the virtual reflection point V

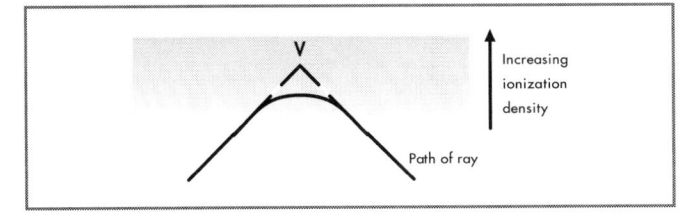

This is known as Snell's or Descartes' Law of refraction. Our experience with light suggests that when it goes from air into any other medium, *eg* water or glass, the wave is always bent towards the perpendicular, corresponding to a reduction in phase velocity. But that is not always the case with radio waves – the phase velocity ***increases*** when entering the ionosphere and thus the wave is bent ***away*** from the perpendicular.

Fig. 2.5 illustrates a variety of situations involving refraction and reflection at a sharp boundary between two media. In the examples in the left-hand column, the phase velocity increases across the boundary. The right-hand column shows the corresponding situation with a decrease of phase velocity. Figs 2.5a and b show the simplest case of perpendicular or 'normal' incidence, in which the wave merely slows down or speeds up but continues in the same direction. Figs 2.5c and d show refraction, away from the perpendicular if the wave slows down (2.5c) and towards it if the wave speeds up (2.5d).

Up to now we have considered only a sharp boundary between two different media. This is usually the case in everyday examples involving the refraction of light, but radio waves more commonly encounter gradual changes in the refractive index of the ionosphere or the atmosphere, taking place over distances which are large compared with a wavelength. As the wave penetrates into a region of increasing phase velocity it is bent further and further away from the perpendicular, and if the change in phase velocity is sufficient for the wavelength concerned, the ray is turned back and re-emerges (Fig. 2.6). Because the boundary of the ionosphere is gradual rather than sudden, a so-called "reflected" signal is in fact returned by this process of continuous refraction.

REFLECTION

Any sharp change of refractive index will produce some reflection as well as refraction. As well as the main refracted rays, Fig. 2.5 shows small reflected rays coming back out of the surface. The amount reflected depends on

VHF/UHF PROPAGATION

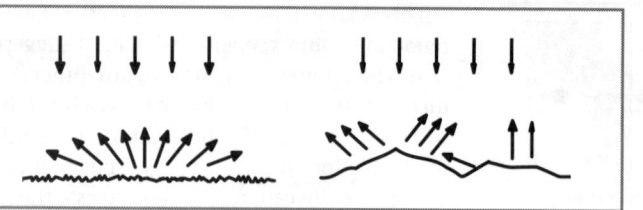

Fig. 2.7. Small-scale roughness gives a 'matt' effect, scattering almost equally in all directions. Larger-scale irregularities act as reflecting surfaces and tend to 'glint' at particular angles

the difference between the two refractive indices, and also on the angle at which the ray strikes the surface. As the angle between the incident wave and the surface decreases, the reflected power increases. To demonstrate this, take a sheet of glass and look at the reflections in it while varying the angle. The reflections will improve as the angle moves away from the perpendicular, and it will become difficult to see through the glass at very shallow angles. This is the situation of Figs 2.5f and h; yet the sheet of glass never reflects as well as a proper mirror.

Complete mirror-like reflection can only occur from incidence on to a medium with a lower refractive index. Compared with air or free space, the phase velocity of light and radio waves is markedly greater in metals and in the ionosphere, which is why both can be good reflectors. As the wave speeds up on entering the second medium, the ray is refracted away from the perpendicular and towards the interface. At a certain angle of incidence it will actually travel along the interface itself (Fig. 2.5e), and this angle of incidence is called the *critical angle*. At shallower angles, reflection is *total* (Fig. 2.5g).

Reflection is not always perfect, of course. In order to be effective, reflection requires a sharp boundary as shown in Fig. 2.5. In contrast, refraction can take place even if the change between two media is gradual. A rough or irregular surface will produce scattering rather than reflection, in a manner which depends on the size of the irregularities compared with the wavelength. Fig. 2.7 shows that a surface with small-scale irregularities will back-scatter in all directions and appears 'matt'. If the irregularities are much larger, the surface will 'glint' back at certain preferred angles determined by its geometry.

THE ROLE OF POLARIZATION

The relative power of the reflected and refracted waves depends also on the polarization of the incident wave. This is a consequence of the dipole radiation pattern of the scattering electrons, which has nulls along the direction of oscillation (Fig. 2.2). If a wave has its plane of polarization (which is the same as its electric-field direction) pointing into a surface dividing two media with different refractive indices, cancellation of the reflected signal occurs when the reflected ray would be exactly perpendicular to the refracted wave. The angle of incidence at which this occurs is known as the *Brewster angle*, and its value depends on the ratio of the two refractive indices. No such cancellation occurs at any angle if the wave is polarized parallel to the boundary surface. If the incident wave is polarized at some oblique angle with the surface, some intermediate degree of cancellation will take place.

Practically all VHF/UHF DX stations use horizontal polarization. That is not an accident or mere convention; horizontally-polarized signals have been shown to suffer less from fading and other undesirable effects. This is because the earth, the atmosphere and the ionosphere are all predominantly horizontal surfaces. Horizontally-polarized waves are thus reflected the best, with no Brewster-angle cancellation. Most ground-based obstacles are vertical in aspect (trees, telephone poles, lamp-posts, chimneys, electricity pylons, etc), and therefore cause less scattering if the waves are horizontally polarized.

ABSORPTION

Absorption occurs only when radio waves encounter matter; there is no absorption in empty space. As explained earlier, waves are propagated through matter by the electrons

INTERACTION OF RADIO WAVES WITH MATTER

oscillating and re-radiating the radio wave. If some of the energy temporarily held by the oscillating electrons is transferred to other particles before it is re-radiated, the propagating wave will be attenuated since some of its energy is absorbed. Usually this energy is converted into a rise in temperature of the absorbing material.

The attenuation in insulating materials is generally small – the electrons bound in atoms are little affected by the passing wave. Air is a good insulator, so absorption in the lower (non-ionized) atmosphere is very small at all frequencies. At 1GHz the loss in traversing the 'standard' atmosphere (page 2-13) by the shortest path – directly upwards – is less than 0·03dB, increasing to 0·23dB at the horizon where the path through the atmosphere is maximum. Even for very weak moonbounce signals, these absorption losses are negligible.

In the upper ionosphere (the F layer) the probability of electron collisions is low, so absorption is again low at all frequencies. Lower down, the atmosphere becomes more dense and electron collisions are more frequent. If the D layer becomes highly ionized, for example by an intense solar flare, it also becomes highly absorptive and prevents HF radio waves from reaching the E or F layers above. Ionospheric absorption is much lower at VHF and above because the probability of energy loss by collision decreases as the frequency increases.

In poor electrical conductors the electrons can move more freely than in insulators, but some of the wave energy will be dissipated as heat by the electrical resistance. In this case, absorption increases rapidly with frequency. Most solid natural materials contain water, which makes them lossy conductors – anyone unfortunate enough to have an antenna in the loft will know how signals are reduced when the roof is wet. Foliage and other green vegetation are also made up largely of water. Wood, although a fairly good insulator in a dry state, is a lossy conductor when filled with sap. The absorption of a 15m thickness of various types of vegetation may be estimated

Fig. 2.8. Approximate absorption due to 15m of foliage of various kinds

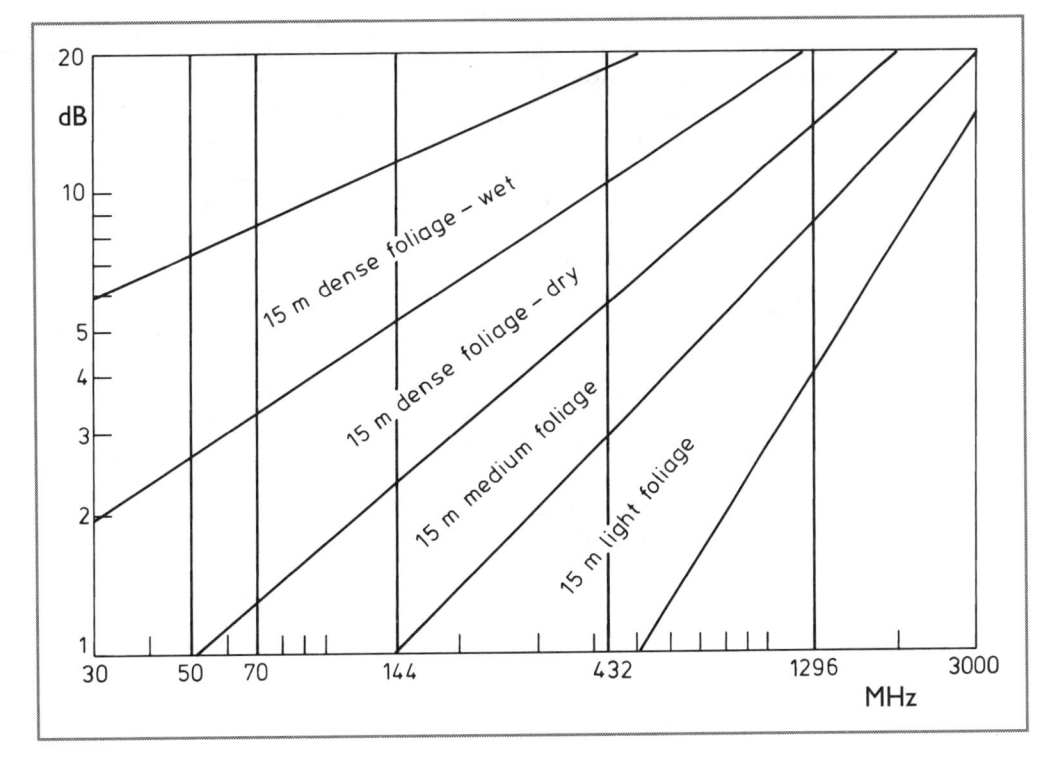

Fig. 2.9. Diffraction, showing the variation in signal strength in the first few wavelengths above and below the edge of the obstacle (not to scale)

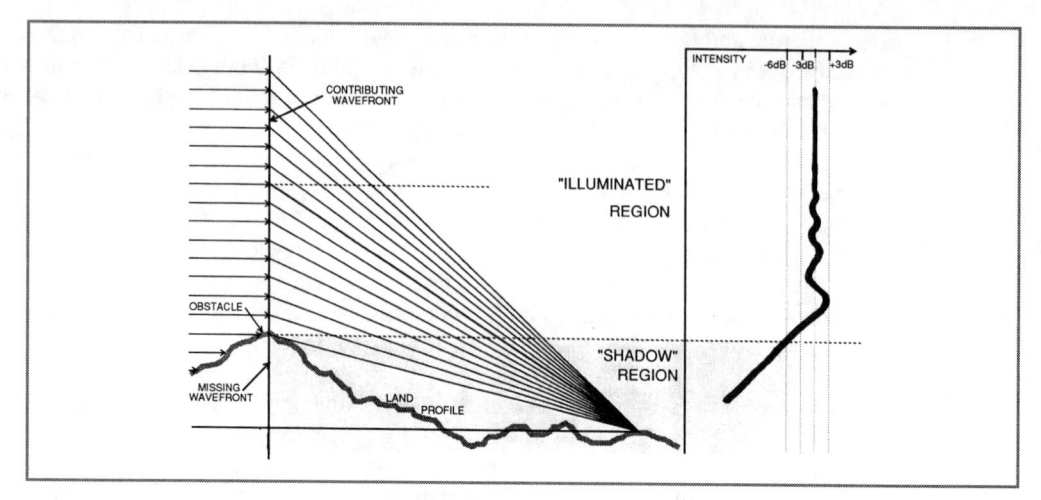

roughly from Fig. 2.8. The data can be scaled for other thicknesses; multiply the attenuation in Fig. 2.8 by $\log_{10}$ (actual thickness in metres divided by 15).

Absorption losses also take place during reflection from a conducting surface. Reflectors re-radiate radio waves because currents are induced in their surfaces, and if they are not perfect conductors there will be resistive losses which increase with frequency. At VHF/UHF, poor conductors such as mountains or buildings act mainly as very lossy reflectors. Losses accompanying reflection from the ground depend upon its conductivity, and this largely depends on its water content – chalk hills and deserts make poor reflectors. On the other hand, the sea is a good electrical conductor and therefore a good reflector.

DIFFRACTION

Reflection, refraction, and absorption are all closely related, but diffraction arises from a different mechanism (Fig. 2.9). Normally the signals re-radiated during the passage of an EM wave through matter will cancel in all directions except straight ahead. If part of a wavefront is removed by an obstacle, cancellation will be incomplete and some signal will propagate into the 'shadow' region. In line with the edge of the obstacle the situation is symmetrical, with half the wavefront missing on one side and the other half present on the other. Not surprisingly, the signal amplitude is halved and hence the power is reduced to one-quarter – a 6dB loss.

Within the shadow region the signal falls off continuously, but there is not the sharp cutoff which one associates with a beam of light (actually, light does undergo diffraction at a knife-edge, but the effect is not normally noticeable). At low angles just above the shadow region, the signal level varies by up to ±3dB.

Diffraction is one of the mechanisms which can aid VHF/UHF propagation well beyond the visible horizon [1], but the loss due to diffraction around an object increases with both the scattering angle and the frequency. Thus a solid obstacle which has a scarcely noticeable diffraction loss on 50 or 70MHz may appear as a significant barrier at 1·3GHz and above, so a clear take-off becomes more and more important at higher frequencies.

That concludes our discussion of the basics of propagation. Now to see how they apply in various propagation media.

THE PROPAGATING MEDIA

Radio waves change direction and travel beyond the horizon because they interact with the medium through which they are propagating. The available propagation media are mainly the troposphere and the ionosphere, as shown in Fig. 2.10, although even so-called "obstacles" can play a useful part.

THE TROPOSPHERE

The *troposphere* is the lowest part of the atmosphere, and what takes place within its boundaries largely determines our weather. Its upper boundary is the temperature-inversion layer known as the *tropopause,* which lies between 8km and 16km high depending on the latitude and prevailing conditions. The very low level of ionization is negligible as far as radio waves are concerned. However, small variations in the physical properties of the air give rise to variations in refractive index (n), which results in several distinct tropospheric propagation modes.

Because the refractive index of air differs very little from that of empty space (n = 1·0) a more practical working quantity is defined as the difference from 1·0 expressed in parts per million. This is the *radio refractive index* (N):

$$N = (n-1) \times 10^6$$

Thus for n = 1·000320 (the mean value at sea level) the corresponding value of N is 320.

A practical expression for determining N is:

$$N = 77\cdot6\frac{P}{T} + 3\cdot73 \times 10^5 \frac{e}{T^2}$$

where P is the atmospheric pressure measured in millibars (mb), T is the temperature in kelvins (add 273 to °C), and e is the water-vapour pressure in millibars. Typical values are P = 1000mb, T = 280K (7°C) and e = 10mb, giving N = 324. Incidentally, you will find that meteorologists are increasingly using units called hectopascals (hPa) for pressure. These are numerically the same as millibars.

The refractive index of the troposphere is almost independent of the frequency, unlike the refractive index of the ionosphere which we consider later. The value of N for air depends only on the three factors mentioned above: temperature, pressure and humidity (or water-vapour content). In the lowest few kilometres of the atmosphere the water-vapour content is the most important factor in determining N. If a mixture of air and water is cooled, its *relative humidity* (the fraction of water vapour compared with the maximum the air can contain) will increase until the air becomes saturated with water and begins to deposit dew. So a practical measure of the water-vapour content of air is its *dew point,* which is the temperature to which the air must be cooled before dew is formed. The drier the air, the lower its dew point. Tables are available to convert temperature and dew point to relative humidity, and also to the water-vapour pressure used in the equation above.

In a well-mixed atmosphere, such as can occur after a rain storm, N decreases almost linearly with height h for the lowest 2–3 km at an average rate dN/dh of –39 units/km, as shown later in Fig. 2.16a. (dN/dh is the difference in N divided by the difference in h; dN/dh is negative because N is decreasing as h

VHF/UHF PROPAGATION

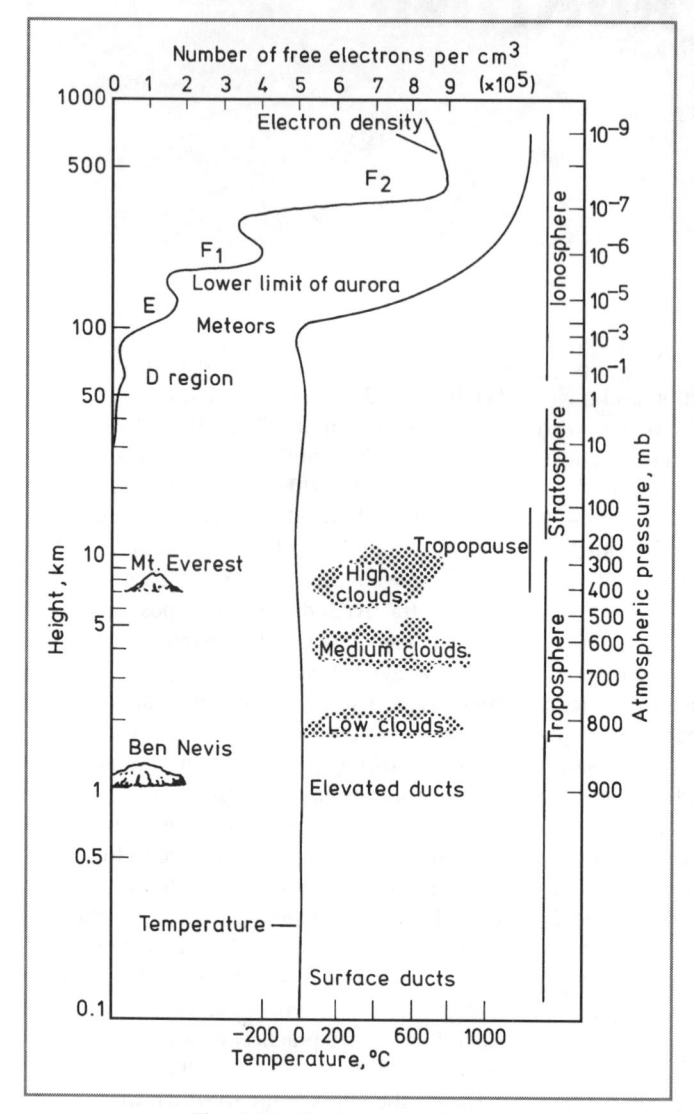

Fig. 2.10. The lower and upper atmosphere, showing the troposphere and the regions of the ionosphere. The logarithmic scale of height compresses the upper portion of the diagram. At the same scale as the lowest part, the whole diagram would be tens of metres high!

increases.) At greater altitudes the rate of decrease of N, and hence the refraction, becomes less and less; mathematically the decrease is exponential. A model based on the averaged atmospheric conditions is known as the *Standard Atmosphere*, and is useful for estimating propagation, though it never occurs exactly in practice – we have weather instead!

The detailed consequences are discussed later in **Tropospheric Propagation**.

THE IONOSPHERE

The ionosphere forms the outer mantle of the earth's atmosphere (Fig. 2.10). It consists of a variable composition of charged atoms (ions) and free electrons, together with non-ionized atoms in an overall electrically neutral mixture known as a *plasma*. The ionization results mainly from solar electromagnetic radiation but is also caused by the impact of atomic and sub-atomic particles of solar and cosmic-ray origin, together with some larger particles of matter in the form of meteors (page 2-5).

Although the total density of matter decreases with height, the number of electrons per unit volume continues to increase up to a peak at 200–400km as shown in Fig. 2.10 on the left. Beyond that height the decreasing density takes over as the outer atmosphere merges into interplanetary space.

Three major divisions of the ionosphere are recognized, known as the D, E and F regions. Within these regions are 'layers', although these are not just slabs of ionization but local maxima within the trend of increasing ionization with height. Moreover, these layers are not fixed but change continuously in position and density with the time of day, season of the year, state of the solar flux, and the solar wind.

The movements of the ionosphere are very complex. In addition to wind patterns, which still occur even in the outermost regions of the atmosphere, there are electromagnetic forces as the charged particles move through the earth's magnetic field, and also electrostatic forces between the charged particles.

THE D REGION

The D region lies between 60 and 90km high. Its main feature is that it absorbs radio waves, particularly at HF, because the electrons are frequently colliding and dissipating the energy of the waves instead of propagating them.

THE PROPAGATING MEDIA

Although the D region is almost transparent at VHF and UHF, it does contribute to ionospheric forward scatter, especially when intense solar activity produces an excess of ultra-violet and high-energy X-rays which penetrate to this depth of the atmosphere. The resulting fade-out at HF is accompanied by an enhancement of scattering from the D region at VHF.

THE E REGION

The E region is defined as that between 90 and 120km above the earth's surface, and it can be regarded as the transition zone from the earth's atmosphere into space; there are discontinuities in pressure, temperature and chemical composition at this height. Above about 90km ions predominate over neutral atoms. Most of the short-wavelength ultra-violet rays and less energetic X-rays from the sun are absorbed by ionizing the E layer. Although the ionization is normally insufficient to reflect VHF radio waves back to earth, some thin and dense layers of ionization occasionally occur, known as *sporadic-E*. This is also the region where auroras and meteors produce ionization capable of reflecting VHF and UHF signals.

THE F REGION

Above the E region lies the F region, which in daytime usually subdivides into two distinct layers denoted F1 and F2. The lower F1 layer plays little part in VHF propagation, except to contribute some refraction. If VHF radio waves are returned back towards the earth, it is from F2-layer heights. The F2 layer has an average effective height between 300 and 350km, though this can increase to between 450 and 480km. The layer varies between 100 and 200km deep. During peaks of solar activity the F2 layer can provide world-wide contacts on the 50MHz band, and propagation also extends occasionally to the 70MHz band. However, the F2 layer does not reach the levels of ionization required to return higher-frequency VHF or UHF signals to the earth except in the special case of trans-equatorial propagation.

PROPERTIES OF THE IONOSPHERE

VIRTUAL HEIGHT

Earlier in the chapter I pointed out that the signal returned from the ionosphere is due to continuous refraction through a medium of changing refractive index. From the ground, however, this looks like a simple reflection from a sharp boundary. If you project the directions of the incident and returned ray, these appear to meet above the height actually reached by the ray, at a *virtual reflection point* marked as 'V' in Fig. 2.6.

The *virtual height* h' of an ionospheric layer can be measured by timing the reflection from an *ionosonde* (a specialized upward-looking HF radar). Since the refractive index of the ionosphere depends on frequency, h' will vary with frequency too.

MAXIMUM RANGE VERSUS HEIGHT

The area on the earth covered by reflection or scattering from an object in the sky depends on its height. From Fig. 2.11 the maximum range D_{MAX} obtainable using a reflector h km above the earth's surface is:

$$D_{MAX} = 2R \arccos\left[\frac{R}{(R+h)}\right]$$

where R is the radius of the earth (6371 km mean)[*]. This is a simple geometric construction assuming straight rays tangential to the earth's surface at both ends, so it applies at all frequencies.

In the case of 'reflection' from the ionosphere (which is actually bending – see Fig. 2.6) the correct geometrical value for D_{MAX} is obtained by using the virtual height h'. The above formula can also be used to calculate the horizon distance for an antenna height h above unobstructed ground or the sea, merely by omitting the factor of 2. For example, the sea-level horizon from a 100m cliff is 36km away, or 50km for an antenna height of 200m.

[*] *This formula assumes that the **arccos**, **acos** or **cos** $^{-1}$ function returns its results in radians. If your scientific calculator only works in degrees, multiply D_{MAX} by (π/180).*

VHF/UHF PROPAGATION

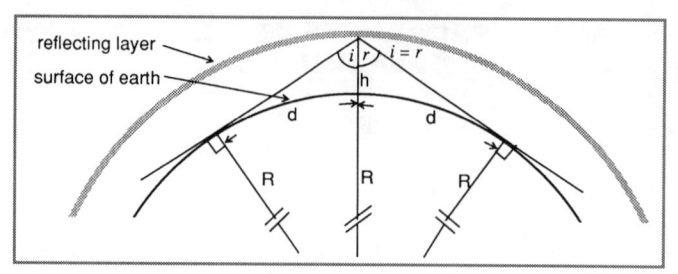

Fig. 2.11. The geometry for maximum range using a change of direction at a height h. Note the possible range extension if the antenna is significantly above ground level (not to scale)

In practice, a number of effects can modify this simple prediction of D_{MAX}. Firstly, no account is taken of the extended horizon due to the height of the antenna (Fig. 2.11). Secondly, refraction in the troposphere will generally extend the radio horizon, although significant bending occurs only at low wave angles and only in the lowest few kilometres of the atmosphere. A third effect, which tends to reduce the range rather than increase it, is that the maximum radiation from most antenna systems is not horizontal but is inclined a few degrees upwards due to ground reflection (Chapter 7). Nevertheless, the sum of all these modifying effects is small compared with ionospheric skip distances.

MAXIMUM USABLE FREQUENCY

The *Maximum Usable Frequency (MUF)* is the highest frequency capable of being returned to earth from the ionosphere. VHF signals require high ionization densities in order to be reflected at all, together with a favourable geometry for reflection. Figs 2.5e and 2.5g showed that partial reflection is easier for a wave approaching at a shallow angle, a long way from the perpendicular. This condition is best met by launching a wave parallel to the ground, to meet a distant ionized layer which is only just visible above the horizon. When there is just sufficient ionization for reflection to take place, you are operating at the MUF.

Fig. 2.12 shows the situation at various frequencies, ranging from above the MUF to well below. Above the MUF (Fig. 2.12a) the ionization is insufficient to return the waves to the earth. At the MUF itself (Fig. 2.12b), only waves taking off at angles grazing the horizon are reflected, and only one distance is workable – the maximum possible for that layer height.

If the ionization density increases, or the operating frequency is decreased, the ionosphere can return signals from more acute angles of incidence, so shorter ranges become possible as well (Fig. 2.12c). Even then, however, there is a minimum skip distance (shown as S) inside which the higher-angle radiation will be lost through the ionosphere.

At a much higher level of ionization – or more practically, at a much lower frequency – even a vertical wave will be reflected (Fig. 2.12d). The highest frequency at which this can occur is called the *critical frequency* (f0). The critical frequency can be measured using a swept-frequency HF ionosonde, and can then be used to calculate the MUF, assuming a smooth reflecting layer. However, this method seriously underestimates the MUF at VHF, where ionospheric propagation is actually making use of small-scale and short-term irregularities in the ionization density. Equally, the relationship between f0 and the MUF is very sensitive to the angle of incidence, and small changes in angle due to refraction or 'tilting' of the ionospheric layer can make a very large difference to the predicted MUF. For VHF DX purposes, the best indications of the true MUF come from a general-coverage VHF receiver.

PROPAGATION BY OBSTACLES

Obstacles are included as a means of making contacts which would not otherwise be possible, because they can reflect, scatter or diffract radio waves. The larger they are, the more useful they can be. Diffraction from the ridge of a large, sharp-edged obstacle obstructing the path may sometimes provide better signals than would be possible over smooth earth or lower terrain, and this improvement is sometimes called *obstacle gain*.

Although reflections from aircraft are mainly

THE PROPAGATING MEDIA

a source of fading (see page 2-20), they have been used to make QSOs over difficult paths of a few hundred kilometres. Man-made satellites at greater heights would provide longer ranges, but they are not large enough to function by passive reflection alone. But one satellite *is* large enough to provide world-wide VHF/UHF propagation: the moon.

Fig. 2.12. Ionospheric reflection in relation to the maximum usable frequency (MUF) and the critical frequency (f0). M is the maximum single-hop distance, and S is the minimum skip distance

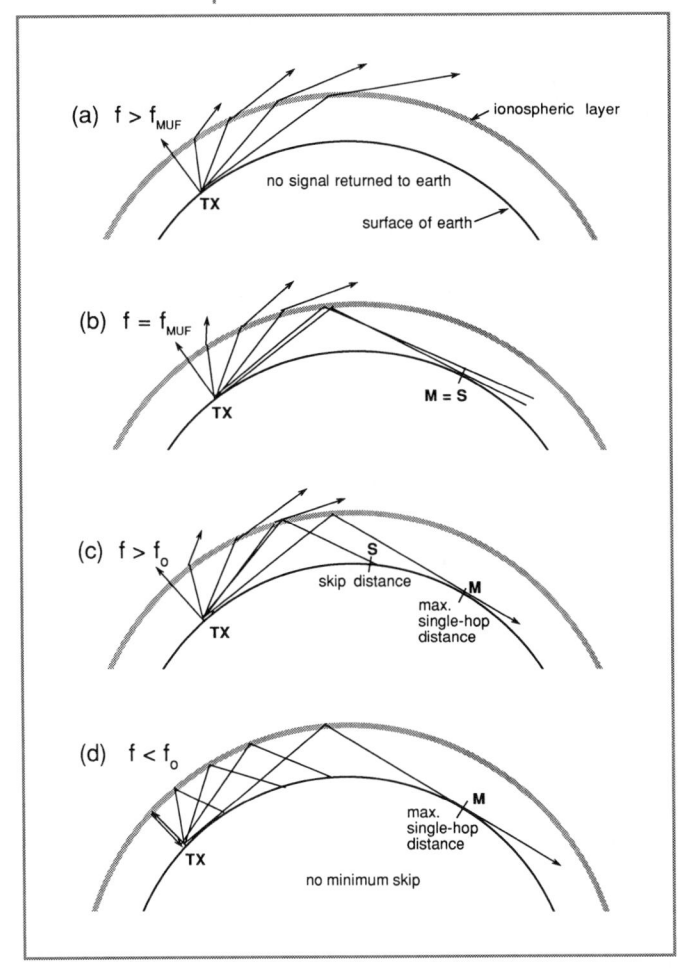

(a) $f > f_{MUF}$

(b) $f = f_{MUF}$, $M = S$

(c) $f > f_o$, skip distance, max. single-hop distance

(d) $f < f_o$, max. single-hop distance, no minimum skip

THE PATH FROM TRANSMITTER TO RECEIVER

PATH LOSS

Path loss is an essential concept in determining whether a usable propagation path exists between two stations. If the transmitters, receivers and antennas of two stations can just manage between them to overcome the path loss, you may manage a weak-signal QSO. Signals will be stronger if the path loss decreases, as it does during an 'opening', and you can also make equipment changes to improve the capability of your station to overcome path losses. Chapter 4 explains how to calculate and improve your station's *path-loss capability*.

FREE-SPACE PATH LOSS

In empty space, radio waves propagate rather like ripples from a stone thrown into a pond, but in three dimensions rather than two. The

Fig. 2.13. Radio energy reduces with distance-squared in free space: the Inverse Square Law

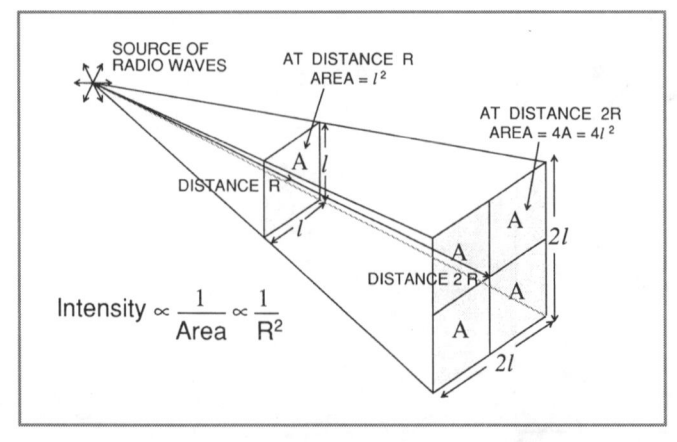

radio propagation path is the route taken by the wavefront as it travels from the transmitting to the receiving antenna. The attenuation along the path is known as the *path loss*.

Consider two stations in free space. The signal radiated from one station decreases in strength with distance as the wavefront spreads out over a larger area. Fig. 2.13 shows that if the distance R is doubled, the energy has to cover an area four times larger, so the signal falls to one-quarter of its the original power. In other words, the signal power reduces by $1/R^2$, which is known as the *inverse-square law*.

The path between terrestrial stations usually involves reflections and other losses, but the free-space loss almost always contributes to path-loss calculations. An important exception is in tropospheric ducts (page 2-24).

In addition, the free-space path loss is used to assess the quality of a given propagation path, by comparing the actual path loss with that of a free-space path of the same length.

CALCULATING PATH LOSS

The received signal obviously depends on the power being transmitted and on the gain of the antenna at each end. The path loss is therefore defined as the ratio of the power received to that transmitted, between stations having perfect *isotropic* antennas, *ie* antennas which radiate power equally in all directions. Since this is never the case (isotropic antennas just don't exist) we will deal with the general case where both the transmitting and receiving antennas have gain.

If an antenna is radiating a transmitted

THE PATH FROM TRANSMITTER TO RECEIVER

power P_T watts with a gain G_T over isotropic in the direction being considered, the power per unit area S available at a distance R metres is $(P_T \times G_T)$ divided by the surface area of a sphere of radius R:

$$S = \frac{P_T G_T}{(4\pi R^2)} \text{ watts per m}^2$$

The signal power P_r intercepted by a receiving antenna of gain G_R over isotropic is equal to the power density S multiplied by the capture area A of the antenna. From Chapter 7 the capture area of an antenna with gain G_R over isotropic is:

$$A = \left(\frac{G_R}{4\pi}\right) \text{ sq. wavelengths}$$

$$A = \left(\frac{G_R \lambda^2}{4\pi}\right) \text{ m}^2$$

So the received signal power P_R is simply S multiplied by A:

$$P_r = G_T G_R P_T \left(\frac{\lambda}{4\pi R}\right)^2 \text{ watts}$$

To obtain the space loss, we must first reduce P_R to the power which would be received using isotropic antennas, which we do by setting G_T and G_R equal to 1. Then the free-space path loss L is simply:

$$L = \frac{P_T}{P_R} = \left(\frac{4\pi R}{\lambda}\right)^2$$

Of course, this formula only applies in the 'far field' of the antenna, where R is much greater than λ, so L will always be a number much greater than 1.

For practical purposes it is more convenient to express the path loss L in decibels and to use kilometres rather than metres for R. Also we tend to use to use frequency rather than wavelengths. Making all these adjustments, the formula for path loss becomes:

$$L = 20\log_{10} R + 20\log_{10} F + 32 \cdot 45 \text{ dB}$$

where R is in km and F is in MHz.

Note that the expressions for free-space path loss involve frequency or wavelength. The term involving wavelength was introduced

with the capture area of the antenna, because the area over which an antenna of fixed gain collects energy depends on the wavelength. However, the properties of the space through which the wave propagates are *not* frequency-dependent.

AN EXAMPLE PATH-LOSS CALCULATION

Although we said at the beginning of this section that free-space loss contributes in most cases to the overall path loss, most types of terrestrial propagation are dominated by other sources of attenuation. For example, assume that you are listening to someone in an orbiting spacecraft at a distance of 1000km. How strong would the astronaut's signal be compared with that from your friend on the hill across town? Let's assume that they are both using 1W hand-held 144MHz transceivers with near-isotropic antennas having 0dBi gain.

Your friend is line-of-sight at 10km, and on your S-meter he is 60dB over the noise level. Assuming his signal strength is governed only by the free-space path loss, this loss is:

$$L_1 = 20\log_{10} 10 + 20\log_{10} 144 + 32 \cdot 45 = 95 \cdot 6\text{dB}$$

while from the astronaut it is:

$$L_2 = 20\log_{10} 1000 + 20\log_{10} 144 + 32 \cdot 45 = 135 \cdot 6\text{dB}$$

The difference is only 40dB, so the astronaut's 1W would be quite a respectable signal, 20dB over the noise level over a 1000km free-space path. In contrast, the additional losses over terrestrial paths of 1000km are so great that it would require an exceptional DX opening to hear a 1W hand-held.

We could have found the answer of 40dB more quickly by returning to basics. Using the inverse-square law, the ratio of the signals is $(1000\text{km}/10\text{km})^2 = 10000$, which is of course 40dB.

PATH LOSS WITH CHANGE OF DIRECTION

Line-of-sight propagation is usually not very interesting to DXers. All but the most local contacts (or contacts with spacecraft) will

VHF/UHF PROPAGATION

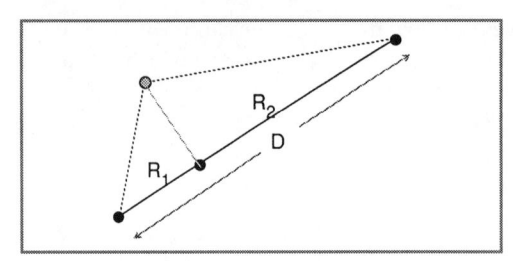

Fig. 2.14. A path with an intermediate reflection

involve at least one point where a change of direction takes place. How is the path loss affected by this change of direction? If we divide the path into two legs of lengths R_1 and R_2 (Fig. 2.14) we have to work out a separate path loss for each leg. The received power-flux density S_R will be the power density at the reflection point, multiplied by the fraction of the power reflected towards the receiver, and then multiplied by the fraction of this reflected power arriving at the receiving antenna:

$$S_R = \frac{P_T}{\left(4\pi R_1^2\right)} \times r \times \frac{1}{\left(4\pi R_2^2\right)} \text{ W/m}^2$$

where r is called the *reflection coefficient*. Its value, which lies between 0 and 1, depends on

Fig. 2.15. Variation of received signal strength with the position of the intermediate reflecting point

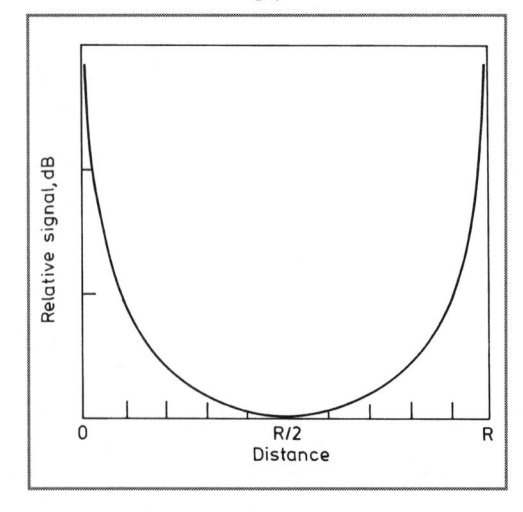

the nature of the reflecting surface and the angle of reflection.

Converting into decibels, the equation for the overall path loss will be:

Path loss (dB) = $20\log_{10} R_1 + 20\log_{10} R_2$
$+ 10\log_{10} r +$ other terms...

where the 'other terms' will be constant for a particular pair of stations.

Fig. 2.15 uses the above equation to show the variation of received signal strength with the position of the reflection point, for the simple case where the reflector is very close to the direct line between the two stations. Note that the signal strength has a broad minimum when the reflector is near the half-way mark; the best signal strengths are obtained when the reflector is close to either end of the path. Even if the reflecting object is not on the direct path between the stations, the half-way mark is still the **worst** place to make a change in direction, from the signal-strength point of view.

This may seem strange, because we generally think of half-way as the 'best' reflection point. But that is strictly from the viewpoint of achieving the maximum possible distance; as already shown, the reflector then needs to be at the limit of range for both stations, *ie* half-way between them. Unfortunately that is also the place where the path loss is at a maximum, but there's no other choice for maximum range.

FADING

Why do signals fade? Well, obviously in some cases because the conditions which led to propagation have gone away. But there are other causes of fading, often regular or quasi-regular in character. The most common is interference between signals arriving from the same station by different propagation paths.

It is often an oversimplification to speak of just one signal path, because rays travelling in slightly different directions can arrive together at the receiving antenna. In general they will arrive with a difference in phase. If the difference between two paths is in the range from 0° to ±90°, the net effect is an increase in

THE PATH FROM TRANSMITTER TO RECEIVER

signal strength (*constructive interference*); if the phase difference is in the range from 180° to ±90° the net effect is a decrease (*destructive interference*). Any change of amplitude or phase will change the situation, causing the net signal to rise or fall. Such changes can be brought about by a variation in the delay along the contributing paths, for example due to a change in the refractive index. Complete cancellation between two paths occurs only when the amplitudes of the waves are identical and also the phase difference is exactly 180°.

Another example of multi-path interference causing periodic fading occurs when the frequency of the waves on one path is shifted slightly compared with the other, due to a motion of the reflecting surface (the *Doppler effect*). The two waves then 'beat' together, the fading rate being the difference between the two frequencies. The most common example of this is the 'flutter' which occurs when a signal arrives by reflection from a moving aircraft, as well as by the more direct path. The flutter slows down and momentarily stops as the aircraft passes directly overhead or directly across the signal path at right-angles, since at that moment the aircraft is moving neither towards nor away from you; it then speeds up again as the aircraft moves away. Aircraft flutter can equally well be explained in terms of changes in path length [2], and this alternative view is entirely compatible with the Doppler-shift explanation; it is simply another way of calculating the phase velocity.

A further mechanism for fading is the changing polarization direction of arriving waves, due to motion of the reflector or changes in the medium along the path. An example of this is Faraday-rotation fading of moonbounce signals (page 2-62). A discussion of how polarization affects fading can be found on page 2-10.

Some propagation modes always involve multiple paths and movements of the reflecting surface, and are therefore intrinsically subject to fading. These modes include meteor-scatter, auroral scatter and moonbounce; refer to the sections dealing with these modes for further details.

One of the striking features of the VHF/UHF bands is the difference in their fading characteristics over both short and long paths. Fading on different bands can be compared by listening to a multi-band beacon. 50MHz and 70MHz are both notorious for long, slow fading, whereas fading on 432MHz and above can be very rapid and fluttery. Some reasons for this are advanced later in this chapter under **Tropospheric Forward Scatter**. Nevertheless, to quote [1]:

"The propagation mechanisms that cause severe fading and phase effects on horizontal and low ... angle paths are not yet well understood."

VHF AND UHF PROPAGATION MODES

Up to this point we have been dealing with the fundamentals of wave propagation. We can now use this information to understand each of the recognized propagation modes. These modes are summarized in the table below which lists the name of each, its common abbreviation and the propagating medium involved.

SUMMARY OF PROPAGATION MODES

Medium	Propagation mode	Abbrev.	Bands (MHz) [1]			
Troposphere						
	Refraction	Tropo	50	70	144	432
	Ducting	Tropo		(70)	144	432
	Forward troposcatter	Tropo	50	70	144	432
Ionosphere						
	Temperate-zone sporadic-E	Es	50	70	144	
	Auroral sporadic-E	Ar-Es	50	70	144	
	Equatorial sporadic-E		50	(70)		
	Field-aligned irregularities	FAI	50	70	144	
	Trans-equatorial	TEP	50	70	144	
	Equatorial FAI	TEP	50	70	144	(432)
	Aurora	Ar	50	70	144	432
	F2-layer	F2	50	70		
	Meteor-scatter	M/S	50	70	144	(432)
	Ionospheric scatter		50	(70)	(144)	
Obstacles						
	Diffraction		50	70	144	432
	Moonbounce	EME	(50)	(70)	144	432
	Reflection from objects		50	70	144	432

[1] Brackets around a band mean that although the mode of propagation may exist in principle on that band, contacts are rare and/or extremely difficult.

TROPOSPHERIC PROPAGATION

Fig. 2.16. Typical profiles of temperature, dew point and radio refractive index with height. (a) normal conditions and the Standard Atmosphere; (b) surface duct; (c) elevated duct

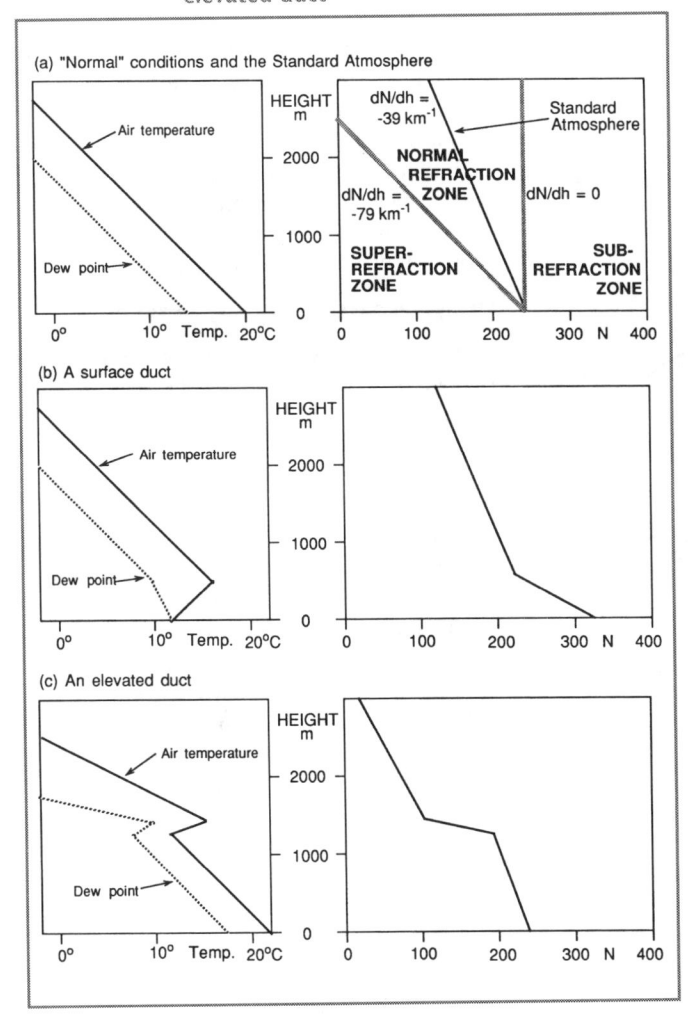

The most common VHF/UHF propagation occurs due to changes of refractive index within the troposphere (Fig. 2.10). There are three types of tropospheric propagation which take signals a significant distance beyond the horizon:

Enhanced tropospheric refraction, caused by a significant increase above the normal value of refractive index

Tropospheric ducting, where the radio waves are trapped within channels bounded by sharp changes in refractive index

Tropospheric scattering, which arises from small-scale variations in the refractive index.

These modes occur almost exclusively in the lowest few kilometres of the troposphere, where the air density and water-vapour content are greatest. In practice, the abbreviation "tropo" is used by amateurs to include all three of these modes, as well as any accompanying diffraction.

ENHANCED TROPOSPHERIC REFRACTION

The refractive index of the troposphere was introduced on page 2-14, where we saw that the radio refractive index (N) normally decreases with height. Since refraction occurs whenever there is a change in refractive index, bending of radio waves around the curvature of the earth is the norm rather than the exception within the troposphere. In the optical analogy, it is like an almost-permanent mirage which enables you to see beyond the horizon. The refractive index profile for the Standard Atmosphere described on page 2-14 is shown in Fig. 2.16a and it extends the *radio horizon* beyond the geometric horizon by approximately one-third (Fig. 2.17a). Within this 'line-of-sight plus one-third' distance, signals are almost as strong as in free space.

In practice, real atmospheric conditions seldom approximate to the Standard Atmosphere. Refractive-index gradients between dN/dh=0 and dN/dh=–79 units/km are considered normal. If dN/dh decreases more rapidly than –79 units/km, *super-refraction* is taking place (Fig. 2.16a) and the radio horizon is extended (Fig. 2.17b). *Sub-refraction* condi-

2 - 23

VHF/UHF PROPAGATION

tions occasionally occur when the refractive index increases with height (dN/dh is positive, Fig. 2.16a), bending the waves away from the earth so that signals become abnormally weak or disappear altogether; in effect, the radio horizon becomes closer (Fig. 2.17c).

To find out whether refractive tropospheric propagation exists between two stations, it is necessary to see if the refracted rays strike the ground or encounter any other obstacles along the path. This can be done by graphical ray-tracing [3, 4], which involves drawing a vertical cross-section of the relief along the path, using an earth radius modified to take standard refraction into account. Short-range ray tracing may be useful at microwave frequencies but it is rather academic at VHF/UHF, where other effects take signals well beyond the refractive radio horizon.

A well-sited VHF/UHF station with reasonable equipment can make contacts over many hundreds of kilometres under average tropo conditions, even though the 'line-of-sight plus one-third' rule would imply a range of only a few tens of kilometres by normal refraction from most locations. Some books imply that under normal conditions only simple refraction is taking place, but this clearly cannot be the case. In fact several mechanisms contribute to this enormous extension of range, including diffraction around obstructions and – more importantly – forward tropospheric scatter which we deal with later.

Similar misconceptions are sometimes found regarding contacts over even longer distances, which are attributed to super-refraction. This too is inaccurate; to make the signal follow the curvature of the earth, dN/dH would need to maintain just the right value over the whole path, which is a very improbable state of affairs. In fact the mechanism responsible for long-distance tropo DX is *tropospheric ducting*.

Fig. 2.17. Radio ray trajectories for (a) Standard Atmosphere; (b) super-refraction; (c) sub-refraction. D marks the optical or geometric horizon

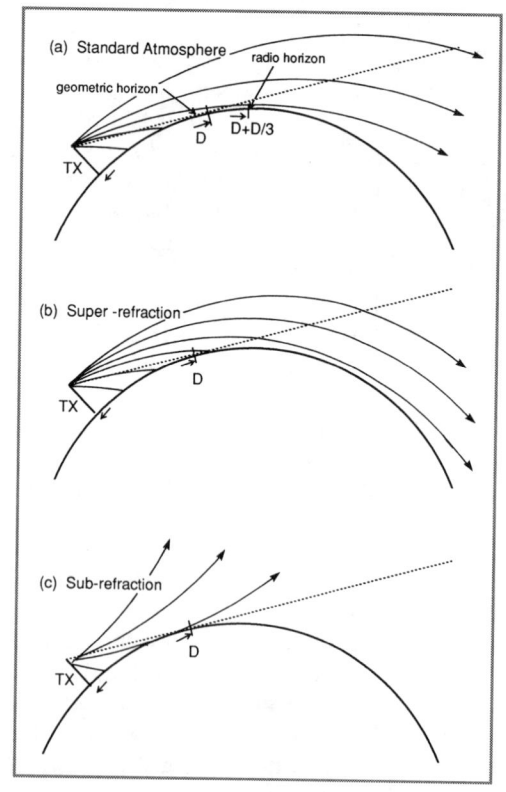

TROPOSPHERIC DUCTING

A duct is formed when a radio waves are trapped between two boundaries. The atmospheric refractive index does not always vary smoothly with height; discontinuities occasionally occur, forming a layer in which refraction is much greater than elsewhere. Such a layer will refract the wave back towards the earth's surface, forming the upper boundary of a duct whose lower boundary is the earth's surface itself, and this is known as a *surface duct*. Alternatively the wave may become trapped between two discontinuities in the troposphere, the upper one bending the wave towards the earth, and the lower one bending it back up again. This is known as an *elevated duct*.

These discontinuities of the refractive index are still very small, of course, so only waves which are at glancing angles to this surface get reflected. In this respect the duct resembles an optical fibre, which is transparent when viewed crossways but will trap light waves launched into its ends. A duct may also be compared to a microwave waveguide, in that a duct will not propagate signals whose wavelength is too long in relation to its vertical depth. The minimum duct thicknesses are approximately:

f	λ	Min. duct
50MHz	6m	400m
70MHz	4m	300m
144MHz	2m	200m
432MHz	0.7m	100m
1.3GHz	0.23m	50m

Although the minimum tropospheric-duct thickness increases with wavelength, the increase is not strictly proportional as it would be in a metal waveguide. The minimum thickness is about 65λ at 50MHz, but about 220λ at 1.3GHz. The difference is due to the irregularities of refractive index which make the boundaries 'fuzzy'. The upper frequency propagated by a duct is determined by scattering losses due to these irregularities.

Because the waves within a duct are confined to a two-dimensional layer, theoretically the signal only reduces linearly with the distance ($1/d$ instead of $1/d^2$ as in free space – page 2-18). In spite of losses due to inefficient coupling of signals into and out of the duct, together with leakage and scattering caused by irregularities, the attenuation at ranges up to 1000km can be comparable with the free-space value and ducted signals are often extremely strong. Propagation is obviously best between stations which are in the duct itself, or can beam directly into its open ends.

SURFACE DUCTS

Surface ducts occur when a steep negative gradient in refractive index forms immediately above the ground or sea (Fig. 2.16b), trapping the radio waves by refraction from above and by reflection from the surface below. In a normal atmosphere, the air temperature and dew point – and hence the refractive index – all decrease with height above ground. When the temperature increases with height instead of decreasing in the normal way, this is called a *temperature inversion*. The effect on the radio refractive index is to make it decrease more sharply than normal at the inversion boundary (see equation on page 2-13), bending the radio waves downwards and trapping them between the boundary and the ground.

Near-surface inversions are often formed when the ground cools rapidly under clear skies, in turn cooling the layer of air in contact with the ground. This is called a *radiation inversion* and usually results in fog or mist. Surface ducts do not propagate well over land, the ground being a relatively lossy reflector covered with obstacles which absorb and scatter radio waves. However, ducts occur quite frequently over the sea, and since the sea (when calm) is a good reflector, sea ducts can propagate over very long distances with little loss. For example, the 2500km path from the Isles of Scilly to the Canary Islands (EA8) is reported to open about once a year on 144MHz. A particular feature known as an *evaporation duct* is very common over the sea, and is formed by the rapid change in humidity which occurs just above the surface. Sea ducts rarely penetrate far inland, so you need to be located either on an island or close enough to the coast to be able to beam into the open end of the duct. Such ducts are often only about 15m high, limiting propagation to the microwave bands, but deeper evaporation ducts capable of propagating lower frequencies do occur occasionally.

ELEVATED DUCTS

Elevated ducts form when a double discontinuity in refractive index occurs (Fig. 2.16c). In this case the waves are bent upwards from the lower boundary of the duct, and downwards from the upper boundary. Thus the waves can be guided over irregular terrain until the duct is broken by mountains or other changes that the duct cannot follow. Elevated ducts form at heights typically between 450 and 2000m, and are the origin of the longest overland tropo openings. Even longer distances can be worked via elevated ducts over the sea.

Although ducts approaching 400m deep are occasionally formed over warm seas, ducts over land in temperate climates are generally

VHF/UHF PROPAGATION

Fig. 2.18 (a). Long-distance tropo contacts may occur tangentially to isobars or across ridges...

Fig. 2.18 (b). ...but also in other directions

much shallower. This is borne out by the observation that while tropo ducting may occur many times a year on the 144MHz band, there are very few records of tropo ducting at 70MHz or lower frequencies in the UK. It also explains why strong ducted signals can be found on the 432MHz or microwave bands at times when ducting on 144MHz is much weaker or even absent.

Well-developed elevated ducts produce a 'skip' effect which carries the signals over the heads of stations beneath the duct. Ducted propagation thus tends to be strongest between specific areas. This is occasionally so marked that the opening resembles sporadic-E propagation. However, a sporadic-E cloud will propagate signals in many directions, while tropo ducts tend to occur in specific directions. Also, ducts can be very stable, lasting for hours or even days, in contrast to the more transient behaviour of sporadic-E. Although less well-developed 'leaky' ducts do not produce the strongest signals, their imperfect boundaries do permit stations at intermediate distances to couple in, thus providing a wider geographical spread of DX.

TROPOSPHERIC PROPAGATION

TROPO WEATHER

Having seen how temperature inversions can give rise to surface and elevated ducts, now let's look at the larger-scale weather conditions in which tropo ducting can occur.

Inversions are formed by two basic mechanisms. As already mentioned, radiation inversions occur under still, clear conditions when the land cools rapidly, thus cooling the air close to the surface but leaving the higher levels relatively unaffected. These conditions occur most often in anticyclonic weather systems, an *anticyclone* being an area of high pressure in which barometers are typically reading "FAIR" or "VERY DRY".

The other mechanism for forming an inversion is connected even more directly with the anticyclone, and is known as a *subsidence inversion*. The air at the centre of a high-pressure system tends to descend at a rate of a few metres per hour. This subsiding air is compressed as it moves downwards into levels where the pressure is higher, and hence it warms up – for the same reason as the air compressed by a bicycle pump gets warm. As a result, the relative humidity of the air decreases (page 2-13) and so does its refractive index. When the subsiding air becomes as dense as the air beneath, it will cease to sink. At the boundary between these two air masses there will be a temperature inversion where the normal decrease of temperature with height is sharply reversed, accompanied by a sharp change in the refractive index.

As a high-pressure system develops, the height of a subsidence inversion will decrease and the inversion may even touch ground level. Elevated ducts have a double bend in the profile of refractive index against height (Fig. 2.16c), caused by a combination of subsidence and radiation effects.

During the day the ground heats and the ducts either disappear or become lossy surface ducts, only to reappear in the evening when the ground cools again. Ducts tend to develop progressively through the lifetime of an anticyclone, so the biggest tropo openings tend to occur after several days of enhanced propagation. The boundary region is typically dome-shaped; at its edges the inversion is lower, so coupling into the duct will be much better.

A major opening affecting the UK generally involves an extensive area of high pressure to the east, giving similar weather over much of western Europe. Ducts extending 800km or more can then appear for several days running, the direction of best propagation changing as the anticyclone drifts across the continent. It has been claimed [5] that the longest paths occur tangentially to the isobars on the weather chart, or across ridges (Fig. 2.18a). But excellent contacts can also be made radially from the centre of a high-pressure area (Fig. 2.18b) so there is no substitute for turning the beam and exploring new directions. An excellent example of a moving tropo opening was that of 29–30 August 1987, which for my station in southern England began with propagation to northern Spain, swinging anticlockwise towards southern France followed by Italy, Austria, Czechoslovakia, East Germany (as it then was), Denmark, Norway, and finally an oil rig off the north-east of Scotland. After the ducting had moved through 180° in the course of two days, the opening abruptly ended.

Unfortunately, anticyclones are not an everyday feature of weather in the UK. Although they can appear at any time of the year, anticyclones are more common in late summer and early autumn, when one or two big tropo openings are very likely. However, the mere existence of a high-pressure system does not guarantee ducting. A short-lived anticyclone will not allow a subsidence inversion time to establish itself, and if there are steep pressure gradients (recognizable on the weather chart by the isobars lying close together) there will be strong breezes which will not permit stable inversions and ducts to form. In other cases, cloud formation at the boundary layer – usually stratocumulus – can cause irregularities which limit the upper frequency of propagation.

The weather in the UK is more generally dominated by cyclones (depressions) and their associated warm and cold fronts. Super-refraction can occur during the passage of either type of front, but those caused by cold fronts are more marked. Enhancements occur because the incoming cold, dry air is more dense than the warmer air ahead of the front and tends to undercut it. The warm air is therefore driven upwards, creating a temperature inversion along the line of the front. So the best direction of propagation is along the front as it passes over your location, but the opening will inevitably be short-lived.

Forecasting tropo openings is no easier than forecasting the weather itself. While it is possible to construct vertical atmospheric profiles from weather observations at specific points, this is not a practical forecasting tool for radio amateurs because the data are not available soon enough. As an aid to catching tropo openings, a barometer is a must – a recording type, if you can afford it. National and continental weather maps are readily available from TV and newspapers, and even more immediate information is available from weather satellites or HF facsimile broadcasts for quite a small investment in equipment. Look for a large, stable high-pressure area with light winds.

Direct observation of the sky – wind, clouds, transparency – can give a lot of information about your local tropo situation. Inversion layers can sometimes be seen as a dark line of trapped smoke just above the horizon. In anticyclonic conditions a layer of stratocumulus with a base below 1000m may indicate the presence of a surface duct, while a base above 1000m could suggest an elevated duct.

Finally, don't ignore the messages given on broadcast TV and VHF radio apologising for interference. This is usually caused by strong tropo signals from other countries – bad news for TV viewers but an opportunity for us!

MAXIMUM DISTANCE SPANNED BY TROPO DUCTS

Fig. 2.19 shows the maximum tropo distances worked by operators reporting in *DUBUS* 1989 for the 144, 432, and 1296MHz bands. The maximum range appears to fall off with frequency, but to some extent this may reflect reduced band occupancy and path-loss capability on the higher bands. From the UK, tropo paths to the north and to Scandinavia are much rarer than towards Spain, southern France, Italy, Switzerland, Austria, and Czechoslovakia. Mountain ranges often limit the distance workable from the UK to around 1000km. On the other hand, sea ducts are limited only by the scale of the weather pattern. The 4000km path from California to Hawaii is regularly spanned by sea ducts, producing world-record tropo DX. During an opening between the Canary Islands (EA8) and the British Isles and Norway on 8/9 September

Fig. 2.19. DX operators' maximum ranges by tropospheric ducting, taken from the 'Top List' in *DUBUS* 1989/1

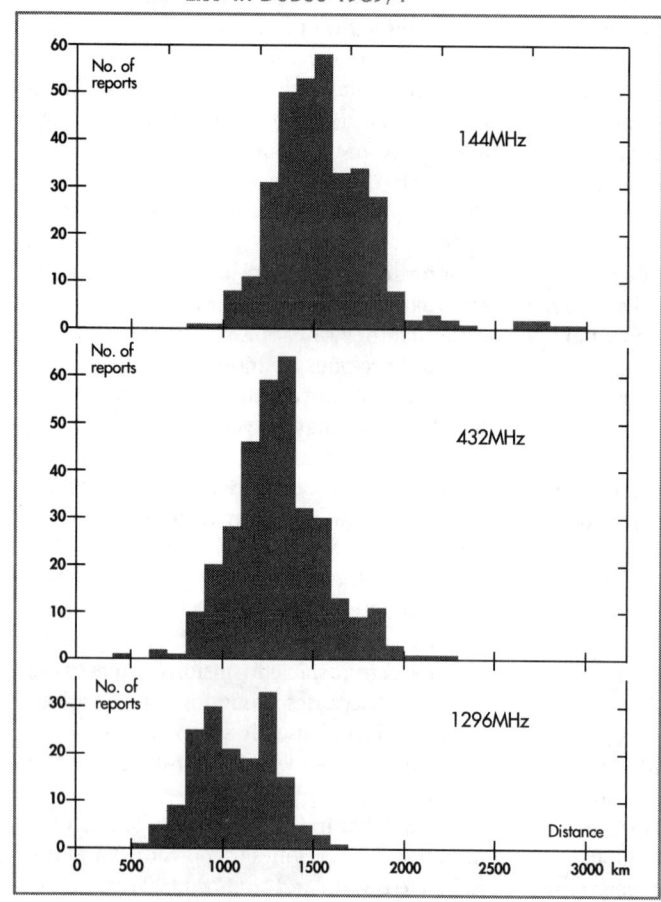

TROPOSPHERIC PROPAGATION

1988, many contacts in excess of 3000km were made on 144MHz. This is approaching the distance between the Isles of Scilly (extreme SW England) and Newfoundland, so it has been suggested that the Atlantic could one day be spanned by a tropo duct.

TROPOSPHERIC SCATTER

Tropospheric scatter arises from small variations in the refractive index of the troposphere. The atmosphere is never completely uniform and differences in temperature, pressure and water-vapour content are always present – if it were otherwise, we would have no weather. Changes in refractive index occur wherever there is turbulence, for example in convection cells and in any strong wind. These irregularities extend beyond the troposphere into the stratosphere (Fig. 2.10).

In the lower troposphere, differences in water-vapour content dominate the changes in refractive index. At greater heights the humidity is much lower, so that temperature variations become the dominant factor and the changes in refractive index are smaller. However, the lower density of the stratosphere leads to physically larger irregularities. The stratospheric irregularities offer a greater range by virtue of their height, but the path loss tends to be so great that only stations with EME-type path-loss capability can exploit them.

The refractive index is effectively independent of frequency throughout the VHF/UHF range, but troposcatter itself is frequency-dependent. This is because the scattering power of an object varies with its size relative to the wavelength, as explained on page 2-7. In the context of troposcatter an 'object' is any mass of air with a different refractive index from that of the air surrounding it.

The ever-changing structure of the troposphere gives rise to the day-to-day variations in 'conditions', and also to the shorter-term variations in signal strength. If the irregularities in the atmosphere are suppressed, for example during or after widespread heavy rain, tropo conditions will be poor on all the VHF/UHF bands.

TROPOSPHERIC SIDE-SCATTER

The conventional 'troposcatter' referred to in textbooks makes use of the irregularities which occur throughout the atmosphere. The classic textbook troposcatter path is obstructed by a very large obstacle, leaving no other choice except tropospheric scattering from the region of the atmosphere that can be seen from both ends of the path. This can involve an appreciable change in direction, so I prefer to call it 'side-scatter' to distinguish it from forward troposcatter which is described in the next section.

A simple formula which successfully predicts the average side-scatter path loss L_{sc} has been given by Yeh [1, 3, 6-8]:

$$L_{sc} = -694 \cdot 5 + 30\log_{10} f + 20\log_{10} R + 10A - 0 \cdot 2N \text{ dB}$$

where
  f is the frequency (MHz)
  R is the distance (km)
  A is the projected scatter angle through which the direction of the signal has to be changed (degrees)
  N is the radio refractive index at the surface.

The $(+30 \log_{10} f)$ term in the formula shows that an increase of 3 times in frequency (144 to 432MHz, or 432 to 1296MHz) increases the predicted path loss by about 15dB. You will recognize the $(20 \log_{10} R)$ term from the free-space loss equation on page 2-19, and the fourth term $(+10A)$ shows that the path loss increases by 10dB for every 1° increase of scattering angle. Another effect of increasing the angle A is that the scattered wave becomes more and more incoherent, resulting in a characteristic 'warble'.

The primary application of this formula is for microwave paths involving scattering from a volume of the troposphere somewhere above the mid-point of the path [3], though it has been applied to lower frequencies [6]. If you wish to estimate path losses using the above formula, don't forget to allow for a shortfall in antenna gain at the receiving end due to the *aperture to medium coupling loss* (Chapter 7) because this is a scatter mode.

VHF/UHF PROPAGATION

TROPOSPHERIC FORWARD-SCATTER

You will not find tropo forward-scatter discussed in the textbooks as a distinct mechanism – yet. However, it is becoming recognized as the principal propagation mode for routine VHF DX contacts, which can be made in most directions on most days and do not require super-refraction.

A very common kind of atmospheric irregularity is the *convection cell*, familiar to glider pilots as an expanding column of rising air which normally signals its presence by forming a cloud at a certain height. On most of our VHF bands a typical cell is many wavelengths across, so it tends to refract rather than scatter (page 2-8). Because the variations in the refractive index of the atmosphere are very small and the size of the cell is not extremely large in terms of wavelengths, its refractive power is very weak and the wavefront will undergo only a very slight change in direction. Also the emerging wave will be almost coherent with the rest of the wavefront which passed by the object. Hence these forward-scattered signals suffer little phase distortion and sound quite 'normal' in speech quality or CW tone, in contrast with other scatter modes.

Picture this process taking place continuously along the propagation path and you will see that the wavefront gradually spreads out, some of it following the curvature of the earth rather than disappearing totally into space. With assistance from diffraction when the path passes an obstacle, and some larger-scale refraction if conditions are favourable, stations can be worked with weak to moderate signals over distances up to about 500km. The range can easily extend to around 1000km with a suitable combination of sites, equipment and propagation conditions.

Like any form of scattering, tropo forward-scatter is inherently multi-path and subject to fading as convection cells and other tropospheric structures form, change and disperse. The fading characteristics will depend not only on tropospheric activity but also on the size of these structures relative to the wavelength. The wavelengths of the 6m, 4m, 2m and 70cm bands span almost a factor of ten, and this difference gives propagation on each band its own distinct characteristics of workable range, depth of fading and timescale of fading. On the 10m band tropo scattering is relatively poor, suggesting that the wavelength is too large for convection cells to act as effective lenses.

TROPOSPHERIC BACKSCATTER

Tropo backscatter is the extreme form of large-angle scattering, and has slightly lower loss than side-scatter. It is a much less useful mode of propagation, though it may perhaps be usable to contact a station when the direct path is blocked by large obstacles. It is most often heard from a station with high ERP, when your two beams are side-on to each other and the direct signal is very weak. Tropo backscatter can be recognized by its rough, almost 'auroral' audio quality, caused by random addition of incoherent wavelets as described on page 2-8.

AURORAL SCATTERING

THE AURORAL PHENOMENON

The *Aurora Borealis* or 'Northern Lights' normally appears as a complex pattern of light in the northern sky. The equivalent in the southern hemisphere is the *Aurora Australis*. Auroral activity is centred in the *auroral ovals*, which are distorted ring-shaped regions surrounding the earth's magnetic poles (Fig. 2.20). At these high latitudes, auroral activity can be seen almost every night as a slowly varying glow in the sky. However, very intense and dynamic displays occasionally occur when the boundaries of the oval expand to both higher and lower latitudes. These intense auroras are unstable, their shape changing continually in patterns of streamers, curtains or rays, while the colours range through greens, blues and reds arising from ionized oxygen and nitrogen. Auroral displays often appear in the northern and southern hemispheres simultaneously, and observations have shown them to be near mirror-images. Visual auroras in southern England are rare, but on occasion a glow has been seen as far south as Rome.

Auroras are caused by an input of energetic electrons into the upper atmosphere. Their erratic behaviour is caused by the chaotic nature of the processes accelerating these electrons, which we will consider later. These relatively high-energy electrons are able to knock out more tightly-bound electrons from an atom than the usual solar ionizing radiations can; when the missing electrons are replaced, the characteristic colours of the auroral light are emitted. The energetic incoming electrons are able to ionize very many atoms before losing their energy, creating showers of *secondary electrons*, and are at their most effective just before being slowed down to a stop. This occurs in the E region where the atmospheric density suddenly increases (Fig. 2.10). Here the secondary electrons far outnumber the original incoming electrons, and it is from the E region that radio signals are scattered.

Auroral propagation works best on the lower VHF bands, 50 and 70MHz. However, most studies of this mode – many of them in the UK – have been carried out at 144MHz owing to the limited international availability of the lower bands. Auroral contacts are more difficult at 432MHz, although contacts have been made at frequencies as high as the 902MHz band in the USA.

The E-layer plasma from which radio waves are scattered is highly agitated due to the flux of incoming secondary electrons. The motion

Fig. 2.20. An auroral oval over the magnetic North Pole, photographed from a satellite in UV light (land outline has been added).

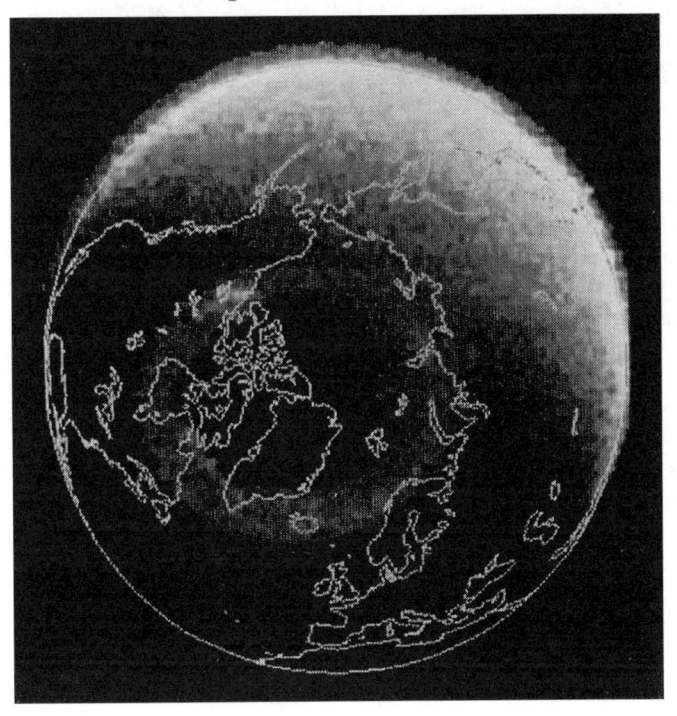

VHF/UHF PROPAGATION

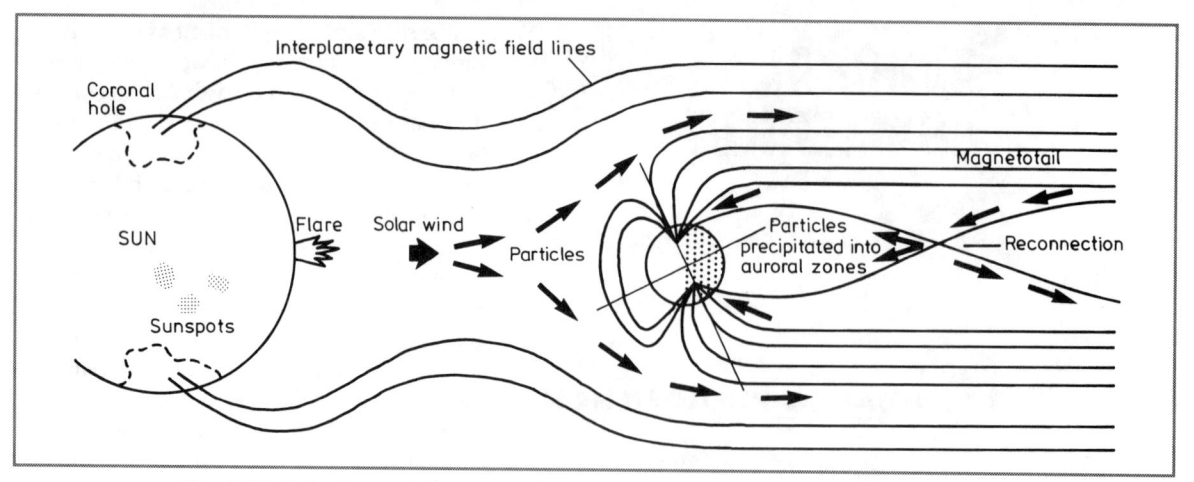

Fig. 2.21. Solar interactions with the earth's magnetic field (not to scale). The earth's magnetic field is pushed away from the day side by the solar wind. Note the magneto-tail containing trapped charged particles, extending away from the earth on the night side

of these electrons modulates the scattered wave, producing a characteristic 'hiss' caused by random Doppler shift. So SSB is difficult to read and you need to speak slowly and clearly (Chapter 3). This technique works well on 50 and 70MHz but the Doppler shift increases in direct proportion to the frequency. CW gives far better communication than SSB for auroral contacts on 144MHz, and is almost essential at 432MHz. In addition there may be a shift in the centre frequency of the signal, due to the collective motion of the secondary electrons which form currents around the hemisphere, so don't be surprised if you need quite a large RIT shift in order to resolve the other station – especially on 432MHz.

The visible and radio manifestations of an aurora may not always be evident at the same time. Visual effects range from 400km in height down to 100km, whilst radio reflections are limited to the bottom of this range. Thus an aurora can be visible while its radio-reflecting part is below the horizon. Sometimes, however, a radio aurora is observed without visual effects being reported. Apart from the obvious reason of clouds or daylight preventing visibility, ionospheric currents transport the secondary electrons around the earth, so that scattering from these currents can occur well away from the position of the original aurora.

ORIGINS OF AURORA

As befits such a wondrous and strange phenomenon, the explanation of aurora is not simple. It is still the subject of research and I can only give a rough outline here. The electrons which cause the aurora originate from the variable streams of atomic particles which are continuously emitted from the sun. This *solar wind* is composed mainly of electrons and protons (hydrogen nuclei) with a small proportion of highly ionized atoms of the elements that make up the sun (chiefly helium), and neutral atoms. Together these form a *plasma* which is electrically neutral overall.

During their journey from the sun towards the earth, the charged particles are influenced by the *interplanetary magnetic field*. This applies particularly to the electrons as a result of their much smaller mass. The path of a moving electron is bent when it travels through a magnetic field because it experiences a force which is at right-angles to the field and also to the direction of motion. This is familiar as the force which makes an electric motor work, or moves the electron beam in a TV tube. As a

AURORAL SCATTERING

result, the charged particles of the solar wind travel in spirals along the field lines linking the sun and the earth. But that is not the whole story – moving charged particles also generate their own magnetic fields which modify those already present. The effect is to distort the pre-existing field lines as if the solar wind was exerting a 'pressure' upon both the interplanetary magnetic field and the field of the earth.

What causes the interplanetary magnetic field in the first place? It is dominated by the field of the sun; but unlike the earth's magnetic field which resembles that of a two-pole bar magnet and is quite steady, the sun's field is more complex and changing. The sun has several 'magnetic poles' called *coronal holes* where the field lines leave the sun and connect into interplanetary space, and these are also where charged particles can escape. The sun's magnetic field therefore switches direction several times per rotation as a number of coronal holes pass by. These changes take place at *sector boundaries*; within each sector, the sun's magnetic field is in the same direction.

For a long time it was thought that auroras were caused directly by electron streams originating from cataclysmic events on the sun such as sunspots, flares or solar prominences. From satellite measurements of the solar particle flux, we now know that this is not correct – but these 'obvious' ideas can take a long time to disappear!

The view now accepted is that charged particles continually emanating from the sun in the solar wind become trapped in the earth's *magneto-tail*, which is formed by field lines which stream from the earth away from the sun (Fig. 2.21). An increase in the solar particle flux stretches the magneto-tail until it 'snaps' and reconnects into a new and more stable configuration. When this happens some of the trapped particles are ejected out into space, while those in the retained part of the magneto-tail are propelled towards the earth by the contracting magnetic field lines. However, this alone is not sufficient to accelerate the particles to the energies necessary to penetrate to the E region where the radio auroral effects are found; the details of the acceleration mechanism are still the subject of debate and research.

It is important to understand the implications of what I have just described. The direct cause of an aurora is *not* solar activity. It is some event which triggers a change of configuration in the magneto-tail. If we wish to detect auroras – or better still, be warned in advance – we need to look for these *triggering events*.

THE WHEN AND WHERE OF AURORAS

The field lines from the magneto-tail connect with the earth to form the auroral ovals around the north and south magnetic poles. The northern oval rotates around the magnetic north pole (presently located near Thule in NW Greenland, although its position is slowly moving) and the accelerated electrons and other particles are guided along these field lines to create an aurora. The pressure of the solar wind pushes the field lines towards the dark hemisphere, so night-time auroras generally extend further south than those occurring in the daytime.

The daily rotation of the magnetic pole about the geographic pole combines with the solar-wind effect to make the auroral zone approach and recede twice a day (Fig. 2.22). As seen from the magnetic latitude of the UK, the oval comes close around 1800 local time (Fig. 2.22d) and then slightly recedes before approaching even closer around local midnight (Fig. 2.22a). This gives two times of day when aurora is most likely to occur, as can be seen in Fig. 2.23.

Fig. 2.23 shows another phenomenon. There is a dip in the probability of finding a radio aurora at local magnetic midnight, *ie* when the longitude of the magnetic pole and the sun differ by 180°. In the UK this falls between 2200 and 2300 UTC. While the aurora is taking place, two huge electron currents (kiloamps) flow away east and west from the point nearest the sun; but the currents cannot meet on the night side of the auroral oval because of the repulsion between like electric charges. Instead both currents turn away

VHF/UHF PROPAGATION

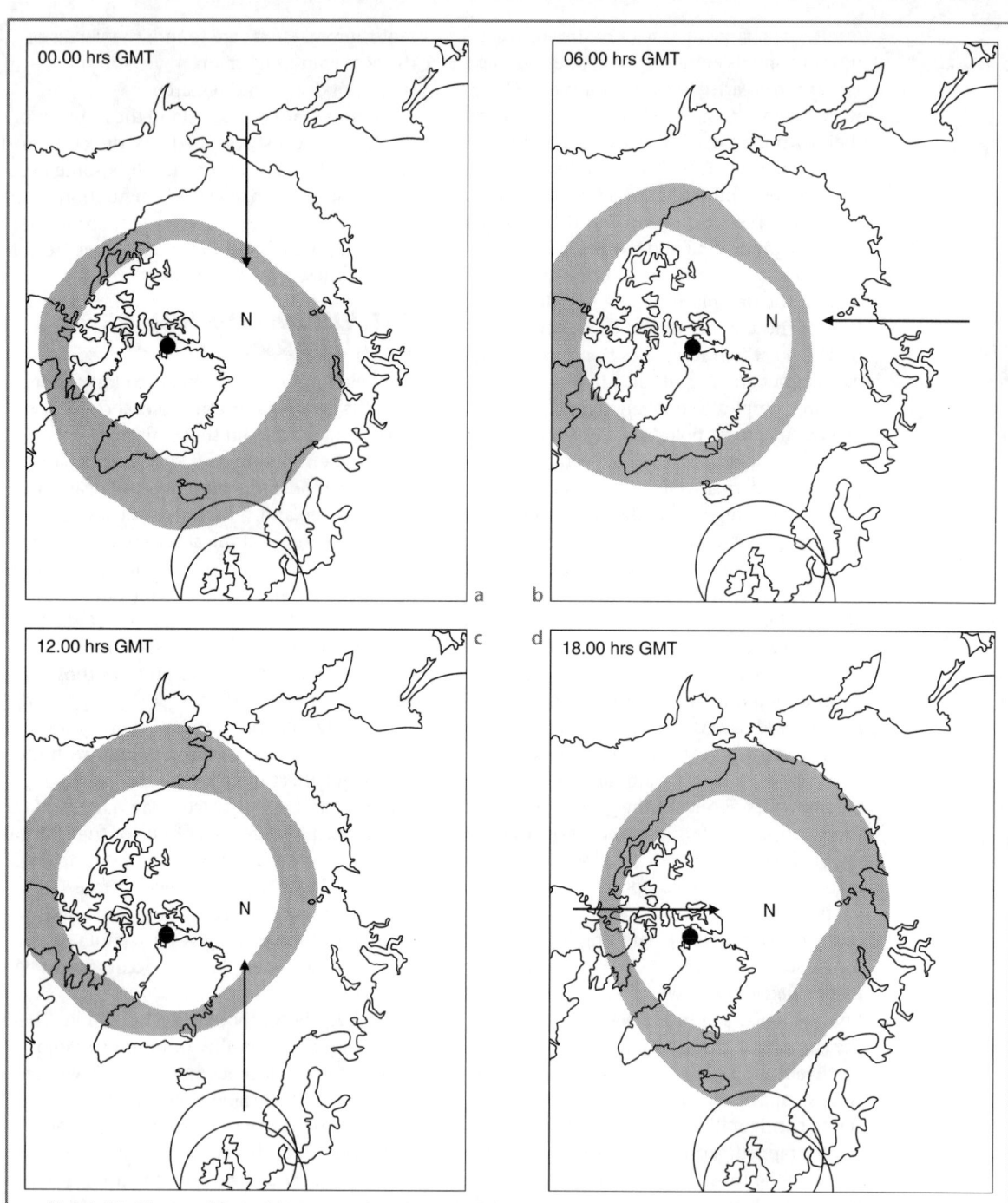

Fig. 2.22. The position of an auroral oval at six-hourly intervals during the day. The geomagnetic north pole (shown by a dot) rotates about the geographic north pole in the centre of each picture, and the direction of the solar wind is shown by an arrow. The result is that the auroral oval approaches and recedes twice a day. Rings show the accessible range from London and Edinburgh

AURORAL SCATTERING

towards the equator, leaving a relatively quiet region of the ionosphere around magnetic midnight. This is the **Harang discontinuity**, named after the Norwegian scientist who first described it. From your viewpoint on the earth, the ionospheric current is flowing from the east before the discontinuity but afterwards from the west, and a magnetometer recording the deflection of the earth's magnetic field will show this reversal [9].

There is also an annual variation of the position of the oval. In the northern winter the magnetic pole is on average tilted further from the direction of the sun, so the oval moves to lower latitudes. This results in peaks in auroral activity around the equinoxes, although Fig. 2.24 shows that auroras can occur at any time throughout the year.

The frequency with which auroras occur decreases very rapidly with distance southwards from the auroral oval. Hence there is a very large difference between the northern and southern UK. As can be seen from Fig. 2.22, at the optimum approach of the auroral oval (around midnight) stations in Edinburgh can access the auroral ionization when it is below the horizon from London.

THE GEOMETRY OF AURORAL RADIO REFLECTIONS

To make contacts *via* aurora, you have to beam at the aurora itself; and its position in the sky will determine the favoured regions for strong signals. Signals are back-scattered from the ionization, which forms a sloping wall because the earth's magnetic field is dipping sharply towards the vertical at high magnetic latitudes. The optimum conditions for a reflection from ionization aligned along a certain direction (in this case the direction of the magnetic field lines) is that both stations 'see' the magnetic field line as being perpendicular to them. This geometrical requirement can usually be satisfied only at one localized region of the auroral ionization, if at all. As the angles between the field direction and the directions of the two stations deviate from 90°, signals will become weaker, but less so if the specular-reflection criterion (angle of incidence = angle of reflection) is still satisfied.

For stations in the UK, the field lines dip approximately along a N-S line. This means that E-W deviations of the path will affect the reflection geometry much less than N-S variations. The potential area which can be worked from each point of auroral reflection is therefore roughly elliptical in shape, with its longest axis east-west. The benefits of a

Fig. 2.23. Occurrence of radio aurora in the UK by time of day, showing peaks in late afternoon and around local midnight, and the Harang discontinuity around 2200-2300h

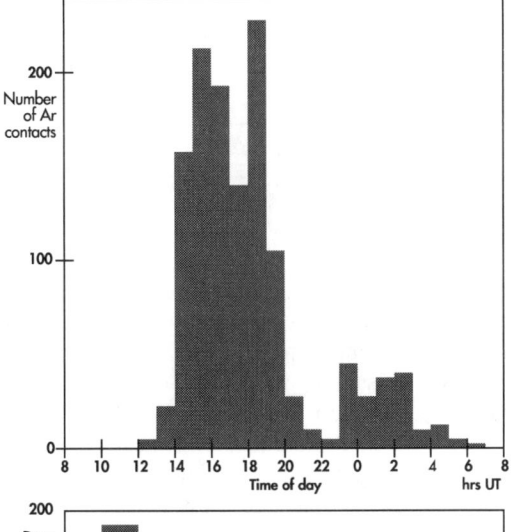

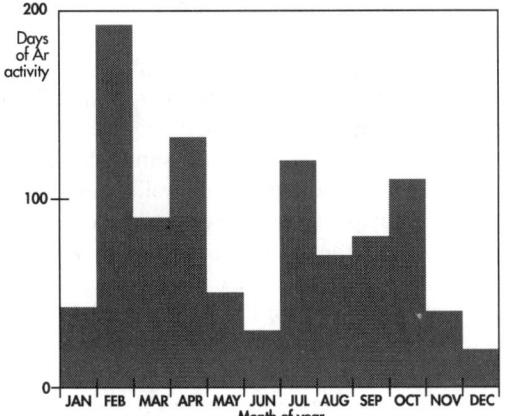

Fig. 2.24. Aurora can occur at any time of year, but peaks around the equinoxes in late March and September

VHF/UHF PROPAGATION

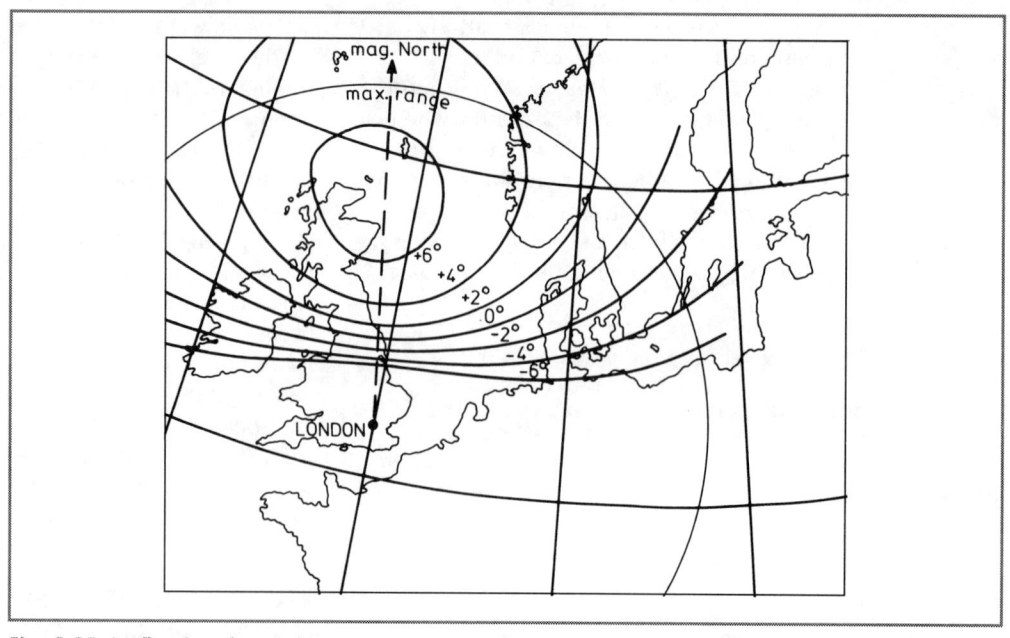

Fig. 2.25. Isoflectional map for a station in London. The contour marked 0° joins locations where the auroral reflection would be most favourable, if ionization is present at E-layer heights. The same map would apply to field-aligned ionization

favourable reflection geometry can often be heard from southern England in relatively weak auroras which provide contacts with Scotland and Northern Ireland. Some stations will regularly be much stronger than others who are using higher power but are less favourably situated.

To explore these effects, in 1979 I developed a computer program for the Scientific Studies Committee of the RSGB (now renamed the Propagation Studies Committee). At certain points in the E layer, a radio beam originating from your station will intersect the earth's magnetic field at right-angles. Contours joining these favoured locations can thus be drawn on maps, which I named *isoflectional maps* [10]. The computer program plots these contour lines as seen from a specified station location. If a similar map for another station is superimposed, the intersection of the contours shows the preferred reflection region, and hence the directions towards which the antennas should be pointed for these two stations to effect a contact. As an example, Fig. 2.25 shows this plot for a station located in London. For E-layer heights of around 100km, only the region closer than 1100-1200km is within range. Notice that you can beam north-east along these contours towards an area over southern Scandinavia where the contours are widely separated. This means that there is a large area over which the field alignment is favourable, and most auroral contacts from the southern UK are indeed made with the beam in this direction. If the aurora comes far enough south, into the region across northern England where the contours bunch together, even local stations at similar latitudes will become strongly auroral. Similar isoflectional maps can be constructed for other locations, but the shape of the contours will be different.

Note that these maps only consider the *geometry* for favourable auroral reflections. They say nothing about the location of the auroral ionization – or even whether there is any! Very often the aurora will not be sufficiently extensive or come far enough south to provide any significant ionization at the optimum locations. In such cases the signals will be relatively weak and operators in

AURORAL SCATTERING

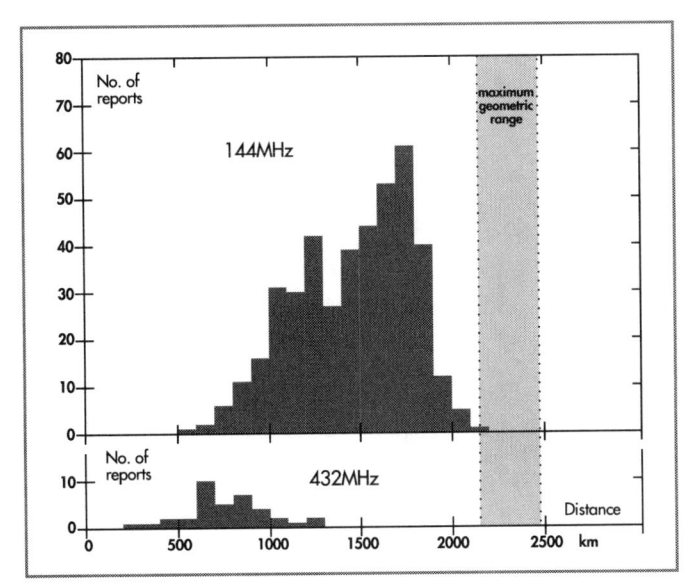

Fig. 2.26. DX operators' maximum ranges by aurora, taken from the 'Top List' in *DUBUS* 1989/1

southern Britain often call these minor events "Scottish auroras". One reason why GM stations are heard is that the aurora is relatively close to Scotland and the inverse-square law favours contacts where one of the stations is closer to the reflection point (Fig. 2.15).

If the antenna directions of pairs of stations are examined after the event, in most cases one station is much closer to the reflection point than the other. However, the reflection points for the longest contacts have to be near the mid-point of the path, with reflection taking place at a glancing angle to the magnetic east-west direction. Because the geometry of aurora does not permit true forward-scattering, the maximum distances achieved (Fig. 2.26) are slightly less than those by other E-layer propagation modes such as sporadic-E and meteor-scatter.

The region within which auroral contacts are possible has been called the "boundary fence". It consists of an approximately oval area, about 2000km to the magnetic east and west of your location, and about 1000km to the magnetic north and south. There is little doubt that this limit exists; very few indeed have worked more than 2000km via aurora (Fig. 2.26), and those extreme-range contacts have always been east-west.

The location of your own boundary fence can be determined by a very simple geometrical construction (Fig. 2.27). Begin by drawing a semicircle of about 1000km radius, facing towards magnetic north from your site; this defines your usable beam headings for aurora. If there happens to be an aurora in the direction you're beaming towards, your back-scattered signal will cover another 1000km semicircle centred on the point you're beaming at, but this time looking away from the aurora. Move that semicircle around to cover all your usable beam directions, always facing away from the aurora, and it will sweep across the entire region which you can potentially reach via this mode.

Fig. 2.27. Derivation of the auroral 'boundary fence'

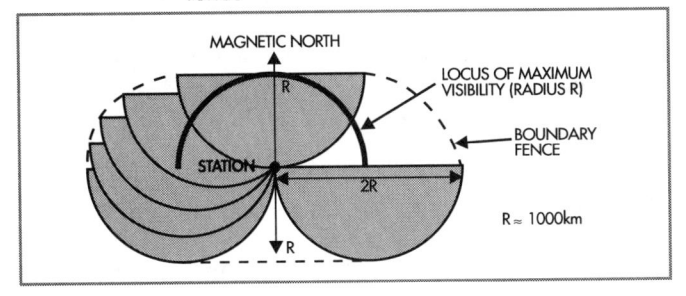

PREDICTING AURORAS

I have already explained that a high level of solar activity will not lead inevitably to an aurora. Although the original particles are derived from the solar wind, they first have to be trapped by the magneto-tail; then the magneto-tail must become extended to the point where the field lines are forced to reorganize and reconnect. This latter event usually needs some form of 'trigger' – which need not necessarily occur immediately after the high solar activity.

Since high solar activity is itself a potential auroral trigger, it is worth monitoring. Sunspots are small areas of the sun with a lower temperature, and therefore appear dark against the solar disc. They are easily observed with a small telescope, or even by the naked eye

VHF/UHF PROPAGATION

through fog or thin cloud*. Coronal holes are not directly visible, but provide another potential source of auroral triggers.

Some of these phenomena associated with the sun's surface may be long-lived, and may thus trigger repeat auroras. These will reappear at intervals approximately equal to the mean *synodic* solar rotation period of 27 days (the synodic period is the time interval between the same point on the sun's surface passing directly opposite the earth). Since the sun is fluid, the rotational period changes with latitude, and 27 days is only a mean value. It is therefore useful to keep a record of auroral events on a calendar made up of lines 27 days long, known as a *Carrington chart* after its inventor. You can draw this yourself, or find a ready-made version [11]. Look for repeats within the few days lying on the next line below a previous auroral occurrence.

We can now consider some of the common triggers for auroral events. Fluctuations in the solar wind, due to changes in the sun's interplanetary magnetic field, are perhaps the most common causes. The passage of the earth through a sector boundary causes some magnetic disturbance as the connections with the earth's magnetosphere rearrange themselves, and this can trigger an aurora. It appears that for an intense aurora to form, the interplanetary field component must be parallel and opposite to the earth's field. The passage between magnetic sectors is difficult to forecast precisely, except by using data from satellites measuring the interplanetary magnetic field – and those data are not usually available in time to be useful for forecasting. In any case, a sector-boundary transition may not trigger an aurora at all.

On a shorter timescale, there are a number of reliable precursors of an auroral event. Prior to the onset of a radio aurora, the incoming high-energy electrons in some cases penetrate down to the ionospheric D region, increasing

---

* NEVER look at the unobscured sun directly, *either with a telescope or the naked eye*. **Your eyesight could be permanently damaged!** *Filters mounted at the eyepiece of a telescope are also extremely dangerous. The safest method of solar viewing is by projection through a telescope on to a white card in shadow.*

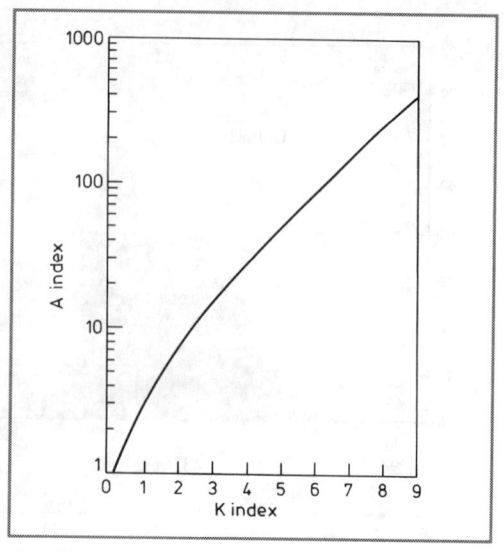

Fig. 2.28. Conversion from A index to K index

its ionization and producing absorption at HF. Monitoring the HF bands for these fadeouts can be rewarding. Other effects are due to the currents flowing in the ionosphere, affecting the earth's magnetic field and inducing currents in the ground; these are discussed in the next section.

MAGNETIC INDICATIONS

The huge electric currents flowing in the E layer significantly disturb the earth's magnetic field, providing one of the most reliable short-term indicators of auroral events. This disturbed magnetic field also has effects on both E- and F-layer propagation, as explained more fully in later sections. A high F-layer MUF is rarely found at periods of auroral activity and the formation of sporadic-E seems to be inhibited; it is probably the occurrence of a number of auroral storms during June 1991 which resulted in so few Es openings that year.

The earth's magnetic activity is monitored at many observatories and a disturbance factor (the *A* index) is calculated. The *K* index, a logarithmic derivative of the A index, is perhaps more useful; the conversion is given in Fig. 2.28. So-called 'planetary' values (Ap and Kp) are computed from weighted averages of a number of the observing stations and they show a close correlation with auroral activity.

The values of Ap and Kp are broadcast by several sources (see page 2-50) but they all suffer from the disadvantage of being some hours out of date. The immediate state of the geomagnetic field can be monitored by building a simple magnetometer, provided that a suitable location can be found. This must be well away from metallic and electrical disturbances, so I would not recommend this approach to anyone living in a block of flats or close to an electric railway! Several designs have been published using Hall-effect sensors [12], though more satisfactory results seem to be obtained using magneto-resistive sensors [9]. A more professional device, which does not depend on a suspended magnet and is therefore more portable, is the fluxgate magnetometer. Although a little more complicated, it is not beyond the amateur constructor [13].

As an alternative, you might try the less well-known method of measuring earth currents [14]. The earth acts as an electrical generator as it rotates through its own magnetic field, which partly arises from the circulating currents in the ionosphere. This develops a voltage in the north-south direction which is affected by ionospheric disturbances. To measure earth currents you need two efficient buried electrodes, separated by at least 25m in the north-south direction. The considerable advantages of this method are a relatively high signal (approximately 1mV per metre of electrode separation) and insensitivity to the local disturbances that can affect the magnetic field sensors.

SPORADIC-E

Sporadic-E (Es) is a spectacular mode of propagation. A previously 'dead' band can suddenly become alive with extremely strong signals from distances ranging from 500km to perhaps 8000km on the 50MHz band, and 800km to nearly 4000km on 144MHz. Signals disappear just as suddenly after a few minutes, or occasionally a few hours – openings at higher frequencies are generally much shorter. The 144MHz band is approaching the highest frequency affected; the upper frequency limit of sporadic-E propagation is around 200MHz, although very occasional openings extending to the 220MHz band have been reported from the USA.

Es is often extremely geographically selective. You may work many stations at 59+, all in one or two locator squares, while an operator a few miles away hears nothing at all! At the same time someone a few hundred miles away may be working a different set of locator squares, probably an equal distance away from the ones you are working.

There are three major categories of sporadic-E. The best-known to UK amateurs is *temperate-zone Es* or *mid-latitude Es*. Being solar-driven, this is mainly a summer daytime phenomenon. *Auroral Es* on the contrary follows the times and locations of auroral activity. The third type is *equatorial-zone Es*. This is a much more regular phenomenon, and provides 50MHz propagation in the region around the magnetic equator, but unlike the other two types it does not reach the 144MHz band. It could possibly contribute to multi-hop Es contacts from mid-latitudes, but since there seems little recorded on this

VHF/UHF PROPAGATION

subject I will not deal with equatorial-zone Es any further.

TEMPERATE-ZONE Es – FREQUENCY BEHAVIOUR

The apparent sudden onset of an Es opening is largely due to our habit of listening on one frequency band. By monitoring the frequency range above 28MHz you can follow the upward trend of the Es MUF as the ionization intensity increases. However, the further the MUF rises, the less likely it is to continue to rise. Fig. 2.29 plots the highest E-layer critical frequency (f0; see page 2-16) achieved at

Fig. 2.29. Critical frequencies f0E measured above Wallops Island (North Carolina, USA) during 1976

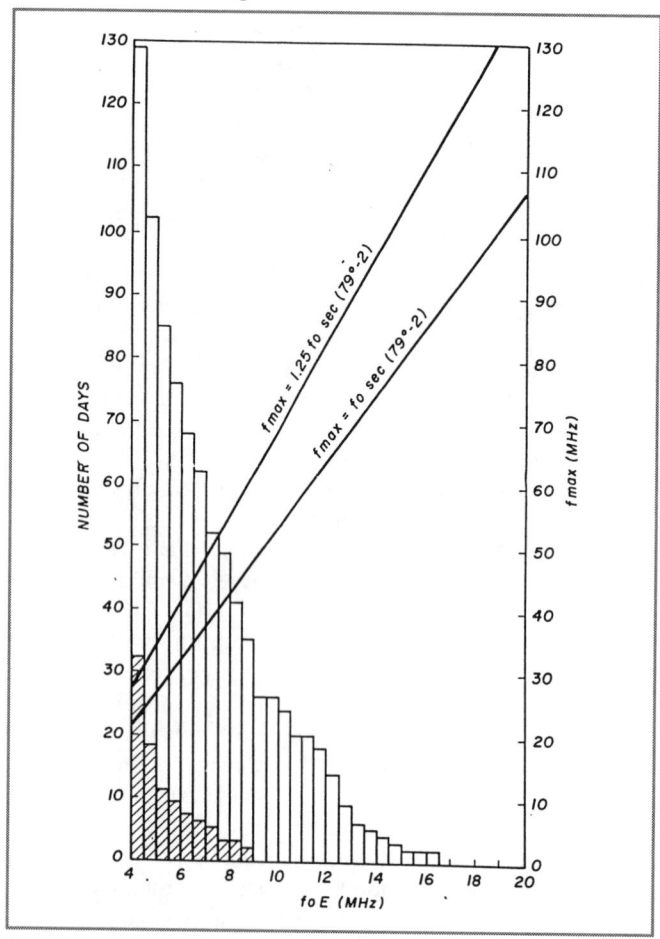

Wallops Island, North Carolina during 1976 for days when it exceeded 4MHz. Most of these days occur between May and August; the rest of the year is represented only by the small hatched area. The data from Wallops Island were used because that particular ionosonde took samples every 15 minutes during that year, unlike most ionosondes which sample at hourly intervals. Even so, the data may still be pessimistic – excursions to even higher frequencies may occur between sampling periods, and the ionosonde only measures the situation immediately overhead while more intense Es may be occurring elsewhere.

Fig. 2.29 shows a rapid tail-off of Es probability as the frequency increases; re-plotting the data on a logarithmic scale suggests that the smoothed probability curve is heading towards a frequency limit. This implies an upper limit to the ionization density that can be produced by the Es mechanism, as the observations confirm.

PATTERNS IN TEMPERATE-ZONE Es

Despite the label 'sporadic', which implies that its appearance is irregular and unpredictable, there are predictable patterns in the occurrence of Es. It is primarily a midsummer phenomenon, occurring in the in the northern and southern hemispheres alternately. The times of year at which Es openings have occurred at 60MHz and 144MHz in north-west Europe are shown in Fig. 2.30, together with the times of reception of a 28MHz beacon for comparison. (The 60MHz data are reasonably representative of 50MHz.) All frequencies show a similar peak between early May and the end of August, the maximum occurring during the first week in June. Indeed, some people say that openings on 144MHz from the UK are "almost guaranteed" within a few days around this peak – though the years 1989, 1990, and 1991 showed that this is not necessarily the case. Clearly the Es season varies greatly from year to year, just like the British summer!

Claims have also been made that the

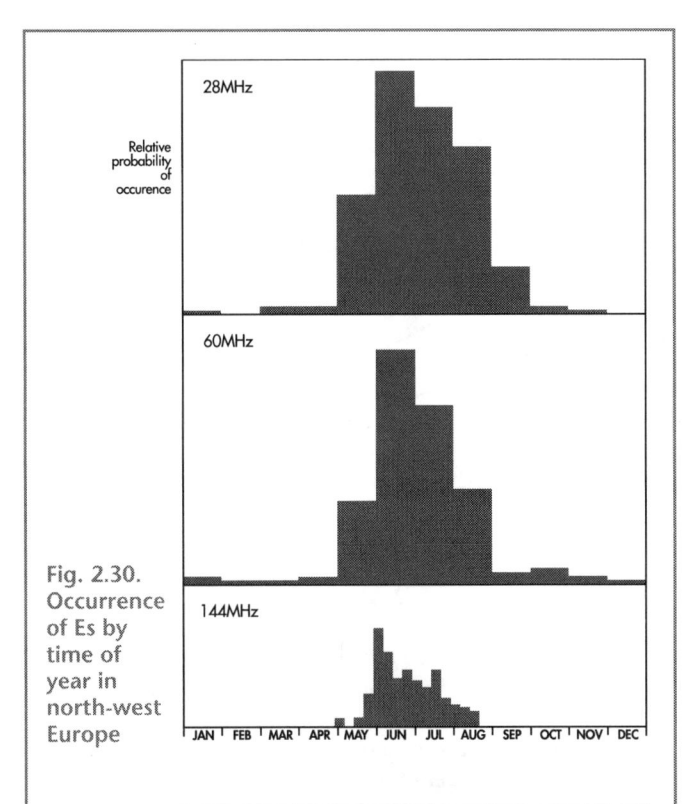

SPORADIC-E

Fig. 2.30. Occurrence of Es by time of year in north-west Europe

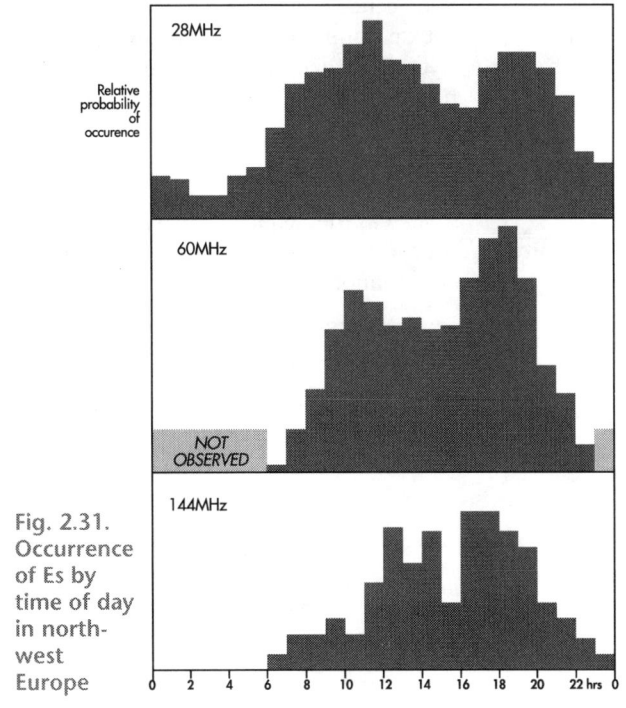

Fig. 2.31. Occurrence of Es by time of day in north-west Europe

144MHz Es season displays a second peak around a month later than the main peak in early June; but when proper statistical tests are applied, this is unconvincing. However, my analysis of the reception of 28MHz amateur beacons showed a weak (10%) periodicity of about fourteen days [15]; this is probably an effect of monthly tidal cycles in the ionosphere and was first noticed in ionosonde data thirty years ago [16].

A few Es events occasionally occur around midwinter; these could be 'spill-over' from the summer Es season in the opposite hemisphere. Although interesting, these events have little effect on the overall annual pattern.

Mid-latitude sporadic E is also characterised by the time of day it is most likely to appear, as shown in Fig. 2.31 from observations in western Europe at 144, 60 and 28MHz. The behaviour is similar on all bands, with a mid-morning peak followed by an early evening peak which seems to be relatively more important at higher frequencies. Probably there is a build-up of ionization throughout the day, which results in the most intense concentrations occurring in the evening. During the night, the ions recombine in the absence of solar radiation, although Es propagation on 28MHz can last throughout the 24 hours at mid-season – remember that the summer night is even shorter at E-region heights than it is at ground level.

Es propagation also tends to occur repeatedly between the same areas, implying that the ionization also occurs again at the same place. This is very likely on a timescale of days, but also seems to occur over a season. For example, during the summer of 1988 many Es openings occurred from the British Isles to the south of Spain. This was fortunate for the ZB2IQ expedition because 1987 and 1989 had favoured the opposite end of the Mediterranean. As an example of the 'repeat phenomenon', Fig. 2.32 shows two almost identical Es openings to north-west Yugoslavia and neighbouring Italy occurring on successive days. On investigating the short-term statistical correlation of Es openings on 28MHz, I found that if it occurs on a particular day,

VHF/UHF PROPAGATION

Fig. 2.32. Two almost identical Es openings on successive days. Dots indicate the region where reflection takes place, near the midpoints of the paths

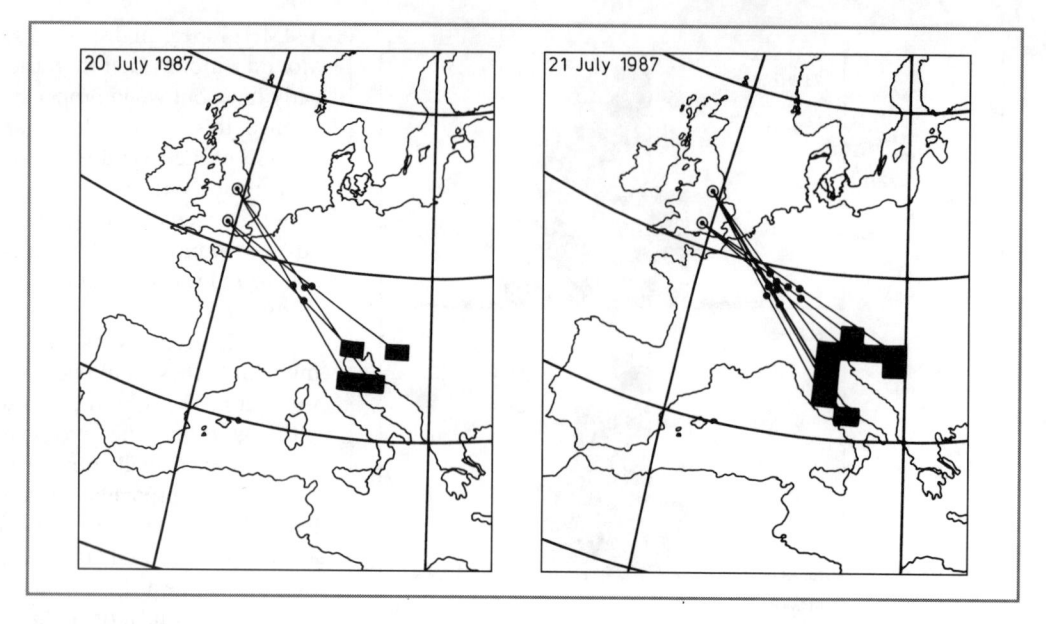

there is an enhancement of about 30% in its probability of occurring the next day over the same path, and an enhancement of about 10% in its probability of occurring on the third day [15]. If we make the analogy of a capacitor discharging through a resistor, this would imply a decay time-constant of one day. At present, however, it is only a guess that similar probabilities might apply at VHF.

SPORADIC-E CLOUDS

Investigations by rocket probes, radar and ionosondes reveal the following picture. Es layers contain a high concentration of ionized metals such as iron and magnesium, rather than the atmospheric gases which form the regular E and F layers. These metal atoms need much less energy to become ionized than do the atmospheric gases, and they also take longer to recombine. Es layers are often seen to descend from about 120km high to about 100km during the day. Often two or more thin layers (less than 1km thick) form at particular heights. The ionization within these layers is not uniform, and if you could look at the E layer at a particular radio frequency you would see discrete 'clouds' which appear to get smaller as the frequency increases. At 144MHz you would see only the highest concentrations of ionization measuring perhaps tens of

Fig. 2.33. Wedge-shaped 'footprint' from a reflecting Es cloud

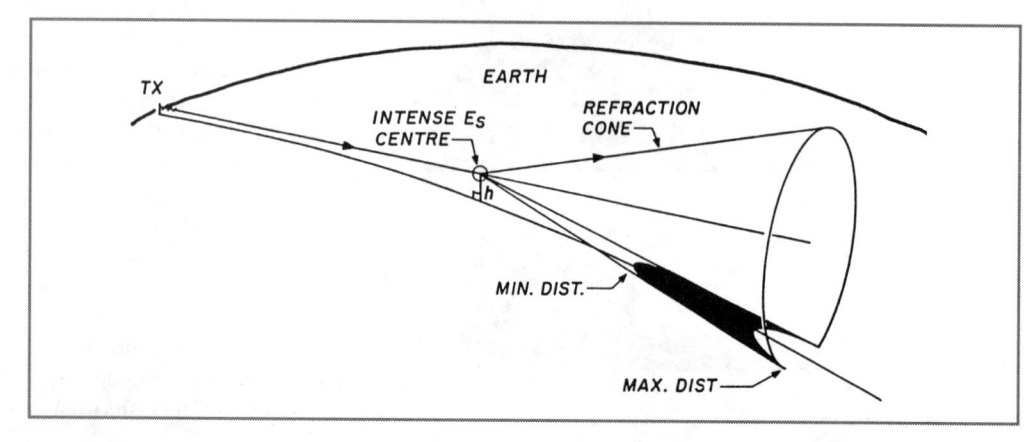

SPORADIC-E

Fig. 2.34. DX operators' maximum ranges by Es, taken from the 'Top List' in *DUBUS* 1989/1

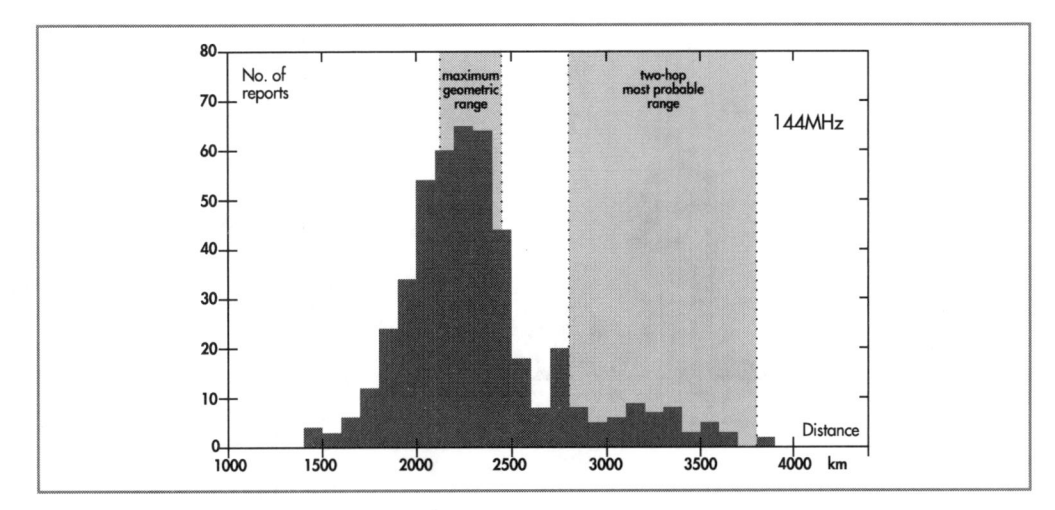

metres across, and this is what makes 144MHz Es so geographically selective.

Since signals at the Maximum Usable Frequency are bent through the smallest angles, they will provide the longest-distance propagation. For this they use clouds on the far horizon (Fig. 2.12b). A distant 'blob' of high ionization will appear almost as a point when viewed from the antenna, and the refracted wave will take the shape of a cone. The intersection of this cone with the ground gives a wedge-shaped 'footprint' of propagation which is familiar from Es openings (Fig. 2.33). These clouds of ionization often move or disperse quite rapidly, so contacts need to be made quickly!

SIDE AND BACK REFLECTIONS

Occasionally Es contacts are reported with large deviations of antenna direction from the great-circle direction. Most often these occur on the lower-frequency VHF bands, particularly 50MHz. This is because the bending angle at lower frequencies is larger for a given ionization density, so the angle of the refracted cone of rays covers a wider region on the ground. See also the section on ionospheric 'tilts' on page 2-51.

Sometimes both antennas need to be pointing in roughly the same direction, and for this type of back-reflection we need a different explanation. Since these conditions seem to occur towards the end of a long Es opening, it may be that the ionization which had become concentrated into Es layers is beginning to spread out vertically again, providing a near-vertical reflecting surface which may also become field-aligned. If so, this back-reflection may be very similar to the post-Es type of FAI propagation (page 2-48).

MAXIMUM Es DISTANCES

The maximum geometric range corresponding to E-layer heights of 90 to 120km is 2130 to 2450km (page 2-15). For comparison, Fig. 2.34 shows the maximum distances on 144MHz claimed by operators in the *DUBUS* DX lists. In good agreement with prediction, the distribution drops rapidly at 2400-2500km. Some extension beyond this limit may be credibly explained by good sites with distant horizons, with a little help from tropospheric refraction. However, this does not account for the significant tail of contacts extending to a minor peak at 3300±250km, which is not seen in other E-layer modes such as aurora or meteor-scatter at this frequency. The *DUBUS* lists and Fig. 2.34 represent people's best DX, while Fig. 2.35 shows the distances of ordinary Es contacts sampled randomly from reports of major 144MHz openings during 1987 and 1988. The most probable distance worked is between 1400 and 1900km, so the 3000km+ peak in Fig. 2.34 clearly represents double-hop Es. There are also a few examples of double-hop contacts (*eg* [17]) at times when both

VHF/UHF PROPAGATION

Fig. 2.35. Distances worked during several Es openings on 144MHz; the most probable range is 1400–1900km

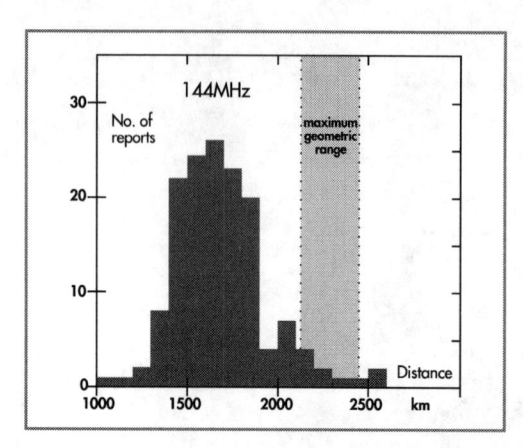

Fig. 2.36. Derivation of the most probable time of day for transatlantic 50MHz Es (curves), and actual times of contacts (histograms). See text for details

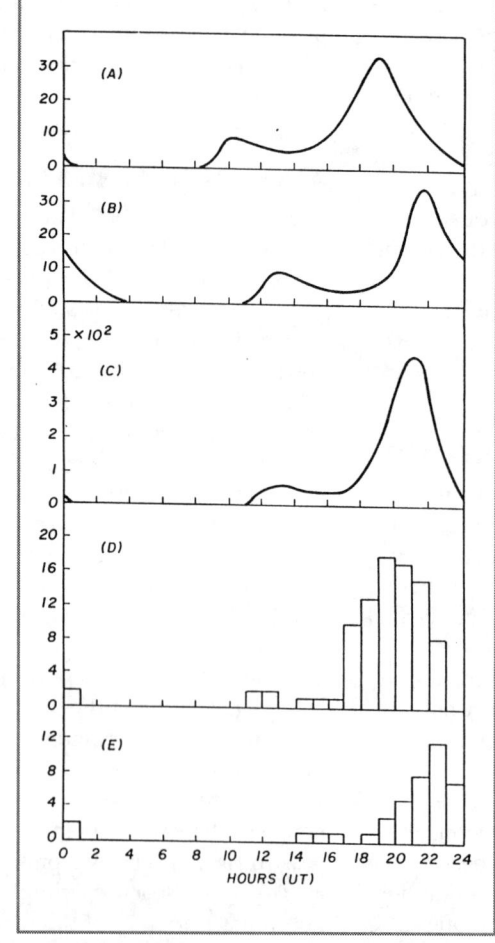

stations could work stations in the middle of the path by single-hop.

On the lower bands (50 and 70MHz) the upper distance limit for single-hop Es is essentially the same as for 144MHz, being determined largely by geometry. However, the minimum skip distance at these lower frequencies is reduced by the higher refraction for a given ionization density. Since usable Es ionization is much more widespread for lower frequencies, multiple hops can extend the range to many thousands of kilometres – enough to span the Atlantic on 50MHz [18] and even 70MHz.

MULTI-HOP Es

The time of year and day when Es is most likely to occur depends on your geographic position. The best time of year is clearly the summer, regardless of whether you live in the northern or southern hemisphere. Fig. 2.31 showed the most likely times of day, which were given in terms of *local* time. In terms of Standard Time (GMT or UTC), the plots must be shifted forward by one hour for every 15° longitude east of the Greenwich meridian, or back an hour for every 15° west. The dependence upon local time shows that temperate-zone Es takes its daily timing from the rotation of the earth, and therefore that solar radiation probably is responsible for generating the ionization. However, since Es occurs sporadically rather than every day, there must be other factors involved too – as discussed in the next section.

Multi-hop paths are likely to be formed from hops of the most common distance, *ie* 1650±250km. By dividing the great-circle path into segments of this length, you can find the most probable number of hops. Read off the latitude of the midpoint of each hop, and shift the characteristic daily frequency curve of Fig. 2.31 to the corresponding times in GMT. Finally, multiply each point on these curves together to produce a curve of the overall probability. In practice you need only work out enough points to sketch in the curve [18].

An example prediction is shown in Fig. 2.36 for a two-hop 50MHz Es path from the UK to the eastern seaboard of the USA. Figs 2.36a and b are the smoothed hourly distributions, moved westwards (*ie* later in the day) to the required positions for the two hops, and (c) is the combined probability curve. Compare this prediction with the actual hourly distributions of transatlantic Es openings: Fig. 2.36d gives data for 1987, while Fig. 2.36e gives the cumulative data for 1982-85. The agreement is good, showing that middle to late evening in the UK is the most hopeful time. A similar prediction for the even longer path from the FY7THF beacon shows equally good agreement with the actual times of reception, and looking in the other direction suggests late afternoon as the best time for the double-hop path to Cyprus and the Middle East [18].

CAUSES OF TEMPERATE-ZONE Es - FACTS AND FICTION

There have been many suggestions about how Es is formed, but not all of them stand up to scientific analysis. In this section I will try to sort out the facts from the fiction.

As mentioned earlier, Es layers are observed to contain high concentrations of ionized metals. These relatively heavy atoms tend to gravitate towards the earth, and it is supposed they are being continuously replaced by debris from meteors, many of which have a high metallic content. From this it has been suggested that the level of Es follows the annual variation of meteor rate, which peaks around the same period as Es in the northern hemisphere. But the Es season in the southern hemisphere is in their summer too, which is six months out of phase with the peak of the world-wide meteor rate. Hence it is not surprising that analyses of meteor and Es data have failed to reveal a convincing correlation.

Another apparent connection between meteors and Es is the enhanced rate of meteor reflections observed on 144MHz when the Es is approaching the critical frequency for that band. The explanation is simple – the pre-existing Es ionization simply adds to that arising from the burn-up of meteors at E-layer heights, producing an enhanced rate of audible meteor bursts. So although it is generally agreed that meteors provide the metallic raw material for Es, they clearly aren't the whole story.

Correlations between Es and many other factors have been claimed. Let's consider a few which have received the most publicity. Firstly, the solar cycle; this is known to have little effect on the E-layer MUF. The radiation absorbed in the E layer is primarily ultra-violet, which like visible light is almost independent of the solar cycle. A recent analysis of many years of data claimed a very weak dependence of VHF Es on the twelve-month mean sunspot number [19]. However, my own analysis of openings on 144MHz between 1972 and 1991 suggests a weak *negative* correlation with sunspot numbers; the probable reason for this is explained in the next section.

There have been suggestions that Es forms preferentially over mountains. If you look at a relief map of Europe, mountains such as the Vosges, Eifel and Jura can indeed be found at around half the most probable Es distance from central England. For longer paths, which maybe deserve higher mountains, the Alps and Pyrénées together with their lower extensions form an almost continuous arc covering all popular directions from the UK. However, this is rather a parochial view, and perusal of *DUBUS* magazine shows spectacular 144MHz Es contacts with no land above 200m anywhere near the path midpoint! Es on 50MHz certainly does not require high ground, as many transatlantic contacts have shown. If mountains do have any effect on Es it is clearly rather small.

What about thunderstorms? This idea looks promising, because it would explain why mountainous regions might be favoured – because storms tend to develop there during the summer months – while not ruling out the possibility of Es over lower-lying terrain. Furthermore there is a mechanism to explain the connection, namely that strong convection cells (cumulonimbus clouds) can develop high upward velocities which break through the tropopause at about 12km, causing gravity waves which propagate up to 100km altitude, where they form standing waves which could concentrate the available ionization into Es layers. But a number of facts suggest that storms cannot be the only cause of Es, or even the primary cause. Firstly, at the height of the season Es exists almost continuously at lower frequencies; second is the lack of correlation on a world-wide scale between the occurrence of Es and that of thunderstorms; and third is the observation of Es when the satellite weather pictures [20c] show no trace of thunderstorms! The advocates of the theory reply that the highest-frequency excursions of Es are sometimes connected with thunderstorms, or perhaps with high-altitude jet streams [20a-c], but none of these connections has yet been demonstrated with scientific conviction.

WIND-SHEAR THEORY OF Es

On a global scale, the distribution of sporadic-E is shown in Fig. 2.37 as contours of the percentage of the time during the summer months for which the Es critical frequency (f0: page 2-16) exceeds 7MHz, corresponding to an MUF of at least 50MHz. The geographic distribution is roughly proportional to the square of the horizontal component of the earth's magnetic field. This lends support to the normally accepted *wind-shear mechanism* for the vertical concentration of ionization into a thin Es layer.

The wind-shear mechanism works as follows [21]. If the charged ions are carried along by winds (which do exist at E-layer height) they will also drift at right-angles to both the

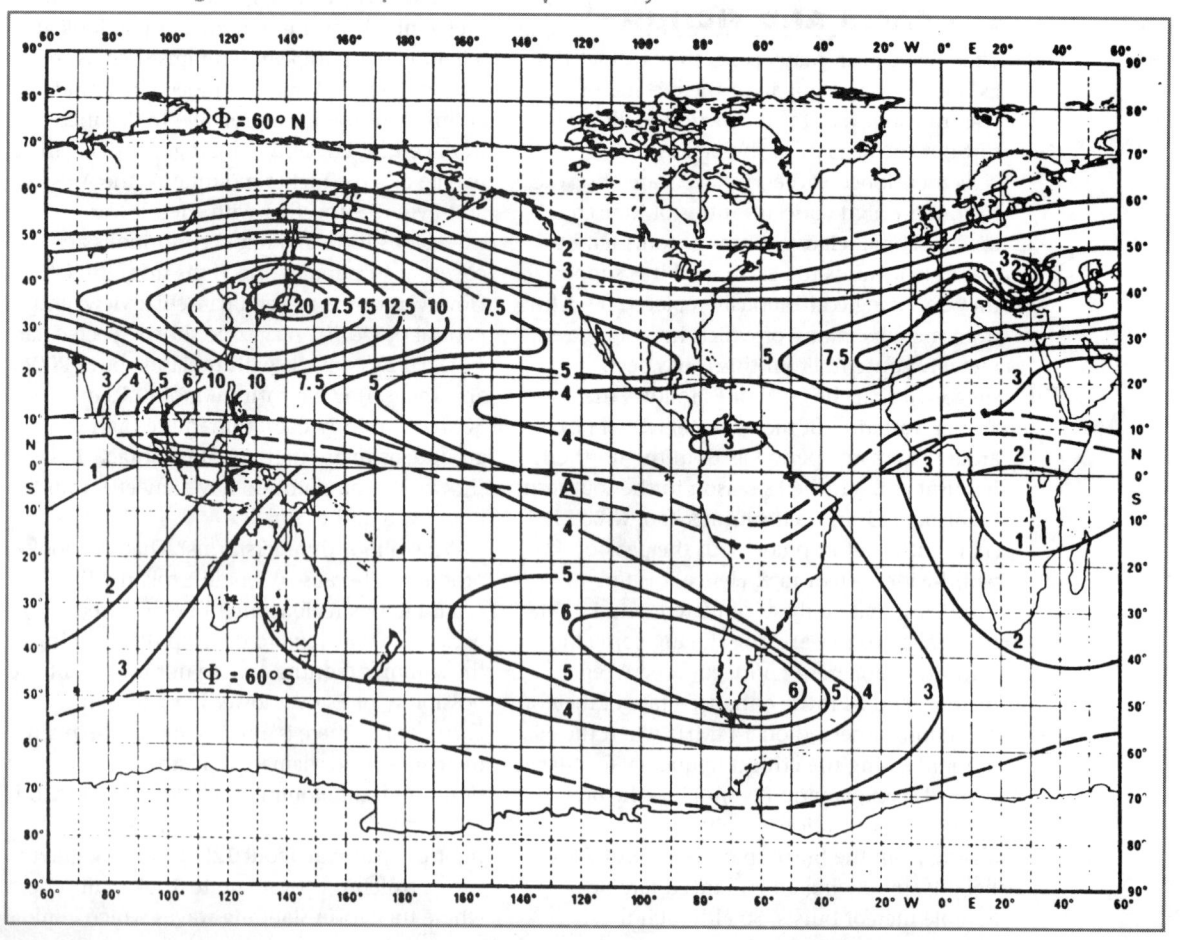

Fig. 2.37. World map of relative Es probability

earth's magnetic field and the wind direction. The force producing this drift is the same as that which deflects the electrons in your TV tube to scan the screen, and the direction of drift will reverse if the wind direction is reversed. The drift in the vertical direction is determined by the horizontal component of the earth's magnetic field. If the wind moves with different velocity at different heights (the technical definition of 'shear'), electrons at different heights will drift at different speeds, and hence will bunch into layers. If the wind direction actually reverses, the layer will be stabilized at the height of the reversal.

This wind-shear mechanism probably applies to all types of Es as a means of concentrating the ionization into layers. The mechanism clearly operates best under stable geomagnetic conditions and will be upset by auroral activity; hence the negative correlation with sunspot number mentioned earlier. Auroral Es (see later) normally occurs immediately before or after an auroral storm when the magnetic field is not so highly disturbed, and is facilitated by the massive ionization spread throughout the E region at such times. However, wind shear at 100km is difficult to detect and measure, and its relationship with more readily observable objects or events has been the subject of much speculation.

To summarize, the best scientific authorities – even including the inventor of the wind-shear theory – still do not have a complete theory for the formation of Es [21]. We know that the raw material to be ionized is provided by meteors; and that the energy for ionization comes directly from the sun in the case of temperate-zone Es, and from auroral particles in the case of auroral Es. These long-lived ions are concentrated into irregular thin layers by wind shear combined with the earth's magnetic field. But we do not yet understand these factors well enough to predict the occurrence of Es. The 'weather' at E-layer height bears little relation to the more readily observable conditions down here in the troposphere, so the prospects for amateurs to predict Es from ordinary weather maps are not good. The very fact that intense Es *is* sporadic suggests that it requires several unrelated factors to be present, all at the same time.

If any one of them is missing, you will search the bands in vain.

AURORAL Es

Auroral-zone sporadic-E differs from the more familiar temperate-zone Es by being generally found at higher latitudes, at night as well as during the day, and at other times besides the summer. This is because the ionization originates from incoming auroral particles rather than solar ultra-violet radiation, hence the time and place of auroral Es tends to follow that of the aurora. Usually auroral Es is formed from the ionization remaining after an auroral storm and its associated geomagnetic disturbance have subsided, though it can precede an aurora if sufficient ionization is already present from particle precipitation. The mechanism which concentrates the ions into a layer sufficiently dense to reflect VHF is probably wind shear, as described above for temperate-zone Es.

Although auroral Es is much more common at more northerly latitudes, after a big storm the ionisation will spread southwards and becomes accessible at mid-latitudes. For example, during the huge aurora of 13 March 1989, 50MHz auroral-E signals from Finland appeared in the UK as the auroral oval was still expanding and passing overhead. This aurora spread so far south that intense auroral Es formed in traditional temperate-zone Es territory, giving almost T9 signals on 144MHz from Italy, Yugoslavia and Bulgaria and the surrounding area as late as 0200 hours! On the same day, very similar auroral Es was observed in the USA [22].

VHF/UHF PROPAGATION

FIELD-ALIGNED IRREGULARITIES

Field alignment refers to the effect of the earth's magnetic field on the electrons trapped in the ionosphere. As with auroral particles, their net motion is aligned along the geomagnetic field lines. As the field alignment becomes more distinct, the scattering of radio waves becomes more coherent (page 2-8). The term 'FAI' (standing for field-aligned irregularity) refers to this type of propagation in the E region, but field alignment also plays its part in the F region during trans-equatorial propagation (TEP) – see page 2-52.

FAI signals are weak and display the characteristic frequency spread of semi-coherent scattering [23].

The levels of ionization necessary for FAI to be useful are mostly only found during the temperate-zone Es season; FAI may thus be thought of as an above-MUF manifestation of Es. The times of day perhaps extend later into the evening, so it is worth checking for FAI after an Es opening. As with Es, FAI requires quiet geomagnetic conditions to avoid disturbing the field alignment.

The strongest signals scattered from aligned structures such as FAI will be when both the incident and reflected directions are perpendicular to the axis. Moreover, the maximum effect will be when the electric field (polarization direction) of the wave is parallel with the axis. These requirements are exactly the same as for aurora and meteor trails; but unlike those two modes, the FAI signals are so weak that you can only expect to work stations in the most favoured locations.

An isoflectional map giving the favoured locations for auroral scattering from London has already been shown in Fig. 2.25, and

Fig. 2.38. Isoflectional map for a station in Geneva. Comparing this with Fig. 2.25, the prospects for FAI contacts between Geneva and London seem unfavourable

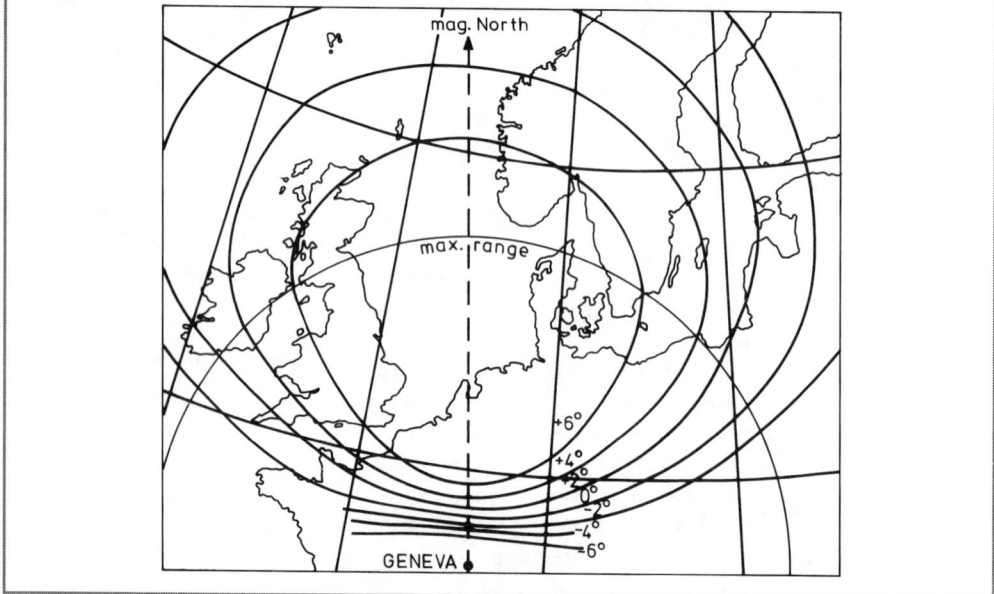

F2 PROPAGATION

applies equally to FAI. Fig. 2.38 is a further example for Geneva. The isoflectional contours are roughly symmetrical about lines of magnetic latitude, which are tilted with respect to ordinary geographic latitude. The isoflectional map is unique for each station's location, and cannot simply be shifted across large distances as some people have tried to do. The circle of about 1150km radius labelled "max. range" represents the maximum distance at which the ionosphere at a height of 100km is visible from Geneva.

By superimposing the isoflectional maps for two different stations, areas of potential FAI propagation may be identified. The necessary condition is that both stations' isoflectional contours must intersect somewhere within the area where their 1150km circles overlap. Unfortunately the contours in Figs 2.25 and 2.38 do not intersect as required within the overlap area, so FAI propagation between the UK and Switzerland is very unlikely because the unfavourable geometry will make the scattered signals too weak. From the UK, east-west paths appear to be favoured, using a north-easterly scattering region, the same as described for aurora in Fig. 2.25. Since Es-type ionization at this latitude is much less likely than further south, FAI is unlikely to be an important mode for amateurs in the UK. If contacts are sometimes made by this mode, they are likely to be attributed to a weak aurora because of the rough tone of the signals.

Even in southern Europe, where the isoflectional maps and higher occurrence of Es suggest that FAI will be more probable, contacts are attributed to FAI which certainly cannot be justified. There have been suggestions that FAI occurs over specific locations, *eg* Brussels, while others express a more general preference for mountainous regions, in particular the Alps. However, if contacts made by beaming at the so-called "FAI hot-spots" fail to satisfy the magnetic criteria for FAI, they must be taking place by some different mode. Perhaps the main virtue of these supposed 'hot-spots' is to focus interest in trying for weak-signal DX!

Ionospheric F2-layer propagation is used by HF operators to contact stations around the world, and its fluctuations are well known to any listener to the short-wave bands. When all the factors are favourable, the Maximum Usable Frequency (MUF) extends to 50MHz and very occasionally to 70MHz. Under these conditions 50MHz resembles the open 28MHz band, although propagation tends to be geographically more selective. Sometimes signals are very strong, because ionospheric absorption is lower at VHF. Factors determining the level of F2-layer ionization are mainly the solar cycle, the season of the year and the time of day. Less predictable variables include changes in solar radiation and geomagnetic disturbances of auroral origin.

THE SOLAR CYCLE

The higher-energy shorter-wavelength radiations from the sun (gamma rays, X-rays and far ultra-violet radiation) are absorbed in the outermost region of the atmosphere to create the F region of ionization. F-layer propagation depends on the average intensity (or 'flux') of these shorter-wavelength radiations, which in turn depends on solar activity. This follows the well-known eleven-year cycle, which is actually a 22-year cycle if the polarity of the magnetic fields associated with sunspots is taken into account [24].

In contrast the longer-wavelength, less energetic radiations which penetrate much further down into the E region are almost unaffected by the solar cycle. There is no 'sunspot cycle' in the levels of visible light or the near ultra-violet.

VHF/UHF PROPAGATION

Fig. 2.39. Widely scattered points show that the correlation between sunspot number and solar flux is only approximate

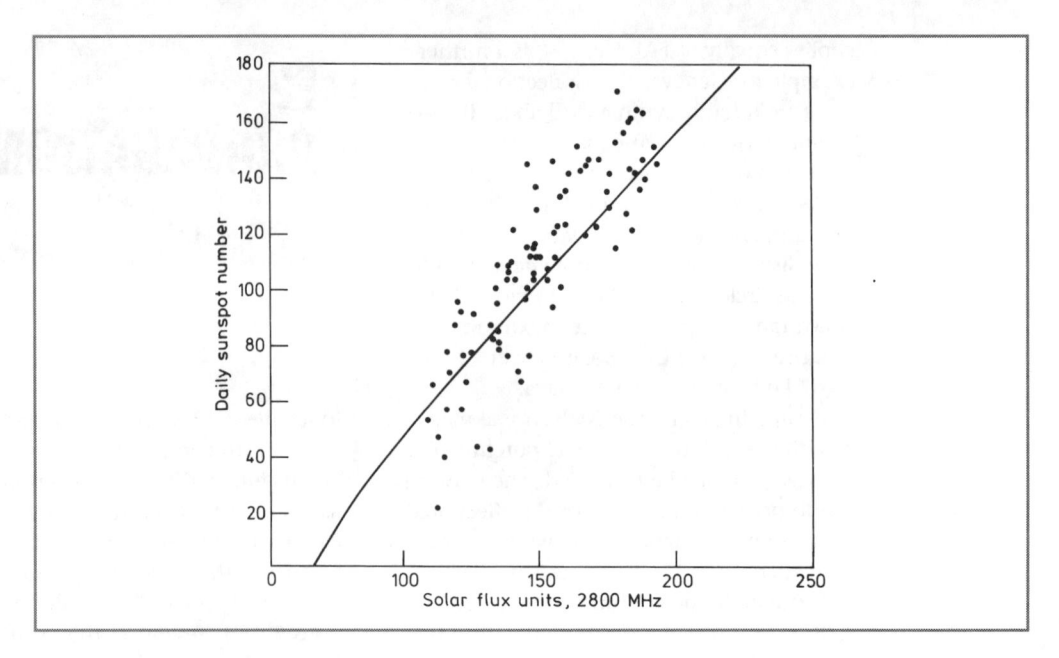

MEASUREMENTS OF SOLAR ACTIVITY

The traditional measure of solar activity is the sunspot number (SSN) which is discussed in more detail in the next section. However, a better measure of solar activity as it affects the ionosphere is the *solar flux unit* (SFU). This is a simple measurement which is immediately available, unlike the SSN which requires processing. The solar flux is measured at several frequencies, the value at 2800MHz being considered the best measure. There is a correlation between the SSN and the flux (Fig. 2.39), but the widely scattered points show that this is an ***approximate*** relationship and not the simple curve shown in Fig. 2.39, which is often assumed in home-computer MUF predictions.

Current solar-flux and magnetic-disturbance data can be obtained by phoning the Geophysical Data Centre at Boulder, Colorado directly on (010) 1 303 497 3235. The information from Boulder is also broadcast on the WWV standard-frequency transmissions, and are usually available on the DX packet-cluster network. The RSGB plans to broadcast solar and geophysical data several times a day from the UK on a frequency just above 3·8MHz.

SIGNIFICANCE OF SOLAR ACTIVITY

Contrary to continuing popular belief, the day-to-day MUF does not follow solar activity up and down. The ionosphere acts like a capacitor, taking time to charge up and also time to decay. The capacitor analogy can be taken further, since the rate of production or recombination of the ions depends on the number already present; this results in an exponential charge or decay curve, just like a capacitor. The value of SSN or flux which is normally used in programs to predict HF propagation is in fact a mean value averaged over several months; three, twelve and thirteen months are popular periods.

Rather than being beneficial, sudden increases in solar activity can produce deleterious effects such as *sudden ionospheric disturbances* (SIDs), *polar-cap absorption* events (PCAs) and fade-outs. The SID and PCA have little effect at VHF, but the fade-out does indicate a decrease in the MUF. Sudden increases in solar activity can also trigger aurora (page 2-37).

SEASONAL VARIATIONS IN THE F2 LAYER

Although the ionization in the F2 layer is

F2 PROPAGATION

caused by solar radiation, the maximum electron density is not found at latitudes where the sun is directly overhead but in regions 10–15° north and south of the magnetic equator. The MUF decreases either side of the maximum by about 0.8MHz per degree of latitude, as stations at higher latitudes know to their cost. These regions of maximum MUF do not follow the seasonal movement of the sun north and south of the geographic equator, but merely change in relative intensity. The F2 MUF peaks around the equinoxes when the two regions are equally illuminated, rather than at mid-summer in either hemisphere. When approaching the solar maximum, the later (autumn) equinox gives higher MUFs, while during the decline towards the solar minimum the spring equinox is better.

DAILY VARIATIONS IN THE F2 LAYER

The largest change in MUF is its daily variation. However, we know that this cannot be entirely due to the increased ionization around the sub-solar point, because changes in solar flux on a timescale of hours do not result in a significant increase in the global MUF. In fact, the daytime increase in MUF results partly from the pressure of the solar wind on the ionosphere, which results in a lowering and compression of the F region. This produces a higher ionization density and also a more favourable geometry for reflection (Figs 2.21 and 2.40).

The MUF may show a peak considerably before or after local noon. Sometimes two daily peaks occur, as with sporadic-E (page 2-41) but closer together. Statistically, however, the paths with the highest MUF are still most likely to be those at which local noon occurs at the midpoint. In other words, look for openings to Australasia and the Far East just after dawn, Africa around mid-day, and the Caribbean and then the USA in the afternoon and evening. If in doubt, point your beam towards the sun!

TILTS, CHORDAL HOPS, AND NON-GREAT-CIRCLE PATHS

At dawn and dusk, when the change in height of the electron concentration is taking place, the ionosphere is 'tilted' from east to west. This tilt can launch *chordal hops* involving two or more ionospheric reflections but no intermediate ground reflection (Fig. 2.40). The MUF for chordal hops is higher by virtue of the shallower angles of incidence and there are no ground-reflection losses, so the chordal hop is probably the favoured mode of long-distance F2 propagation on 50MHz. For a given number of ionospheric reflections, a chordal hop covers no greater distance; for example, two maximum-distance conventional hops joined by a very shallow ground reflection will cover essentially the same distance as a maximum-distance chordal hop passing just above the ground.

Chordal hops are not confined to twilight propagation, however. Every change in electron density with latitude or longitude can be thought of as a 'tilt', in that it changes the effective height of the layer. In this sense sporadic-E clouds are full of tilts due to the lumpy nature of the ionization. Also the F layer changes density with latitude, as described above. These effects are important for

Fig. 2.40. Effect of F2 layer tilt in launching chordal hops which involve no intermediate ground reflection (not to scale)

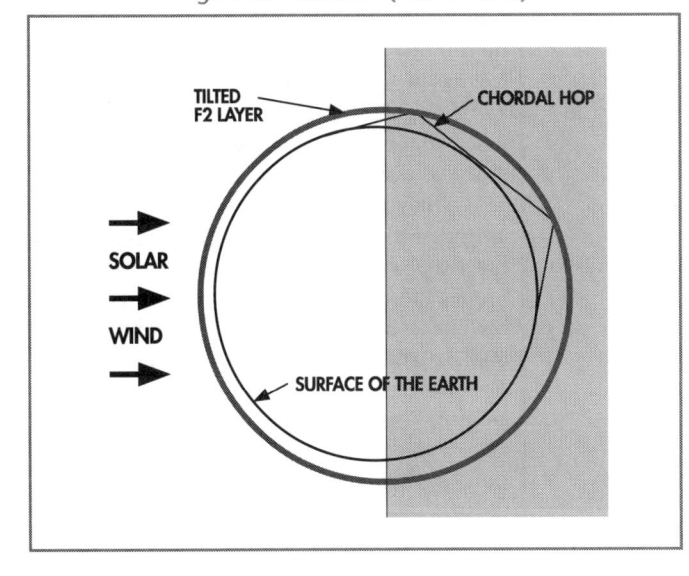

trans-equatorial propagation (see next section), and tend to produce an equatorial chordal-hop.

Six-metre operators frequently observe that F2 signals do not arrive *via* the great-circle path. There are several possible reasons for this. Except on direct N-S paths there will be some 'sideways' refraction through the equatorial F-layer belts which will affect the direction from which signals arrive. Also the concentration of ionization where the sun is directly overhead produces a 'tilt' in all directions, rather like a domed convex mirror. Finally there are many smaller fluctuations in ionization density from place to place. All in all, the idealized picture of a smooth horizontal ionospheric layer at a constant height is entirely false, especially at VHF where only the extremes of electron density give propagation.

TRANS-EQUATORIAL PROPAGATION

Trans-equatorial propagation (TEP for short) has become an interesting mode for UK amateurs since the release of the 50MHz band. The identification and exploration of this propagation mode has been carried out largely by amateurs applying scientific methods, an achievement of which we can be proud [25]. By measuring the time delay along the path, it was demonstrated that TEP involves reflection from the ionospheric F layer.

As mentioned above, the maximum F-layer ionization occurs in two belts located north and south of the geomagnetic equator. These belts form in the morning, are well developed by noon, and decay after sunset to reach a minimum just before dawn. The positions of the ionization belts are independent of the time of year, but they become unbalanced in intensity as the sun favours the one or the other. TEP makes use of both these regions of high ionization, and the favoured regions thus follow the geomagnetic equator around the world (Fig. 2.41). TEP is therefore at its best when the intensities of the two regions are greatest, which is around the equinoxes.

TEP may be classified into three main types, although they are not totally distinct in practice. Normal single-hop or double-hop F2 propagation across the equator shows a higher MUF than in temperate latitudes. So the 50MHz paths to southern Africa and South America can be open when other directions are not. This F-type TEP merely takes advantage of the equatorial regions of high ionization, and signals have normal F-layer characteristics.

The mode of propagation more generally

TRANS-EQUATORIAL PROPAGATION

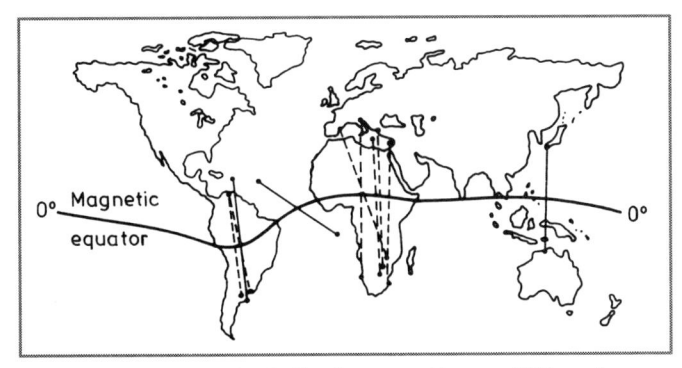

Fig. 2.41. The longest-distance TEP paths on 144MHz have been symmetrical about the geomagnetic equator

recognized as TEP takes place directly across the equatorial region without an intermediate ground reflection. This uses the concentration gradients in the two equatorial regions of high ionization to make a direct chordal hop between them (Fig. 2.42). Because the required angles of bending are smaller, chordal TEP can take place at frequencies well above the MUF for ground-reflected trans-equatorial F2 propagation. A factor of 1·5 is generally quoted, and the MUF for chordal TEP can approach 100MHz during optimum conditions. Paths inclined by up to 45° to the line of maximum ionization are workable. Signal characteristics are again similar to normal F-layer propagation, and signals can be very strong since there are no losses at a ground reflection. Indeed, it might be that some focussing can take place since signals have been claimed to exceed free-space levels by up to 3dB, although it is very difficult to measure path loss to this accuracy.

The third type of TEP appears following strong F2-layer propagation. During the daytime, the pressure of the solar wind compresses the earth's magnetic-field lines. After sunset at a few hundred kilometres above the surface of the earth (which occurs later than sunset on the surface), the magnetic field lines expand again. The result is that the F region becomes more turbulent, and the ionization layers become broken up into a condition known as *spread F*. The forced motion of the electrons causes them to become field-aligned, forming near-horizontal lines of concentrated ionization which propagate VHF and even UHF signals by relatively coherent forward-scattering. However, the motion of the electrons does give the signals a somewhat 'auroral' quality.

These field-aligned F-layer ducts are irregular, and the received signals are weak and arrive by multiple paths from a large angle of the sky. The field alignment makes this type of TEP very sensitive to location; paths are confined to a few degrees around the perpendicular to the magnetic equator, and within a narrow belt about 4000km north and south of it. The way that these signals appear following strong F2 propagation is reminiscent of the weak field-aligned E-layer signals which occur after strong Es.

All these types of TEP occur on the 50MHz band, and the field-aligned type also works on 144MHz and even 432MHz. As the frequency is increased, propagation becomes rarer and geographically more restricted, the duration of the openings becomes shorter and they occur closer to 2000 local time. A typical pattern for a good opening is that 50MHz F-type TEP signals fade out and 144MHz FAI-type TEP signals appear during the period 1930-2030 local time as the F2 region spreads and breaks up into its field-aligned form. Later these 144MHz signals fade out, and 50MHz signals reappear, this time also with FAI characteristics. This indicates that the effective size of the

Fig. 2.42. Transequatorial chordal hop uses the two F2 ionization belts on either side of the geomagnetic equator (not to scale)

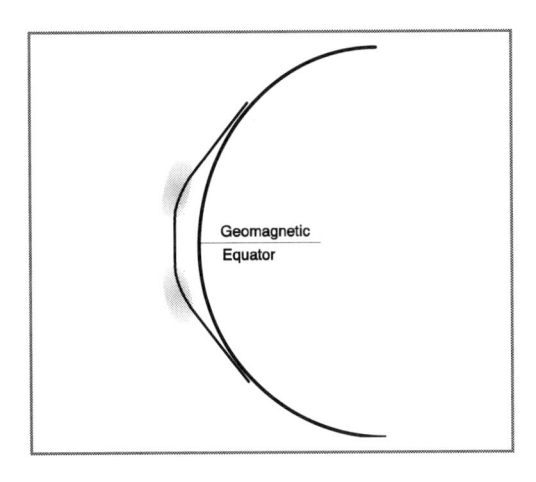

2 - 53

TEP ducts is growing, propagating the longer wavelengths, but that consequently the ionization density has decreased so that 144MHz signals can no longer be trapped within the duct.

It is unlikely that the TEP zones can be accessed from the UK without the aid of another propagation mode. E_s seems the only contender since tropo ducting seldom extends far enough, and it is believed to have assisted in 50MHz chordal TEP contacts from the UK. However, the coincidence of a suitable 144MHz E_s opening from the UK to connect with TEP conditions further south would be very rare, especially at times as late as 2000 hours.

IONOSPHERIC SCATTERING

Ionospheric scattering is similar to tropospheric scattering, except that the variations in refractive index are due to differences in ionization density rather than meteorological properties. Most ionospheric scattering takes place from heights of 100km or below, and it peaks in the D layer at around 85km. This height corresponds to a maximum range of about 2000km. The minimum range is set by the troposcatter signal taking over, which occurs at about 800km. The situation is not as simple as troposcatter, however, because the movements of ionization will be guided by the earth's magnetic field as well as by atmospheric winds and turbulence.

The scattering effect increases with the ionization level of the D layer, so signals are strongest during the peak of the sunspot cycle, in the summer and around noon. *Sudden ionospheric disturbances* (SIDs), which are associated with high D-layer ionization, can enhance signals on 50MHz while causing fadeouts on LF and HF bands. A similar effect may be associated with the precipitation of auroral particles.

Ionospheric scattering decreases rapidly at higher frequencies, the intensity being proportional to about $1/f^{7.5}$. Since the cosmic background noise decreases as approximately $1/f^{2.5}$, the signal/noise ratio varies as $1/f^5$ if receiver noise is negligible. This means that in practice 50MHz is the only band on which this mode is likely to be useful for most amateurs.

Meteor-scatter may be thought of as a special case of ionospheric scattering. Underdense trails from sporadic meteors

IONOSPHERIC SCATTERING

produce very similar signals to those from true ionospheric scatter, but MS peaks at around 0600 local time, much earlier in the day than ionoscatter. Similarly, FAI may be considered as a special case of ionospheric scattering, in regions where the turbulence is small enough to permit field alignment. Since it is sometimes difficult to isolate and identify these modes, there is room for debate about the nature of the propagation. Another obvious potential source of confusion is with tropospheric scattering, which will have the same signal characteristics and may also peak during the midsummer months at around noon.

At the ERPs generally employed by 50MHz DXers in the USA, 'scatter' (of whatever origin) is a reliable mode for weak-signal summer DX, although it is doubtful whether this is usable at the very low ERPs presently allowed on 50MHz in the UK. In spite of the adverse frequency dependence, ionospheric scattering just might be responsible for reported contacts on 144MHz between stations of moonbounce capability on paths such as the UK to Italy and Spain to Sweden which are geomagnetically unfavourable for FAI.

METEOR-SCATTER

The name 'meteor-scatter' is misleading. It is not the meteors themselves which scatter signals beyond the horizon, but the trails of ionization which are left behind as these high-velocity fragments burn up. Meteors capable of producing a useful ionized trail range in size from a grain of dust up to huge boulders. Remainders of the latter very occasionally reach the earth's surface as meteorites, but fortunately the lighter meteors predominate!

A typical meteor is the size of a grain of sand and burns up at a height of about 100km. Variations in mass, composition and velocity result in trails between 20 and 65km long and about 1m in diameter. The trail rapidly expands as the electrons diffuse into the surrounding space; this effect gives rise to some Doppler shift, particularly at the start of a burst.

The velocity at which meteors enter the earth's atmosphere is important, because the kinetic energy of the meteor – and hence its ionizing power – is proportional to the mass and to the square of the velocity. Typical velocities are 10–100km/s, and the actual value depends on three factors:

1 The motion of the meteor itself within the solar system

2 The orbital velocity of the earth round the sun

3 The velocity of the particular point of entry due to the earth's rotation on its axis.

The earth's rotational and orbital velocities add in the morning hours and subtract in the evening (Fig. 2.43a). So the meteors create more ionization in the morning, and there also appear to be more of them because

VHF/UHF PROPAGATION

Fig. 2.43.
(a) Random meteors are most effective from midnight to dawn, when the earth's rotational and orbital velocities add together;
(b) Average rate of random meteors for days in June

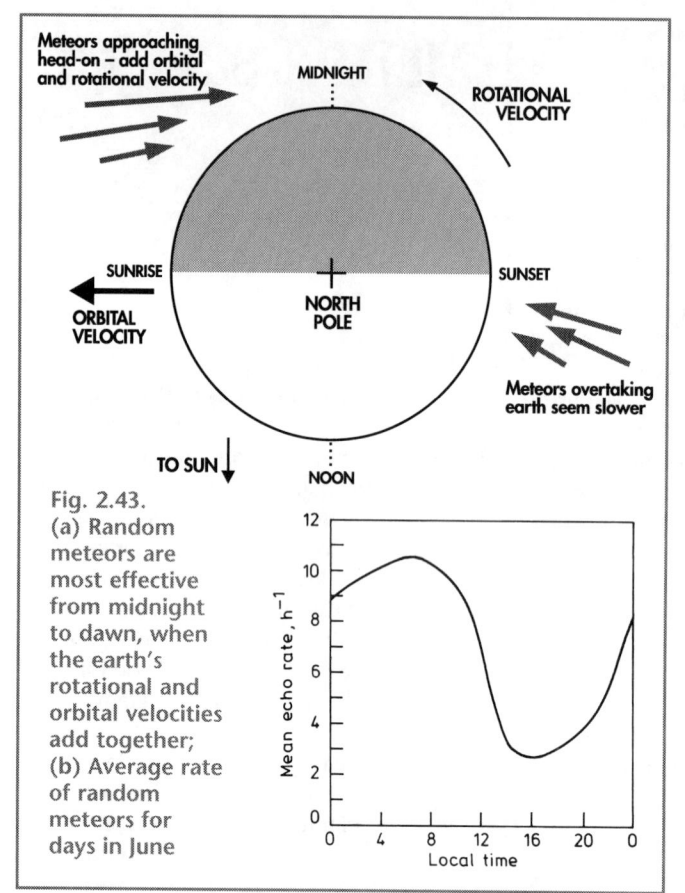

Fig. 2.44. Geometric requirements for maximum reflection from meteor trails. There are two 'hot-spot' regions where these requirements are met, one on each side of the great-circle path

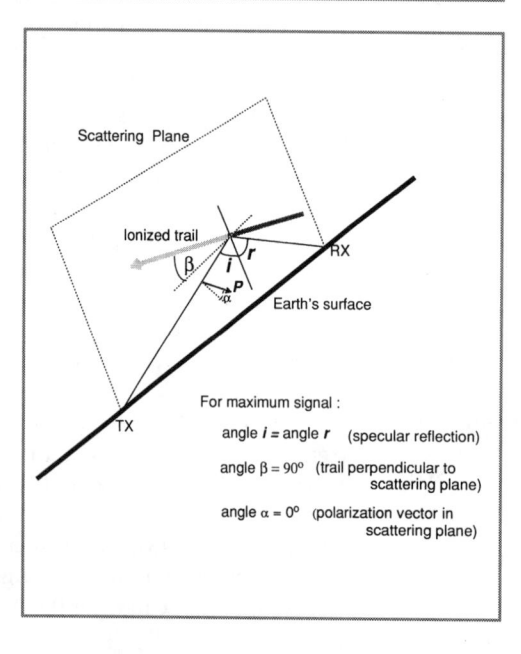

For maximum signal:

angle i = angle r (specular reflection)

angle β = 90° (trail perpendicular to scattering plane)

angle α = 0° (polarization vector in scattering plane)

effective trails can also be produced by meteors of smaller size. At mid-latitudes this gives an approximately sinusoidal daily distribution of sporadic-meteor signals with a factor of four difference in rate between morning and evening (Fig. 2.43b). The hours around 0600 local time are therefore best to try sporadic MS contacts.

The strength of the reflected signal depends not only on the initial ionization density but also strongly on the direction of the meteor trail relative to the two stations. Hence if several stations across the country are listening for the same DX signal, they will in general hear reflections from different meteors, with different message contents. The strongest signals are received when the conditions summarised in Fig. 2.44 are met. For a given distance, these conditions imply that there are actually two areas at E-layer height which give the best reflections, located on either side of the direct path. As the range increases to the maximum, these 'hot-spots' move in towards the direct path.

SPORADIC AND SHOWER METEORS

Sporadic or random meteors arrive all the time and from all directions, although their rate of arrival varies with the time of day and season of the year. Shower meteors are probably the remains of comets which have broken up rather recently on astronomical timescales, and they appear at particular times of the year because they travel in fixed orbits around the sun which the earth intersects annually in its own orbit. A meteor shower therefore arrives from a specific direction in space, called its *radiant*. Each shower is named after the stellar constellation in which its radiant appears. The name of the month or a Greek letter may be added in cases of ambiguity, producing names such as the June Perseids (from the constellation Perseus), or the *eta* Aquarids (named after Aquarius). The radiant moves across the sky as the earth rotates so, if you wish to make contacts via shower meteors, first make sure that the radiant is above the horizon!

The table above gives a list of the principal

METEOR-SCATTER

PRINCIPAL METEOR SHOWERS AND FAVOURED DIRECTIONS

Shower name	Limits	Max	ZHR	N-S	NE-SW	E-W	SE-NW
Quadrantids	1-6 Jan	3-4 Jan	60	02-06(W) 11-16(E)	11-17(SE)	23-03(S) 15-17(S)	00-05(SW)
Lyrids	19-25 Apr	22 Apr	10	22-02(W) 06-10(E)	23-03(NW) 08-11(SE)	03-06(N)	22-01(SW) 05-08(NE)
Eta Aquarids	24 Apr – 20 May	5 May	35	03-04(W) 10-11(E)	04-09(NW)	05-11(N)	08-12(NE)
May Piscids	4-27 May	12 May	10	03-08(W) 12-16(E)	05-09(NW) 14-16(SE)	08-11(N)	04-08(SW) 10-14(NE)
Arietids	22 May – 2 July	8 June	60	04-08(W) 11-15(E)	05-09(NW) 14-16(SE)	08-12(N)	04-06(SW) 10-14(NE)
Zeta Perseids	20 May – 5 July	13 June	40	05-10(W) 13-17(E)	06-11(NW) 15-17(SE)	09-14(N)	06-07(SW) 11-15(NE)
Nu Geminids	9-15 July	12 July	60	06-10(W)	07-11(NW)	10-14(N)	11-17(NE)
Delta Aquarids	15 July – 20 Aug	29 July, 7 Aug	20 10	Equatorial and southern hemisphere only			
Perseids	3 July –	12-13 Aug	75	23-04(W) 09-13(E)	08-17(SE)	11-01(S)	18-04(SW)
Orionids	16-27 Oct	22 Oct	25	00-03(W) 07-09(E)	00-04(NW)	03-06(N)	05-08(NE)
Taurids	20 Oct – 30 Nov	4 Nov	10	02-05(E) 20-22(W)	20-01(NW)	22-03(N)	00-05(NE)
Geminids	7-15 Dec	13-14 Dec	75	04-09(E) 20-01(W)	22-02(NW) 05-09(SE)	01-04(N) 03-07(S)	03-07(NE) 19-23(SW)
Ursids	17-25 Dec	23 Dec	5		07-01(SE)	00-24(S)	16-09(SW)

Notes
ZHR indicates zenithal hourly rate of meteors.
Favoured directions and times are valid for latitudes around 50°N. Local times are valid for any longitude. For optimum reflection geometry on shorter paths, both stations should offset their beams by 5-15° towards the direction shown in brackets.
Table updated and revised from the *RSGB Operating Manual*, using additional data from the *BAA Handbook 1990* (1989) and *Meteor Showers: A Descriptive Catalog* by G W Kronk (Enslow 1988).
Many minor showers are also listed in these sources and may be worth trying.

showers in the northern hemisphere. A number of major showers occur in the months of May to August inclusive, and these dominate the annual totals of meteors. The most effective showers for meteor-scatter are generally the August Perseids and the December Geminids. The astronomical parameters of meteor showers change only slowly, though their intensities can vary considerably from year to year. Meteor rates at the peaks of showers may far exceed the random meteor rate, although the densities of individual meteor streams can vary markedly from year to year.

The values which appear in the annual Handbook of the British Astronomical Association are revised each year.

PINGS AND BURSTS

Two types of received MS signals can be recognized. The first is generally called a 'ping'; the initial signal is abrupt and rapidly dies away. Pings occur from underdense trails which scatter signals semi-coherently rather than truly reflecting them (page 2-8). The longer signals are called 'bursts', although many of these also come from underdense trails. Contrary to popular belief, only the very longest and loudest reflections on 144MHz arise from overdense trails capable of supporting true coherent reflection. The signal from an over-dense trail increases in strength as the trail expands, due to the increased reflecting area, but also oscillates in strength as the trail's diameter passes through multiple half-

VHF/UHF PROPAGATION

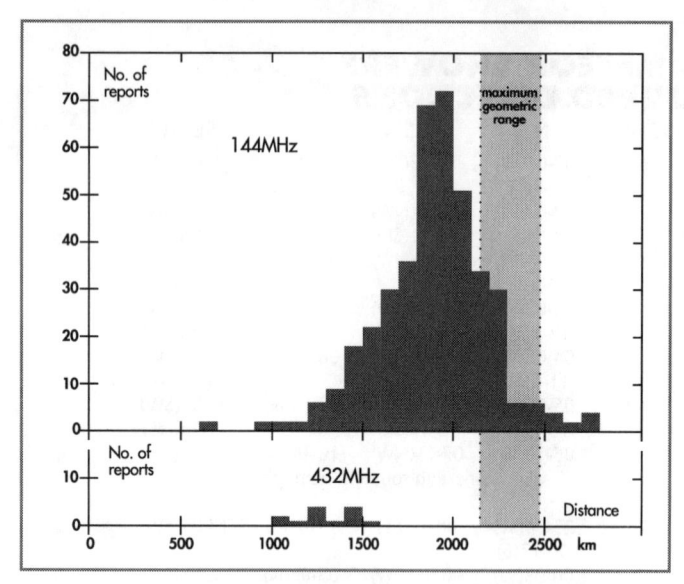

Fig. 2.45. DX operators' maximum ranges by meteor-scatter, taken from the 'Top List' in *DUBUS* 1989/1

wavelengths. After a time, the trail expands so much that it becomes underdense and then the signal decays rapidly. Obviously some trails remain overdense for longer times than others, and on rare occasions the best trails last long enough for a slick operator to make several contacts in succession.

The ionization density necessary to support coherent reflection increases with frequency, so a trail which is overdense at 50MHz will probably be underdense at 144MHz, and will certainly be underdense at 432MHz. Calculations show that for both kinds of trails, the power returned varies as the cube of the wavelength (proportional to $1/f^3$), while the signal duration varies as the square of the wavelength ($1/f^2$). Meteor-scatter is therefore much more effective on the 50 and 70MHz bands, where the longer bursts and smaller Doppler shifts also make SSB contacts much easier. On 144MHz the signals are less frequent and of shorter duration, so random MS contacts tend to take much longer. On 432MHz even more patience is needed, and skeds are better arranged only during the more intense showers (Chapter 3). Although less effective during the spring months when meteors are few, MS provides DX contacts throughout the year when other modes are absent.

BEAM DIRECTIONS AND TIMES

On all frequencies, the weaker underdense returns predominate. This means that high ERP is an advantage, though the narrower beamwidth which accompanies high-gain antennas will reduce the volume of sky from which returns can be heard. Especially for sporadic meteors, the 'hot-spots' on either side of the direct path should both be illuminated to double the chances of a suitable meteor burning up thereabouts. In other words, point a not-too-directional beam along the great-circle path and hope for some meteors. The path geometry strongly favours horizontal polarization at both stations.

The situation is different for shower meteors. Here the direction of the trail is predetermined, so the quality of MS paths in particular directions can be calculated for your location. Some years ago I wrote a computer program to tabulate the optimum directions, and printouts of these predictions for a location in central England can be purchased from RSGB. The calculation makes use of spherical trigonometry rather than the over-simplified 'flat earth' approximation used in other published programs. For shower meteors, one of the two offset scattering regions or 'hot spots' will be better than the other, and the beam offsets from the great-circle heading are given in the program output. For shorter contacts it may be an advantage to elevate the beam, but remember that the antenna radiation pattern is normally tilted upwards already by ground reflections (Chapter 7).

One final note about the path-quality predictions. They are based purely on geometry; even if you are beaming in the right direction, the predictions cannot guarantee that any meteors will arrive! When arranging skeds for reliable showers, check for evidence of last year's peak time from the QSO lists in *DUBUS*. This may not be the same as the peak time for visual observations published in the astronomical magazines. On the basis that the

METEOR-SCATTER

earth intercepts the shower orbit every 365·25 days, simply add six hours to the date and time of the previous year's radio peak. For leap years, you must also subtract one day for all shower dates after 29 February (including the January Quadrantids in the following calendar year).

DISTANCES BY MS

Distances worked by MS are what one would expect from E-layer reflection. On 50 and 70MHz the majority of returns come from between 88 and 100km high, corresponding to a maximum range between 2100 and 2250km. Unfortunately, 70MHz contacts of those distances from the UK must generally be crossband. A plot of maximum claimed distances on 144MHz reported in *DUBUS* implies a similar reflection height, though there are a few claims extending to 2800km (Fig. 2.45). It can be argued that the maximum range should be greater on the lower bands, since they can use the less dense trails commencing at greater heights.

Not shown in Fig. 2.45 is one isolated claim of a distance of 3200km being worked from Europe, and there are a few similar examples from the USA. The propagation mode for these extremely long distances remains unexplained. Double-hop MS does not fit the evidence; nor does single-hop using meteors burning up at extreme heights. The most likely explanation seems to be single-hop MS assisted by some other mode.

MOONBOUNCE

Reflecting signals off the moon – normally called EME (earth-moon-earth) or moonbounce – has always been considered a rather esoteric mode, because it is only just possible using amateur equipment. However, EME is one of the most reliable modes of DX propagation and is the only one to give world-wide coverage on the higher VHF/UHF bands without using satellite repeaters. It is also one of the few modes in which the path loss is fairly well known in advance. When the moon is closest to the earth the round-trip path loss is 251·5dB at 144MHz, 261dB at 432MHz, and 270·5dB at 1296MHz. From this information and the station parameters, the expected signal/noise ratio can be estimated. In practice there are considerable fluctuations in signal strength (mostly downwards!) which we describe below.

TRACKING THE MOON

Obviously the moon must be above the horizon for both stations, although atmospheric refraction may mean that signals may still be heard with the moon a few degrees below the geometric horizon. You then have to locate the moon, point your antenna at it and follow its motion through the sky.

The traditional method of locating the position of the moon – you certainly can't rely on seeing it from the UK – is to use the *Nautical Almanac*. The moon's positions are given in Celestial co-ordinates used by astronomers, known as *Right Ascension* (or Hour Angle) and *Declination*. These have to be converted to the earth-based co-ordinates of azimuth and elevation for your particular

VHF/UHF PROPAGATION

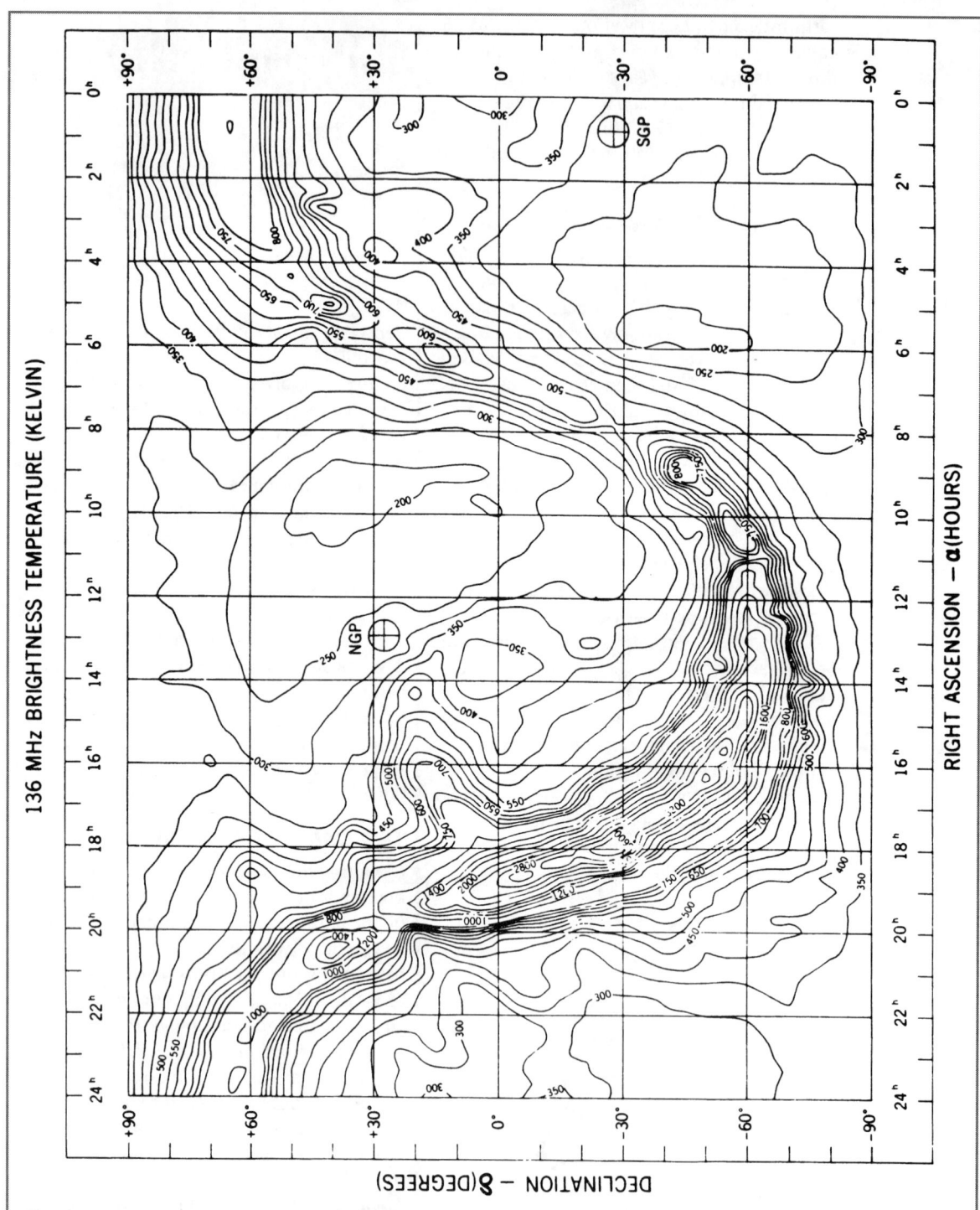

Fig. 2.46. Equivalent noise temperature of the sky at 136MHz, in kelvins. Principal feature is the Milky Way, which appears S-shaped owing to the map projection used

location – an ideal task for a personal computer, which can also be used to predict the Right Ascension and Declination so you don't actually need the *Almanac* at all. As well as telling you the moon's azimuth and elevation at suitable time intervals, a good moon-tracking program will also calculate the times when you have a mutual 'moon window' with any chosen part of the world, the level of cosmic background noise and the Doppler shifts caused by the relative motion of the earth and the moon [26].

The dominant motion of the moon, as seen from the earth, is due to the rotation of the earth itself. One rotation in 24 hours amounts to 2·5° of arc every 10 minutes, with a small correction for the moon's own orbital motion. The angular diameter of the moon seen from the earth is about 0·5°, which is far less than the beamwidths of most amateur antennas so it is generally sufficient to make aiming corrections every 5–10 minutes at most.

At its the highest elevation, midway between rising and setting, the moon moves only in azimuth and no elevation correction is needed for an hour or more. Similarly, the motion close to moonrise and moonset is predominantly in elevation.

GROUND GAIN

Even with no elevation adjustment, EME contacts can still be made when the moon is near the horizon. At low elevations, uncluttered ground in front of the antenna can provide up to 6dB extra gain. For example, I have contacted W5UN many times on 144MHz with the moon as high as 20°, using a single horizontal 16-element Yagi. However, at other angles the ground-reflected signal subtracts to give a minimum, and Chapter 7 shows how the angles of these maxima and minima depend upon antenna height.

Ground gain is mainly useful on 144MHz where cosmic noise usually exceeds the thermal noise from the ground (Chapter 4). On the higher bands the reflection characteristics of the ground deteriorate, and any extra gain is more than cancelled by the ground noise.

BEST TIMES FOR EME

Since EME signals will always be very weak, all possible steps must be taken to reduce noise, both external and system-generated. Night-time is best because sun noise will be absent and man-made noise reduced. This implies a full moon, rather than a new moon with the sun nearby. Apart from this, the phase of the moon is not important – the entire moon is still there, even if you can't see it all.

The other major noise source is the galactic background which follows the contours of our galaxy, as shown in Fig. 2.46 [27]. Although the Milky Way is roughly planar, it appears S-shaped on the diagram because of the projection used to map the celestial sphere onto the page. The noise temperatures in Fig. 2.46 are at 136MHz; multiply all values by about 0·87 for 144MHz. The multiplying factor for 432MHz is about 0·06, though reference [27] also includes a 400MHz map which would be a more accurate starting point. During the course of a lunar month, the moon drifts across the celestial background shown in Fig. 2.46, moving sinusoidally between declinations of approximately +28° and –28°. However, it moves only slightly against that background during a single day. The best times of the month are thus when the moon is in front of a quiet region, towards the top centre of Fig. 2.46. Since the moon will then be high in the sky for stations in the northern hemisphere, the quietest days of the month will also provide the maximum moon-time for the majority of the world's EME stations.

The moon is in a slightly elliptical orbit around the earth, so its distance varies a little. The free-space part of the path loss (page 2-19) therefore varies by about ±1dB between *perigee* when the moon is nearest to the earth and *apogee* when it is furthest away. The period of the apogee-perigee cycle is not quite the same as to the moon's orbital period; and in fact all the various aspects of the moon's motion drift with respect to one another, over periods ranging from one month to several years.

EME conditions are at their most promising when full moon, night-time, perigee and high northerly declination all coincide. This will happen next during 1999-2000 [28]. In any other year and month, optimum times for schedules are something of a compromise. Perigee (when the signal is strongest) is not necessarily the best time; a low sky-noise temperature is more important with today's sensitive receivers.

DOPPLER EFFECT

As a result of the relative motion of the moon seen from the station, the received frequency is shifted due to the Doppler effect. The relative velocity is mostly due to the rotation of the earth, so the Doppler shift is upwards in frequency as the moon is rising and moving towards you, passes though zero when the moon is almost due south, and goes progressively LF towards moonset. The shift is proportional to the operating frequency, amounting to a maximum of a few hundred hertz on 144MHz and up to 1kHz on 432MHz. This effect needs to be allowed for when searching for your own echoes or for weak stations on scheduled frequencies, especially when using narrow receiver bandwidths. The simplest way to deal with Doppler shift is to set the RIT to cancel the frequency offset on your own echoes and then to use the main tuning control for all frequency setting. If you tune in a station to the same pitch as your own echoes, he will hear your EME signal on the same frequency as his own echoes.

POLARIZATION OF EME SIGNALS

Imagine a horizontally-polarized wave, transmitted at the moon from the UK and reflected back without any change of polarization. At a longitude of 90° east or west it will arrive with nearly vertical polarization. This *spatial polarization rotation* depends on the relative longitudes of the two stations and the position of the moon, and could in principle be calculated and corrected by rotating the antenna(s) by this amount. However, the chances are that the signal will not arrive as predicted because the plane of polarization will also be rotated each time the electromagnetic wave passes though the ionosphere. This is called *Faraday rotation*, and is due to the earth's magnetic field which causes the electron oscillators to be twisted away from the electric-field direction of the incident wave, and hence to radiate the new wave with a different plane of polarization (see page 2-5).

When both stations are using linear polarization, what is the likely effect of these polarization changes? The power in the radio wave is proportional to the square of its electric field (page 2-4), so the attenuation A_p due to a linear-polarization offset angle of θ between the incoming signal and the antenna is:

$$A_p = -20\log_{10}(\cos\theta) \text{ dB}$$

So the polarization offset required to cause a loss of 1dB is only 27°. Because Faraday rotation makes all angles of incoming polarization equally likely in the long term, it follows that mismatched polarization could cause a noticeable deterioration (>1dB) in weak EME signals for about 70% of the time, with serious effects on the success rate of marginal contacts.

FARADAY ROTATION

Faraday rotation on the outward and inward passages through the ionosphere is additive, and can combine with the spatial rotation to give apparent one-way propagation [29]. Even if you can hear your own echoes, it doesn't follow that other station can hear you – and the opposite applies too! The number of rotations depends on the lengths of the two slanting paths through the ionosphere, and also on the levels of ionization and geomagnetic field along those paths.

The magnitude of Faraday rotation, and hence its rate of change, decreases with increasing frequency. At 144MHz the time for the polarization to rotate through 90° is typically half an hour, which is not too long to wait if the orientation is unfavourable. At 432MHz, however, Faraday rotation can be more of a nuisance because it changes more slowly, and can become 'stuck' for long periods.

Although stations with fixed linear polarization suffer the most from Faraday rotation, they also usually need its help on intercontinental QSOs to overcome the offset due to spatial polarization. But Faraday rotation is an unreliable ally and can lead to many unsuccessful schedules, sometimes with good signals one way but not the other, until luck and persistence eventually yield a two-way contact. Since linear polarization is the standard on 144MHz and 432MHz EME, the best practical option on these bands is fully-rotatable linear polarization.

However, stations with rotatable linear polarization report that received EME signals sometimes seem spread over a range of polarization angles, and the expected cross-polarization null is not observed. Part of this polarization spreading is due to the geometrical effects of reflections from the irregular surface of the moon; but the larger and more variable component depends on ionospheric conditions, probably because the propagation velocity through the ionosphere depends on the alignment between the plane of polarization and the direction of the earth's magnetic field. Thus signals that were initially linearly-polarized can become spread out, while circularly-polarized signals (the standard on 1·3GHz and above) can lose their true circularity. In both cases there can be a loss in signal strength of up to 3dB.

LIBRATION FADING

As well as Faraday fading, EME signals suffer a much more rapid form of fading called *libration fading*. Although the moon always shows the same face to the earth – because its rotational and orbital periods have become locked together through tidal forces – it still rocks a little on its axis. This motion is called libration, and one consequence is that considerably more than half of the surface of the moon becomes visible from earth over a period of time. Another consequence of libration is that the path lengths of signals backscattered from various parts of the moon's surface are always changing, leading to quite rapid fading and occasional brief enhancements of several dB. The typical timescale of libration fading is a few seconds at 144MHz, and a second or less at 432MHz; however, on all frequencies it can be sufficiently rapid to break up individual Morse characters.

LONG-DELAYED ECHOES

Though not strictly moonbounce, the puzzling phenomenon of long-delayed echoes deserves a mention here. Rare but well-documented cases of echoes being heard with much longer delays than the 2·5-second earth-moon-earth path have been recorded by stations carrying out EME tests [30]. Similar effects are observed more frequently on HF but possible mechanisms have been suggested for these cases, such as trapping of signals in the ionosphere for several turns around the earth. As far as I am aware, no credible explanation has yet been put forward for long-delayed echoes on VHF/UHF.

VHF/UHF PROPAGATION

REFLECTIONS OFF OTHER OBJECTS

A variety of other objects besides the moon have been used for reflecting radio signals, ranging in size from buildings to mountains. Fig. 2.15 showed that signals will be better – other considerations being equal – if the reflector is near one or the other station. A reflector also tends to become more effective when its surface looks large and flat compared with the wavelength. Thus buildings are effective reflectors for 432MHz up to microwaves, but are poor on the lower bands.

Reflections can help get VHF/UHF signals out from very difficult locations. For example, I have worked several stations in western Switzerland who were in the shadow of the nearby Jura mountains and unable to access a tropospheric duct towards the UK. These stations made contact by beaming in the opposite direction towards the Alps and reflecting their signals into the duct.

Signals reflected from rough objects tend to be weak (because of the poor reflecting surface) but constant. If there is variation in signal strength, it is probably due to atmospheric effects. To calculate the path loss, you would need to know the effective reflection coefficient of the reflector. Little data on this seem to be available, giving scope for amateur research! The polarization may change on reflection from an oblique surface, and this should be compensated if possible by polarization rotation or by using circular polarization.

Aircraft make good reflectors on the 144MHz band and higher since they consist of large, smooth metallic surfaces. Unfortunately they do not stand still for us, so the reflected signals tend to be transient. To offer a useful reflected signal, the aircraft should be not too far from one end of the path and hence not too high. This limits the useful range to only a few hundred kilometres. Aircraft reflections are often heard on stations being worked on other modes, and are recognizable by the characteristic periodic fading which changes in rate with the aircraft's direction and distance (page 2-21).

Other aerial objects besides those so far considered have been used as passive reflectors. The most obvious are satellites, but most of these are too small and too far away to produce a useful reflection. An exception was a large reflective balloon satellite launched in the 1960s, which had a brief but scintillating life! Low-orbiting satellites such as re-entering launch vehicles and other types of 'space junk' move very quickly across the sky and are only visible for a short time. Although they could be tracked automatically, it is very difficult to obtain the necessary tracking information ahead of time. Other potential reflectors have been suggested or actually employed; these include copper 'needles' and a release of barium vapour. However, concern about pollution of the space environment means that such experiments are unlikely to be repeated.

Another reported source of reflection, or rather scattering, is the ionization track left after a lightning discharge. Although the limited height of these trails and their uncertain direction would seem to limit their use for working DX, enhancements are possible for a second or two. But when thunderstorms are about, most people sensibly prefer to stay well away from the radio gear and its attached lightning conductor.

RECOGNIZING PROPAGATION MODES

From the descriptions in this chapter, you might be tempted to think that there is seldom any difficulty in distinguishing between the various propagation modes. Most of the time this is true – a full-blown Es or tropo opening on 144MHz can hardly be mistaken for anything else! But sometimes more than one propagation mode contributes to a path, either in series or in parallel, so the situation is not always clear-cut.

As an example, the signals from a very short Es opening or from a very long meteor burst sound very similar – in fact both sources may contribute to the ionization formed. Similarly, very short-lived tropo ducts are difficult to distinguish from Es, particularly since both modes can occur at the same time of day or season of the year. But it is important to know whether a short burst of Italian SSB is merely a good meteor reflection which is unlikely to recur, or a prelude to a full Es opening. This is where good knowledge of propagation pays off. Ask yourself – are we near the peak of a meteor shower? Is there Es on the lower bands? Are the weather conditions suitable for tropo in that direction?

To have all this information at your fingertips requires some organization and perhaps some ancillary equipment, but it needn't be out of anyone's reach. The perfect system would include a weather-satellite receiver together with several scanning receivers connected to dedicated antennas looking at transmitters in different directions and recording meteor pings, with a system to raise an alarm whenever long periods of steady signal appear. It would also be useful to have a recording magnetometer to give advance warning of auroral conditions.

Lacking any these facilities, you still can learn a lot from everyday observation. Monitor the broadcast channels and note when VHF radio or TV signals seem disturbed; announcements are often made when co-channel interference is being experienced, which could mean an opening for you. Watch the weather maps on TV, and buy a barometer. Observe the sky and look for the signs given in **Tropo Weather** (page 2-27). Sharing information in telephone warning nets and the DX packet-cluster network is also extremely useful (Chapter 3). Above all, there is no substitute for leaving the rig switched on, monitoring the beacons regularly and *being active on the band*.

WHAT YOU CAN WORK ON VHF/UHF

Propagation is by far the most important factor affecting whether you can work a particular station. If there is no propagation path open to that station, the sad fact is that site, equipment and operating ability are all to no avail. On the other hand, when the band is wide open, people work all kinds of DX with minimal equipment. So be aware of the possibilities and make use of them when they occur.

The second most important factor on the VHF/UHF bands is your site. A clear, uncluttered horizon is far more important than absolute height. Even if you do live in a poor site, don't despair; choose your band and propagation mode appropriately, remembering that the lower VHF bands are less affected by a screened location. And you may still be able to work the world using moonbounce; a quiet environment with enough space for the antenna is more important for this mode of propagation.

BE YOUR OWN PROPAGATION EXPERT

"Science says the first word on everything, and the last word on nothing." (Victor Hugo)

If you've read this chapter carefully, you have probably thought of a lot of questions which it hasn't answered. Hopefully this will lead you to further reading; it may not be easy to find the references you need, but they can usually be obtained though inter-library loans, a purposeful visit to a specialist library or by borrowing from a friend. However, textbooks won't provide the answers to all the questions you may want to ask. Radio propagation is still not fully understood, and this is particularly true of some of the short-lived VHF/UHF modes.

By frequently observing your own favourite bands – and above all, by *using* them – you can learn more about their practical behaviour than many so-called 'experts'. And you can work plenty of DX at the same time!

REFERENCES

[1] *Recommendations and Reports of the CCIR*, 1986, Vol.5: Propagation in Non-ionized Media. ISBN 92-61-02741-5.

[2] P.W. Sollom G3BGL, A Little Flutter on VHF. *RSGB Bulletin*, (a) November 1966, p.709; (b) December 1966, p.794.

[3] *Microwave Handbook*, Volume 1. RSGB, ISBN 0-900612-89-4.

[4] J. Priedigkeit W6ZGN, Ray tracing and VHF/UHF Radio Propagation. *QST*, November 1986, p.18.

[5] R.G. Flavell G3LTP, Studies of an Extensive Anticyclonic Propagation Event and some Short-term Enhancements Observed at VHF and UHF. *Radio Communication*, February 1984, p.128.

[6] Julian Gannaway G3YGF, Tropospheric Scatter Propagation. *Radio Communication*, September 1981, p.710 (based on reference 2).

[7] G. Roda, *Troposcatter Radio Links*. Artech House, 1984, ISBN 0-89006-293-5.

[8] Commission of the European Communities, *COST Project 210 Final Report*, Annex 2.16: Development of the BTRL/Yeh Troposcatter Interference Prediction Model. EUR 13407 EN (1990).

[9] D.J. Smillie GM4DJS, Plotting of Magnetic Deviation and Aurora. *Radio Communication*: (a) February 1992, pp.51-53; (b) March 1992, pp.31-33. *NB The circuit described can be made much less sensitive to RF and transient interference by reducing the amplifier bandwidth to about 1kHz.*

[10] C.E. Newton G2FKZ, *Radio Auroras*. RSGB, 1991. ISBN 1-872309-0308.

[11] *RSGB Amateur Radio Call Book*.

[12] H. Hatfield, Recording Jamjar Magnetometer. British Astronomical Association note ON3 (February 1983).

Technical Topics: Kilner-jar magnetometer. *Radio Communication*, July 1988 p.522.

Russ Wicker W4WD, Simple Magnetometers. *Proc. 22nd Conference of the Central States VHF Society*. ARRL, 1988. ISBN 0-87259-209-X.

REFERENCES

[13] D.O. Pettit, A Fluxgate Magnetometer. *BAA Journal* vol. 94 (2), p.55 (1984).

R. Noble, A Simple Magnetometer. *Electronics World & Wireless World*, September 1991.

[14] A. Hopwood, Riding the Solar Storm. *Electronics World & Wireless World*, March 1990.

[15] G.H. Grayer and M. Harrison, Spatial and Temporal Correlation in Bistatic Communications via 28MHz Sporadic-E. *Proc. Fifth Int. Conf. on HF Radio Systems and Techniques*, IEE, July 1991. IEE Conference Publication 339.

[16] Ernest K. Smith and Sadami Matsushita, *Ionospheric Sporadic E*, pp.194-214. Pergamon, 1962.

[17] Michael R. Owen W9IP, The Great Sporadic-E Opening of June 14, 1987. *QST*, May 1988, p.21.

Emil Pocock W3EP, Sporadic-E Propagation at VHF: a review of progress and prospects. *QST*, April 1988, p.33.

[18] Geoffrey H. Grayer G3NAQ, Sporadic E and 50MHz Transatlantic Propagation during 1987. *Ham Radio Magazine* (USA), July 1988, p.10.

[19] G.A. Moraitis, Temporal variations of MED FOEs. National Observatory of Athens Ionospheric Institute, preprint (1989).

[20] Jim Bacon G3YLA, An introduction to sporadic E. *Radio Communication*, 1989 – (a) May, p.37; (b) June, p.37; (c) July, p.51.

[21] J.D. Whitehead, Sporadic E layers: History of Recent Observations. COSPAR 1988.

J.D. Whitehead, Recent Work on Mid-latitude and Equatorial Sporadic-E. *Journal of Atmospheric and Terrestrial Physics*, vol.51 (5), p.401 (1989).

[22] Emil Pocock W3EP, Auroral-E propagation at 144MHz. *QST*, December 1989, p.28.

[23] T.F. Kneisel K4GFG, Ionospheric scatter by field aligned irregularities at 144MHz. *QST*, January 1982, p.30.

[24] H.H. Sargent III, A Prediction for the Next Sunspot Cycle. *Proc. 28th IEEE. Vehicular Technology Conference, Denver, Colorado, March 22-24 1978*.

[25] R.G. Cracknell ZE2JV (G2AHU) and R.A. Whiting 5B4WY (G3UYO), Twenty-one years of TEP. *Radio Communication* (a) June/July 1980, p.626; (b) August 1980, p.785.

[26] John Morris GM4ANB, *Amateur Radio Software*. RSGB, 1985. ISBN 0-900612-71-1.

Suitable programs include those by VK3UM, WA1JXN and G3SEK for IBM-compatible machines, available on disks from G4ASR. Some programs are also available for the BBC micro, Spectrum and others – all enquiries to G4ASR.

[27] R.E. Taylor, 136MHz/400MHz Radio-sky Maps. *Proc. IEEE*, April 1973, p.469.

Reprinted in Eimac EME Notes AS-49-25, and also in *Recommendations and Reports of the CCIR*, 1986, Vol.6: Propagation in ionized media. ISBN 92-61-02751-2.

[28] Michael R. Owen W9IP, Orbital mechanics of the moon. *Proc. 21st Conference of the Central States VHF Society*. ARRL, 1988. ISBN 0-87259-066-X.

[29] Tim Pettis KL7WE, Spatial Polarization and Faraday Rotation. *Proc. 22nd Conference of the Central States VHF Society*. ARRL, 1988. ISBN 0-87259-209-X.

[30] H.L. Rasmussen OZ9CR, Ghost Echoes on the Earth-Moon Path. *Nature*, September 4 1975, p.36.

Fig. 2.10 is reproduced by permission of the RSGB, and Fig. 2.20 by permission of the Rutherford Appleton Laboratory.

CHAPTER 3

OPERATING

OPERATING

by David Butler G4ASR

Anyone with an interest in VHF/UHF DX knows about the spectacular 'tropo' and sporadic-E openings, but it takes a more seasoned DX operator to realise that ranges of at least 500km are attainable any evening by weak-signal tropo, and up to 2000km via meteor trails. The horizons of VHF/UHF are expanding beyond Europe too: a growing number of stations are claiming operating awards which were originally conceived as HF-only. At the peak of a sunspot cycle you can work the world on 50MHz; many have earned the Worked All Continents award on 144MHz and above by moonbounce; and W5UN has actually achieved 144MHz DXCC – one hundred countries confirmed!

Yet all of this takes place against a background of a widespread belief that the VHF and UHF bands are for chatting to the locals via the nearest FM repeater, or – sadly prevalent among some HF DXers – that "real DX" is only on HF. But unlike the short-wave DX operator who must rely almost completely on ionospheric F-layer reflection, we VHF DXers can use a multitude of propagation modes: tropospheric refraction, ducting and scatter, sporadic-E reflection, auroral reflection and scatter, to name just a few. These modes of propagation have already been described in Chapter 2.

This chapter is about acquiring the operating skills that make VHF DXing a uniquely challenging and exciting part of amateur radio. I'll begin by describing the general principles and techniques which apply to almost any mode of VHF/UHF DX propagation.

GENERAL OPERATING TECHNIQUES

The essence of good operating is very simple: you just need to do exactly the right thing at any given moment! For that reason, some modes of propagation require more specialised operating techniques, and I'll deal with those later.

WORKING WITH WEAK SIGNALS

Your first big step in effective DXing is a willingness to put up with some noise on the received signal. Some people never even get this far; if they ever progress beyond switching FM channels, they tune the band with great sweeps of the wrist, stopping only for the really big signals. They're missing a lot, because –

If all the DX is S9, you're missing the real DX.

Skill in working with very weak signals comes with practice, and anyone can learn it. In the 'sixties there was a myth that a few individuals were gifted with 'moonbounce ears' which allowed them to work miraculous DX, but that was an illusion: what they were achieving just *seemed* like magic. When the rest of us started to use weak signals too, it soon became clear that 'moonbounce ears' are just another technique which anyone without a clinical hearing problem can develop with practice.

Copying weak-signal information is much easier if you use some conscious intelligence to sift through what you're hearing. There is a certain structure in any amateur contact, especially as signals become weaker and the contact is stripped down to its bare essentials. When an unidentified station calls you, the callsign will have a fixed format and you can easily familiarize yourself with the callsign formats of the countries you're likely to work. Once the QSO gets going, you expect to receive a signal report which will normally only contain numbers, so numbers are all you need to listen for at that stage. Similarly, the QTH Locator has a fixed structure of letters and numbers. If you're familiar with the locator map, and also know that in a large number of countries the prefix or suffix of a callsign may give the geographical location within that country, you can eliminate many impossible combinations of callsign and locator. More detailed descriptions of these tactics are given in a later section.

LISTEN, LISTEN, LISTEN

To make long-distance contacts from an average site requires skill, patience, determination and sometimes low cunning! You've got to find your DX, because unless you live in a rare locator square it won't come looking for you. So the most important lesson when trying to make long-distance or rare contacts is to *listen*. Almost every keen VHF DXer eventually learns that although blasting away with full power to a group of Yagis will net you shoals of medium DX and lots of lovely contest points, it doesn't always produce the really prize DX contacts.

ANSWERING A CQ

But you can't listen all the time – if you hear someone calling CQ and you want to work them, you've got to press the PTT or reach for the Morse key and say something. What should you say? Well, step back a little and think what you're trying to do. You're trying to get into contact with that particular station, and you're trying to get the basic QSO business completed in a clear and efficient way. If it's a contest or a frantic DX pile-up, that's all you're trying to do; when things are a bit more relaxed, you can slide into a more comfortable style of chatty contact. And just one more thing: having started to think what you're

OPERATING

doing, *don't stop thinking!*

Your first objective is to get into contact with the person who's just finished calling CQ. But what if someone else is calling too? To adapt the old saying: "One caller is company, two's a pile-up." Pile-ups sometimes get totally out of hand, but at their best they are a game of friendly rivalry ruled (mostly) by skill. The key to getting your callsign recognized in a pile-up is simply this:

Put yourself in the place of the other operator

What is he or she doing, thinking, right at this instant? (Chances are that it's a he, though CW brings some nice surprises.) He's probably trying to hold the microphone, to finish writing the details of the last contact in the log, and to gulp a cup of coffee, all at the same time. He won't have a fourth hand free to tune the receiver, so call on exactly the frequency he's already listening to.

Call him by spelling only your own callsign, just once. He will assume that anyone on the frequency is calling him, and he already knows his own callsign, thank you. Call as soon as he goes over to receive, then listen. If he comes straight back, your QSO is under way. If not, and you can hear other longer-winded stations still calling, you can slip your callsign in again.

A good operator will probably catch your callsign – or somebody's – first time. But not even the best operator can do that every time, and some people seem to need several repetitions before anything sinks in! In any case, the other operator may be suffering a lot more interference than you and, if you're running less than legal-limit power, your signal may not be so strong as he is with you. Possibly he has a pile-up of stations of roughly equal signal strength, all calling at once. In Europe he will probably be working in a foreign language – and if it's English he's doing you a big favour. So never underestimate the other person's operating ability if he doesn't catch your callsign first time. If he needs to call "QRZ?" (meaning "please call again") then simply do that, same as before.

If an operator can get something out of a pile-up, even if it's only one letter, he can use it as a 'hook' to pull out the station he wants. For example he might say "QRZ the station with 'Kilo' in the call?", or "QRZ the 'Kilo' station?" If that could have been you, call again. If not, keep quiet and wait for your next chance.

And don't call unless you're sure the other station is listening! This sounds obvious, but it's amazing how many stations continue to call a station who is actually transmitting, or worst of all continue to call when clearly a QSO has begun with someone else. In a heavy pile-up for a weak DX station it isn't always obvious what's going on, and everybody makes mistakes – but nobody ever spoiled their on-air reputation by deciding *not* to press the PTT.

Success in pile-ups isn't easy, and there are no guaranteed solutions. If you aren't doing too well, the important thing is to try and understand why not and maybe change your tactics.

THE NEXT CHANCE

If you didn't make it last time, you can always try again at the end of the QSO that's going on now. The station who called CQ generally retains the frequency when the QSO is over, and a competent operator will immediately tell everybody that he's ready for further calls. Let him show you the correct timing. Meanwhile you should have placed yourself accurately on the frequency of the station being worked, ready to drop in your callsign at exactly the right moment. We'll be talking some more about this as the chapter continues.

CALLING CQ = TAKING CHARGE

If nobody ever called CQ, there would be precious few contacts made. But before you do, there are three things you need to consider. First of all, your chances of raising some real, rare DX are pretty slim, especially if you're just one of many stations in your locality. Wouldn't you do better to listen

GENERAL OPERATING TECHNIQUES

carefully and then call a DX station you know you want to work? Secondly, if you do call CQ you are in charge of the proceedings and had better do it right. I'll come to that in a moment, but don't forget the third point. When starting a new CQ call, don't forget to ask:

"Is this frequency in use, please?"
or in Morse:
"QRL?"

The answer may be an earsplitting "YES – but thanks for asking," from a local DXer who is straining for a weak station on that frequency but some other beam heading. Your question doesn't just apply to that exact frequency, either; people can recognize that particular phrase even if you're a kilohertz or more off tune on SSB, and they would suffer interference if you started calling CQ. So even if someone replies on a slightly different frequency, the answer still means "Please find somewhere else to call".

So let's assume that you've found a clear frequency (somehow) and are launching into your CQ call. As with every branch of operating, you've got to make it clear who you are and what you're doing. A reasonable pattern for a CQ call would be:

"Hello CQ, CQ. Golf Nine Zulu Zulu Zulu, Golf Nine Zulu Zulu Zulu, Golf Nine Zulu Zulu Zulu calling CQ."

That's the basic CQ call. For most purposes you'll need to add a little more, *eg* "CQ contest", "CQ eastern Europe", "beaming north", or mention your locator square if you think that might be an added attraction. "CQ DX" is not particularly useful, because it might well attract replies from semi-locals who think that a station two counties away is DX. If you're warming up a new frequency, you might do well to repeat the whole call a total of three times.

When you've finished, give some brief but definite indication that you're going over to receive, like "... and standing by" or simply "Over". ("Over-over", "K please" or "Dah-di-dah" are generally regarded as lacking in style. And CB procedures or your own inventions aren't a good idea either – see **Do It Right** on page 3-7.)

WHAT IS A 'CONTACT'?

The answer seems so obvious that many amateurs never give it a thought. But VHF/UHF DXers are often working so close to the limits of propagation that difficult contacts may only just pass or fail, and the line has to be drawn somewhere.

The basic requirements for a contact are that both stations correctly receive **ALL** of the following –

(1) **Both** complete callsigns – your own and the other station's.

(2) Some other information that was not previously known. This is usually a signal report, but not necessarily; in US contests it may well be the IARU Locator (or 'grid') square, *eg* "N32".

(3) An acknowledgement that all the above information has been received, *eg* "QSL", "roger", or "R" on CW.

Generally speaking in Europe, you are therefore looking for both callsigns, your report and an acknowledgement as the minimum requirements for a successful QSO.

These rules apply – and work – for all types of contact: tropo, meteor-scatter, contest, moon-bounce or whatever.

ANSWERING A CQ CALL

With luck, somebody comes back to your CQ call. What now? Well, it's now up to you to sort out any pile-up, and to direct the course of the contact. What is a 'contact', anyway? Very few people actually know, so I've put the basic rules in the panel above.

With these rules in mind, it's pretty easy to see what you're aiming to do. You want to exchange both callsigns, your report and an acknowledgement as the minimum requirements for a successful QSO.

The first thing to say when replying to a station who's answered your CQ is that station's callsign. This immediately establishes who you're working, and tells any other stations to wait till you've finished. Then say who you are, say hello, give a signal report and pass the transmission straight back. The other station should respond with both callsigns, an

OPERATING

acknowledgement of your signal report to him, and his report to you. Follow this with your own acknowledgement, and the basic QSO is complete.

So a typical bare-bones QSO would go like this (where a callsign is printed in **bold**, it should probably be spelled out phonetically) –

"CQ DX, CQ DX from **HB9QQ, HB9QQ, HB9QQ**, over."

"G4ASR"

"G4ASR from HB9QQ – hello David, you're 5 and 9 in JN47, over."

"From G4ASR – roger, thank you Pierre, you're 5 and 9 also, in IO81**MX**, over."

"From HB9QQ – roger ..."

From the moment I hear that "roger" from Pierre, our basic QSO is complete and valid, and we'd be entitled to exchange QSL cards even if the band went flat at that very instant. Since we aren't super-DX to each other, and have obviously worked several times before, we'd probably have a brief chat about conditions. We'd probably also ask about activity from rare locator squares in each other's locality, wish each other good DX – and then sign off and go looking for some!

That's all very idealized and simplified. What you'd actually do would depend on the precise details of the situation. Obviously you don't need to go through the formalities in such a brisk and businesslike way with Fred in the next town; even Pierre and I started to relax as we each realised that the band was wide open. But at the other extreme, where signals are in and out of the noise and communication is only just possible, there is considerable potential for misunderstandings and mistakes. This applies particularly to meteor-scatter and moonbounce (see later sections), so formal QSO procedures are absolutely mandatory for these modes.

SIGNING OFF

When the QSO draws to a close – all the sooner if there are other stations waiting – the station who called CQ always has the last word. In the example above it was HB9QQ, so he ends with something like this:

"OK, Seven Three, David and thanks for the nice QSO. HB9QQ signing with G4ASR, and **HB9QQ** is calling QRZ DX, over."

Note that there's absolutely no doubt about who is staying on the frequency, or when he's ready for another call, so the foundations are already laid for the next QSO.

As I said, if you called CQ, you generally retain the frequency when the QSO is over; but there are exceptions. If you're called by a station which is much more rare or exotic than yourself, your QSO will probably attracting a pack of locals who also want to work him so it's courteous to everyone if you give up the frequency. If you're being called by more DX, you can always announce where you intend to go, *eg:*

"*G***4ASR** signing with HB9QQ and going twenty-five HF, twenty-five HF".

I'll say more about this and other specialized operating techniques later in this chapter. Now let's look at some other general aspects of VHF/UHF DX operating.

CW OPERATION

The most effective mode for difficult and long-distance VHF/UHF work is undoubtedly CW. Chapter 4 shows that CW gives you far more ability to overcome the losses of long paths than any telephony mode, mainly owing to the narrower bandwidth but also because of the simple nature of the signal – at any given moment you only need to decide whether the carrier is on or off. There is no valid reason why a CW capability shouldn't be part of every VHF/UHF station, but a surprising number of operators make no use of it. Yet CW is the key to success with many weak-signal transmission modes such as meteor-scatter or moonbounce. During an aurora, while SSB operators are struggling to work the semi-locals, the CW DXers are working the rare stations with relative ease. There's a saying that "A little CW goes a long way"; well, a lot of CW goes further still! And once you're comfortable using it, Morse is actually *fun*. Until you've acquired the skill, it needs a major leap of faith to believe that – but it's true.

In general the good CW operator is the one

GENERAL OPERATING TECHNIQUES

whose Morse is easy to read, and doesn't try to impress by sending fast. The optimum speed of sending depends entirely on circumstances, and when conditions are poor it's sensible to send more slowly than when the signals are loud and clear. For DXers, another major virtue of CW is that the abbreviations used on the amateur bands jump clean over the spoken language barrier; fluent amateur CW operators anywhere in the world can understand each other perfectly. For an excellent introduction to CW DXing on any band, read *The Complete DXer* by past-master W9KNI [1].

Many VHF/UHF CW operators find it difficult to put their transmitter or transceiver exactly on another station's frequency. Unless you do, you could well be calling as much as 2kHz off frequency. If the other station is using a 250Hz CW filter on receive, he just won't hear you. The following technique will enable you to zero-beat your frequency to that of the other station. First, turn off the RIT and XIT (incremental tuning or clarifier) and centre the IF passband tuning. In almost any transceiver built today, the audio frequency of the CW monitor is meant to match the offset of the transmitted CW note. So the trick is to turn off the transmitter VOX and send a string of dashes while still receiving. Adjust the transceiver frequency until the audio tone of the station you wish to zero-beat is identical in tone to the CW monitor note. When you transmit, you will be exactly on that station's frequency. If you wish, you may now use the clarifier to adjust the pitch of the received signal to your liking. This technique obviously relies on the offset of your CW carrier being the same as the sidetone note, and in most rigs you'll need to adjust the frequency of the CW oscillator crystal to make certain that it is.

VOICE OPERATION – PHONETICS AND LANGUAGES

SSB is the only serious mode for DX voice communication. All transmission modes travel the same distance, but SSB shares some of the advantage of CW in being a narrow-bandwidth mode (Chapter 4). Also the demodulation process of CW and SSB enables these signals to be recovered with a better signal/noise ratio than FM when the signals are weak.

I've already touched on the question of when to spell letters phonetically. Basically it's a balance between speed and intelligibility, and you judge it best by experience. Appropriate internationally recognized phonetics can be most helpful under certain circumstances, especially in weak-signal work or during sporadic-E openings when conversing with VHF-only stations who don't normally have QSOs in English at all. The ICAO phonetic alphabet (Alpha, Bravo *etc*) should probably be the first you should try, since that's what the other operator is most likely to be expecting, but it isn't optimum for all letters in all languages. It's up to you to devise the best combination of phonetics which works for your own callsign.

Better still, if you find yourself working into a country where English is not frequently spoken, try to use the native language. Even if your accent is terrible and you can only stumble through a contact, give it a go. You don't have to learn the whole language; there are an increasing number of books and tapes aimed specifically at helping you to make amateur radio QSOs [2]. Provided you use the correct language (*eg* French doesn't go down well in Flemish-speaking northern Belgium), your efforts will be very much appreciated and very effective in terms of DX contacts. When TA finally comes through on 144MHz sporadic-E, the first to crack the pile-up will be the operator who can call in pidgin Turkish! There's a VHF DXer in Southern England who is willing to attempt a basic QSO in just about any language. Not only does he have the QSLs to prove it, but he's also made a lot of friends in the process.

DO IT RIGHT

Why all this emphasis on neat, clean and 'correct' amateur-band operating procedures? I'm not trying to regiment you, or limit your freedom of self-expression. My point is simply this: your best chance of making a difficult contact is to make it easy for the other operator. DXing already involves plenty of the

OPERATING

unexpected, so don't add to it by using fancy operating tricks or by borrowing procedures that don't belong on the amateur bands.

For working DX, just keep your operating style simple and predictable, and stick close to the standard pattern and procedures of an amateur-band QSO.

CALLING FREQUENCIES

One feature of VHF and UHF operating in Europe is the use of calling frequencies or centres of activity. This system may sometimes work during poor to moderate conditions, but it produces absolute chaos when conditions are 'up'. Personally, I reckon that the use of calling frequencies has set back the art of DXing by decades. Instead of getting out there, tuning around and building up a mental picture of band conditions, people stick their receivers on a calling frequency and expect the DX to come knocking on their door.

Meanwhile the real DX listener will be tuning the band carefully, pausing at signals that sound interesting, listening to what other stations are working and perhaps checking the beacons. Chapter 2 should help to give you the experienced DXer's in-depth knowledge of the likely propagation modes that could be encountered at any particular time, so you can adjust your listening techniques accordingly.

If you do find some DX on the calling frequency, your best bet is either to make a quick basic QSO and be done, or else to QSY (change frequency) as quickly as possible before you're drowned out by a semi-local at one end or the other. But if you nominate a QSY frequency which neatly rounds off to 10kHz, the chances are that it'll be occupied and then you're both in trouble. Most DXers have transceivers with real VFOs, so why not use them? Selecting a frequency such as 144·164MHz in an uncrowded section of the band is more likely to produce a neat QSY and a successful contact.

The same applies when you're moving frequency after finishing a contact, and hope to take any DX callers away with you. If your QSO has attracted a crowd of local stations who want to work the same station as you just did, you'd better move away some distance to get clear of the pile-up you're leaving behind.

Making a successful CW QSY can be very difficult, especially in crowded conditions close to the CW calling frequency, so that's all the more reason to spread out and tune around when you know the band is open.

BAND PLANS

In common with most amateur allocations, the VHF and UHF bands are split up in terms of usage by voluntary agreement. Within any particular band there will be sub-bands for CW, SSB, FM, beacons, repeaters and satellites, and also specific frequencies for specialist modes such as data, SSTV, FAX etc. Because of the wide variety of activities which take place on the VHF and UHF frequencies, it should be obvious that some international self-regulation is necessary. That's what band plans are for.

In Europe, with many small independent countries close together, the International Amateur Radio Union (IARU) is the vehicle for international band planning. Although IARU band plans are purely voluntary, it's in everyone's interest to keep to them. Running a meteor-scatter schedule in the FM, satellite or beacon sub-bands, although legal within the terms of the UK amateur licence, is hardly conducive to making or keeping friends. What's more, it degrades the reputation of VHF DXers as a group.

Since band plans do change from time to time, it's important to keep up-to-date copies close to hand.

VHF AND UHF BEACONS

Beacon transmitters are in operation on most amateur bands from 14MHz to 24GHz. You'll find them of particular value in determining band conditions, especially if the activity level is low, and they can of course be used for more formal propagation experiments. Even though a semi-local beacon won't tell you anything about DX propagation, it can be invaluable for checking antenna or receiver performance. Lists giving up-to-date information about beacon status should be available from most

GENERAL OPERATING TECHNIQUES

national amateur radio societies or from specialist newsletters. Avoid transmitting within the beacon sub-bands – interference reduces the value of the beacon service to others and spoils your personal reputation.

PLAYING THE GAME

Whenever anyone tries to do something really challenging, there's always a temptation to bend the rules – written or unwritten. And there's always a temptation for other people to accuse those who are successful of having somehow 'cheated'. VHF/UHF DXing is no exception, so let's look at some of these questions.

DID YOU REALLY WORK HIM?

Keen intuition can be as important as keen ears in copying weak signals. There's no dividing line between the role of the ears and that of the brain. But intuition is not the same as guesswork. If you think you know what you're hearing, you still need to check it rigorously, point for point, against what you're actually hearing. Almost all DX operators are scrupulously honest about this, and are quite prepared to scrap a difficult contact unless everything about it is definitely correct.

It follows that everyone will have a certain percentage of failed DX contacts – and the more you try to achieve, the higher that percentage is likely to be. An operator who seems to be 100% successful is either not trying for the real DX or else attracts the greatest suspicion! Anyone can make an honest mistake in claiming a contact, and that's what QSL cards are for. But operators who show – well, let's call it 'excessive optimism' – are well known on the DX grapevine. And even if they eventually grow out of it, there are some very long memories...

ETHICS IN PILE-UPS

Working a new country or a much-wanted locator square can be very exciting. Losing out in a pile-up isn't as much fun. But losing is a part of DXing. We all have to lose sometimes. If we managed to crack every pile-up successfully there would be no challenge left in DXing. Obviously you don't like missing the DX, but take comfort in one thought: nobody goes to the trouble of arranging a good VHF station for only a single evening's operation. There will usually be another chance some other time. Therefore an important virtue a DXer must possess is long-term patience.

Now let's look at another pile-up situation. During an opening, you hear many stations trying to work OK1KRA on 144MHz. You've already worked that square and country, and in fact the White Mountain Radio Club's QSL card has been on the wall for ages. So what do you do? Do you switch on the high-power amplifier and go for it? Have you forgotten the anguish that you suffered when you were chasing your first contact with Czechoslovakia? You have two choices: leave the amplifier on and go looking for something rarer, or turn it off and indulge in a little of what the Russians call "radio sport". You can set your own handicap, and only you know whether you did it with ten watts, four hundred watts or two kilowatts. Taking a multi-dB advantage into a pile-up and winning is nice; but it isn't always fair.

SKEDS – ARE THEY FAIR?

Some people consider it unfair to make VHF/UHF schedules ('skeds') via the HF liaison frequencies or by telephone. But anyone who's tried the really challenging modes like long-distance troposcatter, meteor-scatter and moonbounce will confirm that skeds are legitimate and often necessary. On these modes, the other station is in and out of the noise at the best of times. Although a QSO may be possible when both beams are lined up and both stations are looking for each other on the same frequency, it's stretching coincidence to expect that many such QSOs will happen by chance. In fact it's only through making skeds that we've developed these modes at all.

People who criticize skeds are often ignorant of the stringent requirements for a complete QSO. If you and I make a sked, we may know each other's callsigns in advance and we may even be able to guess what report we're going to send each other. But you and I also know whether we each heard that information correctly during the actual QSO, and whether we succeeded in exchanging the final confirmations. So either of us can say, "Sorry, it wasn't complete; I copied the callsigns and report, but never heard your final Rs. Let's try again tomorrow night, shall we?" Neither of us will be any the worse for having honestly admitted that we didn't make it.

The other surprising thing about attempting difficult skeds is the link that it creates between the two operators. Working VK with extreme difficulty on EME – and perhaps not making it – isn't the same as chatting easily on 14MHz every morning. But when two people are both trying hard to make even a minimal QSO, the human contact is definitely there.

SPECIAL OPERATING TECHNIQUES

Just as each mode of propagation has its own particular characteristics, our operating procedures should be tailored to match. Therefore this section should be read alongside Chapter 2 on **Propagation**. Operating techniques for most propagation modes are just extensions of the general principles and advice I've outlined earlier in this chapter, but procedures for meteor-scatter and moon-bounce are more specialized, so I've devoted a separate section to each.

TROPO DX

Since tropospheric propagation is the most common DX mode, I've already used it in the previous section to outline and discuss general operating techniques. However, there are a few special tricks and topics to tropo as well. Unlike most other VHF/UHF propagation modes, tropo openings can last for days and sometimes weeks. So you can often relax your operating procedures, compared with other modes which may only last for seconds. You can even exchange real locations and talk about the weather, just like 3·5MHz!

Long-distance working on SSB takes place within the appropriate sub-section of the band concerned. In Europe and Africa (IARU Region 1) this equates to 350kHz of sub-band on 144MHz, and in contests and extensive openings this space fills up very rapidly. During openings it won't be necessary to use the calling frequency as you might when band conditions are flat. In fact, you should make a firm resolution not to go anywhere near the calling frequency during an opening, except to listen for DX; and let me tell you in advance

SPECIAL OPERATING TECHNIQUES

that the chances of hearing anything interesting will be slim verging on none, because of the pandemonium in that small part of the band.

If you're an average suburban station in a well-populated locator square and you want to work similar stations a few hundred kilometres away, by all means call CQ in a big tropo opening. But if you want to work anything new or rare, the only way to accomplish this is to listen, listen and listen some more. Tune up and down the band, checking out every interesting-sounding signal, and don't move away until you have confirmed that either it's not DX or it's not in a square you require. It can also be useful to monitor what the well-sited stations in your area are working, to give yourself an idea of how propagation is going and what chance you have of working the really long-distance stations.

PILE-UPS

You may conclude from this that the only way to work rare DX is to attach yourself to a pile-up and rely on your brilliant operating technique. Sometimes that may be true, but it's also important not to fall into the syndrome that HF DXers call "pile-up fixation". You can keep pounding away, determined to work the station come what may. Maybe you make it, maybe you don't – but either way you spend half an hour trying. The experienced operator may make only two or three calls, and if that fails will leave the second VFO on the pile-up and go looking elsewhere. The chances are that other stations located in the same wanted squares will also be heard.

Another facet of pile-up fixation is the '59+ syndrome' – an irrational urge to work all the strong stations. For example, a cursory sweep around the band suggests that the only station audible from Austria is OE3JPC. Since his signal is strong, he has attracted a very large pile-up. However, a more careful tune of the band finds OE3OKS calling CQ at a perfectly workable S4. As all the other local stations are locked in mortal combat over OE3JPC there's no competition for OE3OKS and you bag him first call. He might have only been S4 instead of 'JPC's S9, but so what? Result – one new locator square and maybe a new country as well. It's really satisfying when you try some of these more subtle operating tactics for the first time and find that they really do work.

But what if you have to run with the pile-up? Let's assume you've tuned the band carefully and hear a massive dogfight in progress. Somewhere underneath it is the one station which will give you DXCC on 144MHz. What sure-fire technique can you use to work him? Answer: there isn't one. DX-chasing is very similar to fishing – what works in one situation may be a total failure next time. But even though there are no guarantees of success, knowledge of a few techniques can help. Effective radiated power obviously helps too, but it's nothing like as important a precisely-timed call on the right frequency.

What is the right frequency? In some instances it will be the frequency being used by the operator who's just ending a QSO with your wanted station. Tail-ending – dropping in your own callsign just as he completes his last transmission – is accepted by some operators, but is open to grave abuse if used unskillfully. Unfortunately there are too many operators who know about tail-ending, and too few who know how to do it properly. Also, there aren't enough DX stations who are prepared to take charge of their pile-up and create some semblance of order by telling callers exactly what they want them to do. So tail-ending is best avoided unless that's the only way the DX station is accepting the next call. If he's not accepting tail-enders, a call just off the frequency of the last person contacted can often produce results, on the basis that if everyone else is zero-beat, your call will stand out by being different. This applies particularly to CW operating, though calling on CW can sometimes give good results in an SSB pile-up too. It makes your call stand out, even if the station doesn't want a CW QSO, or can't copy CW!

I take it all back – there is just *one* sure-fire way to beat a pile-up. That is to be in the other station's log before the pile-up starts. In other words, you need to be tuning around so that you're right there when the DX station pops

OPERATING

up. So we keep coming back to the DXer's golden rule – *listen, listen, listen*. If you don't hear them – you don't work them!

CONTESTS

A typical contest exchange may sound something like this (where a callsign is printed in **italics**, it should probably be spelled out phonetically):

"CQ contest, CQ contest from **GD4APA** portable, **GD4APA** portable, contest, over."

"G4ASR"

"G4ASR, you're 59672 in IO74**PD**, QSL?"

"Roger, you're 59084 in IO81**MX**, good luck".

"Thanks Dave, and 73. QRZ contest from **GD4APA** portable, over."

Make it easy for the other station to copy your information onto the contest logsheet. Always give it in logsheet order – report, serial number, locator. The quickest way to get your information accurately into his log is to speak clearly at dictation speed – no faster than you yourself could write in block capitals. If signals are weak, use your best judgment about repeats and phonetic spelling.

Time is at a premium in contests, so people frequently cut corners in QSO procedures if they can. The trick is to understand where you're cutting corners, and what risks you're taking. Good contest operators are accurate and fast – but strictly in that order. What might get left out of a contest exchange? You don't always get your own callsign back from the other station; because he comes back on the right frequency at all the right times, you have to assume that he's working you. This may or may not be true! Similarly, if a contest station wishes you "thanks and 73" and presses on to the next contact, as GD4APA/P did in the above example, you can usually assume that he's copied everything OK from you. If you let things go at that, he'll assume that you've copied his information as well; so if there's anything you're not absolutely certain about, you *must* stop him and make sure. There's no need to read back everything he said to you, but if you do need to confirm something, do it *straight away* before you let the station go – you'll never get a better chance.

A good contest operator never seems to be in a hurry, but is actually making a constant stream of split-second decisions. Normally a good operator can well afford the time to greet callers and thank them for the contact. (That "don't-have-time" feeling isn't actually true; it's merely distracting your attention and spoiling your efficiency as an operator.) Even when the going gets really rough and there's no substitute for 100% efficiency, you can still sound interested and pleased to work someone without actually expressing it in flowery words. Although contests are competitive, outright rudeness is fortunately rare – it usually means that someone's getting rattled because too many things are happening at once!

AURORA

The first time anyone ever hears an auroral signal, they think their receiver's gone faulty. SSB signals are often so badly distorted that you can hardly understand them. CW signals sound rather like keyed white noise, but with a little practice they are very much easier to copy than telephony during an aurora.

Regardless of your station location, beam north initially whenever auroral activity is suspected. When stations are heard, swing the beam either side of north to maximize the signals. From the UK on 144MHz, various auroral 'windows' appear possible. Beam headings between 015-030° from southern England can produce many contacts between 400 and 800km with stations situated in Scotland, Northern Ireland or northern England, especially if the aurora is not particularly strong. Beam headings between 040-060° from England generally give contacts with stations in the Netherlands, Denmark, western Germany and southern Scandinavia. If the event strengthens, contacts into Poland, Latvia, Lithuania and Estonia – amongst others – may be possible. This second window can give contacts in the range 700 to 1800km. During very big auroral openings, beaming between 070-090° can give contacts from 1400-2000km into Austria, Hungary, Yugoslavia and maybe even Spain and Italy. In general, the longest-distance stations peak the farthest away from due north.

Because aurora is basically a scatter mode (Chapter 2), the real DX will be quite weak so CW is a must. With experience the rough-sounding signals can be copied fairly easily. When calling CQ the letter A is added to the call ("CQA") and is also added after the readability and signal strength report in place of the normal tone report – *eg* "52A". Sometimes beam headings ("QTF" on CW) are also exchanged so you can plot where the point of reflection is. You'll generally need to use the RIT control because the Doppler shift of the received signal can be as much as 2kHz at 144MHz; at 432MHz you'll probably need to receive on the second VFO to be sure that you're not missing anything.

SSB signals are normally so badly distorted that it's very difficult to understand what is being said. The amount of distortion increases with frequency, so signals on 50MHz will exhibit far less distortion than those on 144MHz, while 432MHz signals will always be very badly distorted. Therefore if you're using SSB your speech must be delivered slowly and clearly, using phonetics because individual letters cannot easily be distinguished. Pass only the minimum of information – callsigns, report, Locator (preferably only the square, *eg* "IO81") and of course that essential confirming "R".

High power is not vital for aurora operation, but aurora is essentially a weak-signal mode so low power will require a lot of perseverance, especially if the aurora itself is not intense. An increase from 20W to 100W will make your signal vastly more readable when it's close to the noise level, and increasing power from 100W to 400W will be almost equally worthwhile.

OPERATING

SPORADIC-E

Sporadic-E in the northern hemisphere is most likely to occur between April and August, with a possible minor peak in late December. As Chapter 2 explains, Es is very common on 50MHz, fairly rare on 144MHz, very rare on 220MHz and never detected on any higher amateur band. One-hop signals in particular can be very strong, and it's actually possible to miss an Es event by tuning straight through the DX in mistake for a local signal.

The E-layer cloud can be quite small and moving very rapidly, so on 144MHz in particular you may not have much time to snatch a contact and go looking for another. Callsigns, report, locator and "Roger" are all you need to exchange, and with a little practice you can complete the contact within 10 to 15 seconds. On 50MHz you can sometimes be a bit more laid-back, but you should recognize that signals can disappear just as rapidly on this band as on 144MHz. It's also impolite to launch into a long discussion when there's a whole pile-up waiting to work the same station; let the DX operator be your guide to the appropriate length of the contact.

One of the attractions of sporadic-E is that it's a great leveller of station equipment. You don't always need a mega-station in terms of power or antenna system. It's much more important to be in the right place at the right time, and then do the right things.

On days when sporadic-E does occur, there often seem to be several patches across the whole of Europe (or even the northern hemisphere – see Chapter 2). So for example you might be working stations in Sicily at one moment and within a few minutes be working Spain using a different cloud. Since the one-hop path from the UK towards the south crosses areas of low amateur population – the Mediterranean Sea and northern Africa – it's easy to be misled into believing that the opening is over. On numerous occasions DXers have been heard discussing what they had worked in the opening, while it's actually continuing all around them. I've done it myself – but only once...

Operating on 50MHz can present problems unique to the band. Some operators who work 50/28MHz crossband get so engrossed with working stations on 28MHz that they forget that the band could well be open to countries permitted to transmit on 50MHz. It's very worthwhile to keep tuning around 50MHz to see that nothing is being missed. DX stations should take the initiative to make sure that crossband operation takes place well away from the parts of the band used for direct DX. During some 50MHz Es events, there is likely to be propagation path in two or more directions at the same time. If so, be particularly aware of the '59+ pile-up fixation' mentioned earlier in this chapter. One example of this occurred during a sporadic-E opening when a very strong OH station was creating a huge pile-up; while 5kHz up and in a totally different direction there was CT3BX on Madeira at S3. Which would you prefer to work – Europe at S9+ or Africa at S3?

FIELD-ALIGNED IRREGULARITIES (FAI)

FAI is a new mode to most VHF amateurs. However, because the FAI and sporadic-E seasons coincide, it appears that at least some 144MHz openings which have been thought of as due to sporadic-E were in reality FAI. Experience shows that FAI propagation may take place along with afternoon and evening sporadic-E openings, and intensify as the Es opening fades.

Although SSB can be quite intelligible via FAI, signals are sometimes weak and fluttery so CW may yield better results. As with all weak-signal scatter modes, the exchange of information should be kept to a minimum. When calling for contacts it may be prudent to call "CQ FAI" and indicate your beam-heading ("QTF") to give other stations some indication which way to beam.

To aid effective communication, the following sub-bands have been designated by IARU Region 1 for FAI working:
144·140–144·150MHz for CW and
144·150–144·160MHz for SSB.

METEOR-SCATTER

Meteor-scatter is unlike most other propagation modes, in that neither station can hear the other until an ionized meteor trail exists to scatter or reflect the signals. So MS requires specialized operating techniques. The two stations have to take turns to transmit and receive information in a defined format, following the procedures as formulated at IARU Region 1 conferences. These rules may at first seem pretty formidable – see below – but you soon get the hang of them because they're basically just common sense.

Signals may be strong at times, but may only last for a second or less as the ionization rapidly dissipates. Meteor pings and short bursts of signal are very common on 50 and 70MHz, less so on 144MHz and fairly rare on 220MHz and 432 MHz, although a number of successful contacts have been made on the two latter bands. No amateur MS contacts have been made above 432MHz.

Contacts on 50MHz are usually made on SSB because the bursts are long enough, and SSB can also be used on 144MHz. However, serious MS DXers on 144MHz and above use high-speed CW, with speeds generally between 500–2000 letters/minute (100–400 words/minute). A memory keyer is used on transmit, and a variable-speed tape recorder on receive. The received bursts are recorded and then played back at a slower speed to decode the transmitted information (Chapter 12 gives more details of the necessary equipment).

SCHEDULED AND RANDOM CONTACTS

For a scheduled ('sked') contact the two

OPERATING

interested stations will arrange in advance the frequency, timing and duration of the test, as well as the transmission mode and callsigns to be used. Scheduling may be by exchange of letters, one station offering a proposal, or by radio via the European VHF net which is active around 14·345MHz and is particularly busy on Saturdays and Sundays before major shower periods. As the popularity of meteor-scatter is increasing, it's not uncommon to find some activity on this net at many times throughout the week also. Depending on propagation, 3·625MHz and 28·385MHz are also used for scheduling of MS tests.

Random (*ie* non-scheduled) MS contacts are far more difficult. Because you're starting entirely from scratch, it's particularly important for both stations to follow the standard QSO procedures.

Because of the very short burst lengths normally encountered via meteor-scatter, a greater level of station organization and operating skill is required than for normal DX working. In Europe, the standard procedures co-ordinated by IARU Region 1 must be followed to ensure that a maximum of correct and unmistakable information is passed both ways. As with operating procedures in general, the virtues of the MS procedures are mainly that they are standard and are widely understood throughout Europe and beyond. And although the MS procedures are different in the USA, it's obviously equally important to be using the same procedures as the person you're trying to work.

TIMING

Accurate timing of transmit and receive periods is important for two reasons: to maximize the chances of hearing the other station, and to avoid QRM between local stations. MS takes place in relatively narrow frequency segments and it can be very difficult to hear weak DX while a local station's keyer is hammering away on a nearby frequency.

The recommended time period for random CW contacts is 2½ minutes to avoid interference between local stations. With scheduled contacts you can arrange for any time period you wish, *eg* the periods could be reduced to one minute by agreement; but your first priority should be to avoid causing interference to local stations who are using the standard periods.

The recommended standard period for both random and scheduled SSB is one minute. However, time periods of less than that are encouraged during major meteor showers. Quick-break procedure within scheduled contacts can be very effective. This could involve taking a break every 15 seconds, using a pip-tone (Chapter 12) if so desired. Although not defined in the IARU procedures for random SSB contacts, a growing number of stations leave a break on the half-minute in case the QSO can be completed in one long burst. If other local stations adopt this procedure and are able to set their clocks correctly, there should be no problems.

TIMING ACCURACY

Unfortunately a large number of operators seem unable to set their watches or clocks accurately. This is surprising since accurate Standard Time (UTC) is freely available anywhere in the world by the magic of radio. It is absolutely vital that time-pieces are checked prior to any MS activity, eg by checking against the time-ticks on standard frequency transmissions, or by using teletext or the telephone company's 'speaking clock'. For MS activity, clocks need to be set with better than two seconds of Standard Time. Any inaccuracy will result in wasted time, when both stations are either listening to nothing or both transmitting at once, and will also cause totally unnecessary interference to local MS stations. Time it right.

SCHEDULE DURATION

Scheduled contacts are usually in the range 1–2 hours, although during shower periods this can be reduced to 30 minutes or less. Start times should be arranged to be on the hour, *eg* 0000, 0100, 0200 *etc*. Every schedule period must be considered as a separate uninterrupted test. It is not permissible to break off and then recommence at some later time.

METEOR-SCATTER

TRANSMIT/RECEIVE PERIODS

All MS operators living in the same area should agree to transmit simultaneously, to avoid mutual interference. The IARU Region 1 procedures state that if possible northbound and westbound transmissions should be made in periods 1, 3, 5... counting from the full hour, with southbound and eastbound transmissions during periods 2, 4, 6... The location of the UK means that we can work towards both the north-east and the south-west, so the IARU rule is overridden in the UK by a local agreement to transmit in the second period.

PERIOD	SSB	CW
1st	hh00 – 01	hh00 – 02½
*2nd	01 – 02	02½ – 05
3rd	02 – 03	05 – 07½
*4th	03 – 04	07½ – 10
5th	04 – 05	10 – 12½
*6th	05 – 06	12½ – 15 and so on.

* = UK transmit period

FREQUENCIES

For scheduled MS contacts the frequency selected will usually be in the least-used section of the SSB or CW sub-band, as appropriate. For example, schedules on CW could be arranged to run in the section 144·130–144·150MHz, and SSB between 144·430 and 144·500MHz. It makes sense to avoid all known popular frequencies, and also the random MS sub-bands between 144·095–144·126MHz and 144·395–144·426MHz.

For random MS contacts the frequency used for CQ calls should be 144·100MHz for CW and 144·400MHz for SSB. IARU Region 1 has struggled long and hard to find a method to avoid the continent-wide QRM which results from large numbers of stations attempting to make complete QSOs on these calling frequencies during major showers. The current recommendation is that the contacts resulting from these CQ calls should take place in the frequency range 144·101–144·126MHz (CW) or 144·401–144·426MHz (SSB) so as to avoid interference to the calling frequencies. The problem is how to move a beginning QSO off the calling frequency without losing contact altogether, and the recommended procedure for doing this is as follows.

Before you make your CQ call within a few kHz of 144·100MHz (CW) or 144·400MHz (SSB), select the frequency you wish to use for the subsequent contact. The frequency chosen should be clear of traffic and interference. Immediately following the letters CQ, a letter is inserted to indicate the frequency you have chosen. The letter used indicates the frequency offset HF from your actual present frequency, as distinct from the nominal calling frequency.

A = 1kHz *	(CQA)	N = 14kHz	CQN	
B = 2kHz *	(CQB)	O = 15kHz	CQO	
C = 3kHz *	(CQC)	P = 16kHz	CQP	
D = 4kHz *	(CQD)	Q = 17kHz	CQQ	
E = 5kHz *	(CQE)	R = 18kHz	CQR	
F = 6kHz	CQF	S = 19kHz	CQS	
G = 7kHz	CQG	T = 20kHz	CQT	
H = 8kHz	CQH	U = 21kHz	CQU	
I = 9kHz	CQI	V = 22kHz	CQV	
J = 10kHz	CQJ	W = 23kHz	CQW	
K = 11kHz	CQK	X = 24kHz	CQX	
L = 12kHz	CQL	Y = 25kHz	CQY	
M = 13kHz	CQM	Z = 26kHz	CQZ	

* = frequency shift too small to be useful in clearing QRM

At the end of the transmitting period, listen on your chosen frequency as indicated by the letter in your CQ call. If a signal is heard, it may well be a reply to your call; if so, move your own transmitter up to that frequency and make the whole QSO there.

For example, I wish to try for a random CW MS contact. I check my receiver in the range 144·101–144·126MHz and find a clear frequency on 144·113. However, 144·100MHz is already occupied by other G stations so I decide to call CQ on 144·102. Since I shall be listening 11kHz up from my original CQ frequency, I need to call "CQK". Thanks to publicity for the IARU Region 1 MS procedures in national magazines, any reasonably clued-up operator who receives "CQK" on 144·102MHz will know that the reply should be made on 144·113MHz.

The first place where newcomers to 144MHz

OPERATING

DX encounter MS is probably in the SSB subband. The correct calling frequency is 144·400MHz, although 144·200 is still in use despite having been dropped from the Region 1 band plan since 1981. Generally speaking therefore, the people on 144·400MHz know what they're doing, and those on 144·200MHz are largely wasting their time – see the section on **Invalid Contacts**!

Most MS activity on 50MHz is presently via SSB; the IARU-recommended calling frequencies of 50·350MHz for SSB and 50·300MHz for CW are little used. Periods of one minute are utilized for SSB and 2½ minutes for CW, but far less use is made of short pings than on 144MHz; like MS operators on all bands in the USA and Canada, European 50MHz operators tend to chant away until a long burst comes along. There is very little MS activity on 70MHz or 432MHz, though for entirely different reasons; 70MHz is good for MS but under-populated, whereas 432MHz MS is very difficult. Most QSOs or tests on these bands are scheduled rather than random.

CW TRANSMISSION SPEED

Speeds from 200 to 2000 letters/minute or higher are now in use, but in random MS work a speed of more than 800 lpm is not recommended. In scheduled tests the speed should always be agreed upon before the contact. It is wise to check that the message being sent is correct and readable, both before and if possible during the transmission.

M-S QSO PROCEDURE

CALLING: Scheduled contacts start with one station calling the other, at the prearranged frequency and time for a whole period.

"ZB2IQ G4ASR ZB2IQ G4ASR ZB2IQ G4ASR..."

On CW the letters "DE" are not used between callsigns unless required by national licence conditions. For random operation the call is:

"CQ* G4ASR CQ* G4ASR ..."

where * is the letter indicating the receiving frequency.

REPORTING SYSTEM: Meteor-scatter reports consist of two numbers, the first indicating the maximum burst duration and the second indicating signal strength.

First digit	Second digit
2 = up to 5 seconds	6 = up to S3
3 = 5 – 20 seconds	7 = S4 to S5
4 = 20 – 120 seconds	8 = S6 to S7
5 = over 120 seconds	9 = S8 and stronger

Thus the lowest possible report is 26 and the highest (seldom heard!) is 59.

REPORTING PROCEDURE: A report is sent when you have positive evidence of having received either your own or the QSO partner's callsign – in other words, you can start sending a report once you're sure you are involved in the right QSO. The report is given as follows:

"ZB2IQ G4ASR 38 38 ZB2IQ G4ASR 38 38..."

The report should be sent between each set of callsigns, three times for CW and twice for SSB. The report must not be changed during a contact, even though a change of signal strength or duration might well justify it – it's too confusing for the other station.

CONFIRMATION PROCEDURE: As soon as either operator has copied both callsigns and the report – completely – he can start sending a confirmation report. You are allowed to piece the message together from fragments received over a series of bursts and pings, but it's up to you to ensure that you've done so correctly and unambiguously. The confirmation report is:

"ZB2IQ G4ASR R38 R38 ZB2IQ G4ASR R38 R38 ..."

(A station such as mine with an R at the end of the callsign could possibly send –

"ZB2IQ G4ASR RR38 RR38 ZB2IQ G4ASR RR38 RR38 ..."

– so that ZB2IQ could be sure that two Rs together must be a confirmation report, and not just the trailing R in my callsign.) Obviously the '38' part of these example confirmation reports would change according to the burst length and signal strength, so for the rest of this section I'll call it an "R**" report.

When either operator receives a confirmation report, he can't get any further until he himself has received everything he needs.

METEOR-SCATTER

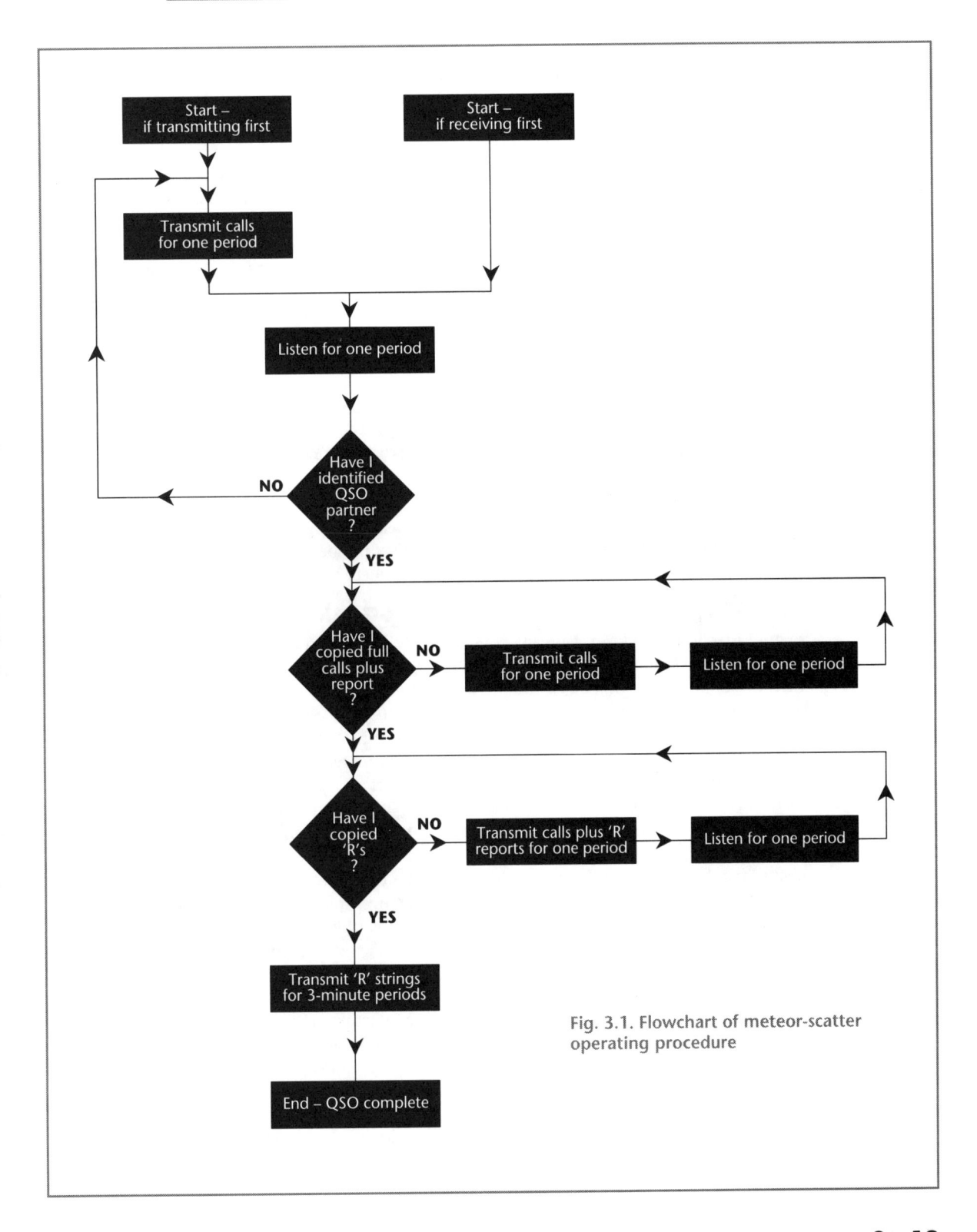

Fig. 3.1. Flowchart of meteor-scatter operating procedure

OPERATING

When he has, he confirms with a string of Rs, inserting his callsign after every 8th R to avoid confusion with any other QSO on an adjacent frequency:

"RRRRRRRR G4ASR RRRRRRRR G4ASR RRRRRRRR G4ASR ..."

When the other operator has received just one of these Rs, the contact is complete. There is some general uncertainty about this, so let me explain in more detail.

In our example, if I hear an "R**" report from ZB2IQ (implying that he's copied both our callsigns and also my report to him) and if I too have copied everything I need, then I'll start sending "RRRRRRRR". When ZB2IQ hears just one of my Rs, the minimum valid QSO will be complete and he can start writing out a QSL card for me.

But I have no way of knowing what he's heard, or whether the sked has succeeded or failed; so to save me the agony of waiting until the QSL arrives, ZB2IQ would usually send a string of Rs back. In fact we'd probably exchange Rs for two or three further periods if there was time. Note that this is purely for our own satisfaction – the QSO is already complete. Alternative (though less positive) methods of indicating that a QSO is complete are to stop sending, or to start calling CQ again if appropriate.

If you find yourselves running out of scheduled time in an MS test and the QSO seems nearly complete, it is conventional to run on for a period or two into the next sked. Conversely, it's unwise and discourteous to abandon a sked if you don't hear anything straight away – the other station may be finishing off with someone else, his alarm clock might be late, he may be making a quick repair, or there may simply have been no meteors in the first few minutes.

FLOW CHART: If you prefer to visualize the MS QSO procedure as a flow chart, take a look at Fig. 3.1 which incorporates the same information as I've been discussing above [3].

MISSING INFORMATION: If you receive an "R**" report, it means that the other station has copied both callsigns and the report from you, yet you may still need something from him. At this stage, you can try to ask for the information you need, by sending one of the following strings:

MMM	my callsign missing
YYY	your callsign missing
BBB	both callsigns missing
SSS	duration and signal strength report missing
OOO	information incomplete – you seem to have left something out
UUU	unreadable keying

The other operator should respond by sending only the required information. Although none of the above strings can be confused with an "R", even if the characters are run together, this technique must be used with great caution to prevent confusion.

SSB CONTACTS: Contacts using SSB are conducted in the same way as for CW. When attempting random contacts, speak the letters clearly, using phonetics where appropriate. It may not be necessary to use phonetics during a scheduled contact, but don't gabble.

LOGGING FOR M-S

Logging is an important aspect of meteor-scatter work. It provides each operator with a record of the received information which can be used to establish whether sufficient has been received to send a confirmation report or signing report. A copy of the logsheet is a nice courtesy if you're QSLing direct.

The example MS logsheet in Fig. 3.2 shows the development of a contact with T70A. After four receive periods, sufficient information had been received to allow a "Roger" report to be transmitted. Two receive periods later a string of Rs was received from T70A and the contact was complete. For our own satisfaction I transmitted a string of Rs and 73s at 0227½, but these were not strictly necessary: the contact was already complete.

It is customary to record the number of separate bursts, pings and duration and signal strength of the longest burst received. This is not part of the QSO procedure but is generally reported on QSL cards, and it is also the reporting format used in specialist newsletters such as *DUBUS* [5].

RAPID REVIEWING OF CW TAPES

To enable a CW contact to proceed smoothly, you need to review the tape recordings of the received bursts quickly so that your transmitted information can be changed if appropriate. When each burst is received, note the reading on the tape counter and keep track of which bursts are the longest or loudest. During your transmit period you can then rapidly return to each burst, starting with the most promising, and play it back slowly. If you get a long, loud burst which is certain to contain everything you need, you can of course play it back straight away. Experience will tell you how much information is contained in a specific length of burst, and even though you can't fully read the high-speed Morse you can often recognize the rhythms of callsigns and reports. As a guide, it takes a little under two seconds to receive "G4ASR SP5EFO 36 36 36" transmitted at 600 letters per minute. At that speed, whole callsigns sound rather like dashes sent at a much slower speed, and reports or Rs sound like dots, so the message above has the rhythm of the Morse character "7".

Fig. 3.2. Example of a meteor-scatter logsheet

METEOR SCATTER TEST | SHOWER/SPORADIC | SKED/~~RANDOM~~

G4ASR | LOCATOR: IO81MX | HEREFORDSHIRE | YEW TREE COTTAGE LOWER MAESCOED HEREFORD HR2 OHP

DATE 5 JULY 1986 TIME 0200 TO 0300 GMT
FREQ 144·165MHz | MODE A1A | SPEED 800 LPM | 2½ MINUTE PERIODS
CALL T70A | WHO STARTS T70A | LOCATOR JN63
DISTANCE 1450 KMS | BEARING AZ 122 DEGREES

NOTES: SAN MARINO
GENERALLY VERY LOW LEVELS BURSTS (S1) AND VERY SHORT PINGS

TIME	MESSAGE INFORMATION		SCATTER INFORMATION		
U.T.C.	SENT	RECEIVED	BURST	PING	LENGTH
0200		NIL			
0202½	T70A G4ASR				
0205		SR T	1	111	
0207½	T70A G4ASR				
0210		39 39 39	1	1111	
0212½	T70A G4ASR				
0215		G4ASR T70A 39 39 39 G4ASR T70A 39 39 39 G4ASR	1	11	4 SEC AT S4
0217½	T70A G4ASR R27 R27 R27				
0220		G4ASR T	1	111	
0222½	T70A G4ASR R27 R27 R27				
0225		T70A RRRRRR 73 73	1	1111	
0227½	G4ASR RRRRRR 73 73				
	5 BURSTS	16 PINGS	4 SECONDS		MAX S4

QSO COMPLETE/~~UNCOMPLETE~~ | 73's DAVID G4ASR

INVALID MS CONTACTS

On many occasions, especially during major shower periods, operators are heard using the wrong procedures and consequently making doubtful or completely invalid contacts.

Here's how ***not*** to work ZB2. You are listening on 144·400MHz during the Perseid meteor shower:

You receive – "CQ ZB2IQ CQ ZB"

You transmit – "G4ASR 38" repeatedly on 144·400MHz

You receive – "ROGER 37 ROG"

You transmit – "ROGER" repeatedly, until –

You receive – "ROGER QRZ ZB2"

That was *NOT* a valid contact! How many things can you find wrong with it?

For a start, G4ASR didn't give ZB2IQ's callsign when calling with the report. The chances are that six other people also think they are working ZB2IQ, but even ZB2IQ can't be sure. Next, ZB2IQ is heard to be sending "ROGER 37" to someone; but it might not be G4ASR, because G4ASR never said who he was calling in the first place. Then ZB2IQ continued the contact on the SSB calling frequency and not on a nominated frequency away from the calling frequency. Finally, G4ASR didn't identify himself while giving the closing Rogers – but it's too late by then, because the whole thing is already a complete mess.

Here's how to do it properly. ZB2IQ has to give a lead by calling for a reply away from 144·400MHz. So:

You receive – "CQM ZB2IQ CQM Z" on 144·400MHz

You transmit – "ZB2IQ G4ASR 38" repeatedly on 144·413MHz

until –

You receive – "G4ASR ZB2" ... "ROGER 37 ROGER 3" on 144·413MHz

Right! You've copied both callsigns, "37" and "ROGER", so –

You transmit – "ROGER G4ASR" repeatedly until –

You receive – "QRZ ZB2IQ"

– and further QSOs continue on 144·413MHz.

This time both operators have copied both callsigns, the report and an unambiguous confirmation. What's more, you've heard ZB2IQ move on to the next QSO, so you can be pretty certain that yours is in the bag.

OPERATING TACTICS

When starting MS, I strongly advise you to listen to other people's skeds – particularly in order to master the technique of rapid review of high-speed CW tapes. When you're ready to try a QSO, it's best to make a sked with a DX station who you've already been hearing well, rather than plunge in at the deep end on the random calling frequencies. Above all, don't get into MS by picking up the bad operating practices which occur around 144·200MHz during showers.

Although it is feasible to work MS with low power, *eg* 50W to a nine-element Yagi, you would have much more success by scheduling tests rather than trying to work random-style. In fact the same could be still said for stations running higher power to larger antennas! Higher power pays off in MS, not so much by making the signals stronger but more by making weak bursts longer as they trail away into the noise and thus increasing the information transfer.

You might think that contacts via SSB should be quicker than CW, because of the shorter time periods. This may be true on 50MHz where the bursts are long and strong, but on 144MHz the signal/noise advantage of CW enables contacts to be made more easily than on SSB, especially if you operate outside major shower periods.

MOONBOUNCE

By using the moon as a passive reflector, any two stations whose antennas can simultaneously 'see' the moon can work each other – maybe. Hundreds of stations are now active on the 144 and 432MHz bands, and a few have very large antenna systems which can make up for the smaller systems of other operators. As noted in Chapter 4, it is now quite easy to work these mega-stations on CW with quite a small system at your end, especially during moonrise or moonset periods.

The best time for EME (earth-moon-earth) contacts is when the moon is nearest to the earth (perigee), when it is overhead at relatively high northern latitudes (positive declination), and when the moon is nearly full (less ionospheric absorption and Faraday rotation at night – see Chapter 2). The one weekend of the month which has the best combination of all these three factors is designated the main activity or sked weekend. The dates of these activity periods can be found in specialist VHF/UHF newsletters available in many countries throughout Europe and North America [5].

Schedules are normally arranged via the EME nets which meet on 14·345MHz every Saturday and Sunday from around 1600 UTC for stations interested in 432 and 1296MHz, followed by the 144MHz net at about 1700 UTC. As an alternative, 432/1296MHz scheduling arrangements can be made via the *432 and Above EME Newsletter* published by K2UYH, or by direct mail or telephone using the address lists in the newsletters.

As the moonbounce procedures are different on 144MHz and 432MHz, the newcomer may

MOONBOUNCE

become somewhat confused. However, as most moonbouncers only operate on one band, the procedures for that particular band are picked up fairly quickly.

TRANSMIT/RECEIVE PERIODS

Time periods are two minutes on 144MHz and 2½ minutes on 432MHz and above. It's easy enough to keep track of 2½-minute periods, but for 144MHz it's advisable to use an operating chart, similar in style to the meteor-scatter log. Unless signals are strong enough to allow the transmission to be passed back and forth with a good degree of certainty, strict time periods are always used.

The station whose callsign appears first in a sked-list always transmits first, and subsequently in the other 'odd' time periods. Otherwise, on 144MHz the station located furthest east generally transmits first, while on 432MHz and above it's the station furthest west.

SIGNAL REPORTING SYSTEMS

Signal reports on EME are used differently from those on MS, and there are also differences between 144MHz and 432MHz.

144MHz	Report	432MHz
Signal just detectable.	T	Portions of call copiable.
Portions of calls copiable, but not complete calls.	M	Both calls fully copied, weakly.
Both calls fully copied.	O	Both calls copied comfortably.
Both calls and 'O' report copied.	R	Both calls and 'M' or 'O' report copied.

An "RST" signal report may be used when signals are loud enough, though the meaning may not be the same as on terrestrial modes. For example a weak tropo signal that might scrape RST319 would probably get a "449" report on EME and be considered very good!

QSO PROCEDURE

When first starting a schedule, both callsigns are sent for the whole duration of your time period. During the middle stages of the contact the callsigns are sent continuously during the first 1½ minutes (144MHz) or two minutes (432MHz), and the report is sent continuously during the last thirty seconds of your time period.

CW speed is very important in EME. You need to send slowly enough for other people to copy your Morse comfortably, but not too slowly; otherwise you are wasting valuable sked time and individual Morse characters may get broken up by fading, making them more difficult to read. In fact the optimum technique is to send at an overall speed of 50–75 letters/minute (10–15 words/minute) but to send the letters and numbers a little faster with more distinct spaces between them. Don't change speed, and take care over letter spacing – in particular, make "T", "M" and "O" sound very different!

When both callsigns have been positively and completely copied, only then may you begin to send a report. On 144MHz the minimum report acceptable for a valid contact is "O", whereas on 432MHz the report can either be an "M" or "O". Reports must of course be acknowledged by the other station for the contact to be complete, and this is done by adding an R to the report being sent, eg "RO RO ...". Note that the criterion for sending reports on EME is different from that used in meteor-scatter work; on MS you can start to send reports at a very early stage, but on EME you have to wait until you've copied both callsigns.

On receiving "RO" (or "RM" on 432MHz), you respond by sending "RRRRR" on its own, much as in meteor scatter. At the end of a contact, you may complete by exchanging "73", "GL" and other abbreviated pleasantries followed by "SK" to end the contact; since these are only sent when everything else is all right, they are generally accepted as a substitute if the closing Rs were lost in a deep fade.

A typical 144MHz EME contact might go something like the log overleaf.

EME contacts are seldom as easy as that! You may not copy the callsigns or report, and repeats will then be necessary. The lack of a report from the other station will tell you that he still needs both callsigns, the lack of an "R" that he still needs a report as well. If absolutely

OPERATING

A TYPICAL 144MHz EME CONTACT

Period	GMT	First 1½ minutes	Last 30 seconds
1	1000-1002	ZB2IQ DE G4ASR	ZB2IQ DE G4ASR
2	1002-1004	G4ASR DE ZB2IQ	O O O O O O O
3	1004-1006	OR OR OR OR OR OR	OR OR OR OR OR
4	1006-1008	R R R R R R R R	73 73 73 DE ZB2IQ
5	1008-1010	R R R R 73 73 73	SK SK SK SK DE G4ASR

nothing is heard from the other station, send nothing in the last 30 seconds; this is a very definite indication to the other station. If you can hear the other station, but not well enough to copy both callsigns then the 'T' report could be used. This may prompt the other station to change the polarization of his antenna system, if he can, but it can also be dangerous since "TTT" could easily be confused with "M" or "O". Sending the callsigns during the period set aside for reports can also be very confusing for the other station if signals are weak, especially if fragments of either callsign could be mistaken for "M" or "O". You have to operate carefully on EME, and it's no wonder that quite large proportions of skeds don't make it. But then again, nobody goes on EME for an easy ride.

When answering CQ calls on EME, you should send only callsigns but not the report because you haven't yet heard your own callsign back from the other station. If the other station can't copy you, he will reply with "???" or "QRZ?" and his own callsign for the whole transmitting period. If he copies his own callsign (which is much more familiar) but cannot make out yours, he may send "YYY"; reply with only your own callsign for the whole period.

Moonbounce is a popular mode in ARRL VHF and UHF contests, so on contest weekends EME stations in the USA and Canada may send "GGG" if they need your locator square as a multiplier. Unlike IARU Region 1 contests in which moonbounce is banned because someone might try too hard, the ARRL contest exchange is both callsigns plus the locator square (or "grid" as they call it, for example "FN32") but no report. So if you receive "GGG", reply with "IO82" or wherever you are.

Finally, there are a number of EME-only contests – notably the annual ARRL event – which attract a lot of interest and often stimulate DXpeditions and other special stations. Contests can be particularly interesting to the first-time listener, although it has to be said that the hurly-burly of a contest is hardly the best way to start your EME transmitting career. Operating procedures are often shortened between stations with good signals, especially in contests.

If you want to cut corners with EME procedures you need to be an experienced moonbouncer and require no lessons from me. If you're a beginner or you want to be absolutely sure of making the QSO, stick to the procedures.

THE DXER'S YEAR PLANNER

To monitor for all the propagation modes described in Chapter 2 you could spend 24 hours a day, 365 days a year, tuning up and down your favourite bands. This is hardly practical (though some DXers certainly seem to do just that) and is more than likely to push up the divorce-rate dramatically. What you really need is some sort of yearly plan that gives an idea when particular propagation modes are most likely to occur. In this way it may just be possible for your amateur radio and family life to co-exist!

The diagram below shows the DX propagation to be expected in the 'DXer's Year'. It's only a generalization, though at least it does show that the best time for planned antenna maintenance is during the Easter holiday period – but don't blame me if propagation springs yet another surprise!

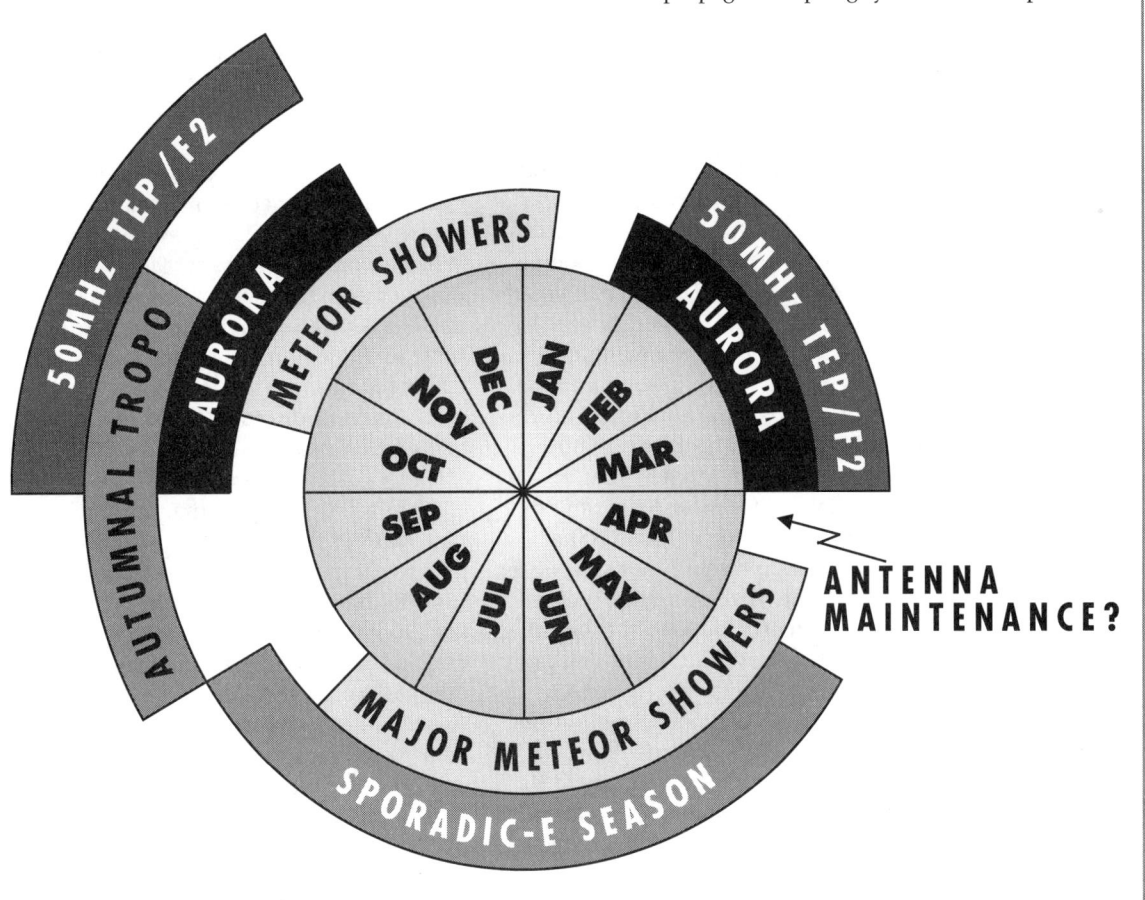

OPERATING

WORKING MORE DX

Having mastered the techniques required to work DX via specific propagation modes, you may now feel that this is all you'll ever need to know. Far from it! You are now on a learning curve which may take years to climb. Other factors and techniques are involved if you intend to work DX stations consistently on the VHF and UHF bands.

AVAILABILITY, OPERABILITY AND RELIABILITY

One of the most important points when trying to work DX via short-lived propagation modes is – are you in the operating room and/or available to go on the air? It's no good spending enormous effort and hundreds of pounds on developing your VHF/UHF system if you're not then prepared to use it. You may need to make yourself available at a moment's notice.

Taking this one step further, you must also make sure that your station equipment is equally ready for a hot session of operating. Can you imagine the panic when a major sporadic-E opening appears and you can't find the microphone? A greater problem is that many people's station layout is not suitable for a long period of intensive operating, such as the celebrated transatlantic 50MHz Es opening in 1988 which lasted over six hours. You can't operate intensively for such a long period while perched on a stool with the table at the wrong height, and with one hand doing nothing while the other has to tune, hold the microphone, write the log and hoist endless cups of coffee. Even if you're not interested in competing, contests can be a test-bed to develop a convenient operating layout.

When you switch on your station, does everything always work first time? If not, fix it – or else a big DX opening will find you out. This applies in particular to any electronics outdoors; for example low-noise amplifiers at the masthead are great until they decide to resign at the most inopportune moment. That's what routine system tests are for. Similarly, thrashing the last microwatt out of the power amplifier may give you a psychological boost, but it does nothing for long-term reliability (and it doesn't do your reputation for a clean signal much good either – see Chapter 6). The time spent in planning and maintaining a reliable and operable system pays DX dividends in the long run.

DX WARNING NETS

If you're not prepared to live in the shack during every possible period of good propagation, why not organize a DX warning net? Setting up a telephone warning chain doesn't involve a great deal of effort. At its simplest level, it can consist of two friends who telephone each other if a band opening is heard. With experience this chain can be expanded to cover a larger geographical area. When a chain becomes very large, some form of operational notes and telephone numbers must be issued to everyone participating in the scheme. These notes will show names, telephone numbers at home and work and special instructions such as "six metres only" or "no telephone calls after 11pm".

For speed of response it's best to organize a chain as a closed ring, so that messages can go two ways around it. When a participant starts off a warning, he makes two telephone calls to his neighbours on either side in the ring. If you receive a call, you need only pass the message to the next in line. If any telephone calls go unanswered, skip over to the next person beyond – don't let the message die with you. This type of organization is very dependent on self-discipline but after a number of practices it should work very smoothly. The important thing to remember is that the chain is for mutual help. If one member of the group only receives but never

WORKING MORE DX

gives, or never takes the trouble to pass the message on, this is easily noticed and obvious steps can be taken to cure the problem!

DX PACKET CLUSTER

Through the use of packet radio, operators can now access a powerful network providing real-time warnings of band openings. Typically up to 30 DX operators are logged into a regional DX cluster node via packet radio, with user access normally being on the 70, 144 or 432MHz bands depending on local arrangements. These individual clusters, 10–15 in the UK, are then networked so that potentially 200 or more DXers are connected together.

Once logged into the cluster, you can make use of its many facilities. The primary function is for spotting DX. Anyone who finds a station worthy of note can announce it to all connected stations. A typical message might read;

50.121MHz VK8ZLX 59 PG66 LISTENING UP

All you do then is turn on the 50MHz rig and work the DX! You don't even have to be near the rig, since your computer can be made to 'beep' every time a DX spot is received. Some computer-aided transceivers will even tune your rig to the correct frequency!

The DX spot is not only broadcast but also stored in the cluster's database. So when you arrive home late from work you can ask the system to list what DX you've missed on any particular band. Fig. 3.4 shows the 144MHz data received on my local cluster (GB7DXC) during an aurora on 13 July 1991. It shows frequency, callsign, date, time, signal strength, locator and the callsign of the station who supplied the information.

Other facilities available on the cluster are too numerous to mention in detail but the following topics may whet your appetite – WWV data, MUF, bearings, sunrise and sunset calculations, QSL details, and mail to other DXers.

CONTESTS AND DXPEDITIONS

Increased operating skills and greater station efficiency are the direct result of amateur-band contesting. If you don't believe this, tune across any VHF/UHF band when a contest is in full swing and listen for the most expert operators. The chances are that they are also well-known DXers. Contests can be viewed in several ways: as an end in themselves, as a way of measuring your own performance and that of your station, as an opportunity to "give a few points away" and see what turns up, or as a chance to search the band selectively for well-equipped DX stations.

The aim of looking for contest stations in specific locations can also be applied to the working of DXpedition stations. They will generally be operating from rare locator squares, counties or countries. These stations usually operate from good sites with high power and large antenna systems, and are therefore equipped to contact large numbers of stations with more modest systems. If you're working towards a certain goal such as an operating award, you are more likely to receive a QSL card from DXpedition station than from a contest group that's only interested in winning a trophy.

LOGGING

During some intense openings, via sporadic-E for instance, the numbers of stations contacted can be enormous. In one such opening, I worked 120 stations in 2 hours, during which

Fig. 3.4. Extract from a DX cluster database. Obviously there were many more DX stations active – users enter only the rarer ones

144022.0	HG8CE	13-Jul-1991	1455Z	55A	KN06EN	<GW4LXO>
144036.0	IW0AKA	13-Jul-1991	1452Z	53A	JN61FI	<G4ASR>
144040.0	EJ7FRL	13-Jul-1991	1445Z	56A	Fastnet!	<G6HCV>
144049.0	RB5PA	13-Jul-1991	1441Z	55A	KO21FC	<G4CVI>
144038.0	SM5EFP	13-Jul-1991	1435Z	53A	JO79WJ	<G4VXE>

time I also answered three telephone calls from the DX warning net, checked the beacon sub-band and had an FM contact with a Hungarian mobile station on 145·5MHz! At times during the opening, contacts were being made at the rate of five per minute. To allow you to work so many stations and not end up with a totally indecipherable mess in the logbook, you should make up scribble sheets. Very simply, these sheets can consist of 20 lines, spaced half an inch apart, with columns for TIME, CALL, REPORTS and LOCATOR. Leave plenty of space on the sheet to allow for corrections and other notes. After the opening, you can then check the scribble sheets for accuracy and write up the main station log in the knowledge that it is an accurate and permanent record. If you feel really enthusiastic, a series of sheets could be produced for different modes such as aurora or tropo.

BACKGROUND KNOWLEDGE

Success in both HF and VHF/UHF DXing comes from having good information before the event, because that's what gets you in the right place at the right time. As well as understanding the kind of background information which this book is intended to provide, you need a continuous supply of information on current affairs on the bands you're interested in.

SHORT-WAVE BANDS AND THE VHF/UHF DXER

The successful VHF/UHF DXer must also be an HF DXer. Well, not strictly speaking, but the real VHF enthusiast must have some sort of HF capability to keep in touch with events beyond normal range. It's also very nice to be able to chat at leisure to people with whom you have a strong common interest, but can only work occasionally and with great difficulty on the VHF/UHF bands. Pursuing the link with HF DX a little further, it isn't too surprising that some of the world's top VHF/UHF DXers are also prominent on the DXCC Honor Roll.

The following HF frequencies are used nationally or internationally for VHF/UHF scheduling and the exchange of operating news.

3·625MHz – Alternative LF frequency for the European VHF Net.

3·718MHz – UK Six Metre Group net. Active daily from about 0730 through to 0830 UTC. The net discusses many topics including propagation forecasts, operational news, DXpeditions *etc*. All UK 50MHz operators are encouraged to call in.

BACKGROUND KNOWLEDGE

7·050MHz – GM to UK talk-back. Sometimes used by expedition stations in Scotland, operating on the 50MHz and 70MHz bands, who wish to arrange schedules with stations throughout the United Kingdom.

14·345MHz (14·335–14·350MHz) – European VHF Net. Not a formal net; used primarily at weekends and during most meteor shower periods for MS scheduling. Expedition news and forecasts of impending propagation events, auroras *etc*, can also be heard. To arrange schedules, simply call "CQ European VHF net" – but avoid 14·345MHz itself after 1500 UTC at weekends because that's when the intercontinental EME nets meet. The VHF Net and the EME nets are an invaluable source of up-to-date operational news. Newcomers to meteor-scatter or moonbounce will find the details of other people's schedules particularly useful when first setting up their receive systems.

28·385MHz – 50MHz liaison, used primarily in Australia and Japan. This frequency is occasionally used as an alternative to 14·345MHz for the scheduling of MS tests.

28·885MHz – 50MHz crossband and liaison, used on a world-wide basis. The activity, particularly throughout Europe, is such that operators are requested to spread out either side of this nominal frequency. Crossband operators will call "CQ crossband" and indicate the frequency, normally 50·185MHz, on which they will listen for calls (please avoid 50·110MHz or 50·200MHz). If you're looking for 70MHz contacts, the frequency to use is 70·185MHz (avoiding 70·200MHz please).

28·885MHz is also used as a meeting-place for 50MHz enthusiasts to exchange operating news and DXpedition information. It's not unusual to have rag-chews about 50MHz with DX stations which would normally attract pile-ups lower down the 28MHz band.

CALLSIGNS AND LOCATORS

Familiarity with callsign structures and the IARU world-wide locator system can enable you to eliminate obvious mistakes in both callsigns and locators when working with weak or partly unintelligible stations. This may be particularly useful in contests.

All amateur callsigns have a fixed format, generally prefix-digit-suffix, the prefixes being allocated to each national administration by the International Telecommunications Union (ITU). The prefix consists of letters and/or one digit, and the suffix after the central digit is usually just letters. However, the ITU allocates large prefix blocks to be used for all kinds of radio services and there's plenty of scope for national self-expression.

Generally the prefix indicates the country, and the suffix belongs to the specific station. But details vary considerably between countries. In some administrations the prefix letters indicate a country or region (*eg* G = England, GM = Scotland), while in others that's denoted by the digit (*eg* SM0–SM7 or I1–I8). The prefix and/or the central digit can also indicate the licence class, *eg* G*4 = UK class A / all bands; G*7 = class B / 50MHz and above.

It should be apparent from these examples that in many cases the callsign can provide invaluable geographical information. By using this knowledge and cross-referencing with locator maps you can eliminate many impossible combinations of callsign and locator. In a number of instances it is possible to identify a specific locator square from the callsign alone. Although you cannot – and should not – work backwards from a locator square to a complete callsign, it may be possible to eliminate certain callsign groupings if initially only the locator information is received correctly. For instance you may have partially received the callsign ???HBR and the locator JN60XS. Checking the locator map indicates that the station is located on the island of Ischia. Instantly you recall that islands situated off the coast of Italy in this area are prefixed with IC8. That's our man – must be IC8HBR. Except that you still need to hear him say "IC8" to the satisfaction of your own conscience!

The release of the 50MHz band to many countries within Europe on a rising sunspot cycle brought the problem of callsign and locator recognition on a world-wide basis. Obviously, instant recall of the world's intricate callsign systems doesn't come easily.

OPERATING

Fortunately, specialist operating manuals are available which catalogue most of the information that you will require [3, 4]. One problem with these books is that callsign systems are forever changing throughout the world, so it is impossible to keep right up-to-date. Band plans and specialist activity frequencies are other items of information which seem to change constantly or require modification.

NEWSLETTERS, CHARTS AND MAPS

The only way to keep up-to-date with changes in operational procedures, or to learn about the latest DXpeditions, is to haunt the bands, swap news with other DXers and listen a lot. Monthly newsletters [5] are a good way to learn about events fairly quickly after they have happened, but the deadlines for the VHF columns in monthly magazines are simply too long for late-breaking news. In other words, to be a successful DXer you need always to use the grapevine for both technical and operating information. More than that, you need to become part of it. Contribute to a specialist monthly VHF/UHF newsletter or write a VHF column in your radio club's magazine. The more information you contribute, the more you're likely to receive in return!

REFERENCES

[1] Bob Locher W9KNI, *The Complete DXer*.

[2] Juka Heikinheimo OH1BR and Miika Heikinheimo OH2BAD *Radio Amateurs' Conversation Guide* (book and tapes)

[3] Ray Eckersley G4FTJ, *Amateur Radio Operating Manual* (RSGB)

[4] *ARRL Operating Manual* (ARRL)

[5] Newsletters are a prolific source of operating information, news and views –

DUBUS is a widely-circulated quarterly VHF/UHF/Microwave DX and technical magazine. Published by DUBUS-Verlag in Germany, it is available through national distributors. (If you don't know anyone who takes *DUBUS*, you really aren't on the DXers' grapevine yet!)

VHF/UHF DXer is a UK-based monthly newsletter covering all aspects of VHF/UHF DX, edited by G8ROU.

Six News is the publication of the UK Six Metre Group. This members' newsletter is edited by G0JJL and G0JHC.

The 50MHz DX Bulletin is issued twice monthly and contains comprehensive details of world-wide 50MHz DX information.

European Moonbounce Edition is a monthly newsletter primarily intended for the 144MHz enthusiast, although it does occasionally give details of 432MHz operation. It is edited by HB9CRQ and HB9DBM.

2-Meter EME News contains all you need to know about 144MHz moonbounce operation. It is edited and published by K0IFL.

The 2 Metre News Sheet, issued monthly, gives details of forthcoming expeditions and other DX information. It is edited by SM6EOC and SM6AFH.

2-Meter EME Bulletin, edited by KB7Q, is obviously for the 144MHz moonbouncer.

432 and Above EME Newsletter is the monthly newsletter for the UHF or SHF EME enthusiast. It is edited by K2UYH and contains news, technical notes and the sked-list for the next month's activity weekend.

OSCAR NEWS is the magazine for satellite enthusiasts. Edited by G3AAJ, it is available to members of AMSAT-UK.

CQ-TV is the definitive magazine for those interested in all aspects of ATV, SSTV on the VHF, UHF and Microwave bands. Only available to members of the British Amateur Television Club (BATC).

All Europe VHF, UHF, SHF Contests Calendar, a yearly publication detailing most contests being held throughout Europe. Written by DH2NAF.

CHAPTER 4

ASSEMBLING YOUR STATION

ASSEMBLING YOUR STATION

by Ian White G3SEK

Why can one station work all the DX, when another with similar equipment can't? It's very easy to blame the site or the operator, but that isn't always true. Far too often, the real reason is that the unsuccessful station is just a collection of boxes plugged together with no thought about how the various items interact. In fact many stations contain equipment which has never worked properly from new. The owners don't even suspect that anything is wrong, because they don't understand how their amateur radio station *should* perform.

That's what this chapter aims to show you: how your receiver, transmitter, linear amplifier, feedline and antenna work together *as a system*, and what kind of DX performance you should expect from it. We read and hear a lot about designing circuits and antennas – and there's plenty of that in other chapters – yet it hasn't been traditional in amateur radio to think about our stations at a system level. However, if you want to get the most from your equipment, that's exactly what you need to do, and this is equally true whether you build or buy your equipment. If you build it, you're responsible for making sure it works as well as possible. And if you buy most of your equipment, system design is your *only* means of turning a collection of boxes into an efficient station.

Chapters 5, 6 and 7 are about the separate parts of the station RF system: receivers, transmitters and antennas. This chapter lays the groundwork about station performance as a whole. You'll learn how to balance receive and transmit performance, and how your antenna and feedline affect the performance of your station. And above all, you'll begin to understand how well your station *should* perform, and where improvements can be made.

We will be working our way towards an idea called *path-loss capability* – this is a way of measuring your station's ability to work DX.

RECEIVER SENSITIVITY AND NOISE TEMPERATURE

There are many ways to specify the weak-signal performance of a receiver, but the basic concept underlying them all is *noise temperature*.

Any electrical conductor contains electrons which are free to move around. At normal temperatures, electrons in a conductor are in random motion, rushing back and forth but on average getting nowhere. This random thermal movement of electrons constitutes a fluctuating current, which can be detected as random noise. Converted to audio frequencies, random noise is the rushing sound your SSB receiver makes when tuned to an unoccupied frequency – as distinct from ignition noise and other forms of man-made pops and crackles.

At any temperature above absolute zero, the thermal noise-power generated in a conductor is proportional to its physical temperature measured on the 'absolute' temperature scale in units of kelvins (K). To convert from degrees Celsius (Centigrade) to K, add 273·16. Thermal noise is spread evenly over the entire electromagnetic spectrum, so the noise power detected by a receiver is proportional to the

RECEIVER SENSITIVITY

spectral bandwidth across which the noise is being collected. The basic formula relating noise power P_N (watts) to noise temperature T (K) and bandwidth B (Hz) is simply:

$$P_N = kTB \qquad (1)$$

where k is called Boltzmann's constant and has the value 1.38×10^{-23} J/K (joules/kelvin).

For example, the thermal noise power generated in a resistor at a temperature of 290K (around room temperature) is:

$$1.38 \times 10^{-23} \times 290 \times B \text{ watts}$$

ie 4.00×10^{-21}W in every hertz of bandwidth B, across the entire electromagnetic spectrum. The noise power is independent of the ohmic value of the resistor, and the value calculated in equation (1) is the power which would be delivered into a matched load having the same impedance as the resistor.

Another example: if a 50Ω resistor at 290K was connected to the input terminals of a noise-free receiver, with a 50Ω input impedance and a bandwidth of 2·5kHz (Fig. 4.1), what would be the noise voltage? The noise power from (1) is:

$$1.38 \times 10^{-23} \times 290 \times 2500 = 1.00 \times 10^{-17} \text{ W}$$

The voltage across 50Ω is obtained simply from $P = \dfrac{V^2}{R}$, or in this case $V = \sqrt{PR}$, and is:

$$\sqrt{(1.00 \times 10^{-17} \times 50)} = 22.4 \text{nV}$$

That's right, **nano**volts – that's what we mean by 'small' signals in modern VHF and UHF receivers!

Resistors are not the only electronic devices which generate noise. Imagine a preamplifier (Fig. 4.2) connected between the 50Ω resistor and the noise-free receiver. The noise power delivered to the receiver now has two components. One is the amplified thermal noise from the resistive noise source, and the other is noise generated in the preamplifier itself. The noise contribution from the preamplifier can be expressed simply as a power in watts, or more subtly – and usefully – as the *equivalent noise temperature* of the preamplifier.

Equivalent noise temperature is one of those ideas that are hard to swallow at one gulp, so let's take a few steps back and look at it. As far as the receiver's concerned, the noise power that it's hearing could be coming either from a noisy preamp, or merely from heating the 50Ω resistor above its actual temperature; the receiver has no way of knowing that in fact it's from the preamp. So we can measure the 'noisiness' of the preamplifier in terms of the increase in the temperature of the resistor that would be required to produce the same amount of noise. That's the meaning of T1 in Fig. 4.2. A perfect preamp would have a noise temperature of zero K, *ie* it would contribute no extra noise. But nothing in this universe is noise-free; any real device has an equivalent noise temperature greater than zero K. The higher its equivalent noise temperature, the more noise the device is making. The word "equivalent" is there simply to remind us that the temperature of the resistor at the input doesn't physically increase – and indeed there needn't be a resistor there at all.

The practical advantage of thinking in terms of *equivalent* noise temperature is that it form a common basis for measuring random noise arising from just about anything, from a GaAsFET to a galaxy. There are many different types of noise in electronic devices, and most of them are not truly thermal in origin – nor is galactic noise, for that matter. This doesn't

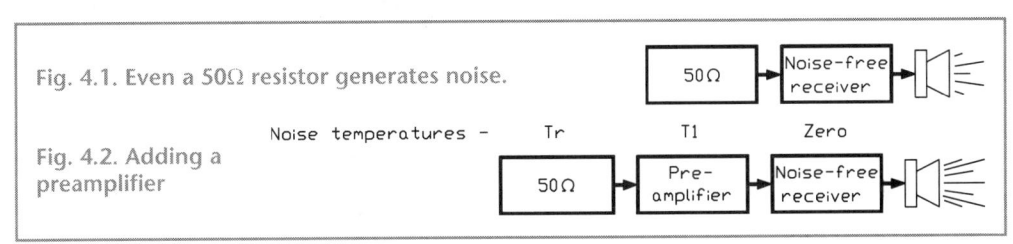

Fig. 4.1. Even a 50Ω resistor generates noise.

Fig. 4.2. Adding a preamplifier

Noise temperatures – Tr   T1   Zero

ASSEMBLING YOUR STATION

matter so long as they can all be expressed as *equivalent* amounts of thermal noise. In practice, people don't always include the word "equivalent", and after a few more paragraphs I'll probably start to forget it as well.

Because much electronic noise is strictly speaking non-thermal, the equivalent noise temperatures of most electronic devices (apart from resistors) are seldom the same as their physical temperatures. For example a modern MOSFET amplifier at room temperature may have a noise temperature as low as 50K at 144MHz, and GaAsFETs can go even lower. On the other hand, a very noisy device can have a noise temperature well above 1000K without glowing white-hot! This means that you can't always cure a noisy device by cooling it towards absolute zero, because that will only reduce the thermal part of its noise. The reason cooling works for parametric amplifiers used in radio astronomy is because most of their noise *is* thermal. Cooling also works for GaAsFETs in the right circuit – but don't get too excited, because we'll shortly see that low noise isn't everything as far as VHF and UHF DX is concerned.

Unlike true thermal noise, which is uniform across the whole electromagnetic spectrum, noise from an active device like a transistor (or a galaxy) is frequency-dependent. Although noise temperature may be taken as constant within an amateur band, it will probably be different on another band. Since there are probably several different noise-generating mechanisms operating within the same device, the noise temperature may well increase at both the HF and the LF end of its useful frequency range. At the 432MHz preamplifier noise-measuring session in a recent Moonbounce Convention, it was instructive to see a cheap consumer GaAsFET almost beating an expensive microwave High Electron Mobility Transistor. The reason was that 432MHz was too low a frequency for the HEMT device!

Compared with other ways of measuring noise, noise temperature has some very useful properties. It is independent of bandwidth (unlike noise power, which is proportional to bandwidth), and it's also independent of the gain or loss of the device generating the noise. Most useful of all, the noise temperatures of different components in a total system can be added directly to give a *system noise temperature*.

SYSTEM NOISE TEMPERATURE

CASCADES OF AMPLIFIERS

So far we've looked at a simple RF preamplifier fed from a 50Ω resistor (Fig. 4.2). We've seen that the noise temperature of this combination is the sum of the noise temperature of the resistor (T_R, equal to its physical temperature because it is a resistor) plus the equivalent noise temperature T_1 of the preamplifier. This sum is actually a simple form of system noise temperature.

Now let's add to the system another stage whose noise temperature is T_2 (Fig. 4.3). We can calculate the noise temperature of this new system by finding out what excess noise the second stage contributes, in terms of excess equivalent noise temperature at the input. The excess noise temperature is simply T_2/G_1, where G_1 is the power gain of the first stage, so the system noise temperature T_{SYS} is given by:

$$T_{SYS} = T_R + T_1 + \frac{T_2}{G_1} \qquad (2)$$

Dividing T_2 by G_1 is called 'referring T_2 to the input'. Power gain G_1 must be a pure ratio

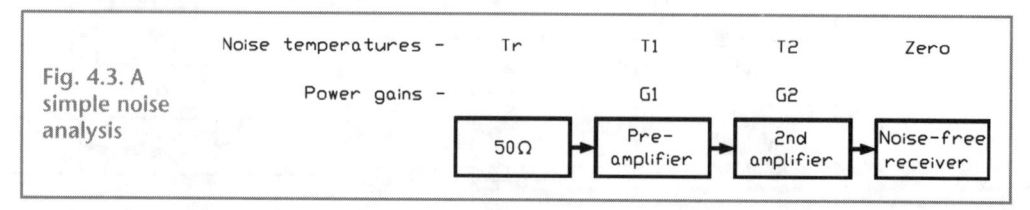

Fig. 4.3. A simple noise analysis

Noise temperatures – Tr, T1, T2, Zero
Power gains – G1, G2

50Ω → Pre-amplifier → 2nd amplifier → Noise-free receiver

SYSTEM NOISE TEMPERATURE

(eg 'a power gain of 15 times'), and not in dB.

We can continue in this way, adding on extra stages (Fig. 4.4A) and referring their noise contributions back to the input, so that:

$$T_{SYS} = T_R + T_1 + \frac{T_2}{G_1} + \frac{T_3}{(G_1 G_2)} + \frac{T_4}{(G_1 G_2 G_3)} + \ldots \quad (3)$$

The noise temperature T_3 of the third stage is referred back to the input by dividing it by the cumulative gain up to that point ($G_1 \times G_2$ of the two previous stages) and so on. Once all the noise temperatures of the later stages have been referred back to the input in this way, their contributions to the system noise temperature are simply added together.

There are limits to how low the system noise temperature can get. Go back to the simplest two-stage system (Fig. 4.3) and look at equation (2) in more detail. For a start, the system noise temperature can't be less than T_R, because that's the noise temperature of the source outside the receiver where the signal is coming from. This is the most basic limit, which many people don't appreciate: there is *always* some noise coming into the receiver, along with the signals. It forms part of the overall system noise and, no matter how low-noise your receiver, that external noise is still there.

The noise temperature T_1 of the first stage adds directly to the system total, but the situation is different for the later stages. If they were completely noiseless, T_{SYS} would come down to just $(T_R + T_1)$. And T_{SYS} can be brought close to this lower limit if the last term in equation (2), namely (T_2/G_1) can be made small. We already know how to do this in practice: by either using a low-noise device, which makes T_2 small, or by increasing the gain G_1.

CUMULATIVE GAIN

In a multi-stage system we can minimize the effect of later stages by using low-noise circuitry or adding more gain. In practice it's usually the gain effect which dominates: in equation (3) the cumulative effect of G_1, G_2 *etc* on the bottom lines of the fractions can quickly make the noise contributions of later stages extremely small. To show this effect in practice, let's put some typical performance figures into the system of Fig. 4.4A, to give Fig. 4.4B. In a real multi-stage system we can say that the last stage is 'the rest of the system', and give it a noise temperature, so we no longer need the fictional noise-free receiver which appears in Figs 4.1 – 4.4A. Then using equation (3) we have:

Fig. 4.4. Noise analysis of more stages: (A) in symbols, (B) with numerical values

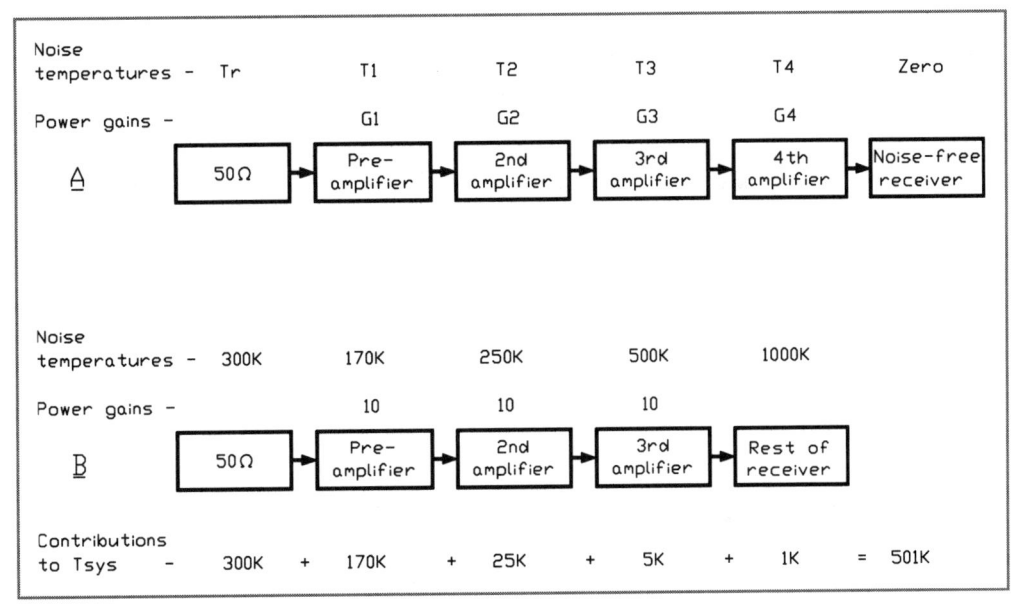

A

Noise temperatures	Tr	T1	T2	T3	T4	Zero
Power gains		G1	G2	G3	G4	
	50Ω	Pre-amplifier	2nd amplifier	3rd amplifier	4th amplifier	Noise-free receiver

B

Noise temperatures	300K	170K	250K	500K	1000K
Power gains		10	10	10	
	50Ω	Pre-amplifier	2nd amplifier	3rd amplifier	Rest of receiver
Contributions to Tsys	300K +	170K +	25K +	5K +	1K = 501K

$$T_{SYS} = 300K + 170K + \frac{250K}{10} + \frac{500K}{(10 \times 10)} + \frac{1000K}{(10 \times 10 \times 10)}$$
$$= 300K + 170K + 25K + 5K + 1K$$
$$= 501K$$

So you can see how the cumulative gain makes the noise contributions of the later stages dwindle away. Mixers behave exactly the same as amplifiers in these equations, and even lossy mixers such as diode-ring devices can be treated in the same way (see Chapter 5).

NOISE FROM LOSSY DEVICES

We also need to know how to deal with such loss-introducing devices as cables, attenuators and filters. These degrade the system noise temperature, mainly by attenuating the wanted signal. But there's also another, less obvious source of noise: thermal noise generated in the resistive part of the losses. This second effect is little-known and often neglected, but it can be important in the ultra-low-noise systems we are dealing with here.

The complete formula for the effect of lossy components on system noise temperature is:

$$T = T_F + (L-1)T_P \qquad (4)$$

where:
- T is the noise temperature at the input of the lossy component
- L is the loss expressed as a power ratio (ie L = 1 for a lossless system, going up to infinity for infinite loss)
- T_F is the cumulative noise temperature of all following stages, referred to the output of the lossy device
- T_P is the physical temperature of the lossy device.

The first term in equation (4) is the attenuation term and the second is the term for extra thermal noise, which depends on the physical temperature of the lossy device. The latter term is small, except in extremely low-noise systems. Note that we have been assuming that the losses are resistive and occur in a matched system. If the losses are solely due to an impedance mismatch, only the first term in equation (4) is relevant.

Let's try an example. If the receiver noise temperature is 180K, how much does a 10m length of URM67 (RG213) with a loss of 0·9dB add to the noise temperature? Well, 0·9dB represents a loss factor L of 1·23, and let's assume the cable is at a physical temperature of 290K. Plugging these numbers into equation (4), we have:

$$T = (180 \times 1 \cdot 23) + (1 \cdot 23 - 1) \times 290$$
$$= 221 \cdot 4 + 66 \cdot 8$$
$$= 288K, \text{ near enough}$$

So you can see that 0·9dB of cable loss has changed the noise temperature from 180K at the receiver to 288K at the top of the cable. And 66·8K of that is due to thermal noise generated in the cable itself.

NOISE FACTOR AND NOISE FIGURE

Noise temperature is the fundamental concept underlying every aspect of noise in receiving systems. So how are the more familiar-sounding terms 'noise factor' and 'noise figure' related to noise temperature?

Noise factor F is related to noise temperature T by a standard definition [1]:

$$F = 1 + \frac{T}{290}$$

290K is about 17°C or 62°F, close to room temperature, so noise factor is a measure of how the device or system noise compares with thermal noise under typical conditions of use. This definition gives the perfect system a noise factor of 1, and all practical noise factors are greater than 1.

Even more familiar than noise factor is *noise figure*, which is simply noise factor expressed in decibels:

$$NF = 10\log_{10}\left(1 + \frac{T}{290}\right) \qquad (5)$$

Since the minimum possible noise factor is 1, the minimum possible noise figure is 0dB. Unfortunately, noise factor and noise figure are easily confused, and there's little need to become familiar with both. In this book we

SYSTEM NOISE TEMPERATURE

use either *noise temperature* (T – and always with its units of K) or *noise figure* (NF – and always with its units of dB).

Noise temperature and noise figure each have their own distinct advantages. The noise temperatures of cascaded stages add up in a very simple way, as we've seen, whereas the NFs don't. The handy thing about NF is that the attenuation of a lossy component (in dB) can be added to the NF of the following stage to give a new, larger NF for the two stages combined. To see how, let's go back to the example I gave a few paragraphs ago, about the effect of a length of cable with 0·9dB loss. If you convert the initial noise temperature (180K) into NF, and do the same for the final noise temperature (use the more accurate value of 288·2K), you'll find that the NF has increased by precisely 0·9dB. But there's a catch: this trick only works accurately if the lossy component is at the standard physical temperature of 290K.

Finally, there's one more serious problem about noise figure. It can only deal with noise in receivers – and receivers are not the whole story.

ANTENNA NOISE TEMPERATURES

The receiver is only one part of the receiving system as a whole. An equally important contributor to the system noise temperature is the noise picked up by the antenna, along with the signal. This 'antenna noise' can be brought into the system noise analysis by giving it an equivalent noise temperature.

The noise temperature of an antenna (T_{ANT}) is determined by the RF environment that it can 'see', not by its physical temperature. For example a 144MHz beam pointed at the sky for satellite work may be picking up only a very small amount of galactic noise, which can be represented by an equivalent noise temperature as low as 150-200K [2]. At the other extreme, if the antenna is picking up thermal noise from its earthly surroundings at physical temperatures of 270-300K, that will also be its noise temperature. The usual situation is somewhere in between, and depends rather upon the frequency.

At 144MHz, flat ground is a fairly good reflector of radiation and a correspondingly poor emitter, *ie* the ground acts mostly like a mirror. So a horizontal antenna sees either 'cold' sky above the horizon or the sky image reflected by the ground, and the warm ground itself makes only a small contribution to the antenna's equivalent noise temperature. Pickup of man-made noise may be much more important than thermal or galactic noise, but so long as the man-made noise is virtually random this too can be expressed as an equivalent noise temperature. The noise temperature of a 144MHz antenna can therefore range from less than 200K in satellite work to well over 1000K in an urban environment full of man-made noise. For terrestrial working on 144MHz, 200K is a reasonable *minimum* estimate for the antenna noise temperature.

At higher frequencies, galactic and man-made noise tend to decrease but the ground and nearby objects become poorer reflectors and better emitters of thermal noise. At 1GHz and above, the noise temperature of an antenna pointed at the ground will be quite close to the physical ground temperature. A normal horizontal antenna will see approximately half warm ground and half cold sky, so its noise temperature will be about 150K. For design purposes, you could also reckon on a *minimum* noise temperature of 150K for 432MHz antennas aimed at the horizon. With a little care, you can use 'ground noise' to estimate receiver performance (see box on page 4-15).

At lower frequencies, man-made and atmospheric noise become considerably greater than at 144MHz or 432MHz. Although the pops, crackles and lightning flashes from many different sources tend to merge together into something very much like random noise (like hand-clapping applause from a large crowd), a good noise-blanker to remove the worst excesses can be a distinct advantage at 70MHz, 50MHz and below. At 50MHz, the minimum galactic noise-temperature which your antenna sees, either directly or mirrored in the ground, is about 4000K [3], and subur-

ASSEMBLING YOUR STATION

ban man-made noise on top of this brings a typical 50MHz antenna noise temperature up to more like 300,000K [4]. Typical figures for 70MHz would be about 3000K for galactic noise and 150,000K including man-made noise.

USABLE SENSITIVITY

Knowing that system noise temperature can be divided into two contributions, from the antenna and from the receiver, we can now tackle the question:

How much sensitivity do we really need at VHF/UHF?

In Days of Old, when RF devices were noisy and the VHF/UHF bands were empty, the answer was clear: we always needed more sensitivity, simply to improve the chances of ever hearing a signal at all. Modern low-noise devices can give us all the sensitivity we need, and more. Yet we can no longer afford to aim for the highest possible sensitivity, because that also entails excessive front-end gain which brings overload problems (Chapter 5). So we have to *choose* what system noise temperature we want, and *design* our receivers to achieve it.

Very well, what noise temperature *do* we want? Even with a perfectly noiseless receiver, the system noise temperature cannot possibly be lower than the antenna noise temperature.

Noise from any practical receiver must degrade the signal-to-noise (s/n) ratio of a weak signal to some extent, so the choice of receiver noise temperature can be framed in terms of how much degradation of s/n is allowable.

Let's try a few numbers. Consider a 144MHz receiver connected to an antenna with a noise temperature of 200K, the 'typical minimum' figure suggested earlier. If the receiver noise temperature is also 200K (NF = 2·28dB), the system noise temperature is 400K, and the noise power is doubled. Thus the signal/noise ratio of an incoming signal would be degraded by 3dB, compared with the un-achievable noiseless receiver. Fig. 4.5 shows the relationship between the receiver noise temperature and s/n degradation for a range of antenna noise temperatures. Clearly, by aiming for the lowest possible receiver noise-temperature, the s/n degradation can be made very small. But these improvements in weak-signal performance will also imply higher front-end gain, which will ruin the strong-signal performance. Unless you carefully balance sensitivity against front-end gain, on a crowded band you can easily end up hearing *less*, not more!

If we aim for a receiver noise temperature which is equal to the minimum probable antenna noise temperature, the scope for further improving the signal/noise ratio is at most 3dB; more often it will be only 1-2dB.

Fig. 4.5. Effect of receiver noise temperature on signal/noise degradation. Curves are for a range of antenna noise temperatures from 50K to 5000K.

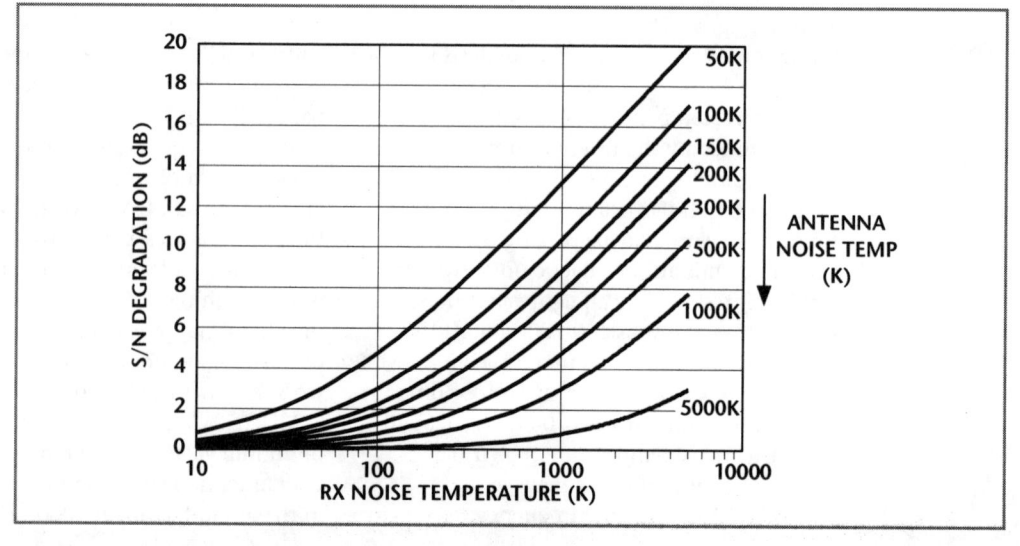

SYSTEM NOISE TEMPERATURE

Such a small improvement may scarcely be noticeable on the air, except at the very limits of weak-signal working such as moonbounce. In other words, if the receiver noise temperature is as low as the minimum probable noise temperature of the antenna, the receiver is about as sensitive as it needs to be for terrestrial work. On the lower VHF bands, man-made noise can exceed galactic and ground noise by 10dB or more, so there's absolutely no need for very low noise figures on 50 or 70MHz.

Referring back to the typical minimum antenna noise temperatures, the design-target receiver noise temperatures and noise figures for ordinary terrestrial DX-chasing would thus be:

50MHz	4000K	12dB
70MHz	3000K	10dB
144MHz	200K	2·2dB
432MHz	150K	1·8dB

These are not hard-and-fast rules, merely design targets in round figures. When you set up your receiver, you will have an opportunity to balance sensitivity against strong-signal handling, to suit your own particular noise environment. Thus you can optimize the *overall* performance of your receiving system.

NOISE FLOOR

A concept related to sensitivity is *noise floor*. This is simply the power level corresponding to a receiver or system noise temperature, calculated by $P = kTB$. For example, the noise floor of an SSB receiving system meeting our target noise temperature of 200K for 144MHz is:

$$1·38 \times 10^{-23} \times 200 \times 250 = 6·9 \times 10^{-18} \text{ watts}$$

To make this estimate, I had to assume a bandwidth of 2500Hz. The *noise bandwidth* used in these calculations is not quite the same as the more familiar –6dB filter bandwidth. Imagine a filter with a perfectly flat passband and perfectly steep sides, sometimes called a 'brick wall' filter because that's what its passband would look like: straight up, flat top, and straight down. Practical IF filters aren't like that, of course. Their skirts are not perfectly steep, and their passbands are round-shouldered and by no means flat. The noise bandwidth of such a filter is simply the bandwidth of a brick-wall filter which would let through the same total amount of noise power. For most IF filters with reasonably flat passbands and steep skirts, the –6dB bandwidth is a fair approximation for the purposes of calculating noise floor.

Notice that if we change the bandwidth, the noise floor changes too. Narrower bandwidths result in a lower noise floor, *ie* a 'quieter' receiving system. This is a very real effect, and is largely responsible for the vast improvements in weak-signal communication obtainable by using CW.

In order to talk about s/n ratios in the usual units of dB above the noise floor, it's conventional to convert the noise floor into dB relative to one watt (dBW) or one milliwatt (dBm), by the usual formulae:

$$dBW = 10 \log_{10}(\text{watts})$$
$$dBm = 10 \log_{10}(\text{watts}) + 30$$

In our previous example, $6·9 \times 10^{-18}$W corresponds to –171·6dBW or –141·6dBm.

One final point about noise floor, until Chapter 5, is that its definition can depend on whether we're talking about weak or strong signals. When we're asking whether a weak signal is audible, the definition of noise floor **must** include the **actual** antenna noise temperature; we calculate noise floor using a noise temperature of $(T_{RX} + T_{ANT})$. But when we're thinking about performance with strong signals (Chapter 5), an accurate definition of noise floor is less important; for that application the noise floor is frequently calculated assuming a nominal antenna noise temperature of 290K.

POWER, GAINS AND LOSSES

In this short section we'll assemble the rest of the ideas we need in order to think about *path-loss capability*. When it comes to stringing together a series of power gains and losses,

ASSEMBLING YOUR STATION

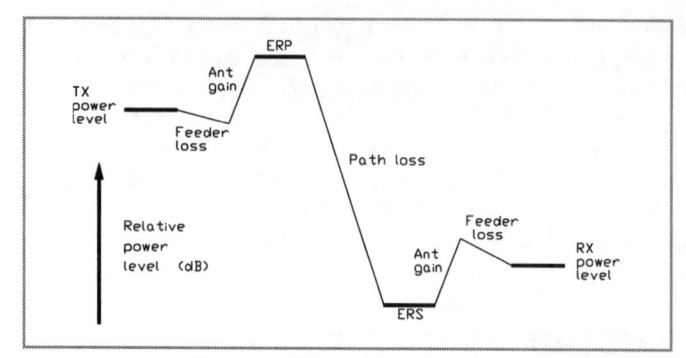

Fig. 4.6. Changes in relative power levels, from transmitter to distant receiver. (Receiving antenna gain does not change the actual power level, but represents collection of power from a wider capture area – see Chapter 7.)

decibels are much easier to work with than power ratios because you can simply add and subtract them to create a power and path loss 'budget'. Working in dBW, decibels relative to one watt, we'll start at the transmitter output and see what happens to the signal levels *en route* to the receiver.

The basic formula connecting all aspects of a transmitter-receiver path is:

$$ERP - PL - ERS = s/n \text{ (dB)} \qquad (6)$$

where
ERP = effective radiated power
PL = path loss (dB)
ERS = effective receiver sensitivity (dBW)
s/n = the resulting signal/noise ratio (dB).

Fig. 4.6 shows how these quantities are related, as the power levels go up and down *en route* from transmitter to distant receiver. Let's look at each in turn.

TRANSMITTING STATION – ERP

Effective radiated power is a simple and convenient way of combining transmitter power and antenna gain, not forgetting loss in the feedline.

Quite simply:

ERP(dBW) = transmitter power (dBW)
 − transmitter feeder loss (dB)
 + transmit antenna gain (dBi) (7)

Transmitter power is easily expressed in decibels above a watt (dBW):

$$P_{TX} = 10 \log_{10}(\text{watts}) \text{ dBW}$$

This is familiar from the UK licence conditions, *eg* 400W = 26dBW.

A beam antenna concentrates the transmitted power in the direction of interest, instead of spraying it around equally in all directions as would an *isotropic antenna* (see Chapter 7). To take account of this enhancement in the wanted direction, add the antenna gain in dBi (decibels relative to isotropic). For clarity, an ERP calculated using isotropic antenna gain should be called EIRP to distinguish it from the value calculated using gain over a dipole (dBd) which is more conventional below 1GHz. For path-loss calculations – at any frequency – EIRP is the one we want.

ERP is much misunderstood. You can easily generate some enormous-sounding figures, *eg* 100W plus 20dBi gives an awesome 10kW EIRP. But don't forget that the actual RF power is still only 100W! The step-up in EIRP due to antenna gain merely means that power is being concentrated in one particular direction; in all other directions the radiated power is *less* than it would be from an isotropic antenna. That's why beam antennas are everyone's most potent weapon in the war against unwanted interference.

PATH LOSS

On its way from the transmitter to the receiver, the signal suffers considerable attenuation; Chapter 2 deals with this in more detail. For the present, let's simply subtract the path loss, a large number ranging typically from 150 to 250dB.

RECEIVING STATION – ERS

At the other end of the path, the power collected by the receiving antenna depends on its gain. More accurately, it depends on the capture area (Chapter 7). *Effective receiver*

POWER, GAINS & LOSSES

sensitivity (ERS) is very similar to ERP, in that it combines the receiver's noise-floor power level with the antenna gain.

$$\text{ERS(dBW)} = 10\log_{10}\{k(T_{RX} + T_{ANT})B\}(\text{dBW})$$
$$-(\text{RX antenna gain, dBi})$$

(8)

The plus and minus signs require a bit of explanation. Although the 'kTB' term looks positive, it works out as a large negative number in dBW; you can receive signals way, way below one watt. The antenna gain allows you to receive even weaker signals, so it is subtracted to give an even more negative number for ERS. As with EIRP, strictly speaking we should say EIRS to remind ourselves that it was calculated using antenna gain in dBi.

Another similarity between ERS and ERP is the possibility of misunderstanding what the antenna and feedline are doing. A beam antenna will collect more of the RF energy coming from the wanted direction, and hence will improve the signal/noise ratio, but it won't actually make your receiver any more sensitive. Although feedline loss doesn't come into the ERS equation (8) because we measured T_{RX} at the antenna terminals, don't forget that it can seriously damage your ERS as well as your ERP.

SIGNAL/NOISE RATIO: WHAT CAN WE REALLY HEAR?

The ability to copy a weak signal depends primarily on the signal/noise ratio. This is simply the difference in dB between the received signal level P_{RX} and the noise floor of the receiving system. Weak-signal working obviously implies signals "right down in the noise", *ie* signal/noise ratios around zero dB. Since the signal is received against a background of receiver noise plus antenna noise, both of these contributions to the system noise temperature must be included when calculating the receiving-system noise floor.

A machine can measure signal/noise ratio, but how well do human beings copy weak signals? The answer ultimately depends on our own ears and brains, but for the purposes of calculating station performance we can assume that a minimal SSB signal will require a s/n ratio of at least +3dB in a bandwidth of 2·5kHz, whereas a minimal CW signal is copiable at 0dB in a bandwidth of 100Hz.

These figures are highly debatable because they are crude mathematical representations of the abilities of the human brain, so don't take them too literally. In particular, the bandwidths I've just quoted are not necessarily the physical bandwidths of the IF or AF filters. Although typical IF bandwidths for SSB reception are reasonably close to the bandwidth of the speech signal itself, the CW receiving bandwidth is more closely connected with how well the brain can distinguish a single tone in the presence of noise. This varies from person to person, and also depends on the amount of practice one gets. Experience suggests that the above estimates of aural s/n ratio and bandwidth are about right for many people, give or take a few dB. If you feel your ears don't behave that way, you can simply plug your own figures into the formulae. But do remember that we're talking about copying **very weak** signals, which is where CW really shines compared with SSB. As the signal/noise ratio improves the two modes become equally readable, and it then becomes more a matter of personal preference.

In order to obtain a desired s/n ratio, the gain/loss budget has to look like this –

$$\text{EIRP} - \text{path loss} - \text{EIRS} = \text{s/n (dB)} \qquad (9)$$

Path loss is outside our direct control, but each of us can *do* something about the ERP and ERS of our own station.

PATH-LOSS CAPABILITY

The *path-loss capability* of your station is the sum of its transmitting and receiving performance.

$$\text{Path-loss capability (dB)} = \text{EIRP (dBW)}$$
$$- \text{EIRS (dBW)} \qquad (10)$$

In other words, path-loss capability is the path loss that you and an identically equipped station could overcome between you. Echo-testing on moonbounce gives a very clear illustration of path-loss capability. You press the key, and about 2½ seconds later you hear the result in your own receiver – maybe!

Coming back to earth, let's work out the path-loss capability of a minimal 144MHz DX station. Let's give it a single long Yagi, a 100W amplifier, a sensitive receiver and fairly low feedline losses. Inserting some reasonable figures, the path-loss capability of this station comes out as follows.

On the transmit side, we have 100W = +20dBW. Assume the feeder is about 12m of URM67 (RG-213U), with a loss of 1·0dB at 144MHz. A medium-length 144MHz Yagi will have a gain of about 15dBi, so the EIRP is

+20dBW − 1·0dB + 15dBi = 34·0dBW

On the receive side, assume a receiver NF of 2·0dB; add the feeder loss of 1·0dB (it adds directly to the NF) and convert to noise temperature. Work this backwards through equation (5) in the earlier part of this chapter, and you should get an answer of 289K. To be optimistic, assume a rather low antenna noise temperature of 200K to give a system noise temperature of 489K. Convert this to the noise-floor power in a 2·5kHz SSB noise bandwidth using equation (1), and convert to dBW; if you're still with me, the answer is close to −168dBW. Subtract the antenna gain of 15dBi, and the EIRS comes out at −183dBW.

So the total path-loss capability is:

EIRP − EIRS = 34dBW − (−183dBW) = 217dB

If you get confused with juggling all the plus and minus signs, just remember a few obvious facts. The overall path-loss capability should be large and *positive*. The receive noise-floor should be large and negative, but the double minus sign turns this into a *positive* contribution to path-loss capability. Each of the following also makes a positive contribution: higher power, lower feedline loss, and higher antenna gain.

Try working out the path-loss capability of your own station, and list the contributions of all the individual parts of your RF system. Having done so, you have two ways of using this information. One is to ask how far you could work, and the other is to study how you could improve your station.

WORKING RANGE

Forecasting your working range is not an exact science, unless you're prepared to work with long-term statistical trends, because path loss on any mode of propagation depends on what we loosely call "conditions" – and of course these can change markedly from day to day, or even minute to minute. With these limitations in mind, how far do you think a path-loss capability of 217dB will get you? According to the so-called 'standard' tropospheric propagation curves for 144MHz, you might expect your SSB signals to disappear into the noise at about 500km. More to the point, in 'average' conditions from an 'average' site you could hold an SSB QSO with a similarly-equipped station about 400km away at a tolerable s/n ratio of +10dB. Although these figures are pretty vague, they do accord with experience in 'normal' conditions, as distinct from an opening or those 'dead-band' conditions when it feels like a struggle to work across town.

As a flight of fancy, what about moon-bounce with a path-loss capability of 217dB? Well, the minimum path loss for 144MHz EME is 251·5dB, so your SSB echoes would come storming in with a s/n ratio of... oh dear, 34·5dB *below* the noise. OK, so you already knew that 100W and a single Yagi won't let you chat with the Ws on 144MHz EME, but now you know precisely *why* not!

CHANGING TO CW

All the previous calculations were for SSB. If you rework the receiver sensitivity part with a bandwidth of 100Hz instead of 2500Hz, you'll see an immediate improvement of

$$10\log_{10}\left(\frac{2500}{100}\right) = +14\text{dB}$$

If you also take account of the fact that weak-signal CW will get a message through at a much lower s/n ratio than SSB, you've gained the best part of 20dB in path-loss capability. In 'average' tropo conditions this

STATION IMPROVEMENTS

might extend your working range by another 200km. So switching to CW and being prepared to use weak signals will considerably increase your potential for working DX, especially when you remember that you're more likely to find the Eastern European DX on CW than on SSB.

Changing to CW has also made considerable inroads into that 34dB deficit on the moon-bounce path. All you need now is a little help from a mega-station like W5UN, whose array of 48 long Yagis boasts a gain about 17dB greater than your own 15dBi. Throw in a few dB of ground-reflection gain at one station's moonrise or moonset, and you're in with a shot at working the States!

The final question is: does it work? Take a listen around 144·010MHz at moonset, on a weekend when the moon's been high in the sky. If you have a single Yagi of reasonable size, the chances are that you *will* hear W5UN and other big stations like his. All around the world, there are dozens of single-Yagi stations running 100W of CW who now have the EME QSL cards to prove that path-loss budgets *do* work!

STATION IMPROVEMENTS

Path-loss budgeting can give you a very clear idea of what you ***ought*** to be able to hear and work. And if the performance of your station is regularly falling short of what it should be, then maybe yours is one of those stations with equipment which isn't working properly. The key to making your station work properly is to understand how the various parts contribute to the overall station performance. And that's the second major role of path-loss capability calculations.

The antenna gain and feeder loss each appear twice in your station's path-loss capability, once on transmit and once on receive. So there's no point in having superb equipment in the shack and then connecting it to a dipole in the roof space or an HB9CV on the chimney. The main improvements that can be made to any station will always be in the antenna and feedline – Chapter 7 tells all.

The next most important aspect of overall station performance is your receiver. Chapter 5 is about optimization of receiver performance, although as far as sensitivity is concerned we've already seen that the achievable improvements will ultimately be limited by antenna noise. The best way towards good receiving performance is not through micron-sized gallium arsenide devices in the preamplifier; it's through large-sized aluminium-alloy devices in the sky.

The remaining area for improvement is in transmitter power, but you need to balance this against your own receiving capability and that of other stations. Suppose you're testing your new 400W amplifier in 'flat' band conditions on 144MHz, and someone 350–400km comes back to your CQ call. Let's imagine that you have a middle-aged commercial 144MHz transceiver with a front-end noise figure of perhaps 8–10dB, while his receiver has a noise figure of 2dB as I recommended earlier. He'll probably be giving you a report of 5-and-2 but you'll only be copying him at about 3-and-1. Your 400W amplifier has turned you into an 'alligator' – all mouth and no ears! To make proper use of high power, you need a good receiving system too. Unless you actually enjoy low-power (QRP) operating for its own sweet sake, it's equally frustrating to be an 'elephant' – all ears and no mouth – in which case Chapter 6 should provide some inspiration.

MEASUREMENTS AND RECORDS

From the system-engineering point of view, you need to keep performance of your station in balance so that it's about equal in all areas and without any notable weaknesses. As you progress, make measurements and record them, either in your station logbook or preferably in a more detailed technical notebook. Making notes and keeping records will be a recurring theme throughout this book, for two very important reasons –

(1) So that you can understand what you're doing without having to keep everything in your head; and

(2) To keep track of station performance by

repeating the same tests over the months and years.

Chapter 12 describes a wide variety of system tests using home-made instruments. Two additional tests for the antenna and receiving system are measurements of ground noise and sun noise – see the box oppposite. Ground noise measures receiver performance, and sun noise tests your receiver and antenna together.

PLANNING YOUR IMPROVEMENTS

You can *plan* the best way to upgrade your station, by estimating how the various changes would improve your path-loss capability. By looking at your station from the 'system' point of view and doing a few simple calculations, you'll soon begin to see what's going on – where the weaknesses are, where you can make real improvements, and where best to spend your hard-earned money.

The biggest rewards come first, which is encouraging if you're a beginner. They come from simple changes like putting up an outside beam, improving a deaf receiver, and increasing power from 3–10W to around 100W. These improvements will raise a beginner's station to one with moderately successful DX performance. After that, further improvements become more difficult and expensive; instead of increasing your path loss capability by leaps and bounds, you're reduced to picking up a dB here and there. But you'll still find it worthwhile – especially if the bug has bitten!

Bearing in mind the need to keep every area of station performance in balance, the next steps onwards from 100W and a small Yagi should probably come in the following order –

(1) Improve your antenna; doubling its size will gain you about 2·5dB on both transmit and receive.

(2) Improve the feeder; every 1dB reduction in loss is as good as 1dB more antenna gain, or 1dB off your receiver NF.

(3) Finish improving the receiver sensitivity (not much more to be gained here).

(4) Last of all, upgrade to a 400W power amplifier (+6dB improvement over 100W).

None of these improvements may seem very worthwhile on its own, considering some of the difficulties involved. But look – adding up all the bits and pieces, you'd have improved your path-loss capability by at least 13dB! Even 1dB can make all the difference to the success or failure of a weak-signal QSO, so every improvement will make a real, positive difference to the DX performance of your station in the long term.

Finally, remember that lots of other people are reading this book too. If both you and they upgrade your stations along the lines I've suggested, we'll *all* be working a lot more DX.

REFERENCES

[1] *IEEE Standard Dictionary of Electrical and Electronic Engineering Terms*. Wiley-Interscience.

[2] R. E. Taylor, 136/400MHz radio-sky maps. *Proc IEEE*, April 1973.

[3] Ray Rector WA4NJP, 6 meter EME. *Proc 22nd Conference of the Central States VHF Society*. ARRL, 1988. ISBN 0-87259-209-X.

[4] Jim Fisk W1DTY (W1HR), Receiver noise figure, sensitivity and dynamic range – what the numbers mean. *Ham Radio*, October 1975.

STATION IMPROVEMENTS

GROUND AND SUN NOISE

The overall noise temperature of your receiving system is a combination of receiver noise plus noise picked up by the antenna from the surroundings (see page 4-7). By pointing your antenna down towards the warm ground and then up towards an area of quiet sky, you can measure the ratio of noise powers received. This is very like a hot/cold measurement of noise temperature (Chapter 12), except that you don't know the two temperatures exactly. A slight problem with this measurement is that you don't know to what extent the ground is acting as a perfect emitter of thermal energy, or is merely mirroring the cold sky. But none of this matters if you're mainly concerned with relative measurements, and with making sure that they don't deteriorate with time.

The general formula for ground noise is:

$$\frac{P_{GROUND}}{P_{SKY}} = \frac{T_{GROUND} + T_{RX}}{T_{SKY} + T_{RX}}$$

where

P_{GROUND} = noise power aiming at ground (W)
T_{GROUND} = effective temperature of ground (K)
T_{RX} = receiving system noise temperature at antenna (K)
P_{SKY} = noise power aiming at cold sky (W)
T_{SKY} = 'cold' sky temperature (K)

Note that ground noise does not depend on antenna gain.

As a rough guide, with a receiver noise temperature of about 200K, looking at a ground temperature of 290K and a sky temperature of 200K (a reasonable value for quiet sky at 144MHz) should give about 1·8dB of ground noise. You should do rather better on 432MHz because quiet sky can be as cool as 20K, although pickup of ground noise from sidelobes of the antenna can be a problem. An EME system with a low-noise receiver and a clean antenna pattern would show over 4dB of ground noise at 432MHz. For long-term measurements, you need to be sure that you're always pointing to the same quiet part of the heavens as the seasons change, which may involve you in some astronomy.

Sun noise is rather more complex than ground noise, because it also depends on antenna gain and beamwidth. When your antenna points towards the sun, it mostly sees the cool surrounding sky with the sun as a small hot object in the centre of the beam. Once again you're measuring relative noise power, using an AF voltmeter or similar; this time it's the ratio of noise power levels with the antenna first pointed directly at the sun and then pointed towards a quiet area of sky.

The general formula for sun noise is:

$$\frac{P_{SUN}}{P_{SKY}} = \frac{0 \cdot 288 S \lambda^2 G + T_{SKY1} + T_{RX}}{T_{SKY2} + T_{RX}}$$

where

P_{SUN} = noise power aiming at sun (W)
S = solar flux (solar flux units, 10^{-22} W m^{-2} Hz^{-1}, at the frequency in question)
l = wavelength (m)
G = antenna gain over isotropic (ratio, not dB)
T_{SKY1} = sky temperature behind sun (K)
T_{RX} = receiving system noise temperature, at antenna (K)
P_{SKY} = noise power aiming at cold sky (W)
T_{SKY2} = 'cold' sky temperature (K)

With so many unknowns in the equation, it isn't easy to make meaningful absolute measurements of sun noise unless your real interest is radio astronomy. What you can do is to use sun noise as a more general indicator of antenna and receiver performance.

As a guide, a decent 144MHz tropo system should be able to see 1-2dB of sun noise, and the equivalent 432MHz system somewhat more (mostly because the sky background is quieter at the higher frequency). For comparison, EME systems with large antennas and low-noise preamps should give anything upwards of 10dB of sun noise. These figures relate to a quiet sun, so you need to make several measurements over periods of days, and don't get over-excited if your whole antenna and receiving system suddenly appears to have improved by two or three dB. It hasn't, of course, it's just a temporary increase in solar flux – but there might be an aurora or ionospheric DX on the way!

Even if you don't have an elevation rotator, you can probably still make ground and sun noise measurements. Any antenna on a tilt-over support can be used for ground noise measurements, and possibly also for sun noise if the mast lowers towards the north (in the northern hemisphere). Simply tilt the mast part-way and then rotate the antenna towards the ground, the cool sky or if possible the sun.

If your antenna has no elevation adjustment, you can make measurements on the rising and setting sun. The problem then is the effect of ground reflections which can either add to or cancel the direct signal according to the angle of the sun above the horizon. To understand what's happening, it's best to follow the sun for a long period as it rises or sets through the vertical pattern of you antenna, making sun noise measurements every few minutes, and then to plot the results against the vertical angle of the sun.

CHAPTER 5

RECEIVERS & LOCAL OSCILLATORS

RECEIVERS & LOCAL OSCILLATORS

by Ian White G3SEK

For VHF/UHF DX we want the same kind of performance from our receivers as we would for HF DXing. In particular we want to be able to receive weak DX signals, even if there are plenty of strong local signals on the band at the same time. As a matter of fact, our requirements on VHF/UHF are even more stringent than on HF. The strong signals are just as strong, yet very much weaker signals can be copied because the background noise on the VHF and UHF bands is lower.

THE RECEIVER'S FRONT-END

What makes a receiver good (or bad) at handling both weak and strong signals is its 'front-end', the first few stages after the antenna [1-4]. The front-end includes the RF amplifier(s), the mixer to convert signals to the intermediate frequency (IF), and maybe an IF head amplifier (Fig. 5.1). The local oscillator (LO) converts the wanted signal frequencies to the fixed IF, eg for a 144-146MHz transceiver with an IF of 10·7MHz, the LO would be a variable-frequency oscillator (VFO) covering 133·3-135·3MHz. Although not in the direct signal path, the VFO can have a profound effect on front-end performance, so I shall return to that subject later.

An alternative approach to VHF and UHF is the transverter (Fig. 5.2), which is an add-on box for an HF transceiver. A fixed-frequency local oscillator is substituted for the VFO, and a typical 144-146MHz transverter would have a 116MHz LO which converts the VHF band down to a tunable IF of 28-30MHz for the HF rig. (A transverter also incorporates a transmit converter, although that does not concern us here.) Apart from the fixed LO and the

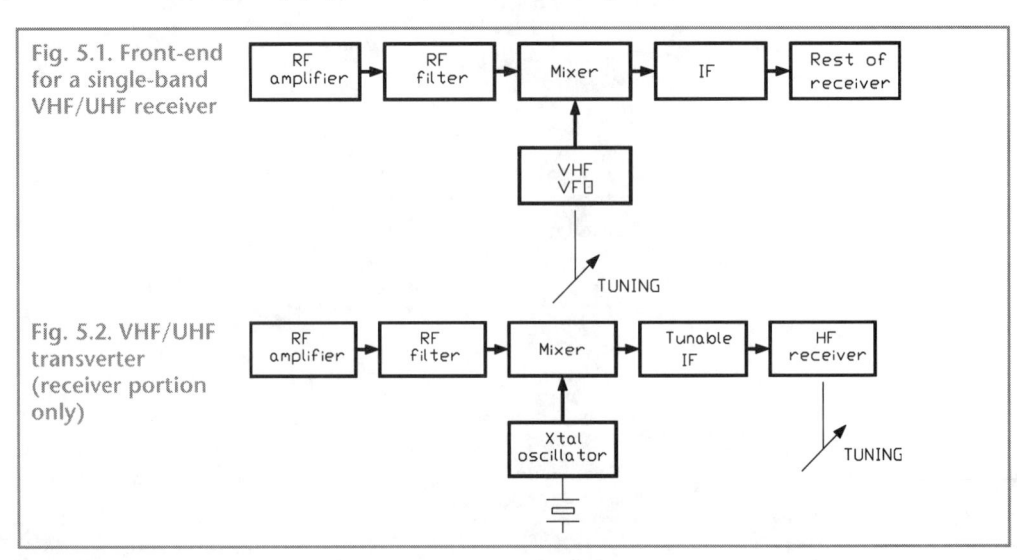

Fig. 5.1. Front-end for a single-band VHF/UHF receiver

Fig. 5.2. VHF/UHF transverter (receiver portion only)

bandwidth of the IF circuits, a modern transverter is very similar to a transceiver front-end for the same band. The advantage of transverting is that it provides all the facilities of the HF rig on the VHF or UHF band, and allows extra bands to be added at modest cost compared with a range of new single-band rigs. The disadvantage is that receiver performance may be compromised, especially with strong signals.

The receiver front-end has to do two things. One is to amplify and frequency-change the weak signals without introducing too much extra noise; the other is to cope with any unwanted off-channel signals, however strong they might be. When all the incoming signals reach the IF filter, only the wanted signals pass through – together with any on-frequency interference, of course. All other unwanted signals should fall outside the filter passband and cease to bother you. But before they are filtered out, strong off-frequency signals may already have caused trouble by overloading an earlier stage in the front-end, perhaps generating spurious signals on the wanted frequency. The IF filter cannot remove these signals, so they can interfere severely with the weak DX that you want to hear.

The problems arise when off-frequency strong signals become amplified to such high levels that they drive one or more stages in the front-end into some kind of overload. A receiver with good strong-signal performance needs to have *as little gain as possible between the antenna and the IF filter*. But a certain amount of front-end gain is essential in a sensitive receiver. We thus have to make a trade-off between sensitivity and strong-signal handling – a very careful trade-off indeed, aimed at achieving the optimum balance of performance in *all* operating conditions.

The optimum balance between weak-signal and strong-signal performance is not a matter of luck. It has to be achieved by careful planning and *design*, based on sound understanding, optimized by trial calculations, and backed up by performance testing.

This chapter can be read at two levels. Just by skimming over the surface, you can learn quite a lot about what's important in front-end design. At a deeper level, you can really get to grips with the subject by working through the example calculations. Calculation is the key to modern RF design, and a scientific calculator or home computer can be just as useful as a soldering iron. If you do decide to delve into the worked examples, don't start using a computer program straight away. Unless you work through the detailed calculations yourself, thinking about what you're doing, you'll never truly understand the subject.

SENSITIVITY AND NOISE TEMPERATURE

In Chapter 4 we looked at system noise temperature as part of a station's overall path-loss capability. Since the lowest possible noise-temperature for the whole receiving system is limited by the noise 'seen' by the antenna, we concluded that a *receiver* noise temperature equal to the lowest probable antenna noise-temperature would result in an acceptably small signal/noise degradation. To be on the safe side, we made estimates of antenna noise temperatures for each band which excluded any contribution from man-made noise, and these became our design targets for receiver noise temperature.

50MHz	4000K	12dB
70MHz	3000K	10dB
144MHz	200K	2·2dB
432MHz	150K	1·8dB

NOISE/GAIN ANALYSIS

Now let's try a more complete example of noise and gain analysis than we managed in Chapter 4. The system we shall consider is G4DDK's *Suffolk* transverter described in Chapter 8. The *Suffolk* is a transverter for the European 144–146MHz band, with a tunable IF of 28–30MHz.

We begin by drawing a block diagram of all the separate stages of the front-end, with the signal flowing from left to right (Fig. 5.3). Don't worry about what kind of circuitry goes inside the blocks; what we're concerned with at present is simply the performance of each

RECEIVERS & LOCAL OSCILLATORS

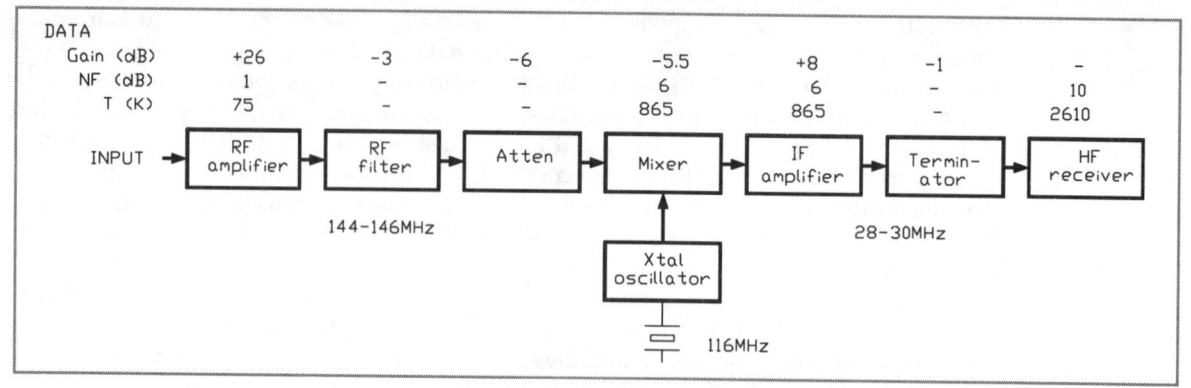

```
DATA
Gain (dB)    +26      -3      -6      -5.5     +8      -1      -
NF (dB)       1        -       -       6        6       -      10
T (K)        75        -       -      865      865      -     2610

INPUT → [RF amplifier] → [RF filter] → [Atten] → [Mixer] → [IF amplifier] → [Terminator] → [HF receiver]
              144-146MHz                  ↑                    28-30MHz
                                  [Xtal oscillator]
                                       116MHz
```

Fig. 5.3. Block diagram of the *Suffolk* transverter for 144-146MHz (receiver portion only) with input data

block in terms of noise temperature and power gain. This we write in the space above each individual block. Since the *Suffolk* is a transverter, its block diagram ends at the HF transceiver.

The performance data for the individual stages of the *Suffolk* come from such sources as manufacturers' data sheets and G4DDK's measurements on prototypes. As we shall see, the HF rig may be a crucial factor, but for the present we only need assume that it has good sensitivity.

The block diagram is shown in Fig. 5.3, with the data for each individual stage above. Below the block diagram will go the results for the system. If you want to follow though this worked example, draw up a similar worksheet for yourself. All the data were originally obtained in decibel units of noise figure and gain. For our analysis, these will have to be converted to noise-temperature and gain ratio.

The formula for converting noise figure (NF) to noise temperature (T) comes from equation (5), Chapter 4, and is:

$$T = 290 \left(\text{antilog}_{10} \left(\frac{NF}{10} \right) - 1 \right)$$

To calculate the system performance, start at the IF filter at the right-hand end of the block diagram, and work backwards stage-by-stage towards the antenna, using equations 1–5 of Chapter 4. To avoid getting confused, work right through in terms of noise temperature alone, and then go back and convert all your system results to NF. Similarly, calculate the cumulative front-end gain first in ratio terms, and then go back and convert all the results to decibels. (A more detailed approach to this kind of worksheet is given in references 1–4.)

The results of the noise/gain analysis are entered below the block diagram as shown in Fig. 5.4. If you're following this worked

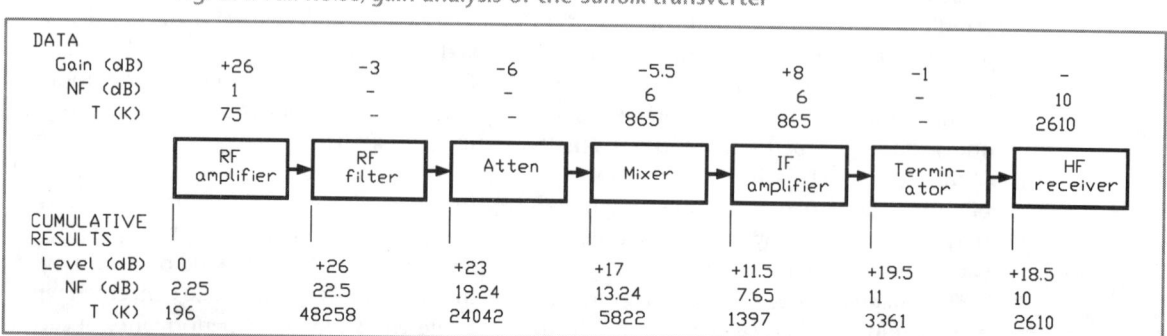

Fig. 5.4. Full noise/gain analysis of the *Suffolk* transverter

```
DATA
Gain (dB)    +26      -3      -6      -5.5     +8      -1      -
NF (dB)       1        -       -       6        6       -      10
T (K)        75        -       -      865      865      -     2610

        [RF amplifier] → [RF filter] → [Atten] → [Mixer] → [IF amplifier] → [Terminator] → [HF receiver]

CUMULATIVE
RESULTS
Level (dB)    0      +26      +23     +17     +11.5    +19.5    +18.5
NF (dB)      2.25   22.5     19.24   13.24    7.65      11       10
T (K)        196   48258    24042    5822    1397     3361     2610
```

RECEIVERS & LOCAL OSCILLATORS

example through in detail, check that you get the same results. The gain can be worked out in two directions; starting from the IF and working backwards towards the antenna, or starting at the antenna and working forwards into the receiver. In the latter case, you're actually calculating the relative levels of an incoming signal as it passes from stage to stage, and that's what I've done in Fig. 5.4.

Having done all that, what can we see? First of all, the noise temperature is very close to our design target of 200K for the 144MHz band. The total cumulative gain is 18·5dB, though signals build to their highest level at the output of the RF stage.

Having once worked through the system analysis, you can try changing the gains or noise temperatures of individual stages, or even the order of the stages, and see how the system responds. The computer program TCALC, described later, makes it very easy to do a range of trial calculations and thus to obtain the desired noise temperature while using a minimum of gain. That's how the numbers in Fig. 5.4 were juggled to make the receiver noise temperature almost exactly equal to the design target. G4DDK will take this topic further when he describes the *Suffolk* design in Chapter 8.

STRONG-SIGNAL PERFORMANCE

The first part of this chapter was mostly about performance with weak signals – which are probably the ones we most want to hear. This part is about the strong signals which get in the way! On today's crowded bands, strong unwanted signals on the same frequency as weak wanted signals are a fact of life. The solutions to QRM are well-known: good operating to avoid the problems in the first place, good IF selectivity, a sharp beam with low sidelobes, and a good pair of ears with something between them.

But there is also a different type of problem caused by strong signals which are **not** co-channel yet which still somehow manage to interfere with the wanted signal. As I mentioned in the introduction to this Chapter, the general problem is called *front-end overload*. Actually this blanket term covers several quite different overload effects. The three most important are –
- Intermodulation
- Gain compression
- Reciprocal mixing.

There are many other front-end overload effects, but they are of less practical importance in SSB/CW receivers. Designing and testing for good performance in those three main areas should take care of the other overload problems too.

INTERMODULATION

Intermodulation occurs when two or more signals mix together to produce extra signals which weren't there before. The normal process of mixing between the signal frequencies and the receiver's local oscillator (LO) to produce the IF is itself an example of intermodulation – and that's the **only** kind which ought to occur in a receiver front-end. All other kinds of intermodulation which ought **not** to occur will produce spurious signals which may interfere with a weak wanted signal. It's important to realise that these unwanted signals didn't exist until they were generated *within the receiver*.

Your receiver front-end has to cope with far more incoming signals than you ever hear. In a typical amateur VHF/UHF receiver the front-end is receiving all the signals present within a bandwidth of several megahertz, while you are hearing only the ones that get through the selective IF filter. At low signal levels all front-ends behave in a linear manner. This means that each incoming signal is processed separately, with no interactions between signals. But if incoming signals are very strong they can easily drive a poor front-end beyond its linear range. Then the signals begin to intermodulate, *ie* they mix together to produce new signals. So intermodulation is caused by non-linearity in the front-end.

If two signals on frequencies f_1 and f_2 mix, they produce three sets of new frequencies:
- second harmonics $(f_1 + f_1)$ and $(f_2 + f_2)$
- sum product $(f_1 + f_2)$
- difference product $(f_1 - f_2)$ or $(f_2 - f_1)$.

These new signals are called intermodulation products – 'intermods' or 'IPs' for short. More specifically, they're known as 'second-order intermods' because they were created by mixing *two* original 'parent' frequencies. If the parent frequencies are close together, the sum product will be up in the second-harmonic

STRONG-SIGNAL PERFORMANCE

region and the difference product will be down at LF. Even a simple filter will separate these widely spaced groups of signals. As I mentioned earlier, your receiver relies on second-order intermodulation to convert from the signal frequency to the intermediate frequency. The signal is at f_1, f_2 is the LO and $(f_1 - f_2)$ is the IF; in this particular case the filter is intended to select only the IF. The mixer stage has to be a non-linear device, or else it wouldn't work; but its non-linearity can often let in a host of further intermodulation problems.

If the two strong signals at frequencies f_1 and f_2 are increased in strength, another set of IPs appears. These are the third-order products, so called because they involve mixing between three signals [5]. These three signals can be totally independent, or the same effect can be generated by two parent signals, counting one of them twice. Since these two frequencies can either add or subtract, the full range of possible third-order IPs from two signals includes:

- third harmonics $\quad (f_1 + f_1 + f_1)$ and $(f_2 + f_2 + f_2)$
- sum products $\quad (f_1 + f_1 + f_2)$ and $(f_1 + f_2 + f_2)$
- difference products $(f_1 + f_1 - f_2)$ and $(f_2 + f_2 - f_1)$.

If both of the parent frequencies f_1 and f_2 are close to the wanted frequency, the third-order sum products must be somewhere up in the third-harmonic region and won't trouble us further. But the two difference products containing a minus sign are somewhere close to their parent frequencies. For example, if f_1 = 144·240MHz and f_2 = 144·260MHz, their third-order IPs fall on 144·220MHz $(f_1 + f_1 - f_2)$ and 144·280MHz $(f_2 + f_2 - f_1)$. Fig. 5.5 shows how the two parent signals and their IPs would appear on a spectrum analyser display.

If the receiver front-end is allowed to generate these in-band spurious signals, we're in real trouble. No practical kind of RF filtering can take them out, because they're too close to the frequencies we want to listen to. If one of these IPs just happens to fall on the frequency of a weak wanted signal, it will cause QRM. And if that *can* happen, then sure enough it *will* – right when you least want it.

In the example above, the two strong signals were spaced 20kHz apart, and note how their third-order IPs were evenly spaced 20kHz above and below them (Fig. 5.5). As the strength of the two equal parent signals increases, the strengths of their third-order IPs

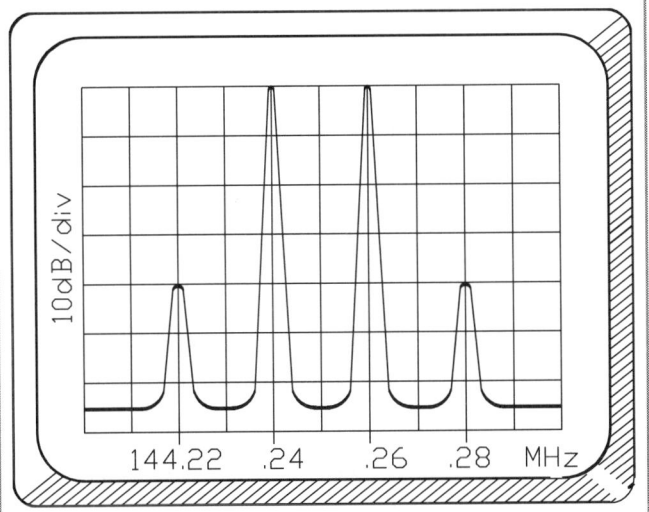

Fig. 5.5. Spectrum analyser display of third-order intermodulation between signals on 144·240MHz and 144·260MHz. The intermodulation products appear on 144·220MHz and 144·280MHz

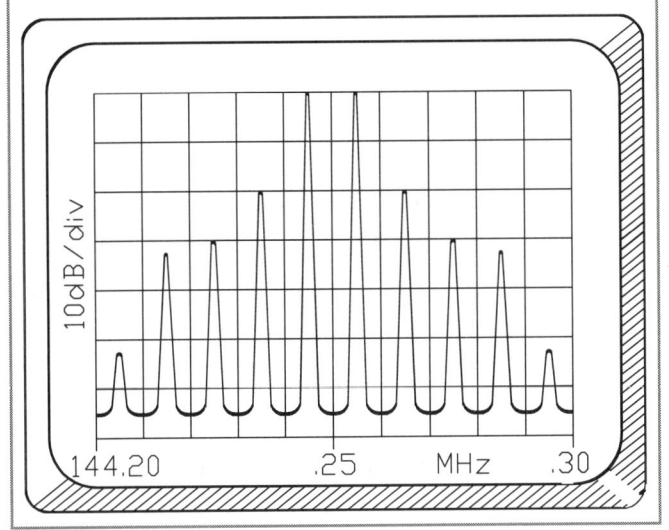

Fig. 5.6. Higher-order intermodulation products, on an expanded frequency display. Note the uniform frequency spacing, equal to that of the parent signals

(also equal) come up three times as fast, increasing by 3dB for every 1dB increase in the two parent signals. So the onset of intermodulation can seem quite sudden. Further increases in the strengths of the parent signals will produce new generations of higher-order IPs. Thrashing through the algebra [5] shows that the odd-order (3rd, 5th, 7th, *etc*) IPs are the ones that cause the trouble, because some of them can fall close to the parent signals.

These higher-order IPs appear above and below their parents, at uniform frequency spacings (Fig. 5.6). Fifth-order IPs require stronger parent signals than third-order before they rise above the noise, but then they increase in strength five times as fast as the parent signals; and *pro rata* for seventh- and higher-order products. Severe high-order intermodulation can thus appear very suddenly indeed, covering the band with spurious signals. High-order intermodulation in transmitters is the cause of 'splatter', as discussed in Chapter 6.

Intermodulation requires at least two strong parent signals. If either one goes off, all their IP offspring will disappear. But relief will be short-lived, because the remaining strong signal can intermodulate with any other strong signal that comes along, to make a new set of third-order IPs. And if the first signal comes back, making three in all, there will now be *four* (work it out) different sets of IPs.

So far we've only been talking about single frequencies, *ie* steady unmodulated carriers. On SSB, it's far worse than that. An SSB signal itself contains several different frequencies at once, all of varying amplitudes. When these intermodulate, the effect is to broaden the bandwidth of the signal. If the intermodulation takes place in the transmitter, the broadening is real and you'll have to ask the other station to fix it (see Chapter 6). But what happens if two perfectly clean SSB signals intermodulate in your receiver? As well as each seeming broader, their in-band third-order IPs together wipe out a further slice of spectrum, totalling typically 20kHz. A more severe case of intermodulation generating appreciable levels of higher-order products will wipe out even more of the band. Remember, this is not necessarily the fault of the transmitters: all your intermod problems can be generated right inside your own receiver. With the low end of the band full of strong SSB and CW signals during a contest or an opening, intermodulation in the average commercial VHF/UHF receiver front-end can be ghastly. Listening on the band is like wading waist-deep through a shifting swamp of inter-modulated mush.

Yet it *is* possible to build VHF/UHF front-ends that have virtually no noticeable intermodulation, even at 'open contest' signal levels. Compared with an ordinary receiver, the effect is startling. The strong signals are just as strong, but there are also spaces between them where the weak DX comes through in the clear. Also, many local signals that appear 'broad' on an ordinary receiver turn out to be quite respectable – and it becomes equally obvious whose signals really *are* broad!

GAIN COMPRESSION

Gain compression occurs when a strong incoming signal drives some stage in the front-end so hard that it can barely produce any more output. The gain of the 'saturated' stage is thus reduced, taking with it all other signals including the wanted one. In an extreme case, gain compression simply makes the receiver collapse in embarrassed silence whenever a strong signal comes on. Gain compression requires only *one* strong signal, unlike intermodulation which requires at least two.

Gain compression occurs with any kind of signal provided it's strong enough, and is particularly easy to recognise when produced by a strong carrier. When the carrier comes on, the background noise goes down, together with all other signals. If the carrier is keyed, the background noise appears as 'negative' Morse code, the dits and dahs being the silences in between.

LOCAL-OSCILLATOR NOISE

Reciprocal mixing is a large-signal effect

STRONG-SIGNAL PERFORMANCE

caused by imperfections of the receiver's local oscillator (LO). Although reciprocal mixing has nothing to do with the non-linearities of stages in the signal path, which are the causes of gain compression and intermodulation, it is still a front-end problem. To understand how reciprocal mixing occurs, we need to take a short detour into the workings of oscillators and frequency synthesizers.

An ideal oscillator produces only one frequency, the one corresponding to the resonance of the tuned circuit or crystal. Energy at this frequency goes round and round the oscillator's feedback loop, with the active device (usually a transistor) providing just enough gain to make up for circuit losses round the loop [5, 6].

Unfortunately, oscillators never produce only one frequency. Besides the main carrier and its harmonics, they produce energy at all – yes, **all** – other frequencies as well! Far away from the carrier frequency, there is a level background of 'thermal' noise due to the oscillator device and other resistive circuit losses. Other sources of wideband noise can include noise on the DC power supplies, and switching transients coupled in from nearby digital circuitry (especially, of course, in digitally-synthesized LOs). All these sources of noise modulate the main carrier, and thus produce noise sidebands on either side of the carrier frequency. The closer to the carrier frequency, the higher the level of the noise sidebands, while at frequencies further away the noise sidebands sink below the level of broadband noise. Seen on a spectrum analyser, the result is a so-called 'noise pedestal' standing out above the broadband noise and centred around the carrier (Fig. 5.7).

Detailed analysis of the factors governing the levels of noise in oscillators is beyond the scope of this book. If you're interested, have a look at the references at the end of this Chapter [5, 6, 7]. Suffice it to say that noise sidebands are better suppressed (relative to the carrier level) if the carrier level itself is high, and also if the loaded Q of the frequency-determining tuned circuit is high. Oscillator design is far from simple, but it's quite easy to build a noisy oscillator by accident.

Because of the difficulties of obtaining adequate frequency stability in a variable-frequency oscillator at VHF or UHF, transceiver VFOs usually involve either mixers or phase-locked loops. Mixer VFOs typically add the frequencies of an HF VFO and a VHF/UHF crystal oscillator (Fig. 5.8A), and amplify the

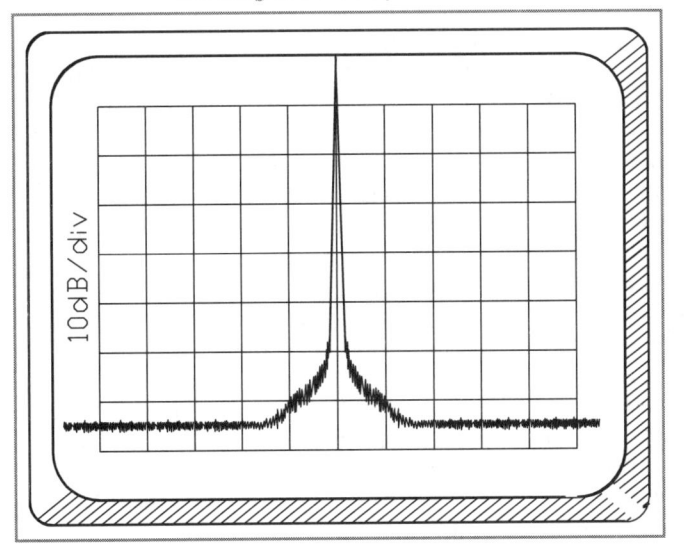

> **OSCILLATOR NOISE LEVELS – dBc/Hz**
>
> Oscillator noise has both AM and FM components, known as *amplitude noise* and *phase noise*. Phase noise is the more serious, and we often speak of 'phase noise' when we actually mean oscillator noise in general. Synthesizer noise is not necessarily pure noise, since it can sometimes involve digital transients, though for many purposes the distinction doesn't matter. So let's use the shorthand and talk about 'phase noise', regardless of the actual type of LO involved.
>
> The level of phase noise is specified in decibels relative to that of the carrier: dBc. Since the noise is broadband, we also have to specify the bandwidth across it's being collected, and conventionally that is one hertz. So the units of phase noise level are dBc/Hz.
>
> A further item which needs to be specified is the frequency offset at which the measurement is being made, because that too affects the phase noise level (Fig. 5.7).

Fig. 5.7. Spectrum of a typical noisy oscillator, showing the 'noise pedestal'

RECEIVERS & LOCAL OSCILLATORS

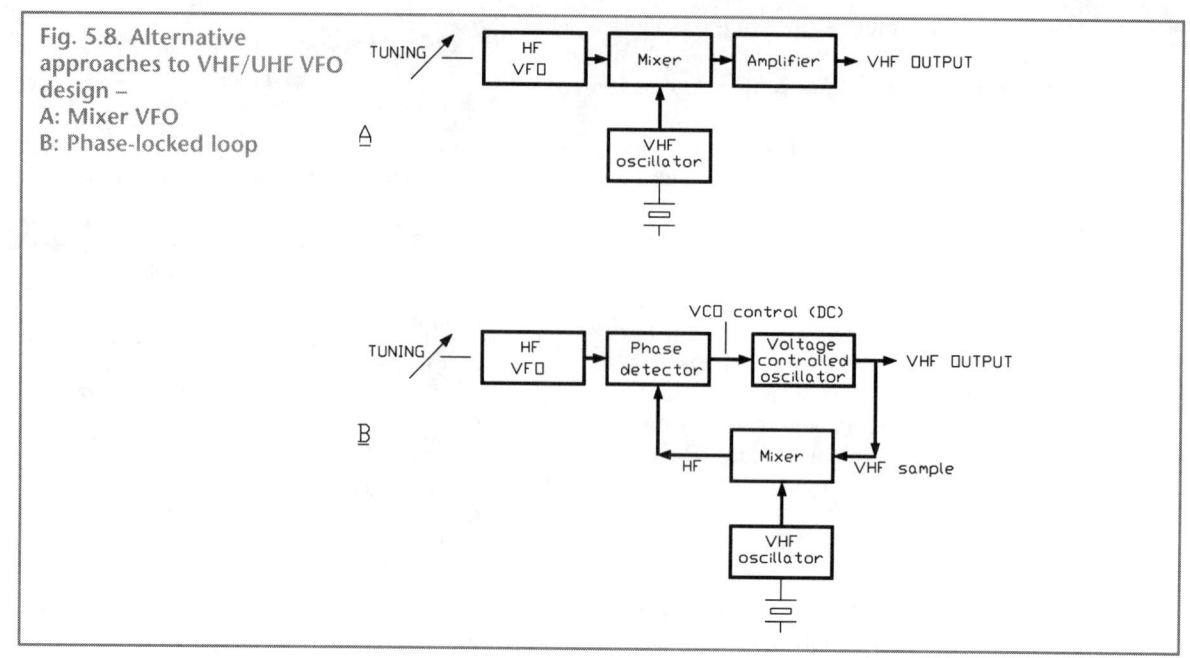

Fig. 5.8. Alternative approaches to VHF/UHF VFO design –
A: Mixer VFO
B: Phase-locked loop

resulting output to a suitable level. Examples of transceivers using this technique are the old TS700 and the IC201. Mixer VFOs offer many opportunities for oscillator noise. The mixer cannot be run at too high a level without producing spurious products, and at the output of the mixer the carrier level can sink perilously close to the circuit noise. Any subsequent amplification of the low-level carrier also brings up the accompanying noise.

Other transceivers of similar vintage, such as the FT221 and FT225, use similar frequencies for the HF VFO and the VHF crystal oscillator, but add them using a phase-locked loop (PLL). A PLL is essentially a feedback control system. In Fig. 5.8B the output signal is generated by a voltage-controlled oscillator (VCO), whose frequency is also heterodyned down to HF using a VHF crystal-controlled oscillator. The phase of this HF signal is compared with that of the VFO, and any error voltage is fed back to control the VCO. Tuning the HF VFO thus makes the VHF VCO follow in step, at a constant frequency difference set by the crystal oscillator. That sounds fine: it produces the same output frequency as if the VFO and the crystal oscillator had been mixed directly, and the PLL avoids the problems of spurious mixing products and noise in low-level circuitry. Or does it?

Phase-locked loop systems certainly *can* give very low noise sidebands if designed and constructed with care. But a bad PLL can be very noisy indeed. In addition to all the noise problems that can beset ordinary oscillators, noise can easily be coupled onto the VCO control line, leading to FM noise sidebands. Also, some designers have skimped on the normal precautions for generating a good-quality VCO signal, relying instead on the phase-locked feedback loop to clean up the noisy, hum-laden, drifting signal for them. It seems to work, if you look only at the carrier, but there's a penalty in the form of high levels of noise sidebands. It's a shame when just a little more care in design and layout could have produced a very quiet LO.

More recent VHF and UHF LOs are usually of the digitally-synthesized PLL variety. These too are capable of good noise performance – but digital synthesis offers the careless designer even more opportunities to get things wrong! The digital parts of these synthesizers generate high-speed, high-level switching transients which can stray onto the VCO control line unless careful precautions are taken. The frequencies fed to the phase detector in digital loops are quite low, and any leak-through to either the supply or control lines of the VCO can cause severe modulation sidebands. These digital modulation sidebands may sound like distinct carriers or like noise, depending on the method of synthesis and the source of the leakage. If unwanted LF signals get onto the VCO control line, it's very difficult to get rid of them because filtering affects the ability of the loop to lock quickly and stay locked as you tune or switch frequency. The design of low-noise PLL synthesizers thus requires some very careful trade-offs between various aspects of performance, plus extreme care in the detailed design and physical layout [6].

A good local oscillator needs to have a noise level of −100dBc/Hz at 1kHz offset, rolling away to −140dBc/Hz at 10kHz and ultimately to better than −160dBc/Hz at offsets of 100kHz or more. We'll return later to the significance of these figures.

To give synthesized local oscillators their due, they have brought us the same standards of frequency stability and readout accuracy at VHF/UHF as we enjoy at HF. And although 99 digital memories are something of a mixed blessing (you still have to remember what's stored where!), a synthesizer with dual VFO capabilities and intelligent programming of the RIT and split-frequency facilities can be a real boon for cracking pile-ups and keeping skeds. Yes, we *have* seen some progress during the past decade. It's just a pity that much of the 'progress' as far as oscillator noise is concerned has been backwards.

So much for our digression into oscillator noise. Now then, what about reciprocal mixing?

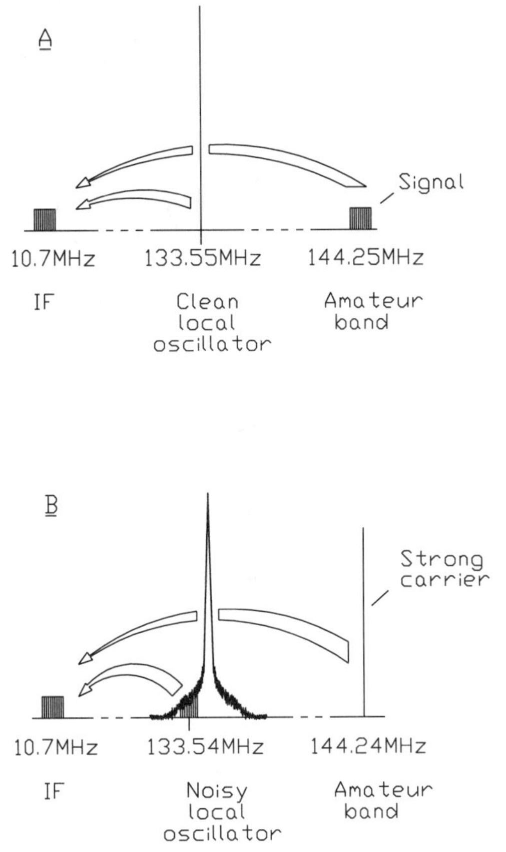

RECIPROCAL MIXING

Normally, the mixer is fed with a strong signal from the LO, and frequency-shifts a fairly weak wanted signal into the passband of the selective IF filter. But the mixer will do the same for *any* pair of signals whose frequencies differ by the IF. Reciprocal mixing gets its name because the roles of LO and signal are reversed: the strong signal is an unwanted one off-frequency, and the weaker signal is part of the LO noise sidebands.

For example, if you're listening to 144·250MHz with an IF filter centred on 10·700MHz, your LO is tuned to 133·550MHz (Fig. 5.9A). The intention is that any signal appearing on 144·250MHz is mixed with the LO and delivered into the IF passband. Reciprocal mixing occurs when a strong incoming carrier mixes with LO noise sidebands, delivering that noise into the 10·7MHz IF and raising the background noise level. Fig. 5.9B shows the example of a carrier on

Fig. 5.9. Mixing processes –
A: Normal mixing: wanted signal + LO = IF signal
B: Reciprocal mixing: strong off-frequency signal + LO noise = IF noise

144·240MHz which mixes with the LO noise sidebands around 133·540MHz, so that the reciprocally-mixed LO noise interferes with the wanted signal being received on 144·250MHz.

The offset between the wanted frequency and the strong unwanted signal – 10kHz in our example – is the same as the offset between the LO carrier frequency and the interfering part of its noise sidebands. This means that the severity of reciprocal mixing will have the same frequency dependence as the LO noise spectrum. Thus the main problems usually occur at close-in frequency offsets corresponding to the LO noise pedestal (Fig. 5.7) although, at very close frequency spacings, reciprocal mixing is often masked by leakage of the unwanted signal through the IF filter. A particularly bad LO with a high broadband noise-floor may cause reciprocal mixing problems across the entire amateur band.

On the air, the effect of reciprocal mixing is opposite to that of gain compression. When a strong carrier appears, the background noise goes *up*. In a transceiver, the LO is used for both receive and transmit functions: thus, if the receiver suffers from reciprocal mixing, the transmitter will also radiate noise sidebands. It can be very hard to distinguish reciprocal mixing in your own receiver from the corresponding problem in someone else's transmitter. The only way to be sure is to test your own receiver independently, using a low-noise signal source.

DYNAMIC RANGES

We've seen in this Chapter that there are three main problems of front-end overload. They are all different, so each problem starts to be noticeable at a different level of strong unwanted signal. Two strong signals are needed to cause intermodulation, but only one to cause gain compression or reciprocal mixing. Because intermodulation and gain compression are both due to some stage in the front-end being driven beyond its linear range, both problems can be cured by keeping front-end gain as low as possible.

Our true objective is not to 'cure' overload, but to make sure that it hardly ever happens, even at the strongest amateur-band signal levels. For all practical purposes intermodulation is non-existent if the products never rise significantly above the background level of system noise. Similarly, the receiver can be considered immune to gain compression or reciprocal mixing if these effects never significantly change the noise level. In other words, what you don't hear won't trouble you.

This leads directly to the idea of *spurious-free dynamic range* (SFDR). The bottom end of all the dynamic ranges is the noise-floor power level of the receiver (Chapter 4). The top end of each individual dynamic range is the power level of the strongest off-frequency signal that the receiver can tolerate without the particular overload effect becoming noticeable. SFDR is simply the difference – in decibels – between these two power levels. The overall SFDR of the receiver is the smallest of its various dynamic ranges.

Although spurious-free dynamic range relates directly to the things you hear when using a receiver, it needs some further definition to turn it into something measurable. I've already defined 'noise floor' in Chapter 4. Another name for this power level is *minimum discernible signal* (MDS), though this is an HF-inspired term – VHF/UHF DXers have very different ideas about copying really weak signals! Formal definitions for the points at which the various overload effects become significant have been proposed by Wes Hayward, W7ZOI, and these lead directly to methods of measuring the various types of SFDR [8]. The W7ZOI dynamic range tests are based firmly on reality – what you measure is also what you hear on the air – so it's not surprising that they have become the standard for HF equipment reviews in *QST* and *Radio Communication*. The test methods described by Hayward were developed for HF receivers, and have needed further development for use with very sensitive receivers at VHF and UHF; we will leave that topic to Chapter 12.

Another aspect of dynamic range which is worth a mention is its frequency dependence.

STRONG-SIGNAL PERFORMANCE

We've already seen that reciprocal mixing is frequency-dependent, according to the noise spectrum of the LO. The same can apply to intermodulation and gain compression, but in a different way. If the receiver front-end is excessively wideband, there is a possibility of gain compression or intermodulation caused by strong FM and TV broadcast signals, not to mention less powerful but closer transmitters on other non-amateur bands. Since the DX portions of the VHF and UHF amateur bands are only a few hundred kHz wide, or perhaps 1MHz including the beacon band, there is no reason to make the RF bandwidth any greater than 1–2MHz. The same applies to the output bandwidth of transverters, which is often grossly excessive. The general principle is to let no greater bandwidth of signals into your receiver than necessary.

The same principle applies in the IF part of the front-end. Almost all receivers now include some kind of 15–30kHz-wide 'roofing filter' ahead of the main SSB IF filter. This is true both for single-conversion VHF or UHF transceivers and for the HF transceivers which we use with transverters. But some front-end designers have forgotten that their task doesn't end at the roofing filter; signals inside the roofing filter but outside the SSB filter passband can cause overload unless careful strong-signal design extends all the way to the narrow IF filter.

This means that dynamic range measurements need to be carried out at a number of different frequency offsets from the frequency to which the receiver is tuned. Preferably these should be combined into a continuous plot of all three kinds of dynamic range against frequency, as pioneered by RSGB reviewer Peter Hart G3SJX.

Such thorough dynamic range measurements are valuable because they test the entire receiver as a single system. But they still leave two questions unanswered. What about the overload performance of the separate parts of a receiver (or accessories like preamplifiers)? And how do you develop and optimize a receiver system in the first place?

INTERMODULATION ANALYSIS

Third-order intermodulation is the most common and annoying front-end overload effect. Fortunately it can be analysed, and we can use the analysis to design better front-ends. In general, a front-end designed to be resistant to third-order intermodulation will also be resistant to intermodulation at higher orders and probably also to gain compression.

If you feed a 'black box' (*eg* an amplifier or a mixer) with two equal signals, and gradually increase their level from zero, Fig. 5.10 shows what happens at the output. With no input signal, the output is merely noise at the noise-floor power level. At low input levels (*eg* point A), the only output signal is the amplified pair of input signals, increasing linearly, *ie* at 1dB per dB of increase at the input. At higher input levels, the third-order intermodulation products appear out of the noise floor (point B), and then increase at 3dB/dB. Ultimately, it looks as if the level of the IPs would catch up with the main output signal. The point where they would do so is called the *third-order intercept point* (point I).

The strength of the intercept-point concept lies in its simplicity. Its weakness is that intermodulation intercept points do not really exist. Before the IPs can catch up with the parent signals, the amplifier or mixer goes into gain compression and its output level flattens off (Fig. 5.10). Thus the intercept point can never be measured directly; it can only be determined by a theoretical extrapolation of measurements made at lower signal levels. There's no guarantee that real systems will behave in the obliging manner that the simple theory demands, and often they don't. However, agreement between theoretical predictions and accurate measurements can be quite good, and one thing's for sure: a system designed with the aid of the intercept-point theory will be a lot closer to optimum than a system that hasn't been designed at all!

In Fig. 5.10 the intercept point (I) can be specified in two ways. If I is specified by reference to the corresponding input power, we call it the *third-order input intercept* (IP_{IN}).

Fig. 5.10. Increase in intermodulation products with input signal levels. Higher-order products increase even more rapidly than the third-order products shown

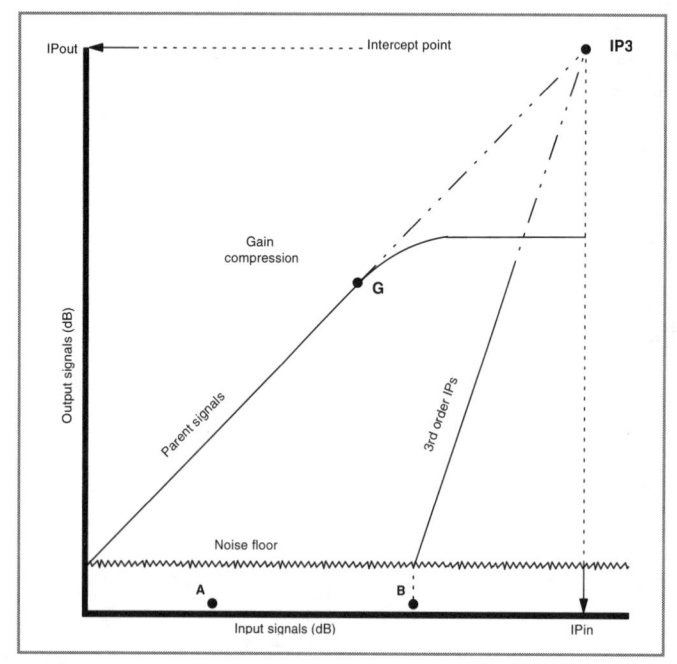

INTERMODULATION ANALYSIS

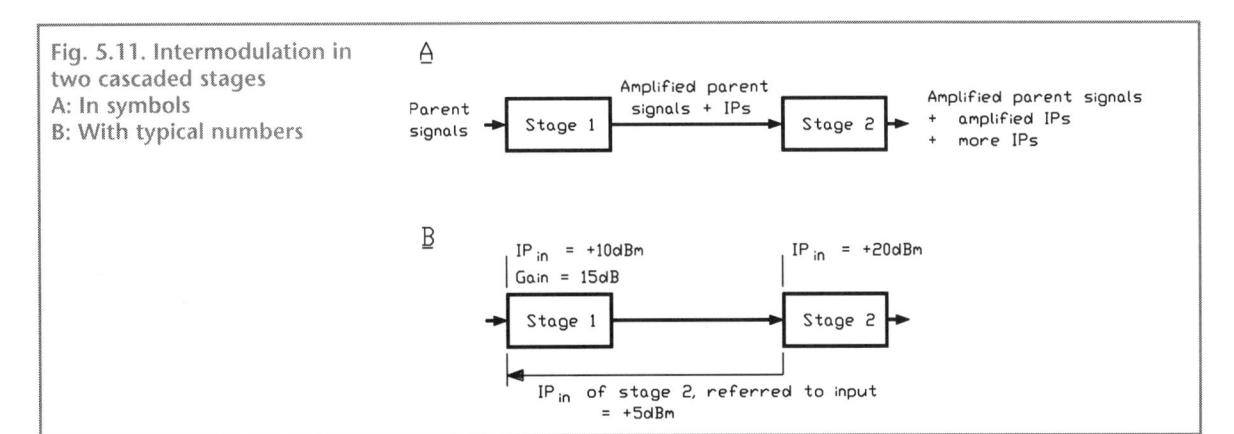

Fig. 5.11. Intermodulation in two cascaded stages
A: In symbols
B: With typical numbers

A
Parent signals → Stage 1 → Amplified parent signals + IPs → Stage 2 → Amplified parent signals + amplified IPs + more IPs

B
IP_{in} = +10dBm, Gain = 15dB → Stage 1 → IP_{in} = +20dBm → Stage 2
IP_{in} of stage 2, referred to input = +5dBm

If we speak in terms of output power, we're talking about the *third-order output intercept* (IP_{OUT}). The two are equivalent, but not equal: they differ simply by the small-signal power-gain of the stage in question. If an amplifier with a power gain of 19dB has an output intercept of 30dBm (30dB over one milliwatt, *ie* 1W), its input intercept is (30 – 19)dBm, *ie* 11dBm. On the other hand, if a diode-ring mixer with a loss of 6dB has an output intercept of 10dBm, its input intercept is 6dB greater, *ie* 16dBm. Since amplifiers have gain, their output intercepts always **look** bigger than their input intercepts, merely because they're higher. Always check which one you're being quoted, especially if someone's trying to sell you something!

The third-order intermodulation intercepts of individual amplifiers, mixers *etc* can usually be estimated either from manufacturers' data sheets, or failing that by some simple rules-of-thumb which I'll describe in the Appendix to this chapter. Given the necessary information, you can predict the intermodulation performance of a multi-stage front-end, in much the same way as we predicted its small-signal performance earlier in this chapter.

To predict intermodulation in a complete front-end, we need to know how IPs build up in cascaded stages. If two equally strong signals are fed into the input of a two-stage system (Fig. 5.11A), the first stage with gain will produce an amplified output, and also some internally generated IPs. The next stage will accept all these signals as inputs, and as well as amplifying them it will produce yet more IPs of its own making. The two sets of IPs will be on the same frequencies, and at small frequency separations or in wideband systems they will usually add in phase at the output of the second stage. If the input intermodulation intercepts of stages 1 and 2 in Fig. 5.11A are IP_1 and IP_2, the system input intercept is given by:

$$\frac{1}{IP_{SYS}} = \frac{1}{IP_1} + \frac{1}{\left(\frac{IP_2}{G1}\right)}$$

Let's look at this formula more closely. It may seem vaguely familiar, because it's rather like two formulae you've seen somewhere before. One is the formula for adding noise temperatures in cascaded stages:

$$T_{SYS} = T1 + \left(\frac{T2}{G1}\right)$$

That was equation 2 from Chapter 4, where I mentioned that the action of dividing T2 by G1 was called "referring T2 to the input", so that it could be added directly to the noise temperature T1 of the first stage. In the intermodulation equation above, we're referring IP_2 to the input in just the same way when we divide it by G1.

The other familiar aspect of the intermodulation equation is that after you've referred all the individual intermodulation intercepts to the input, you can then combine them like resistors in parallel. As you know, the resistance of a parallel combination is always less than the lowest of the individual resistances.

5 - 15

It's exactly the same with intermodulation intercepts: once you've referred everything to the input, the input intercept of the whole system is less than the lowest of the individual stage intercepts. Unlike the chain in the proverb, this chain of stages is **weaker** than its weakest link.

Time to try some numbers. Fig. 5.11B puts some values into the general system of Fig. 5.11A. Using the equation above:

$$\frac{1}{IP_{SYS}} = \frac{1}{(10dBm)} + \frac{1}{(20dBm - 15dB)}$$

This equation cannot be evaluated as it stands. The intercept of the second stage can be referred to the system input simply by subtracting the 15dB gain of the first stage, but power levels in dBm (dB relative to 1mW) must be converted to milliwatts before they can be combined. So, more correctly:

$$\frac{1}{IP_{SYS}} = \frac{1}{(10mW)} + \frac{1}{(3.16mW)}$$

IP_{SYS} = 2·40mW or 3·80dBm

Note that the result, 3·80dBm, is much...

lower than either of the individual stage intercepts of 10 and 20dBm.

What happens if one stage has a much higher intercept than the other (after they've both been referred to the system input)? If the second stage is the more 'robust', the system input intercept still cannot be greater then the 10dBm limit imposed by the first stage. Conversely, no matter how robust the first stage, IP_{SYS} cannot exceed (20−15) = 5dBm. This is because of intermodulation in stage 2, which is made worse by the 15dB gain ahead of it in stage 1. Time after time, we keep coming back to the need to *keep the system gain down!*

WORKSHEET METHODS

Earlier we used a worksheet (Fig. 5.4) to analyse the noise temperatures, noise figures and gain distribution of the *Suffolk* 144MHz transverter front-end. Now let's use the same worksheet to analyse its third-order intermodulation performance. Fig. 5.12 shows the details, with typical intermodulation intercepts for each stage. The stage gains are exactly the same as in the earlier noise analysis, of course.

Since the *Suffolk* is a transverter, you can't usefully analyse its intermodulation performance in isolation from the receiver that follows it. We'll go further into this topic towards the end of this chapter, and also in Chapter 8 where G4DDK describes the *Suffolk* in more detail. For now, let's assume you're using one of the modern generation of HF transceivers with a 28MHz IP_{IN} of +20dBm.

Having written down the input intercept for every stage, including the HF receiver, the next step is to refer each intercept to the 144MHz input. Do this by subtracting the cumulative gain up to that point, remembering that here we're working in decibels: check Fig. 5.12. To combine the stage intercepts into a system value, convert everything into

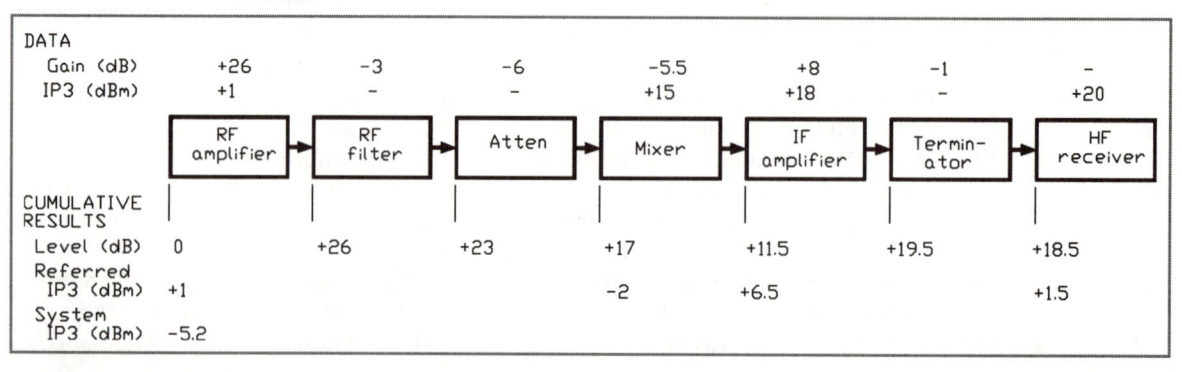

Fig. 5.12. Intermodulation analysis of the Suffolk transverter

DATA							
Gain (dB)	+26	−3	−6	−5.5	+8	−1	−
IP3 (dBm)	+1	−	−	+15	+18	−	+20
	RF amplifier	RF filter	Atten	Mixer	IF amplifier	Terminator	HF receiver
CUMULATIVE RESULTS							
Level (dB)	0	+26	+23	+17	+11.5	+19.5	+18.5
Referred IP3 (dBm)	+1			−2	+6.5		+1.5
System IP3 (dBm)	−5.2						

INTERMODULATION ANALYSIS

milliwatts, and add all the values as if they were resistors in parallel – yes, it *is* tedious, but try doing it at least once with a calculator, just for the experience.

The third-order input intermodulation intercept of the *Suffolk* front-end is thus –5·2dBm. What else does the analysis tell us? As soon as you've referred each of the stage intercepts to the input, you can see which are the vulnerable stages. The mixer is the single weakest link, but note that we have assumed an HF receiver with an excellent IP3 of +20dBm, which represents reasonable performance in the 1990s. Many older or 'budget' HF rigs have poorer intermodulation performance, and such receivers would limit the overall performance of the system. Therefore G4DDK has tried to protect the HF receiver by keeping the overall gain of the *Suffolk* to the minimum consistent with meeting the target noise temperature. And although he hasn't spent an excessive amount on the mixer, he's left the option of upgrading various parts of the system as described in Chapter 8.

PREDICTING DYNAMIC RANGES

INTERMODULATION

Another thing that the intermodulation/gain analysis can predict is the dynamic range for third-order intermodulation. Recalling from Fig. 5.4 that the receiver noise temperature was 196K, its noise floor in a typical SSB bandwidth of 2·5kHz is –141·7dBm (from the 'kTB' equation in Chapter 4, which should be getting pretty familiar by now). The spurious-free dynamic range (SFDR), also known in this context as the *intermodulation-free dynamic range* or the *two-tone dynamic range*, is given by:

$$SFDR = (\text{input intercept} - \text{noise floor}) \times \frac{2}{3}$$

everything being measured in dB, so:

$$SFDR = (-5 \cdot 2dBm - \{-141 \cdot 7dBm\}) \times \frac{2}{3} = 91dB$$

Measurements on the *Suffolk* front-end agree quite well with predictions, and so does experience on the bands. Assuming a decently calibrated S-meter (a big assumption!), with 6dB per S-point and S0 at the noise floor, absolutely no third-order intermodulation should be noticeable until signals reach almost S9+40dB. *That* is the standard of intermod performance you're entitled to demand from a modern front-end.

RECIPROCAL MIXING

As I remarked earlier, a receiver actually has several dynamic ranges. What are the maximum tolerable levels of local-oscillator noise before the overall spurious-free dynamic range becomes limited by reciprocal mixing rather than intermodulation?

When used with a good HF transceiver, the *Suffolk* has a predicted intermodulation-free dynamic range of 91dB in a 2·5kHz bandwidth. So we want to ensure that no discernible reciprocal-mixing noise will be produced unless the carrier responsible is at least 91dB above the noise floor. So the LO noise level must itself be –91dBc in a bandwidth of 2500Hz. What does that mean in terms of dBc/Hz? The equation relating noise levels, in dB terms, for various bandwidths is:

$$N_1 - N_2 = 10\log_{10}\left(\frac{B_1}{B_2}\right) dB$$

So –91dBc in 2500Hz corresponds to –125dBc in a bandwidth of 1Hz.

An LO noise level of –125dBc/Hz would actually raise the noise floor level by 3dB at the signal levels we're assuming, which would be totally unacceptable for weak-signal working. So we really need more like –135dBc/Hz. At close frequency separations, that's asking quite a lot, though it should be perfectly well achievable at offsets beyond a few kHz. And we're still entitled to expect that LO noise levels should roll away towards –160dBc/Hz and better at wider offsets.

FRONT-END ANALYSIS USING TCALC

Up to now, I've taken you through the calculations of front-end noise/gain and intermodulation performance 'by hand', because that's the only way to explain it properly. Now is the time to introduce an easier way of doing the same things. TCALC is a microcomputer program which simply automates the calculations I've just been describing. It is freely available as a BASIC program listing, and also on disk for the BBC Micro and IBM PC compatibles [9].

Noise/gain and intermodulation analyses using TCALC work exactly as I have already described. You divide the system into a block diagram and type in the performance data for each block. TCALC then calculates the cumulative and system performance, taking care of all the tedious details like conversion between noise temperature and NF, or between decibels and ratios.

By making your input as simple and convenient as possible, TCALC frees your attention to think about the real problem – the design of a receiver front-end. For example, when TCALC asks you for the noise temperature or noise figure of an amplifier, it makes intelligent decisions about which one you mean. Enter a number less than 25, and TCALC will assume you mean noise figure; a larger number is taken to mean noise temperature. Having made its decision, TCALC confirms what it thinks you meant and then continues by asking you for the gain.

When you have completed the calculation, you can then go through it again to see the effects of making changes. To save you the trouble of retyping, TCALC remembers all the data you want to leave unchanged, and this can be recalled by a single keystroke.

TYPES OF STAGES

TCALC requires you to label each stage, to help you remember which is which, and so that TCALC knows what data to ask for. If you go back through Chapter 4, you'll find that there are really only three types of stage which need to be considered.

(1) *Amplifiers* have gain and noise temperatures, and the calculation for cascaded stages uses equation 2 from Chapter 4. Since this analysis takes no account of input or output frequency, mixers can be treated exactly the same as amplifiers. Diode mixers with negative gain are no problem in this regard. TCALC permits the labels AMPlifier or MIXer for this type of stage.

(2) *Attenuators* include any type of passive device which has an essentially resistive insertion loss. The calculation uses equation (4) from Chapter 4, and TCALC assumes a physical temperature of 290K.

TCALC permits the label ATTenuator or CABle for this type of stage. When entering data, passive filters also use the ATT label.

(3) *Antenna* denotes the end of the block diagram. On encountering the label ANT, TCALC asks for the antenna noise-temperature, adds it to the receiver noise-temperature and displays the system results. When analysing a receiver in isolation from its antenna, the label END serves the same purpose.

NOISE/GAIN ANALYSIS

The easiest way to describe how TCALC

FRONT-END ANALYSIS

Fig. 5.13. Noise/gain analysis for the *Suffolk* transverter using TCALC. Compare with Fig. 5.4

STAGE PARAMETERS	SYSTEM RESULTS		
	T(K)	NF(dB)	G(dB)
Stage 1 (rest of system)			
NF = 10dB	2610	10	0
Stage 2 Type: ATT			
Loss (dB) : 1	3360.9	11	-1
Stage 3 Type: AMP			
NF = 6dB Gain (dB) : 8	1397.2	7.65	7
Stage 4 Type: MIX			
NF = 6dB Gain (dB) : -5.5	5821.9	13.24	1.5
Stage 5 Type: ATT			
Loss (dB) : 6	24042	19.24	-4.5
Stage 6 Type: ATT			
Loss (dB) : 3	48258.8	22.24	-7.5
Stage 7 Type: AMP			
NF = 1dB Gain (dB) : 26	196.3	2.25	18.5
Stage 8 Type: END			

performs a noise/gain analysis is to show you the output for the *Suffolk*. Compare the screen output in the panel above (Fig. 5.13) with the worksheet of Fig. 5.4.

Notice that TCALC only produces a gain analysis working towards the antenna, rather than a display of relative signal levels working forwards from the input as shown in Fig. 5.4. This is because all the necessary information for a signal-level presentation of gain is not available until the details of the last stage have been entered. However, a relative signal level display is available as part of the intermodulation analysis.

INTERMODULATION ANALYSIS

Having made at least one pass through the noise/gain analysis, TCALC offers the option of intermodulation analysis. By now, the identity and gain of each stage is known to the program, so it lists all the stages and asks you for an intermodulation intercept where appropriate, *ie* for AMP/MIX stages only. If you wish to disregard intermodulation in that stage, you just press 'Enter'.

Having collected all its information, TCALC now produces a table of signal levels, the intermod intercept of each stage referred to the input, and the system results including noise floor and dynamic range. Fig. 5.14 shows the results for the *Suffolk*, which can be compared directly with Fig. 5.12 (below).

Whenever it completes an analysis, TCALC offers you the opportunity to go round again, switch to the other type of analysis, or quit.

OTHER POSSIBILITIES

TCALC is a simple program written in BASIC

Fig. 5.14. Intermodulation gain analysis for the *Suffolk* transverter using TCALC. Compare with Fig. 5.12

STAGE PARAMETERS	SYSTEM RESULTS			
	T(K)	NF(dB)	Level(dB)	IPi3(dBm)
Rest of system IP3(dBm): 20	2610	10	18.5	1.5
Stage 2 (ATT)	3360.9	11	19.5	
Stage 3 (AMP) IP3(dBm): 18	1397.2	7.65	11.5	6.5
Stage 4 (MIX) IP3(dBm): 15	5821.9	13.24	17	-2
Stage 5 (ATT)	24042	19.24	23	
Stage 6 (ATT)	48258.8	22.24	26	
Stage 7 (AMP) IP3(dBm): 1	196.3	2.25	0	1
Whole receiver:				-5.2
Noise floor (2.5kHz) = -141.7dBm				
Dynamic range = 91dB				

for a variety of microcomputers. Ease of translation to different machines and dialects of BASIC was a major criterion when I wrote the program, so I tried to avoid using special screen-handling techniques. An alternative approach would be to model the front-end using a spreadsheet program; such programs are available for almost all microcomputers, but they tend to be machine-specific and the programming facilities that they offer vary widely. One of the major virtues of spreadsheets is in 'what-if' analysis; change any of the input data, and all the consequences are immediately changed on the screen. This would allow rapid optimization of gain distribution, for example.

The disadvantage of a spreadsheet is the difficulty of changing the system configuration by inserting and deleting stages, or by changing the types of stages. All such alterations require changes in the structure of the spreadsheet itself, and are difficult to do automatically. In this respect a simple 'scrolling' program like TCALC is far more flexible. If anyone is willing to tackle the problem of a self-configuring TCALC-style spreadsheet, I'd be interested to hear about it for the second edition of this book! Meanwhile, for my own purposes I'd continue to use TCALC for simple general-purpose jobs, but might be tempted to move a specific problem over to a spreadsheet for more intensive optimization once the basic sequence of stages has been determined.

THE ROLE OF MEASUREMENT IN DESIGN

If you have both the test equipment and the knowledge to use it, you could make all the relevant measurements on your new receiver without having to do all this theoretical analysis. But amateur radio has moved on from the old days of suck-it-and-see. We're demanding high performance at low cost, so we can no longer afford the luxury of taking chances and cutting corners. Careful design is the one area where amateurs can reap big rewards at little or no cost.

Of course, a fully professional approach to RF system development uses the even more powerful combination of both design *and* measurement. But it's still surprising how many ill-designed products reach the amateur radio market, for lack of a morning's effort with a worksheet and a microcomputer.

To sum up the case for careful system design and analysis:
● You get the best possible performance at minimum cost
● You avoid wasting your time building something you *could have known* would be inferior.
● Time, understanding and a home computer can go a long way towards substituting for professional test equipment.

PRACTICAL FRONT-END SYSTEMS

In this final section I will look at some typical amateur VHF/UHF receiving systems. Some of my overall conclusions may not go down too well in certain quarters, but they're all based on data from published equipment reviews and backed up by calculations which you can check for yourself.

I have already said that most commercial transceiver front-ends are noticeably deaf – as any VHF/UHF DXer knows! Reviewers have found that the noise figures of most 144MHz transceivers run in the range 3–7dB, though they are gradually coming down as a result of consumer pressure. Since many receiving systems still fall short of the target noise-figures I recommended in Chapter 4, *eg* about 2·2dB for 144MHz, what can we do about it?

HOW TO CHOOSE A PREAMPLIFIER

One of the most common solutions to a deaf transceiver front-end is to add a preamplifier. The first thing to do when choosing a preamp is to decide why you want it. Here are some possible reasons:
- Your existing transceiver is deaf – the most common reason
- You want to be able winkle out the weakest DX while the band is quiet and there aren't any strong signals about
- You have large, permanent, unavoidable feedline losses
- You work satellites or moonbounce, so you can take advantage of the lower antenna noise temperatures when pointing skywards.

In the last two cases you can generally get away with a permanently connected preamp. However, in the first two cases, which are far more common, the preamp will make your system vulnerable to overload by strong signals. What generally happens is that the gain of the preamp provokes overload in the existing transceiver front-end, or in the HF transceiver that follows a transverter. It's actually quite rare for the preamp itself to overload.

There are two cures for overload provoked by a preamp and you can use them both. The first is to use a switchable preamp, so that you can get rid of it when there are strong signals about. With a bit more effort you will probably still be able to copy all but the weakest DX, which is far better than having the whole band wiped out by receiver overload! The second cure is to use the least possible amount of preamp gain – I'll come back to that later.

Having decided whether you need a switchable preamp (and you probably do) the next step is to look at some advertised preamp specifications, particularly of noise figure and gain. When you add a preamp to an existing transceiver, the new system NF obviously depends to some extent on how good or bad the transceiver was in the first place. Since the preamp manufacturers don't know the NF of your particular transceiver, they have no choice but to quote the NF of the preamp alone, as if it was connected to an idealized noiseless receiver. So the quoted NF is always better than you can get in practice when the preamp is installed as part of your system. Nobody's trying to deceive you; you simply need to be aware of what the advertised specification really means.

RECEIVERS & LOCAL OSCILLATORS

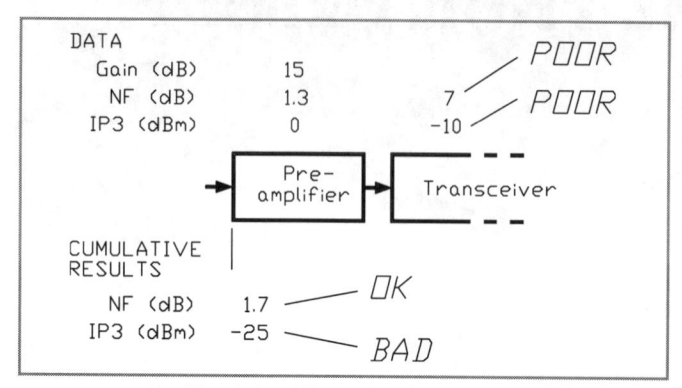

```
DATA
   Gain (dB)      15                  POOR
   NF (dB)        1.3      7
   IP3 (dBm)      0       -10         POOR

         →  [Pre-     ]  [Transceiver]
            [amplifier]

CUMULATIVE
RESULTS
   NF (dB)        1.7              OK
   IP3 (dBm)     -25
                                   BAD
```

Fig. 5.15. Effect of a preamplifier on a typical 144MHz transceiver. Sensitivity is improved, but the intermodulation intercept becomes much worse

The second thing to look at in preamp specifications is the gain. As I keep on saying, *too much gain is BAD*. A typical 144MHz preamp with an NF of 1·3dB and gain of 15dB, as advertised, will improve the 7dB NF of a poor transceiver to about 1·7dB (Fig. 5.15). Fair enough, if it transforms a deaf receiving system into one which, allowing for feedline losses, comes pretty close to our target NF of around 2dB. But 15dB of preamp gain will also transform the strong-signal performance from mediocre to downright bad. That is the penalty of improving the system NF by adding something on the front of the transceiver rather than tackling the real problem: the existing poor front-end.

Reputable preamp manufacturers are actually somewhat apologetic about the fact that their products need to have some gain! Manufacturers who boast about the gain of their preamps either have little regard for your technical knowledge, or else they don't understand what they're doing. Unfortunately there are always some customers who won't believe a receiver is any good unless the S-meter is perpetually bouncing about and the speaker pours out noise like Niagara Falls.

Since the preamp itself is unlikely to overload, you don't need to pay too much attention to its strong-signal performance. Even so, it's better to buy from a manufacturer who has specified the preamp's intermodulation performance. If you do run into difficulties, you're far more likely to receive useful support from such a manufacturer than from one who's never even considered the intermod problem.

When reading a preamp manufacturer's specifications, remember these points:
● *Input* intermodulation intercept is what really matters;
● Output intercept may look more impressive, but you have to subtract the gain to get the input intercept
● Input intercepts significantly above 0dBm are unlikely, unless the preamp uses push-pull or negative-feedback techniques.
● Choose a preamplifier which offers an adjustable gain facility, so that you can adjust your total front-end gain to suit your own local noise environment.
● Don't be besotted with low noise-figures; don't be swayed by this year's fashion; and don't believe in magic!

WHAT'S WRONG WITH COMMERCIAL TRANSCEIVER FRONT-ENDS?

The front-ends of transceivers from the 'Big Three' manufacturers are not sensitive enough for serious VHF or UHF DX working, and they are none too happy with strong signals either. We know this from equipment reviews and from personal experience. If you (or the manufacturer) fit a preamplifier to improve the sensitivity, the strong-signal performance becomes even worse. Yet we also know that it is possible to build front-ends which outperform those of commercial transceivers on both weak and strong signals.

Until fairly recently, reciprocal mixing was virtually unknown (or at least unrecognized) in VHF/UHF receivers for the amateur bands. This was partly because front-ends were even more vulnerable to intermodulation and gain compression than they are today, but also because the LOs in older rigs were generally quieter then their modern synthesized counterparts. Now that front-ends are getting better, while standards of LO noise performance in the amateur VHF/UHF market have

PRACTICAL FRONT-END SYSTEMS

worsened, reciprocal mixing is becoming the limiting factor in strong-signal performance.

In the past few years there have been marked differences in the ways that various manufacturers and importers have tackled this problem. Some have steadily improved their products. But others have done nothing, or have even brought out new equipment whose LO noise is **worse** than in older models! The RSGB VHF Contest Code of Practice sets a target of –90dBc for all wideband spurious products from the transmitter, and if applied to LO noise this would stand our receivers in very good stead too. That target was set at the end of the crystal-controlled AM era when some particularly nasty SSB transceivers were making their first appearance; the figure of –90dBc was chosen because it was being routinely achieved by the older AM equipment, mainly thanks to crystal oscillators and high-level Class-C stages. It's a lot more difficult to achieve that order of noise performance along with the kind of frequency mobility we require for today's operating conditions, but the technology does exist and the design principles are well established [5, 6, 7].

So why can't we buy transceivers with truly excellent VHF and UHF front-ends from the Big Three manufacturers? In their HF models they have recognised our requirements for strong-signal performance as well as sensitivity, and 'high dynamic range' is now a strong selling point. But at VHF and UHF all the recent design effort seems to have gone into digital features like scanning, memories, voice synthesizers and built-in packet TNCs. These are all very well in their way, but what about the *radio*?

As purchasers and users of amateur radio equipment, we need to resist the notion that other people aren't entitled to put strong signals into our receivers; or that we shouldn't expect equipment manufacturers to produce clean local oscillators; or that any rig is acceptable merely because the importers have sold a lot of them. So read the equipment reviews carefully, look out for reports of poor sensitivity, dynamic range or reciprocal mixing performance, and then use your power of choice. If it isn't good, don't buy it.

TRANSVERTERS REVISITED

If you already have an HF rig, the alternative to buying a complete VHF or UHF transceiver is to buy a transverter – or to build one from this book. VHF and UHF SSB got started in the 'sixties when HF SSB enthusiasts added transverters to their HF transceivers. This opened up the VHF bands to the advantages of transceive operation, VFO control, stable and selective receivers, and of course to the advantages of SSB as a DX mode. But by the late 'seventies transverters had acquired a bad reputation for receiver overload, and on 144MHz they have been almost entirely superseded by single-band, single-conversion transceivers for serious DX and contest work. However, transverters have always been more widely used than single-band rigs on the other VHF/UHF/microwave bands, and now they've even made a comeback on 144MHz. So what went wrong with transverters in the 'seventies, and what's changed since then?

A transverter followed by an HF transceiver is a multiple-conversion system. The receive signal path in the transverter is very like the front-end of a normal single-conversion transceiver, except for one thing. The LO in a transverter is crystal-controlled on a fixed frequency, and the HF transceiver does the tuning. That means you can't put a narrowband IF filter directly after the transverter front-end, so the HF transceiver front-end is vulnerable to overload from the amplified wideband output of the transverter.

Transverters are worth a second look today because the front-ends of commercial HF transceivers have improved a lot in the past few years, far more than those of VHF/UHF transceivers. Also the IF performance of HF transceivers has always been a few steps ahead; few VHF/UHF transceivers even nowadays can offer adjustable noise blankers, switchable filters with really good stopband rejection, variable bandwidth, IF shift or notch filtering. All these are very handy on today's crowded VHF and UHF bands.

RECEIVERS & LOCAL OSCILLATORS

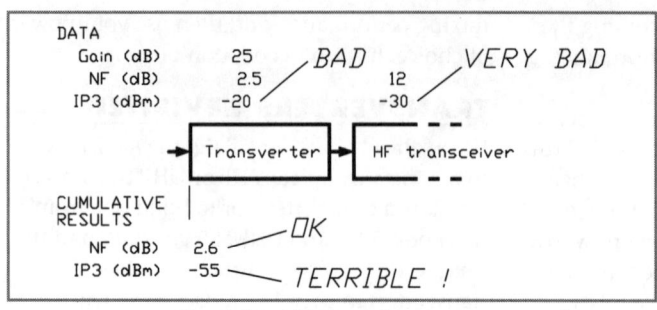

```
DATA
   Gain (dB)    25      BAD       VERY BAD
   NF (dB)      2.5              12
   IP3 (dBm)   -20              -30

        →  [Transverter] → [HF transceiver]

CUMULATIVE
RESULTS
   NF (dB)      2.6  ── OK
   IP3 (dBm)   -55  ── TERRIBLE !
```

Fig. 5.16. 1970's style transverter and HF transceiver

Let's use our techniques for analysing front-end performance to see what went wrong with transverters in the 'seventies. At that time, transistorized HF transceivers were just coming in and we all thought they were wonderful. Actually, their strong-signal handling was dreadful: their third-order intermodulation intercepts were –30dBm or worse. A 1970s-style transverter had about 25dB of gain, which dropped the system input intercept at 144MHz to –55dBm or lower. With a system noise floor of –140dBm the dynamic range would be less than 60dB (Fig. 5.16). In practical terms, signals over about S9 anywhere on the band would cause havoc!

Now do the same analysis for a more modern system. A typical NF for a modern high-quality HF transceiver at 28MHz would be 10dB, with a third-order input intercept of +20dBm. Using a transverter like the *Suffolk*, we can achieve a creditable –5dBm system input intercept and dynamic range of 90dB, along with the required noise figure of around 2dB. That's not quite as good as a single-conversion front-end designed with equal care, but it's a sight better than an ordinary 144MHz transceiver plus preamp. And unlike a single-band 144MHz transceiver, the HF-plus-transverter approach equips you for the HF bands and allows you to add further transverters for 50MHz, 70MHz and 432MHz. That's why we chose the transverter approach for the designs in this book.

Turning to the bands other than 144MHz, a target noise temperature of 150K at 432MHz and above could be achieved using a transverter receive gain of no more than 25dB. To keep the overall gain as low as this, a masthead preamplifier would be essential for 432MHz. Antenna noise temperatures on 70MHz and 50MHz are much higher than on 144MHz or above, so the requirements for low receiver noise temperatures are considerably relaxed and the transverter receive gain can be reduced accordingly. With a good HF transceiver, transverters for 50MHz and 70MHz should have input intercepts significantly better than 0dBm.

ADJUSTING FRONT-END AND TRANSVERTER GAIN

There is no point in using more front-end gain than is necessary to achieve reasonable sensitivity in your particular RF noise environment. Every decibel of unnecessary front-end gain knocks 1dB off your input intermodulation intercept. Considering how difficult it is to get IP_{IN} into the region above 0dBm, it's surprising how many people then waste their efforts by failing to set up the whole receiving system on the air.

Fig. 5.17. Fitting an adjustable attenuator to the output of a preamplifier or transverter. (A) 15-30dB 'constant impedance' design; (B) Wide-range non-constant impedance. RV1 is a miniature cermet or carbon trimpot, and all other resistors are miniature carbon film or metal-oxide.

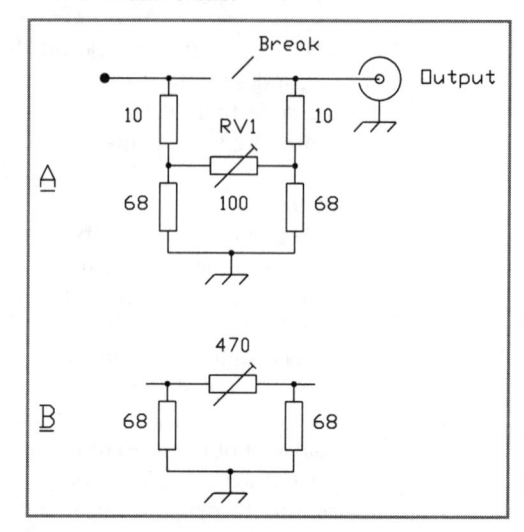

PRACTICAL FRONT-END SYSTEMS

Any system you buy or build will probably start out with too much gain. Commercial preamps need enough gain to produce some improvement on the deafest transceiver. The same applies to transverters: even the ones in this book are designed with enough gain to suit a transceiver with a fairly high noise-figure at 28MHz. Your transceiver may be more sensitive, so you can get rid of the excess gain in the preamplifier or transverter.

Preamplifiers and transverters generally have sufficient gain to perform well in a low-noise environment. The target noise temperatures given for each band in Chapter 4 refer to the *lowest* probable levels of antenna noise. Your own RF environment may be significantly noisier, and this too will allow you to eliminate some excess front-end gain.

Some preamps already include a variable attenuator at the output, so that's a good start. If yours does not, you can easily fit one following the circuits of Fig. 5.17. The attenuator must not carry RF or DC power, so take care if the preamp includes RF switching or has a DC power feed up the coax; if in doubt, check with the technical guru at your local radio club. Some transverters are likewise equipped with a variable attenuator at the 28MHz output. The ones in this book happen not to be, but each has a 50Ω attenuator in the signal path which can be used for the same purpose. More details are given in the appropriate chapters.

To adjust your preamp or transverter gain down to the minimum your system needs, you simply turn the gain down until the signal/noise ratio on a weak signal ***just begins*** to deteriorate. By using a real signal on the band, you are automatically taking account of your own particular RF noise environment, including galactic noise and the high man-made noise levels typically encountered in city environments and practically everywhere on 50 and 70MHz.

To avoid the risk of reducing the sensitivity too much, leaving your system unnecessarily deaf when the band is quiet, you should do the following test at the quietest possible time, *eg* the small hours of Sunday morning, at the beam heading that gives the lowest background noise.

Signal/noise ratio is not easy to judge by ear, so the easiest way to do it is to take advantage of the 'threshold effect' of an FM detector. As you've probably noticed, FM signals tend to be either virtually noise-free or very poor, and the change takes place over a narrow range of incoming signal strength, or rather signal/noise ratio.

Setting-up is a simple three-step process.

1 Tune in a weak and slightly noisy FM signal. If you don't have FM capability, use SSB, preferably with the AGC off.
2 Reduce the front-end gain until the audio starts to become noisier.
3 Then increase the gain *very slightly* until you no longer notice the deterioration.

That's all it takes to optimize your front-end gain. If your system still gets crunched by strong signals, at least you know you've done your best!

APPENDIX

ESTIMATING INTERMODULATION INTERCEPTS

The whole concept of third-order intermodulation analysis rests on three assumptions:
- that intermodulation in each stage can be characterized by an intercept;
- that intercepts add up as shown earlier; and
- that you know what values to assume.

I've dealt with the first two assumptions in the main part of the chapter. They may not be totally accurate but they're far better than nothing at all. But if manufacturers' data sheets don't give you the information you need for analysis and design, you'll have to estimate some intermodulation intercepts for yourself.

Intermodulation intercepts in amplifiers can be estimated by thinking first about gain compression. The maximum output power available from a Class-A amplifier depends on the relationship between supply voltage V (collector-emitter, drain-source, anode-cathode or whatever), standing bias current I, and load resistance R_L. The limiting factor can be either the voltage swing or current starvation [3, 5]. The approximate maximum power output available from a Class-A amplifier is:

$$\frac{V^2}{2R_L} \text{ for voltage limiting}$$

$$\frac{I^2 R_L}{2} \text{ for current limiting.}$$

The lower of these two power levels is the maximum undistorted power output of the amplifier. For a bipolar amplifier biased to 20mA at 6V, and terminated in 50Ω, the voltage limit would occur at $(6^2/100)$W = 360mW; but current limiting occurs at the lower level of $(0.02^2 \times 50/2)$W = 10mW and is thus the operative factor. Any device under the same DC operating conditions would current-limit at 10mW output in the same way. In low-level RF applications, voltage limiting hardly comes into the picture at all, except perhaps for FETs or valves terminated in high RF load resistances.

Subtracting typical stage gains of 15–20dB or more for modern low-noise devices, the input power at the output compression point will be −10dBm or worse. As a rough rule of thumb, the third-order input intercept of a good bipolar amplifier is about 15dB above its compression point, and the same *may* be true of some FETs. Badly biased or non-linear devices may be much worse! Thus the 3rd-order intermodulation *input* intercepts of present-day low noise amplifiers are most unlikely to be above +5dBm, and 0dBm is far more typical.

Negative feedback gives some improvement in intermodulation performance, mainly by eliminating surplus gain and thus reducing current and voltage swings at the output. However, negative feedback does tend to detract from low-noise performance, and even with heavy feedback the fundamental limitation on *output* voltage or current swings still apply. An effective technique (revived for the semiconductor age by G4DGU) is to use a pair of devices in push-pull. This doubles the allowable voltage and current swing, and if applied correctly can improve the inter-modulation intercept by 6dB compared with an equivalent single-ended stage using the same device.

Measurements on diode-ring mixers show apparent variations in third-order input intercepts, *ie* the simple theory doesn't work too well for these mixers. As a rough guide, the third-order input intercept of a ***properly terminated*** passive ring mixer is about 5dB above the LO power level [3]. To obtain optimum intermodulation performance from diode mixers, proper termination of all ports in 50Ω *at all frequencies* – from LF to SHF in the case of VHF/UHF mixers – is vital. Active mixers with gain are most likely to limit at their outputs, and thus can be assessed like amplifiers.

It isn't safe to assume that *any* so-called 'passive' circuit device is immune from intermodulation without careful examination of how it works. Resistors are a fairly safe bet, according to Ohm's Opinion, and so are most capacitors at low signal levels. Diodes are a

APPENDIX

notorious source of intermodulation problems – *all* kinds of diodes, especially varicaps, and not forgetting the 'parasitic' diodes on transistor chips. Unless you have very good reason to do otherwise, diodes in RF systems must be firmly biased either on or off so that normal signal levels cannot disturb their operating conditions. Any kind of ferrite or dust-iron cored device can cause intermodulation unless the magnetic material is operated a long way below saturation. And even such apparently harmless devices as crystal filters can cause intermodulation [3, 10].

Finally, remember that all these methods of estimating intermodulation performance are *approximate*, and that they generally apply to idealized devices of their type. The performance of badly-chosen or badly-used devices will be worse!

REFERENCES

[1] Ian White G3SEK, Modern receiver front-end design. Part 1: Weak-signal performance. *Radio Communication*, April 1985.

[2] Ian White G3SEK, Modern receiver front-end design. Part 2: Strong-signal performance. *Radio Communication*, May 1985.

[3] Ian White G3SEK, Modern receiver front-end design. Part 3: Intermodulation analysis. *Radio Communication*, June 1985.

[4] Ian White G3SEK, Modern receiver front-end design. Part 4: Practical front-end systems. *Radio Communication*, July 1985.

[5] W. A. Hayward, *Introduction to Radio Frequency Design*. Prentice-Hall, Inc., Englewood Cliffs, New Jersey, 1982. ISBN 0-13-494021-0

[6] W. F. Egan, *Frequency Synthesis by Phase Lock*. John Wiley & Sons, New York, 1981.

[7] U. L. Rohde, *Digital PLL Frequency Synthesizers*. Prentice-Hall, Inc., Englewood Cliffs, New Jersey, 1983.

[8] Wes Hayward W7ZOI, Defining and measuring receiver dynamic range. *QST*, July 1975. See also Wes Hayward W7ZOI and Doug DeMaw W1FB, *Solid State Design for the Radio Amateur*. ARRL, 1977.

[9] TCALC was written by Ian White, G3SEK. The program is available from David Butler G4ASR, as a printed BASIC listing, or on disk for the BBCMicro and the IBM PC and compatibles.

[10] Peter Chadwick G3RZP, Dynamic range, intermodulation and phase noise. *Radio Communication*, March 1984.

CHAPTER 6

TRANSMITTERS, POWER AMPLIFIERS & EMC

TRANSMITTERS, POWER AMPLIFIERS AND EMC

by John Nelson GW4FRX

"Hokey religions and ancient weapons are no match for a good blaster"
(Han Solo in *Star Wars*)

The interesting thing about *Star Wars* is this: although the blasters were certainly good, it was the "hokey religion" that finally won the battle. You might care to ponder this paradox if you're intending to build a VHF/UHF transmitter or power amplifier, because you too are going to need more than technology and brute force. You're also going to need a preoccupation with *quality* – with *excellence*, if you like – which is an attitude of mind that can only come from yourself. So here's a hokey suggestion: alongside this chapter, spend a few days reading another book called *Zen and the Art of Motor Cycle Maintenance* by one Robert M. Pirsig. Don't be put off by the title – it has some intriguing things to say about 'quality'. And as the lawyers say before telling you that you've just inherited a fortune, "You might hear something to your advantage".

Achieving reasonable or pretty good high-power VHF/UHF SSB/CW is really quite easy. Achieving *excellent* high-power SSB and CW transmissions, however, isn't. In a predominantly black-box age it may be a shock to find that you can't buy a truly excellent legal-limit PA off the shelf. Instead, you're thrown back on your own creative resources, skill and old-fashioned dedication. We can teach you something about how to tackle transmitters and amplifiers, but we can't tell you how to mobilize your own resources. We can only hope to inspire them and give them something to bite on.

May the Force be with you!

CLEAN SIGNALS

Before we wade in and get ourselves thoroughly involved in the minutiae of transmitters and power amplifiers, it's worth taking the time to examine the basic requirements for SSB and CW transmitters at VHF and UHF in the amateur service. First of all, what do we actually mean by 'SSB'?

In principle, SSB is derived from double-sideband full-carrier AM by suppressing the carrier and one sideband. By convention the 'frequency' of an SSB transmitter is taken to be the frequency of the suppressed carrier – which is not, of course, actually radiated unless there is a fault. Carrier suppression is achieved in the SSB transmitter's balanced modulator. Most commercial equipments achieve 40–60dB of suppression, which is adequate, but from time to time it needs checking. You can do this yourself by listening on another receiver, or alternatively getting someone to listen to the transmission. Equally, the carrier suppression of equipment 'fresh out of the box' is often less than optimum. In all cases one or (occasionally) two adjustments are provided in the transmitter to adjust the suppression. The balanced modulator produces a double-sideband signal with suppressed carrier. The unwanted sideband is then removed by a filter, and in a transceiver the same filter provides selectivity on receive. In principle we can have as much filtering of the unwanted sideband as we desire, but in practice that isn't necessary because other processes in the power-amplification chain tend to refill the unwanted sideband (and beyond, alas) with other spurious signals.

CLEAN SIGNALS

Every SSB transmitter ever built produces intermodulation products. Since good intermodulation performance is crucial to achieving clean SSB transmissions which take up an absolute minimum of band space, I'll spend some time considering what is meant by this concept. In many ways this discussion parallels Chapter 5 which deals with intermodulation in receivers – not surprisingly, because the basic principles are exactly the same. Let's take the case of a truly perfect SSB transmitter tuned to 144·250MHz (*ie* the frequency of the suppressed carrier is 144·250MHz) and switched to 'transmit'. In the absence of any input into its microphone socket, there is no RF output whatsoever – no carrier because it's perfectly suppressed, and not even a single modulation sideband because there's no modulation. Now take two perfect audio-frequency generators set to 700Hz and 1700Hz, combine their outputs and apply them to the transmitter. Fig. 6.1 shows what you'd see on the screen of a spectrum analyser connected to the rig. Assuming that it's transmitting 'upper sideband', two radio frequencies will appear at the output and be seen on the screen; they will be 144·250700 and 144·251700MHz, *ie* the suppressed carrier frequency plus each of our two modulating frequencies. In a perfect transmitter, these would be the *only* outputs.

INTERMODULATION DISTORTION

In real life, however, examination of the output of an SSB transmitter fed with two audio-frequency tones in the above manner would show many other frequencies as well; indeed there would be a veritable family of them (Fig. 6.2). Even very good SSB transmitters will produce some power at spurious frequencies, but only at low levels which decrease rapidly as you tune away from the main transmission – certainly not enough to upset other band users. On the other hand, very bad transmitters will produce enough spurious outputs to cause interference across the entire SSB area of the band for tens of miles around. This is the sort of transmission produced by someone trying to squeeze 60W out of his so-called '50W linear'.

Back to Fig. 6.2: since we only put in two audio tones, where have all the other output frequencies come from? From the transmitter itself; or rather any part of the transmitter's circuit which has a non-linear relationship

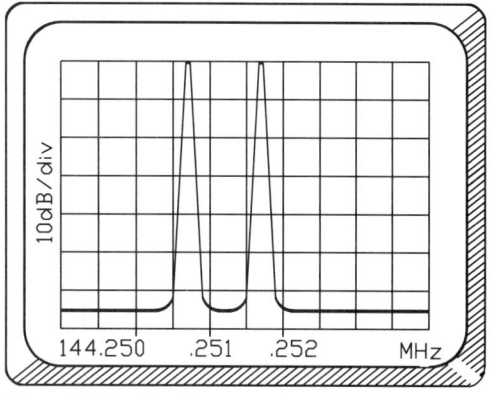

Fig. 6.1. Idealized output from an SSB transmitter with a suppressed carrier frequency of 144·250000MHz, fed with pure sine-wave tones of 700Hz and 1700Hz

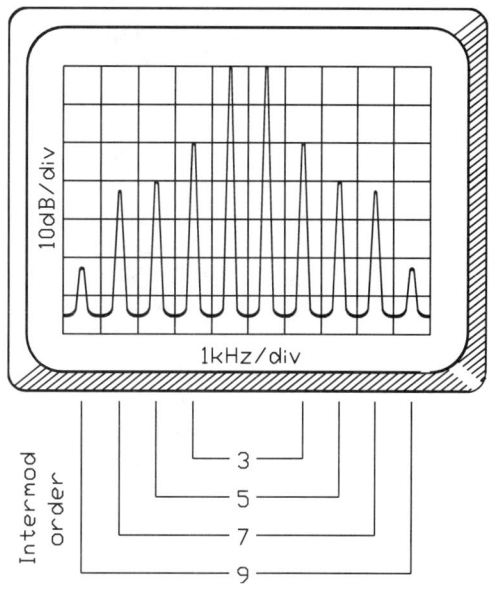

Fig. 6.2. Intermodulation products from a real-life SSB transmitter grossly overdriven by the same tones as in Fig. 6.1. IPs beyond 9th-order will undoubtedly be present too, further increasing the signal bandwidth

TRANSMITTERS, POWER AMPS & EMC

between input and output. You'll recall from Chapter 5 that if two signals at frequencies f_1 and f_2 are fed into an amplifier and increased in strength sufficiently to drive the amplifier beyond its linear range, a set of 'third-order' intermodulation products appears. Their frequencies are $(2f_1 - f_2)$ and $(2f_2 - f_1)$ and they're troublesome because they lie close on either side of the parent frequencies f_1 and f_2. In our example, they will appear on 144·2497MHz and 144·2527MHz – and it's important to note that their level will increase by 3dB for every 1dB increase in the two parent signals. Further increases in the strengths of the two tone signals will produce 'fifth-order' intermod products of $(3f_1 - 2f_2)$ and $(3f_2 - 2f_1)$ on 144·2487 and 144·2537MHz, increasing at an even worse 5dB/dB, followed by 7th-order products which increase even more rapidly, and so on as shown in Fig. 6.2 for odd-order products up to the 9th. The overall bandwidth of the signal on the spectrum analyser's screen has increased to **_nine times_** the original tone spacing of 1000Hz. To make matters worse, there are bound to be even higher-order products off the edges of the display!

Suppose we replace the two tones by speech whose bandwidth is carefully controlled to be 200-2400Hz. In the absence of any intermodulation, the output spectrum would simply consist of a group of radio frequencies spanning some 2·2kHz, the bandwidth of the audio frequencies fed to it. Nothing else – not a scrap. In other words, with a suppressed carrier frequency of 144·250MHz all the RF output would lie between 144·2502 and 144·2524MHz.

As you might be beginning to suspect by now, intermodulation broadens the bandwidth of the SSB signal – which is why it's Bad News for other band users. Bear in mind, by the way, that the intermodulation process usually takes place in the driver and PA stage of the equipment, so the SSB filter bandwidth has very little effect in the overall bandwidth of the eventual signal. Bandwidth is determined by the various kinds of non-linearity exhibited by the amplifier, together with the constantly varying interplay between the range of frequencies which combine to make a human voice. Low-order distortion is not a major problem; for example third-order intermodulation between low-frequency audio components gives rise to products within the wanted upper sideband, which are heard as 'distortion'. They also fall within the area where the lower sideband would be. So an SSB transmitter with poor third-order intermodulation performance but no other defects would produce very distorted speech with a characteristically 'hard' quality, but the transmission would not be particularly wide because the third-order intermod products do not fall very far from the frequencies which produce them in the first place.

Higher-order intermodulation is a different story. If the tones producing intermodulation up to 9th order in Fig. 6.2 are replaced by speech under the same conditions, the width of the transmission would be about seven times the original audio bandwidth – which is an utter disaster for the poor souls who happen to want to be using the band at the same time. When you hear the 15kHz-wide special-event station set up by the local club one fine weekend, with Simon's rig driving Fred's amplifier off Rob's power supply into Bill's antenna and with Walter (who else?) bellowing into the microphone, you're hearing significant levels of intermodulation products up to about thirteenth-order. Not good, but quite typical of what can happen when a station isn't put together properly as a *system*. What's even more irritating is that the wider the signal, the more splattery it sounds. This is because the levels of high-order intermod products are extremely sensitive to the level of drive, *eg* 21st-order intermods (which are spreading something like 25kHz) increase by 21dB for only a 12% voltage change in the AF signal. So the high-order products snap on and off very suddenly, with that characteristic and tiresome sound of buckshot hitting a barn door. Few things in life are as tedious as trying to winkle out some weak DX when a station 30kHz up the band is splattering away on top of you.

CLEAN SIGNALS

As you probably have reason to know, these are real problems – not abstruse and abstract measurements confined to the laboratory. Listen on 144MHz any evening and you're sure to hear someone running an overdriven solid-state amplifier, or a pair of 4CX250Bs which he doesn't know how to handle. You can hardly miss them – they're the ones of which you could say, "Never mind the quality, feel the width". Tune 20kHz either side of the intelligible speech and try to form a mental picture of the intermodulation process which is taking place. Suddenly my little dissertation on intermod generation will make total and awful sense...

Although all SSB transmitters produce some intermodulation products, what matters is how much power goes into them, and how much of *our* band is occupied by *your* signal. So the next logical step is to decide what sort of levels are acceptable. Professional intermodulation specifications for SSB transmitters are much more demanding for seventh-order products and above than for the third and fifth orders so beloved by reviewers, principally because – as we have seen – the majority of the spurious energy from these falls in adjacent channels and would cause interference to other stations. Current requirements for seventh-order products seem to be between –46dB and –57dB for professional SSB equipments, although for third- and fifth-order a frequent requirement is only a modest –26dB. For the amateur service these figures can be taken only as a starting-point; we need something better. Because amateurs do not operate in defined channels when using SSB, and because the dynamic range of VHF and UHF signal levels considered usable by amateurs is likely to be much greater than that encountered in commercial service, it is arguable that even –57dB seventh-order products do not represent a stringent enough design target for our purposes.

Consider the case of two urban amateurs a few miles apart running average 144MHz stations with perhaps 100W PEP (see panel on this page) into a 10dB gain antenna/feeder system. Each is likely to be somewhere in the

---

PEAK ENVELOPE POWER

Peak envelope power (PEP) is *the RMS RF power level at the peak of the modulating waveform*. Let's take this definition apart, understand the pieces, and put them back together.

On VHF and UHF, the RF cycles take about a hundred-millionth of a second. Each sine-wave cycle has its own peak and trough, but when we talk about RF power we're not interested in events occurring inside the RF cycles. RMS (root-mean-square) describes a particular method of averaging power, and when we measure RF power in amateur radio we *always* mean power averaged over very many cycles by the RMS method.

Peaks of speech modulation last for hundreds of microseconds. This is long enough for thousands of VHF/UHF RF cycles, so we can take a good average of the RF power level while the audio peak persists. So that's PEP – *the RMS RF power level at the peak of the modulating waveform*.

PEP is actually very simple. If you have a problem understanding PEP, it's probably because you've already been subjected to other people's misunderstandings. For example, some people speak of 'average PEP', which is simply technical nonsense (though even members of the DTI Radio Investigation Service get confused!). Likewise the number 1·414 ($\sqrt{2}$) has absolutely nothing to do with the subject, unless you're measuring PEP using an oscilloscope. Anyone who talks about 'input power' or 'DC input' is living in the past; modern RF power measurement is universally about *output* power.

The UK licence limits your output power to 26dBW (400W) PEP on SSB, measured at the antenna, on all HF/VHF/UHF bands except 1·9MHz, 50MHz and 70MHz where lower limits apply; check your licence schedule for the details. Since the beginning of 1991, you are also allowed the same PEP output on all other modes including CW – a 6dB step forward for VHF/UHF DXers. Just to be clear about this, the 'peak of the modulating waveform' on CW is simply when the key contacts are closed.

---

region of 80dB above noise with the other, implying that intermodulation products down to about –80dB will be audible and could conceivably interfere with DX signals down around the noise level. For contest stations and others running higher ERP than our urban friends, the demands upon intermodulation performance are greater still. The RSGB VHF Contests Code of Conduct calls for spurious products to be suppressed to –90dB because that's exactly what's needed for peaceful

TRANSMITTERS, POWER AMPS & EMC

coexistence. So the design target for *all* serious VHF/UHF DXers should be something like –80dB for intermodulation products of seventh-order, and even greater suppression for higher orders. That isn't easy; but it *is* very necessary, for the sake of other users of the band and for the sake of your own reputation.

If you're planning to run the full legal limit, these requirements add up to a powerful argument for a grid-driven tetrode linear amplifier. Such an amplifier requires very little drive, allowing the solid-state exciter to operate well within its own linear range: small clean signal in, big clean signal out. Mind you, even with valve power amplifiers, a high standard of intermodulation performance can only be achieved by using high-quality power supplies and taking considerable care with the drive levels and loading. Commercial solid-state transceivers and power amplifiers are only likely to reach levels of –80dB for inter-modulation products higher than seventh-order if their output power level is reduced considerably from the manufacturer's specification (see panel opposite this page). Those containing contemporary hybrid PA modules won't even reach this level, as we've seen elsewhere in this book, although hopefully the situation will change at some point as manufacturers make better components available.

In making these criticisms of solid-state amplifiers, it's important to draw a distinction between the inherent intermodulation performance of the linear amplifier and what happens when it's maladjusted. The best valve linear amplifier ever made will produce splattery signals if it's too lightly loaded, as will the best transistor amplifier if it sees a high output VSWR. And *any* linear amplifier will splatter if it's overdriven or has an inadequate power supply. Even so, some awkward facts remain. Most commercial solid-state amplifiers for the amateur market are extremely non-linear at the manufacturers' rated power levels, and their performance can vary greatly between different samples of the same model. Worst of all, they are chronically overdriven in normal use. It may sound arrogant to say it, but the vast majority of radio amateurs using solid-state amplifiers don't know how to use them properly. That's one reason why the 144MHz band often sounds more like the unacceptable face of 27MHz. The solid-state add-on PA is the worst thing to happen to amateur radio since... (insert whatever *you* think marked the end of the Good Old Days).

It seems to me that the reviewers of commercial linear amplifiers could do much more to raise standards, by presenting more relevant test results and comment. The current practice is to present the results of static two-tone tests of low-order intermodulation performance; typical figures might be –30dB for third-order products and –40dB below the wanted output for the fifths. Considered in isolation, these levels are not seriously anti-social because they are close to the main signal. But it would be far more relevant to measure the higher-order products, up to the nineteenth-order or thereabouts, to exercise the dynamic regulation of power supplies using real speech signals, and also to comment on high-order intermodulation performance under circumstances such as overdrive, mistuning or operating into badly matched loads. Dare I suggest that the majority of reviewers aren't very clear about their criteria in this area?

OUT-OF-BAND SIGNAL SUPPRESSION AND OTHER REQUIREMENTS

Every transmitter ever made produces harmonics and other out-of-band spurious signals, at least to some degree. Although our power levels aren't in the same league as broadcast stations, we commonly use higher effective radiated power than the military and professional services which also operate between our amateur bands. So we definitely do need to take some precautions. The suppression of harmonics and other out-of-band spurious signals can only be checked on a spectrum analyser; if you can't gain access to one, it's best to assume that bandpass and low-pass filtering is required as a matter of course.

The place to put bandpass filters is in the exciter, and further details are given in

CLEAN SIGNALS

LINEARITY AND THE 10% RULE

A 'linear amplifier' gets its name because the graph of output power against drive power is a perfect straight line – the solid line in Fig. A. This line describing the relationship between input power and output power is called the 'transfer characteristic'. For an amplifier to be truly linear, what comes out must be purely a larger-scale replica of the drive signal.

You can easily plot the transfer characteristic of your own PA using a pair of wattmeters and a variable single-tone source (Fig. B1). If your PA has a bypass switch you can manage with only one wattmeter (Fig. B2).

Any distortion of the straight-line transfer characteristic will give rise to intermodulation products. The most common form of non-linearity is gain compression; the transfer characteristic bends downwards until eventually no further output can be achieved, no matter how hard you drive the amplifier (the broken line in Fig. A). Some solid-state amateur PAs exhibit gain compression even at very low drive levels, and almost all are well into the region of severe intermodulation distortion and splatter at the manufacturer's rated output. Valve PAs, on the other hand, tend to be much more linear unless incorrectly biased, tuned or loaded.

A reliable method to stay out of gain compression and keep your signal clean is to follow...

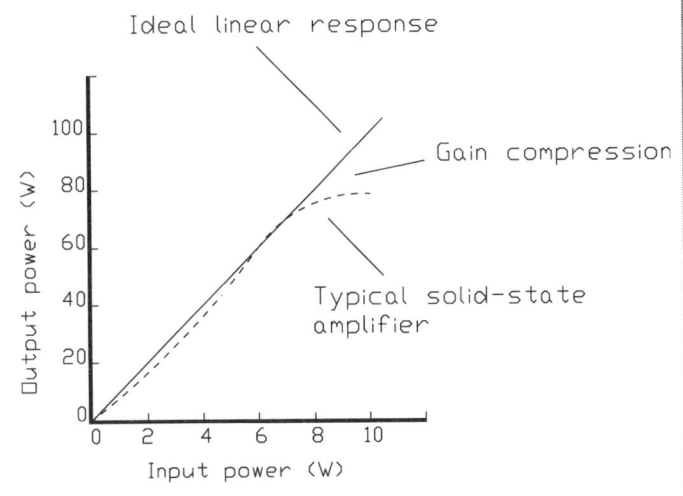

Fig. A.
Transfer characteristic of a perfect linear amplifier (solid line) with a power gain of 10 times. The broken line shows a more realistic transfer characteristic with gain compression at 80W output, and some non-linearity even at low drive levels

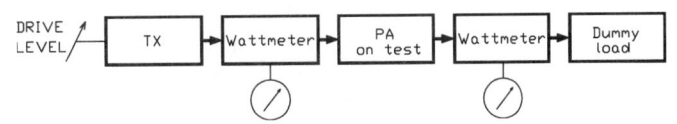

Fig. B. (1) Test arrangement for plotting a linearity curve like Fig. A

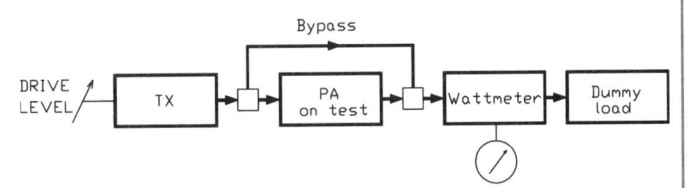

Fig. B(2) Alternative arrangement for an amplifier with a switchable bypass, using only one wattmeter

The 10% Rule:

1 Advance the drive level until the output is hardly rising any more.

2 Back off the drive until the output falls by 10%.

3 Then make sure that the drive level on speech at no time rises beyond the value you've set.

This method will work with any amplifier that is reasonably linear until it reaches a point of gain compression. It won't work if your amplifier has a transfer characteristic that starts to bend even at low drive levels – the broken line in Fig. A shows some tendency to do this. If the problem is serious, your only course then is either to try realigning the amplifier (preferably with a 4lb adjusting tool) or to sell it to a dedicated FMer who doesn't need a true linear and promises never to venture on to SSB...

TRANSMITTERS, POWER AMPS & EMC

Chapters 8–10 and 12. The only spurious signals which power amplifiers can generate (apart from intermodulation products, which are too close to filter out anyway) are harmonics. So there's no point in placing a bandpass filter after your linear amplifier; what's needed there is either lowpass or harmonic notch filtering, or perhaps both. The practice of putting an extremely high-Q bandpass filter or 'high-Q break' at the output of a linear amplifier is particularly dangerous, although several authors who should know better have recommended such practices. The slightest mistuning can result in an extremely high output VSWR, which will certainly wreck the intermodulation performance and can easily destroy solid-state output devices. If you need a filter after your power amplifier – and you probably do – always make it a lowpass filter.

Harmonic filtering is particularly important for 50MHz transmitters since the second harmonic falls in the FM broadcast Band II. Note also that the second harmonic of a 144MHz transmitter falls in a military VHF allocation used principally by aircraft; and higher harmonics fall in the UHF TV Bands IV and V, as does the second harmonic of 432MHz. And of course the third harmonics of 144MHz and 432MHz fall right into our own amateur bands. In all cases, it's a whole lot better to be safe than sorry. You'll find designs for lowpass filters in Chapters 8–10 and 12.

Turning now to other requirements of the SSB transmitter, most of them can be dealt with at considerably less length. Frequency stability and resettability are clearly important, especially if you're proposing to make successful MS and EME schedules; most modern commercial equipments are just about adequate in this respect but only if the various crystal oscillators in the synthesizer and counter have been adjusted accurately to frequency within the last few months. Check your own frequency errors before making skeds.

Some synthesized equipments produce rather large amounts of wideband white noise. Bad cases are easy to detect on another receiver – just key the PTT with the microphone gain turned to zero. Ideally the output should be zero at all frequencies, but you may well be able to hear synthesizer noise close to the nominal output frequency. If you own such a rig, not only will your near neighbours be suffering whenever you transmit but you yourself will also be suffering from reciprocal mixing when your neighbours are transmitting and you're trying to receive weak signals (Chapter 5). Unfortunately there is usually little that can be done to improve transceivers which suffer from this problem. The best plan is not to buy equipment which is known to have a propensity to produce wideband noise. Many earlier Yaesu equipments such as the FT480R and some samples of the FT726 and FT767 are prone to it, for example, and should be avoided for serious VHF/UHF work. Icom equipment is generally much better, and Kenwood somewhere in between – but it can vary between different models from the same maker, or even different samples of the same rig.

CW TRANSMISSIONS

Any investigation of problems involving SSB transmissions should begin by examining the CW signal before adding the complications of modulation and intermodulation. Keyed wideband noise (otherwise known as 'DIY aurora') is sometimes heard on a CW transmission; this is usually because of deficiencies or a fault in the prime mover's synthesizer. However, noisy screen supplies in tetrode amplifiers have also been known to cause this effect: it can be recognized by a somewhat 'phasey' quality and the fact that it is usually gone within about plus and minus 20kHz of the signal. In severe cases the close-in keyed noise is only some 20–25dB weaker than the main CW note! *All* newly-built power supplies need to be checked with an oscilloscope to ensure that they are clean and completely free from spurious noise or modulation, and then checked again when they're actually powering an amplifier and therefore under the influence of RF fields.

The keying characteristics of a CW signal must be beyond reproach, since key-clicks

CLEAN SIGNALS

make a large part of the band unusable by neighbouring stations – and remember that the UK licence also has something to say about the quality of a CW signal. The effects of key-click suppression measures can be evaluated before making on-air tests with the locals, simply by connecting a low-frequency oscilloscope to the output of the diode detector in the station VSWR meter and sticking the latter in the antenna feed. A string of dots from the el-bug will display the keying waveshape, and spikes and sharp edges can easily be observed and corrected. Another useful test is to connect a variable resistor across the key socket, and listen on another receiver for spurious signals at levels between zero and full power. The keyed stages in some rigs (*eg* the FT221 and FT225) can become unstable during turn-on and turn-off, producing transient 'sweepers' which sound like virulent key-clicks even though the waveshape looks perfect on the 'scope.

Unfortunately, many VHF/UHF commercial equipments – perhaps the majority – have poor keying characteristics. The Yaesu FT225 – which with a muTek front end is amongst the very best 144MHz transceivers – is notorious for its very hard and clicky CW, caused by a combination of poor waveshape and the transient instability noted above. The FT221 and FT726 are also deficient in this area. The Icom IC211/251/271 family has keying which, although hard, is acceptable when the transceivers are used by themselves. Some Kenwood products, such as the venerable TS700 series, have excellent keying characteristics; later rigs like the TS711/811, TR9130 and TR751E are usually worse. Although many modifications to keying circuitry have been published, they appear to have very variable effects on individual equipments.

High-speed keying for meteor-scatter is particularly tricky because rise and fall times have to be short, perhaps only a millisecond or two. Good results may be achieved by generating a keyed audio tone whose rise and fall times are carefully controlled, and injecting it into the mic socket. This technique is satisfactory provided the carrier and opposite sideband suppression are adequate; however, it is applicable only to MS operation in which you've chosen your schedule frequencies and transmission periods so that no-one locally needs to operate close to your signal. Keyed tone is definitely **not** recommended for other types of DX CW operation.

However you do it, the keying waveshape will be preserved through a linear amplifier but sharpened by an amplifier in Class C. Since UK licensing conditions relate to power supplied to the antenna, the lower efficiency in Class AB doesn't matter, and the extra on the electricity bill is a small price to pay for helping to keep the bands clean. I would strongly recommend forgetting about Class C forever unless you need to use the amplifier for occasional high-power FM working.

Transverters have been discussed in a receiving context in Chapter 5, and opting to use an HF transceiver as the prime mover is likely to be a Good Move from the CW point of view. Firstly, the keying characteristics of most HF equipments are considerably better than those of most of the radios mentioned above; the designers of HF rigs treat CW as an important mode which should be properly catered for, whereas CW usually seems to be added to VHF and UHF multimode rigs as something of an afterthought. Secondly, most HF transceivers incorporate proper CW filters in the IF stages, which can be worth their weight in gold when there's a good opening to Eastern Europe or during a good aurora. Most VHF/UHF multimodes just use the USB or LSB filter for CW, which can cause you big problems when the band is busy. Incidentally, a potential drawback of using transverters is that the 28MHz low-level outputs of some HF transceivers aren't as clean as they might be, so Chapter 12 includes a design for a suitable filter.

Having looked briefly at all the requirements for transmitters, how well does commercial equipment measure up? The short answer is probably "reasonably well", especially if you use it sensibly and are prepared to make some modifications. But in some areas (in the words of the archetypal school report) there is still "room for decided improvement"; manufacturers definitely "could try harder".

'BAREFOOT' SSB

Many of the problems with SSB transmissions arise from improper use of external linear amplifiers, but even with a basic 'barefoot' rig there is plenty of scope for poor-quality transmissions. As with add-on amplifiers, most of these problems are caused by the operator – yes, *you* – trying to make the rig deliver more RF output than it's capable of.

You get home, unpack the box and plug it all in; it's sitting there in the shack all shiny and new and exuding a faint smell of silicone. You have a tune around – yes, the beacon is S9 and there are some stations on the band. Do you pick up the mic and give them a call to tell them about your new pride and joy? No, not yet. You have some things to do first. First you need to establish the right microphone gain setting for *your* rig, *your* microphone and above all *your* voice.

To do this you need a dummy load and a power meter. If you haven't got a dummy load capable of handling the peak power output of your station, you'd better change your name to Walter; I bet you're also one of those people who whistle and go "Waaaaahlo" on the calling frequency. Come on – if you can afford to buy a VHF rig, or if you're skilled enough to build an amplifier, you have no reason not to possess a dummy load. Buy one, read Chapter 12 and make one, borrow one for now – but for the sake of the rest of us who'd like to use the band occasionally without the station we're trying to work being obliterated by splatter, get hold of a dummy load somehow and *use it*.

Connect the rig to the dummy load *via* some form of power meter. It doesn't have to be a super-accurate laboratory instrument – anything will do at this stage because you're mainly interested in a relative measurement. Chapter 12 contains plenty of ideas for power meters. You should also have some form of single-tone or two-tone source in order to set your drive levels using a constant audio signal. There's no need to collapse your lungs by trying to whistle at a constant level for minutes on end, by the way; take a look at Chapter 12 and save your breath.

Set the mic gain to maximum and apply the tone or whistle. Note the power meter reading and then back off the mic gain control until power level reading reduces by about 10% (see panel on page 6-7). You can then forget all about this control, since its only function is to establish the level at which your audio drives the transmitter. The microphone gain should *not* be used as a general-purpose RF power control; later in this chapter I'll describe some proper ways of winding the power down.

So what's next? It might be worth just checking the VSWR of the antenna before proceeding further. The designer of the solid-state PA in the rig will have worked on the basis that it will always see a load impedance close to 50Ω. Things like the output filters and – most importantly – the ALC system will only work properly if the load impedance is reasonably close to what the designer intended. A VSWR of 1·5 or less should be good enough for this purpose.

Nearly there now – let's just have a quick think about the power supply. If it's internal to the rig it should be all right, but how about fitting a mains plug with integral overvoltage spike protection? The microprocessors in some rigs take offence at spiky mains supplies and are liable to 'crash', and although this is no problem (just switch off, wait a few seconds and switch on again) it can be tiresome – and would be disastrous if it happened in the middle of a short Es opening to SV9. You might also consider some filtering in the main 240V AC line to the shack. Some of the 10–15 amp filters obtainable cheaply at rallies seem to have excellent HF and VHF attenuation.

If your rig – or more particularly an add-on solid-state amplifier – requires an external power supply, you do need to give the subject some careful thought before firing up. The supply will need to be in the region of 12–13·8V and it *must* have excellent regulation right up to the maximum current drawn on speech peaks, which is where the regulation matters the most. Assuming you've

CLEAN SIGNALS

CHECKING THE WIDTH OF A TRANSMISSION

How do you check if someone's transmission is 'clean'? Basically, you're being asked to measure and comment on the intermodulation performance of his transmitter. You do this by using your own receiver as a piece of test equipment – a narrowband tunable microvoltmeter, or if you like, a hand-tunable spectrum analyser. As with any other piece of test gear, in order to give some meaningful answers you need to know its limitations. So you need to test your test equipment first.

It's essential to know how strong a signal you can trust your receiver to handle without introducing various problems of its own as a result of overloading – see Chapter 5. You need to know where this level lies on your S-meter, and also in terms of dB above your noise floor – see Chapter 4. Having thus established your receiver's dynamic range, you'll be in a position to comment on any signals which lie within it.

Find a strong SSB signal which you know to be of good quality. Let's suppose the transmission is on exactly 144·400MHz, and you swing your beam away until the peak signal lies within the range you know your receiver can handle. Now tune HF; you'll hear all sorts of unintelligible gulping noises and the S-meter reading will start to decrease. By the time you get to 144·403MHz you've probably lost all trace of him. Go back to 144·400 and tune away LF. The dulcet tones turn into the usual monkey-chatter and the S-meter reading again starts to decrease, and by the time you get to 144·395MHz you've again lost all trace of him if the transmission is clean.

It may seem from the above figures that his transmission is wider on the lower sideband than the upper, but actually it isn't. The reason for the apparent extra width on the LF side is merely that you started out with your own receiver passband overlapping his upper sideband, so you needed to tune LF by an extra 2–3kHz before your receiver passband totally ceased to overlap the distortion products in his lower sideband. Making this allowance, you can say that you've effectively lost him in 'plus or minus 3kHz'. Any further away than that, and you can't tell that he's on the band – which is exactly as it should be for any signal of S9 or below.

Suppose now that you hear my friend Walter with his FT290 driving a 35W transistor amplifier. You can detect his transmission within something like 10kHz either side. The first thing to check is how strong he is with you. If he's bending the S-meter needle, swing your beam away until he comes down to the level you know your receiver can handle, and then have another tune round his signal. If he now disappears in plus or minus 4kHz, that isn't too bad and you've learned something about the imperfections of your own receiver. On the other hand, if he's only S8 and still 20kHz wide (which is far more likely, knowing Walter) it's time to have words with him.

If you ask someone to check the width of your own transmission, he'll be too polite to do the job properly if you're continuing your conversation with him. So tell him that you're going to count slowly up to twenty and back, which will also give a clear indication of how long and carefully you expect him to tune over each side of your signal. Ask for comments on how soon he loses your signal on the HF and LF sides, and also on carrier suppression and any other odd features. Also try tests each way at a variety of signal levels, and see whose receiver collapses first!

As a technically competent DX operator, your mission is to start a chain of amateurs who know how to comment on each other's signal quality. If we help one another in this way, and each of us trains at least one or two more, our overall level of enjoyment of VHF/UHF DXing will increase dramatically. Poor-quality signals will not persist among operators who *expect* everyone's signal to be good, and are prepared to spend time helping people to get their signals cleaned up.

But even after you and I have sorted out Walter's problems, that won't be the end of the saga. There will always be plenty of newcomers, new equipment and new problems, so Walter never goes away for long. You know what they say:

"Just when you thought it was safe to go back on the band..."

bought or built such a supply, make sure that the leads which connect it to the rig are hefty enough not to drop any volts when high currents are drawn. Unless your power supply is really rock-steady, you and your solid-state PA will find yourselves in deep trouble with other users of the band.

Having checked all those things, now is the time to go on the air and ask the locals who know your voice what they think of it. Your best plan is to ask someone to note your signal level and then to measure the width of your signal (see panel above). At the same time, ask your friend to measure your carrier suppression if he can. Having got your answers, make a note in your log. This means that from time to time you can ask that same station for a routine test (which you are required by your licence to do, by the way) and you can also check the effects of any changes you might make, such as adding a power amplifier.

If you've done everything mentioned above, your transmission shouldn't be over-wide; most modern commercial rigs have reasonable intermodulation performance unless something is amiss. If a new rig is seriously bad, take it back and ask for another. If it's a second-hand rig, try checking the driver and PA standing currents – you can usually do this by removing a link and putting a meter across the pins or connectors. Other than that, one possible cause of peculiar-sounding transmissions is RF feedback – see later.

Just to recap, what points do you need to pay attention to when running your prime mover on its own?
- The match to the antenna (the VSWR) so that the final stage, ALC and output filters can work properly.
- The mic gain – too much and the transmission will sound nasty, although it may not actually spread. **Don't** try using it as a power or drive control.
- The power supply, especially if it's external to the rig.

HOW MUCH POWER DO DXERS REALLY NEED?

So far I've mentioned power amplifiers more or less in passing, but now the time has come for us to do some serious thinking in depth and detail about the various factors associated with these devices. First of all, what do we mean by an "amplifier" in this context? Basically, one or more interconnected units whose function is to take the transmitted signal from the prime mover and deliver a larger and more powerful replica of that signal to the antenna. At present in the UK we may not use a PEP greater than 400W, as measured at the antenna's driven element.

It is pertinent to ask why anyone might need that much power. After all, over 90% of VHF and UHF operators seem to manage quite well with power levels somewhere between 2·5 and 25 watts, and when the band is in good enough shape they can work many hundreds of kilometres. Equally, Chapter 4 placed a high-power amplifier last in the list of ways to improve path-loss capability. So why go to all that trouble and expense?

There are various levels of answer to this question. From the operating point of view, as we pointed out in Chapter 3, high power is no substitute for good operating – but sometimes you need both. If you have to join a pile-up, it's better to get through decisively and be done, rather than sit there contributing to the QRM until your luck turns. From the constructor's point of view, a well-made valve amplifier is a particularly satisfying project because there is no other adjunct to the station which is so easy to make whilst adding so much to the station's potency. A home-built power amplifier represents a very good return on the investment in time and money required to make it.

There are also serious technical reasons for wishing to run high power at VHF and UHF. First of all, some modes of propagation at these frequencies need high power in order to be exploited at all. More accurately, what's needed is high effective radiated power, *ie* both large antennas *and* high power. The challenge of moonbounce is that it is *only just* possible, even for amateurs who are prepared to do everything that's necessary to bridge the half-million-mile path – which includes running the UK legal limit of power, or more if you can obtain a Special Research Permit for this kind of work. Taking another example, there is still a great deal of work to be done with ionoscatter and troposcatter on the VHF/UHF bands, and those modes need lots of power to exploit them. Another interesting point is that high-ERP stations seem to do much better than lower-powered stations during auroral openings, in a way which is not wholly consistent with what would be predicted by reference to theory and textbooks.

Or maybe it's simply because high-ERP operators try harder? In Chapter 1 we made

CLEAN SIGNALS

the distinction between serious and casual DX operation. It amounts to the difference between active and passive – going after the DX rather than waiting until the bands are open wide enough to bring it to you. Serious DXing is often carried out at marginal signal levels, and often calls for full legal power as well as the maximum possible antenna gain.

Anyone can work long distances with milliwatts when conditions are suitable, and good luck to those who choose to do it that way; I have every regard for the serious QRP operator. But casual DXers and QRP evangelists generally forget to mention *who* they worked with their low power when the band opened. The signals which attract pile-ups of casual DXers almost always belong to the kind of dedicated amateurs who take the trouble to build equipment capable of holding up *both* ends of the QSO. In other words, the distant counterparts of your local QRO-and-big-antenna DXer, who in subsequent discussions has to suffer comments of the form "What do you want all that power for? Look what I worked on two watts to a piece of wet string. You only need a few watts at VHF, you know!"

The hard fact is that the stations which casual DXers can only work when the band is open are probably workable any evening by serious weak-signal DX operators running the legal limit. So when the band does open, the serious DXers are taking full advantage of the good conditions to push their range out even further. They're *still* working with weak signals, and *still* need high power.

It's true that a few people do use RF power like a flint club, but they're a minority. The objective of responsible high-power users is to attract the attention of DX, not that of our neighbours. Most of us who'd count ourselves in the QRO category are acutely aware of the effect of amplifier abusers upon the general attitude towards high power – which is precisely why I'm writing this chapter!

To sum up – low power is fine for social radio and casual DX. For serious DXing you need high power at your disposal, and for any kind of experimental work with propagation, you definitely need to *use* it.

WHAT KIND OF AMPLIFIER?

Let's assume that you've decided that the time has come to put more decibels into the DX's receiver. You now have three decisions to make:
1 Valve or transistor?
2 Build or buy?
3 What power level?

In fact these decisions are somewhat interdependent. If you wish to run the full UK legal limit on SSB – which will be 26dBW (400W) PEP plus the feeder loss in dB – you're really going to have to use valves, regardless of whether you buy or build. If we're insisting on *clean* SSB transmissions, the valve/transistor breakpoint presently comes somewhere around 100W PEP. At the risk of irritating sundry manufacturers, this implies using a transistor amplifier rated to give a good 200W flat-out, and carefully under-driving it to obtain good intermodulation performance.

At the risk of being boring, it might be worth reiterating at this point that solid-state amplifiers aren't very good at being *linear* amplifiers. Solid-state amplifiers running off 12–13·8V DC have become very popular on 144MHz and 432MHz, and the 'average' 144MHz station nowadays probably contains one – which is the main reason why the 144MHz band is distinctly dirtier nowadays than it was ten years ago. In simple terms, the transfer characteristic of a bipolar transistor is nowhere as linear as that of a valve. The situation can be improved by using specialized and expensive high-voltage bipolar RF power transistors or MOSFETs in conjunction with fairly elaborate circuitry but, as explained in Chapter 9, this can't be done on the cheap.

One thing is for sure – you won't find such techniques used in commercial amplifiers for the amateur market because they cost more than most amateurs are prepared to pay. And it isn't merely that solid-state amplifiers are inherently non-linear; most people overdrive them as well, and all too many are powered from DC supplies which aren't up to the job. No wonder the bands sound polluted.

I'd firmly recommend derating *all* solid-state amplifiers by at least 50% from the manufacturer's power output figure. If that upsets you, read on...

READING THE CATALOGUES

It's important to be clear about the power ratings of commercial PAs for the amateur market, and one way to approach the subject is by way of rather lower frequencies. Radio-frequency linear amplifiers and high-fidelity audio amplifiers are quite similar in several ways. In both cases we're interested not only in power output but also in distortion (in the form of both harmonics and intermodulation) and how that distortion varies with power. In the world of hi-fi, amplifier power output is always specified at a given distortion figure – in terms such as "50W per channel at 0·01% total distortion". Any hi-fi amplifier is capable of producing more than its rated power if you don't mind the increased distortion, but no reputable hi-fi amplifier manufacturer would dream of specifying the power output of its product without referencing it to a particular distortion level.

Unfortunately the average radio amateur doesn't think of an RF linear as a hi-fi device,

WHAT KIND OF AMPLIFIER?

so the manufacturers can get away without any reference to the amount of intermodulation distortion produced by their so-called "linears" at their rated output powers. To make matters worse, some manufacturers and their dealers completely abandon any claim to professional respectability by specifying their amplifiers in terms of input power – the volts and amps taken from the supply – rather than RF output power, presumably because it makes the product look more 'powerful' and hence more attractive. Given that truly linear amplifiers cannot be much more than 45% efficient, it's obvious that a so-called "500-watt (input) amplifier" may well produce little more than 200W of RF flat-out, and probably less than 200W PEP at acceptable intermod levels.

You've already seen my opinions on acceptable intermod performance, and how test results should be reported. This is an area where someone – preferably national societies like RSGB, ARRL and DARC – should lay down some standards for reviewers and advertisers, for two reasons. One is that claims made by some advertisers suggest that either they should be gently led away by nice men in white coats or their advertisements should pointedly not be accepted by respectable magazines. The other is that there are magazine reviewers who themselves seem none too clear about what to expect from an RF power amplifier.

GOING FOR THE LIMIT

Chapters 8–10 include designs for transistor amplifiers up to about 10W PEP output but – as G3XBY makes very clear in Chapter 9 – more powerful solid-state linear amplifiers of high quality are strictly for the dedicated experimenter who is prepared to spend some serious money. The 10W PEP amplifiers featured in Chapters 8–10 are based on no-tune RF power modules (except on 70MHz for which no modules are available). In spite of the mediocre intermodulation performance of these things, such an approach at least offers less prospect of really appalling signals. There's also some hope that the manufacturers might eventually produce a range of replacement RF power modules with improved intermodulation performance.

Although the 10W PEP level may be adequate for a beginner, a dedicated QRP operator or a casual DXer, you and I probably regard 10W transmitters merely as drivers for something much bigger. Again, let me remind you that if you want to go after the real DX, as distinct from merely accepting whatever DX comes your way, you need more than 10W at your command. Indeed, more than 100W. At these power levels, you may as well forget solid-state and start thinking about valves – or "vacuum tubes" as they say in the United States.

At this point, especially if you cut your teeth on solid-state components, you may experience a sharp stab of pain; your instinctive response may be that those weird thermionic things are emphatically not for you and that financial outlay is called for. Before you raid the coffers, I ought to point out that there are *very few* commercial valve power amplifiers for 144MHz or 432MHz which will reach the UK legal power limit with good intermodulation performance. The *only* ones I'd currently recommend are the American 'Tempo' range which now use 3CX800A7 triodes. Commercial amplifiers such as the Dressler, Nag, Heatherlite *etc* which use a single ceramic tetrode of the 4CX250/350 variety will not – repeat *not* – produce the UK legal limit at anything like acceptable intermodulation levels, for reasons which will become blindingly obvious in the rest of this chapter. And there's no point in anyone, especially equipment reviewers, pretending that they can. Single-4CX amplifiers can perform tolerably well at lower power levels (although some still possess remarkable deficiencies which lead you to believe that their designers never so much as glanced at the valve data sheet) but 'legal-limit' they definitely are not.

This means that for socially acceptable legal-limit power output you only have two choices: buy a Tempo, or build your own. I'm sorry to have to spell it out again, but whatever some of these companies proclaim in their advertise-

TRANSMITTERS, POWER AMPS & EMC

ments, and despite the best efforts of some reviewers to collude with them, the single 4CX250 or 4CX350 simply **will not do the business**.

Valve amplifiers for the UK legal limit are not especially difficult to build. There are several reputable designs for 144MHz and 432MHz which usually present little difficulty from the RF point of view. However, to obtain the performance of which the valves are capable, some care needs to be taken with various aspects of the power supply; I'll be delving into that side of things in Chapter 11. Valve amplifiers for lower powers are also not difficult, although it is tempting to suggest that if you're prepared to go to the trouble of getting a largish valve linear together, you may as well build one that's capable of the full legal limit anyway.

The vast majority of modern designs for high-power VHF and UHF amplifiers use forced-air-cooled ceramic triodes and tetrodes of American origin or design. Since many younger amateurs brought up on solid-state have no background in thermionic devices at all (and the oldsters' knowledge about receiving and small transmitting valves is not all that relevant) let's start at the beginning with an overview of the available triodes and tetrodes for legal-limit VHF and UHF working. I'll list some of their advantages and disadvantages, and discuss in detail how to use them.

TRANSMITTING VALVES AND BASES

We'll take the ceramic tetrodes first, since the majority of both home-brew and commercial amplifiers capable of something approaching the legal limit of power use them. The first so-called "external-anode tetrode", given the type number 4X150A, was developed by the American Eimac company in 1947. As with Eimac products to the present day the initial '4' implies a tetrode and the '150' gives the anode dissipation in watts. 'X' indicates an 'external' anode, meaning that the majority of the anode surface is external to the valve so that it can be cooled by a stream of air blown over it. The letter 'A' simply means that it is the first production version of that particular valve. More recent valves have a four-digit 'EIA coding' as well as Eimac's own 4CX number.

The 4X150A founded a dynasty of external-anode devices. The 4X250B was introduced in 1953, and was similar to the '150A except that it featured ceramic anode/screen insulation

THE 4CX TETRODE FAMILY

Type	Heater (V)	Va (kV)	Pa (W)	Fmax (MHz)	Base	Notes
4X150A	6.0	2.0	150	500	9-pin	Original 1947 valve
4X150D	26.5	2.0	250	500	9-pin	Aircraft version
4X150G	2.5	1.25	150	1250	Coax	UHF/video
4X150R	6.0	2.0	250	500	9-pin	Ruggedized 4X150A
4X150S	26.5	2.0	250	500	9-pin	Ruggedized 4X150D
4X150B	6.0	2.0	250	500	9-pin	Ceramic shell '150
4X250F	26.5	2.0	250	500	9-pin	Aircraft version
4CX250B	6.0	2.0	250	500	9-pin	Classic!
4CX250BC	6.0	2.0	250	500	9-pin	Premium-quality version
4CX250BM	6.0	2.0	250	500	9-pin	Low secondary emission
4CX250K	6.0	2.0	250	1200	Coax	UHF/video
4CX250M	26.5	2.0	250	500	9-pin	Aircraft version
4CX250R	6.0	2.0	250	500	9-pin	Ruggedized '250 with more gain
8930	6.0	2.4	350	500	9-pin	'250R with larger anode cooler
4CX350A	6.0	2.5	350	150	9-pin	Class AB1 only. Very linear
4CX350F	26.5	2.5	350	150	9-pin	Class AB1 only. Very linear
4CX350FJ	26.5	2.5	350	150	9-pin	Class AB1 only. Very linear

WHAT KIND OF AMPLIFIER?

instead of glass, permitting higher operating temperatures and hence the higher anode dissipation. In the following year the glass base insulation was changed to ceramic too, and the device became the familiar 4CX250B, with the 'C' implying all-ceramic construction. And with only slight internal modifications, largely for production reasons, this valve is still very much with us today. It remains an industry standard and users such as the Ministry of Defence, NATO, the Civil Aviation Authority, the Home Office, the Foreign Office and the broadcasters all testify to the long life and ruggedness of the '4CX' in full-time professional service.

Essentially the 4CX250B is a tetrode with an anode dissipation of 250W and an upper frequency limit of 500MHz. In general terms it is delightfully easy to work with, and "a pair of 4CXs" is an excellent choice for a full-legal 144MHz or 432MHz amplifier.

88 & 3CX SERIES TRIODES

Eimac produces VHF/UHF triodes as well as tetrodes, and these are numbered either in the 3CX series (rather like the 4CX family) and/or given a 4-digit code.

The 8873, 8874 and 8875 are high-*mu* power triodes intended for use in zero-bias Class-B amplifiers; they differ only in their anode construction and hence their method of cooling and anode dissipation. The 8874 is like a 4CX tetrode in that it requires axial-flow forced-air cooling; it has an anode dissipation of 400W. The 8874 requires the same order of anode voltage and current as a 4CX250B and can produce rather more output power. Several American commercial amplifiers such as the earlier Tempo models have used 8874s and the like, often in pairs, but there have been relatively few VHF/UHF designs for home construction. It is interesting to note that these valves appear to be seldom used in professional equipment, and it has been suggested that the reason is their rather short life. Users of 8874s have often complained of useful lives of less than 1000 hours, with a long period of poor performance because of low cathode emission. It is not clear whether these problems are due to inadequate design or bad practice by the end users. What is beyond doubt is that these valves are considerably more expensive than 4CX tetrodes to replace, and a 20,000-hour life is by no means unknown for a '250B run at full ratings.

The 3CX series is numbered according to anode dissipation, *eg* the 3CX100A5 has a dissipation of 100W. The 3CX100 family stems from the glass-insulated 2C39 and is occasionally found in amplifiers for 432MHz, but even used in twos or threes the 3CX100 is rather small for our purposes. The more recent 3CX800A7 is becoming popular in amateur service; several amplifier designs have appeared in print, and the 3CX800A7 has superseded the pair of 8874s in the current range of 'Tempo' VHF/UHF amplifiers. Rated to 300MHz, this valve will perform well at 144MHz and even at 432MHz with some careful design and attention to detail. For UHF service the coaxial-based 3CX800U7 would be better still – but at a price, alas.

OTHER VALVES

Other transmitting triodes and tetrodes appear on the surplus market from time to time, and are at least worth a mention here if only to prevent you from repeating other people's bad experiences.

Starting with the tetrodes, and working upwards in anode dissipation, glass-envelope double tetrodes such as the QQV06-40A (aka 5894 or CV2797) and QQV07-50 are far too small for serious power amplifiers. They were never intended for Class-AB1 use and in consequence their intermodulation performance is not good; they can also be unstable at some drive levels, especially if incorrectly mounted through the chassis. This valve family dates back to 1953, and most of the ones you hear on 144MHz sound like original examples! The 8122 family of ceramic tetrodes was introduced by RCA in the 'fifties as a rival to the Eimac products. A few commercial HF amplifiers used them and they sometimes crop up at rallies for reasonable prices. On paper they look an attractive alternative to 4CX250Bs but in practice they seem to have an amazing propensity to flashover and self-destruction. So avoid the 8122 series unless you're looking for a sure-fire way to test your flashover protection circuitry. The 4CX600J/JA/JB family was developed by Eimac some years ago, principally for very linear Class-AB1 applications; their intermodulation performance is incredibly good although you mustn't run them with grid current. Expensive, but highly recommended.

If they ever increase the UK power limits to the levels used in other civilized countries, we could think about some of the larger tetrodes. The 4CX1000A is a pig to get going on 144MHz unless you have a suitable base and are also very clever or very lucky, although the more modern 4CX1500B is much more tractable. You can run even two in a W1SL configuration if you wish to solve persistent pile-up problems and have a suitably mammoth HV transformer. Other large tetrodes occasionally encountered in the UK are the RCA 7213/7214 and its British cousin the DOD006, all having anode dissipations of 1500W. The 7213 and 7214 will perform well at both 144MHz and 432MHz, but the less rare DOD006 is tricky to get going above 144MHz. Even larger tetrodes are to be found in amplifiers of some Continental moonbouncers...

As with the tetrode family, larger triodes are occasionally used by amateurs. The 3CX1000A7/8283 has occasionally been seen on the surplus market; with an upper-frequency limit of 220MHz, it might be worth trying on 144MHz if a suitable 'breech-block' base could be found (or made) for it. The YD1332 and YD1336 are sometimes available as surplus and, being intended for ultra-linear TV transposer use in Class AB1, they will deliver a very clean legal-limit SSB signal on 144MHz or 432MHz. The 8877/3CX1500A7 is a 1500W anode-dissipation triode designed for Class-AB1 or B use, and is popular wherever in the world the licensing conditions allow the power levels it can produce on the amateur bands from 1·8MHz to 146MHz. Its coaxial-based cousins, the 8938 and 3CX1500U7/8962, are much sought-after by the 432MHz EME brigade.

TRIODES VERSUS TETRODES

Having considered both families, which should you go for? Let's look at the advantages and disadvantages of each.

Tetrode advantages
- Low (theoretically zero) driving power required for a grid-driven tetrode, making it easy to achieve good linearity from the prime mover.
- Potentially better linearity than similar triodes.
- Probably much longer life, assuming the valve wasn't worn out before you got it.
- Probably easier and cheaper to obtain than similar triodes.
- Bases probably easier to obtain than for equivalent triodes.
- Many published amplifier designs.
- A few good published power supply designs.
- Simple to tune and load; the screen current meter tells all.
- Internal feedback much lower because of shielding action of screen grid.

Tetrode disadvantages
- Needs a more elaborate power supply and control unit.
- Many surplus tetrodes are suspect even though they look new – see later.
- May need neutralizing, which for some reason terrifies people.
- Prone to flashover, especially when getting a little tired and too lightly loaded – can destroy valve, base and a fair bit of the PSU unless proper protection is built-in.

Triode advantages
- Simpler power supply, metering etc.
- A cathode-driven triode is inherently very stable and easy to set up.
- Some types can use very simple bases and home-made grid rings; no screen bypass capacitor required.

Triode disadvantages
- Some triodes have a short life relative to equivalent tetrodes, and tend to cost more to replace.
- Intermod performance of smaller triodes (8874 *etc*) is not brilliant.
- Fewer published designs for modern VHF/

WHAT KIND OF AMPLIFIER?

'Tempo' range use triodes and work well. They will deliver the UK legal limit of power with quite reasonable intermods, so if you want to run the full whack but don't want to build, big triode amplifiers are presently the only way. You do, of course, need the collateral.

Altogether, these add up to excellent reasons for building your own legal-limit linear for 144MHz or 432MHz. A pair of 4CX250Bs is easy to get going, provided you take a modicum of care and are willing to pay attention to detail. They'll repay you by producing the legal limit with no trouble at all, and once you've got it all working to your satisfaction the amplifier should run for years and years with no fuss.

BASES FOR THE 4CX FAMILY

If you do decide to build a valve amplifier using any of the devices discussed so far, you'll need to give some thought (and probably some cash, unfortunately) to the base into which the valve is plugged. To those brought up on small valves, a valveholder is just something you bought for a few pence or salvaged from another piece of gear. Unfortunately, it isn't that simple when it comes to ceramic triodes and tetrodes. As I've already said, high anode dissipation in a reasonably small size is bought at the expense of forced-air cooling, and this means that the valve base must be physically constructed to allow air to be blown through it. The screen grid of a tetrode also needs an extremely low-inductance bypass capacitor, which must form part of the base itself. The final complication is that the base pinout of the 4CX series is non-standard – the pin layout is ordinary B8F but the control grid is brought out to the centre spigot. All this means that if you want to use a 4CX at VHF or UHF, you need a special base. You're also likely to need a chimney, which is a ceramic or PTFE tube that slips over the valve anode and fits on to the base. Its function is simply to duct the air from the blower through the anode finning in the proper fashion. A few amplifier designs mount the valve in such a way that a special chimney is not required, merely an exhaust tube which

Fig. 6.3. 4CX250R valve in Eimac SK600 base, suitable for frequencies up to 144MHz, with SK606 chimney

UHF triodes compared with those available for 4CX tetrodes.
- If required, special bases are poor value and difficult to get.
- Triodes need a lot of driving power; the average VHF/UHF rig doesn't have enough power to drive a cathode-driven triode to anything like full power, so an intermediate amplifier is usually required – more clutter in the shack, more expense, worse intermods...

So what's the verdict? Personally, I'd take the 4CX tetrodes every time, basically because they offer a combination of performance, convenient drive power, long and consistent life and relative cheapness. However, many American commercial amplifiers such as the

TRANSMITTERS, POWER AMPS & EMC

Fig. 6.4. Eimac SK630 base with shielded screen bypass capacitor, suitable for 432MHz and below

can be rolled by hand from PTFE sheet.

Taking the small 4CX tetrodes first, there are only two regular sources of these bases and chimneys in the UK. One is Eimac (marketed through EMI-Varian), who manufacture a wide range of bases with various pin configurations and appropriate values of screen decoupling capacitance, together with the proper chimneys for each base. The other is the surplus market, which occasionally comes up with bases made by Associated Electrical Industries (AEI) in the 'sixties. These are excellent if you can find them – they have "AEI" written on the top and three small lugs on the upper shell into which the chimneys twist and locate.

If you need to buy new bases for 4CX250-type valves, the Eimac part numbers for use on frequencies up to 146MHz are the SK600, SK600A, SK610, SK610A and SK612, which differ only in having more or fewer of the cathode and heater tags already grounded to the metal body. It's easy enough to ground the cathode and heater pins by stuffing the gap between the tag and the body with tinned copper braid and running in solder with a big iron. For all these bases, the Eimac chimney you need is the SK606 (Fig. 6.3).

For 432MHz you need a different base, because the screen bypass capacitors in the bases discussed above have too much series inductance for proper stability at UHF. For a 432MHz amplifier, therefore, you need the SK620, SK620A or SK630 (Fig. 6.4). These have an internal capacitor of 1100pF as opposed to 2700pF in the lower-frequency versions, but its series inductance is approximately a factor of 3 smaller. Any of these sockets will in fact work on all the VHF/UHF bands down to 50MHz. The chimney for the SK620 and SK630 is the SK626.

The crunch comes when you find out the prices of these nice bases and chimneys. Would you believe that, if you buy new, the bases cost more than the valves? It's difficult to quote prices because it depends where you go; and regardless of the fortunes of the pound and the dollar, UK prices only seem to rise. In general terms, however, you're likely to be paying in excess of £90 for a base and £20 for a chimney at 1992 prices.

Is there an alternative? You can't make one because the screen bypass ring is just too tricky. You still might find the AEI bases, but they haven't been made for many years and there can't be many left unsold. There were two main families of AEI base, for HF and VHF use, which look very similar but only the VHF version has a built-in screen bypass capacitor (you have to look closely to see the mica sheets on the inside edge). The VHF types work well at 144MHz, and seem satisfactory in single-valve designs for 432MHz. But *don't* try to use AEI bases in the K2RIW 432MHz amplifier (Chapter 10). It's impossible to stabilize the beast unless you have access to a network analyser and know exactly what you're doing. The K2RIW *needs* SK620s or SK630s, and that's that.

BASE DESIRES

There's still a lot of confusion about 4CX bases, despite the fact that the subject's been given a good airing in print in recent years. But some of the cowboys at rallies have read and understood only too well, and have learned to offer you the most unlikely-looking *objets d'art* with comments along the lines of "Oh yeah, mate, this is a genuine UHF base". Be warned – if you want your VHF or UHF amplifier to work well, go for the Eimac or AEI

WHAT KIND OF AMPLIFIER?

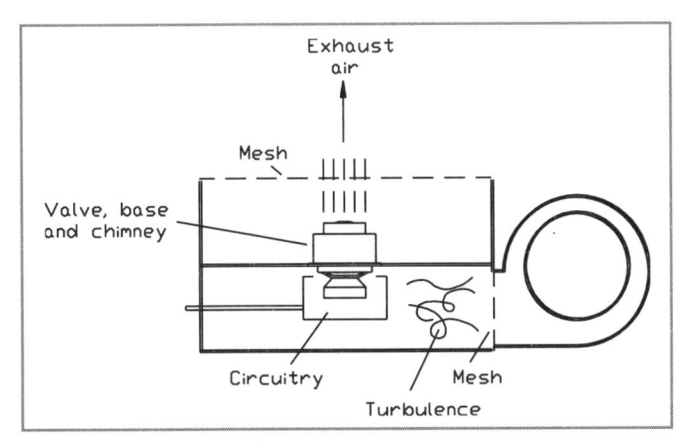

Fig. 6.5. As well as the valve in its airflow socket and chimney, sources of pressure drop in an airflow system include screening mesh, obstructions, turbulence and sharp changes of direction

article and accept no substitute. One particular kind of 4CX base is very often seen at rallies. It consists of the usual pins at the bottom and a receptacle in which the valve sits, with a Y-shaped clamp across the top of the anode. It's offered as a "UHF base" and indeed it is – it's out of a 1962-vintage UHF airborne barrage jammer which used four 4CX250Bs and was deliberately designed to be unstable! Because of all this confusion, unsuitable bases may be offered for sale via the small ads in totally good faith, so always find out exactly what they are before parting with your money.

Bases for the small Eimac triodes are actually made by the E.F. Johnson company in the USA and marketed by Eimac, but they can be very poor value at UK prices. Fortunately triode bases don't require a screen bypass capacitor, so you can actually make your own using phosphor-bronze finger strip for the grid ring, and contacts salvaged from old connectors for the heater and cathode pins.

FORCED-AIR COOLING

On a couple of occasions so far I've referred to these ceramic triodes and tetrodes as 'forced-air-cooled'. All this means is that some form of blower or fan is used to drive a cooling airflow through the around the anode – basically so that it can dissipate lots of heat while allowing the internals of the valve to be small enough to work well at VHF and UHF. We need to think about cooling before considering specific amplifier designs, because you can't get the metalwork together until you've decided on the blower you need.

The folk wisdom about bases for 4CXs is exceeded only by that about blowers, passed on from generation to generation of amplifier builders. The usual approach is to ask Fred at the club what he used when he built his big PA and then perhaps to pay him a visit to have a look at it. Next comes a trip to the local rally; blowers of various shapes and sizes are common enough at these gatherings and the intending amplifier-builder buys one which looks about the same size as old Fred's. He takes it home and hooks it up to the mains – sure enough, it produces a miniature whirlwind in the shack, blows all the QSL cards off the wall and frightens the cat. With a vague memory of having read somewhere that it's important to see what happens when there's a bit of resistance to the airflow, he puts his hand over the outlet, leaving a small gap through his fingers. Not much air seems to come out and the blower sounds as though it's slowed down rather drastically, but even so he reckons he's got it made. Six months later the amplifier is finished and he switches on the blower and thinks "Funny, there doesn't seem to be much air coming through these valves. I'm sure there ought to be more than that. Maybe there's more to this blower business than meets the eye." There is – quite a bit more. He's probably about to find that the correct blower for his new amplifier is too big to fit inside the cabinet...

To avoid that happening to you, let's start again at the beginning. The data sheet for the Eimac 4CX250B has the following interesting things to say about the airflow required:

"Sufficient forced-air cooling must be provided for the anode, base seals and body seals to maintain operating temperatures below the rated maximum values of 250°C. Air requirements to maintain anode core temperatures at 200°C with an inlet air temperature of

TRANSMITTERS, POWER AMPS & EMC

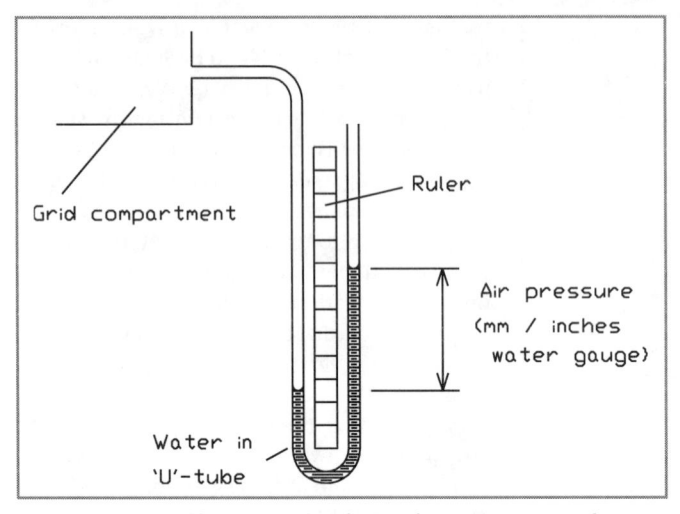

Fig. 6.6. A simple U-tube water gauge for measuring air pressure

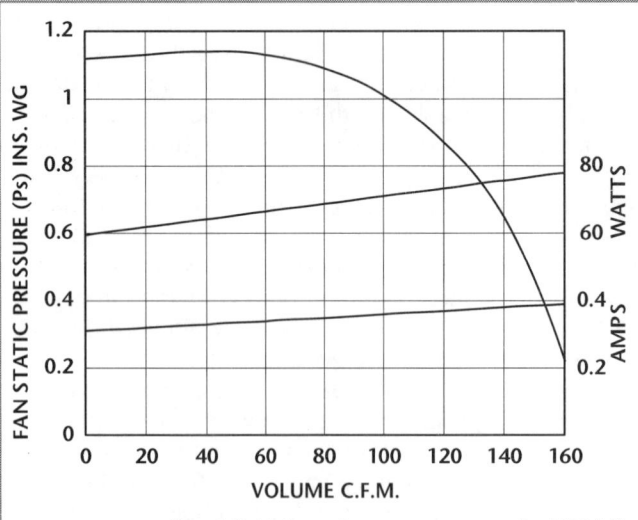

Fig. 6.7. Volume/pressure curve of a VBL5/3 blower. Motor power and current consumption are also shown. (Reproduced by kind permission of Air Control Installations (Chard) Ltd.)

Air Flow Requirements for 4CX250B
(at Sea Level)

Anode dissipation	Airflow (cu.ft/min)	Back pressure (inches of water)
200	5.0	0.76
250	6.4	0.82

"The blower selected in a given application must be capable of supplying the desired airflow at a back pressure equal to the pressure drop shown above plus any drop encountered in ducts and filters. The blower must be designed to deliver the air at the required altitude".

In other words, before you can choose a blower for your 4CX amplifier, you need to know something about the job it is trying to do. An airflow rate of 5 cubic feet per minute (cfm) doesn't sound like much – you probably move twice that amount when you languidly fan yourself on a hot day. But what's this "back pressure"?

Back pressure is the resistance presented to the flow of air by obstructions in its path. In this case the back pressure is caused by all that close finning within the anode structure of the valve, which represents a substantial obstruction to the free flow of air. Think of it as an electrical circuit. The blower could be likened to a battery and the anode fins to a resistor connected across it; the 'potential drop' across the 'resistor' represents the back pressure. The sum of the back pressures in the air system – represented by the valve or valves together with any back pressure presented by the base, the chimney, bends and turbulence within the airtight chassis, RF screening mesh and so on – must all add up to the total pressure of air supplied by the blower (Fig. 6.5)

One way of measuring air pressure is in terms of the height of the water column that the pressure can lift. This means that pressure can be measured using nothing more technical than a length of transparent plastic tubing, some water and a ruler (Fig. 6.6). A back pressure expressed as "0.82 inches of water" would produce 0.82" of displacement between the columns of water. (For comparison, a pressure of 1lb per square inch corresponds to 27.7" of water.) Looked at in another way, the

50°C are tabulated below. These requirements apply when a socket of the Eimac SK-600 series and an Eimac SK-606 chimney are used with air flow in the base to anode direction.

WHAT KIND OF AMPLIFIER?

figures in the data sheet suggest that a 4CX250B presents about 0·82" of back pressure to an airflow rate of 6·4 cubic feet per minute. For a two-valve amplifier you require double the airflow rate but the back pressure stays the same.

The problem with small blowers is that they aren't very good at generating the pressure required to deliver enough air to cool a 4CX, which is where so many people come unstuck. Going back to the electrical analogy, a blower is a bit like a power supply with a high output impedance: the output voltage looks fine without any load but it sags as soon as you ask it to deliver some current. Conversely, choosing a blower on the basis of the gale it creates when discharging into free air is like choosing a power supply on the basis of its short-circuit current; not very useful. Unless the blower is designed to generate sufficient air pressure it won't cool your 4CXs. So you need to **design** your cooling system.

As a first step, look at the specifications of the blowers you might consider using. What you want to know is how much airflow the blower delivers at a given back pressure. Blower manufacturers supply this information in the form of a graph – by way of an example take a look at Fig. 6.7, which shows the output characteristic of the Air Control Installations model VBL5/3. Airflow rate in cubic feet per minute (cfm) is plotted against 'fan static pressure' – basically the air pressure the fan can produce. Could we use this blower to keep a pair of 4CX250Bs cool? Well, first of all, let's examine the requirements again. The specification says that at 250W anode dissipation the valve needs 6·4cfm through it, and at that flow rate the anode cooler produces 0·82" of back pressure. That's just the valve, though, and a good rule of thumb for the '250 series based on experience and a lot of messing about with pressure gauges is to assume that the valve, base, chimney and other impedimenta in the grid compartment (Fig. 6.5) all add up to 1" of back pressure. For a pair of '250Bs that means we need 12·8cfm at 1".

Taking the VBL5/3 graph again, you'll see that at 1" static pressure the blower is producing 100 cubic feet per minute, whereas we said that we needed 12·8. That's always the way of it: any blower capable of generating enough pressure for the 4CX series of valves will deliver far more than the minimum airflow requirements, even for two valves. In any case, plenty of cooling air is a good thing. Professional transmitter manufacturers tend to allow something like a factor of 3 in this area to cater for things like low mains, minor blockages in the anode cooler and so on; in this particular application about 40cfm would be a reasonable minimum figure to shoot for. However, valve life depends very much on the operating temperature, so using a VBL5/3 blower with a pair of 4CX250Bs would be a very good move from that point of view. When you're doing initial tests into a dummy load you may well want to run the valves at more than 250W anode dissipation for a period, and they can cope with this if you blow them hard. Basically, you want to blow any ceramic triode or tetrode as hard as you can, provided your headphones and microphone will keep the noise out and the valves don't get blown out of their sockets!

The two major sources of blowers in this country are Airflow Developments Ltd of High Wycombe and Air Control Installations Ltd of Chard, Somerset. Both will be delighted to supply you with catalogues and data sheets, and you'll probably discover that the blower you need to do the job properly is much larger than the one on Fred's linear. For instance, a common blower found at rallies is the Air Control Installations VBL4/3 – one down from the 5/3 we discussed a moment ago. You might think from looking at it that it was just the job, but it can't get anywhere near 1" static pressure – and sure enough, if you try one of these on a 4CX250 linear you'll find that almost no air comes out of the anodes. A new VBL5/3 cost about £47 in 1992, and an Airflow Developments Type 45CTL (also good for about 100cfm at 1" static pressure) was about £40. Isn't it worth that much to know that you've done the cooling properly?

If you're absolutely determined to go to a rally and pick up a cheap blower, here's what to look for. First of all, axial Muffin-type fans are no use whatsoever; they are designed only to move air and generate hardly any pressure. You're looking for a centrifugal 'snail' blower with a wheel of at least 10cm (4") internal diameter. Expect to see a hefty induction motor, and avoid anything rated below 2000rpm. Pressure-generating capability depends on the square of the speed of the blower blades, *ie* on (diameter x rpm)^2, and is severely affected by a shortfall in either the diameter or the rpm.

When you see the size of a proper blower for a pair of 4CXs, you'll realise why all commercial amateur-band 4CX amplifiers get hot – they're **all** grossly undercooled. One manufacturer of a single-4CX amplifier fits a Muffin-type axial fan, claiming that centrifugal blowers are "...too noisy for consumer use". A 4CX250B in that amplifier seems to have a life of somewhere around 100–200 hours, which must be something of a record low. Like it or not, part of the price of having a 250W anode dissipation tetrode (or triode) which can work efficiently at 500MHz is that you need a large blower to keep it cool. Moving several tens of cubic feet of air a minute through an inch of back pressure is an inherently noisy process. Put it this way – if the blower isn't noisy, it probably isn't big enough!

About a third of the cooling air requirement of a ceramic tetrode is to cool the grid spigot and seals on the valve's base, leaving the other two-thirds to cool the anode. Certain amplifier designers seem to assume that they can cut down on the blower requirement by either blowing air backwards through the valve and thereby avoiding all that clutter at the bottom, or alternatively blowing the air the right way but ducting through the anode only. According to Eimac's original data sheet, you must blow a 4CX only in the base-to-anode direction, with air blowing on to the base seals as well as through the anode. Having said that, there is plenty of favourable amateur experience with the K2RIW airflow system (Chapter 10) which blows air into the output compartment, cooling the tuned circuit before most of it passes out through the valve anodes and up a chimney. Part of the airflow is bled downwards through the valve socket and out via a restriction in the grid compartment, and this is where the weakness lies: there is no air blast directly on to the base seals. The 'RIW supporters reply that because the airflow paths up through the anode and down through the base are in parallel rather than in series, this reduces the back pressure and allows the blower to deliver more air to cool the valve as a whole. Maybe so... it's your decision. By 1992 Eimac had rather confused the issue by admitting that they saw nothing wrong in principle with the K2RIW method of cooling, to which I would reply that I've personally wrecked three 4CX250Rs by using it.

Blowers can fail, and their power leads can drop off or break. Maybe I've been unlucky, but in the last eighteen years I have had one blower which seized solid, another in which the fan became uncoupled from the motor shaft and a third which had a fatigue fracture of an internal wire to the motor – and all failed when the associated amplifier was in use. You might think it would be impossible to miss the deafening silence when the blower stops. But if you're wearing headphones and you're out on a contest, with the wind howling and rain lashing against the side of the tent, it's perfectly possible to miss the fact that the blower has stopped and the amplifier is quietly doing a China Syndrome. It's easy enough to provide some form of airflow switch or sensor which will shut down the whole amplifier if the air ceases to flow. Try the local industrial central-heating boiler spares and repair depot for adjustable low-pressure switches, or make your own by adding a vane to a low-torque microswitch.

One more thing: before you commit yourself to expensive metalwork, spend a pleasant evening building a cardboard mock-up of the proposed airflow system. It doesn't have to look pretty, but it should include all the main features of the airflow path: the valves in their bases, the mesh used for RF screening, and any lengths of air ducting. Power up the blower and check that the water gauge shows sufficient air pressure.

WHAT KIND OF AMPLIFIER?

WHICH AMPLIFIER? WHICH VALVE?

Let's see if we can define what we're looking for in an amplifier. How about this:

"The amplifier I will build is one which is capable of generating the maximum power allowed by my licence. At that power it will have third- and fifth-order intermodulation performance which is better than 26dB below one tone of two at maximum power and preferably quite a bit better. All other intermodulation products will be at least –55dB and preferably much lower than that. Far away from the main frequency, they must roll off to better than –90dB."

If you prefer something simpler:

"The design target for my amplifier is an intermodulation performance 10dB better at the full legal limit than my prime mover is at 10W output".

Or to put it yet another way:

"My SSB transmission must be *narrower* at the full legal limit than it is when I am running barefoot".

These aren't perfect specifications but they do give us some targets to aim for. In the UK, the legal-limit output power that you're going to require from the amplifier is 26dBW plus whatever the feeder loss is in dB. For a 1·5dB feeder loss, a reasonable figure in a high-class DX-chasing station, that means that your amplifier needs to generate 560W while still maintaining the good intermod performance we're aiming for.

A single 4CX250 or 4CX350 won't produce *linear* output at anything near that power level, and don't you believe anyone who says it will! Extensive measurements on various test amplifiers have been taken over the years, and here are some PEP output levels which can be achieved on 144MHz from a single valve. These were measured with reference to the conditions outlined in the first specification above, namely third- and fifth-order intermodulation levels products better than –26dB and anything above that being kept at better than –55dB below one tone of two. At all times the working conditions of the valves remained within the limits given on the manufacturer's data sheet. These figures were established using virtually brand-new valves operating under the best possible conditions: a very high-grade power supply with stabilized HV, and two professional signal generators and amplifiers followed by a high-grade combiner.

Maximum linear output – one valve

4CX250B	285W PEP at 2000V
4CX250R	310W PEP at 2000V
4CX350A	290W PEP at 2000V
	370W PEP at 2500V
4CX350F	280W PEP at 2000V
	360W PEP at 2500V

These figures clearly show why I'm very much in favour of two-valve amplifiers, at least if we're confining discussion to the 4CX250 or 4CX350 family. Building a tetrode amplifier represents quite a lot of work if you plan on doing the job properly, and given that a single 4CX won't produce the legal limit with reasonable intermod performance whereas a pair will do it comfortably, there doesn't seem to be much point in going for a singleton.

The results for the '250B and '250R may be a lot lower than you expected. Remember that the main constraint is the stringent intermodulation level we are defining, especially for high-order products. The 4CX250B and '250R were not designed primarily for linear service and to some extent it shows. However, they are rugged, reliable, available and easy to get going – and they're still vastly better than transistors. The '250R gives a bit more power output because it has a slightly different cathode from the '250B and some mild internal differences; it also has a shade more gain than a '250B, especially at 432MHz. For SSB use, the 4CX250R is a somewhat better valve all round.

Note the very similar power outputs obtained for '250s and '350s for the same intermod performance at 2·0kV anode voltage. The '350s only deliver higher power if you take advantage of their 2·5kV anode voltage rating. This makes one wonder why some commercial amplifier manufacturers offer them as an extra-cost option in an amplifier designed (if that's the right word) for a 4CX250B and with no other changes to suit the '350. The 4CX350 was designed for linear service in Class AB1, and is capable of extremely good high-order intermod performance, *but only if the power supply is good enough to bring out that performance.*

These power output figures mean that you need to use a *pair* of valves to achieve 500W PEP output at the intermod performance which we've specified as the *minimum* desirable – and please note that it's a pretty modest minimum, not some high-grade unattainable pie-in-the-sky figure. If you have the choice and the money, go for 4CX350s at 144MHz. The 4CX350FJ is the best, closely followed by the F or the A. However, do bear in mind that in order to get the performance of which these valves are capable, you'll need to build a very high-grade supply for the screen and control grid, and keep the anode voltage up around 2·5kV with good regulation. If you can manage all this (and it isn't too difficult – see Chapter 11) you can achieve better than –80dB for all intermods beyond seventh-order; if you can't, there's no point in bothering with 4CX350s. Also remember that '350s cannot be used in Class C (which is no loss) and they really don't want to know about the 432MHz band.

Overall, the best buy for either 144MHz or 432MHz is probably a pair of 4CX250Rs. Next best would be a pair of 4CX250BMs if you could find them; they were a special version characterized for low secondary emission and they sometimes show up new or slightly used. After that it's a pair of 4CX250Bs. For 144MHz only, a pair of 4CX350s or '350FJs will do a superb job if the power supply is up to it.

QUIRKS OF THE 4CX FAMILY

A tetrode such as the 4CX250B contains four electrodes: a heated cathode, a control grid, a screen grid and an anode. In all the amplifiers I'm going to describe, the cathode is simply connected to DC and RF ground but the other three electrodes need some kind of DC voltage applied. Before we go further into that, we need to look at the most peculiar feature of the 4CX family – the dreaded 'secondary emission'. This is what makes screen current meters do funny things (ending in an almighty bang if you're not careful), and makes the design of the screen supply about fifty times more complicated than most designers appreciate.

What's supposed to happen in a tetrode is that electrons emitted from the cathode are controlled by the electrostatic field of grid 1 and are whipped away towards the anode by the positive voltage on grid 2, the screen grid. Most then pass through grid 2 to be collected on the anode proper. The balance between electrons intercepted by grid 2 and by the anode depends on the design and internal geometry of the valve, and also on the relative voltages of the screen and anode. When the anode is very positive, either because the DC voltage is high or because we're at the positive peak of the RF cycle, the screen grid intercepts very few electrons. On the other hand, if the anode voltage is not much greater than the screen voltage, a considerable proportion of the electrons emitted from the cathode will be collected by the screen grid.

An electron emitted from a cathode at 4CX250B-type temperatures and attracted towards the anode and screen by a positive voltage on both is doing about 50,000 miles a second by the time it hits the anode. When a stream of electrons moving at this speed smashes into the anode, it dislodges electrons from the anode material itself; this phenomenon is called *secondary emission* and it results in a negative contribution to the net current. In a tetrode there is also secondary emission from the screen and the fate of these secondary electrons depends on the instantaneous anode and screen-grid potentials.

It is extremely difficult to design and manufacture a tetrode which is physically small enough to work at VHF/UHF, has low inter-electrode capacitances, is rugged and reliable, and yet is capable of extremely high power for its size. The trade-off for these good features of the 4CX series is that the overall screen-grid current can be positive at some drive levels and negative at others, and is

WHAT KIND OF AMPLIFIER?

*hyper*sensitive to the applied DC voltage.

This poses lots of problems to the designer of the screen supply. The 4CX family needs between 250V and 350V on the screen depending on what you're doing with them; in linear service you'll probably do best with 350V. For good linearity the screen supply needs to be very well stabilized, preferably to within a small fraction of a volt because 4CX tetrodes are very sensitive to screen-voltage changes; in stately technical language their grid-to-screen transconductance is very high. In CW operation the anode, screen and control grid currents remain steady so just about any screen supply could cope. But with complex SSB modulation, the anode and screen currents swing around all over the place. Most of us (but not Walter) are familiar with the fact that the needle of the RF wattmeter or anode-current meter does not indicate the true peak reading. Less familiar is the fact that although the screen-current meter may appear to move only a few milliamps, it's actually swinging between something like plus and minus 50mA with audio transients.

See the problem? The output of the screen supply needs to be about 350V (the exact voltage isn't critical) but it needs to be held constant to within something like ±10mV, even though one instant it's supplying 50mA and then moments later it's required to sink a similar amount coming back into it due to secondary emission. That's why really good screen supplies for the 4CX family need to be more complicated than you'd expect. The excellent fifth-order intermod performance of the 4CX350A and FJ (with 2·5kV on the anode and 350V on the screen) deteriorates by about a dB for every 10mV of variation in screen voltage. 4CX250Bs and Rs start out with higher levels of intermods but are slightly more tolerant of screen-voltage fluctuations; the deterioration is more like 0·5dB for the same 10mV variation. So if you want to get the best possible intermod performance from the 4CX family, steam-age screen supplies containing Zeners or VR tubes have to be ruled out forthwith.

Before I leave the topic of screen-supply design for now (it's considered in great detail in Chapter 11) we ought to take note of one last tiresome fact. The 4CX family's propensity for secondary emission tends to worsen with time, especially if the valve is allowed to get too hot. Basically this is because small quantities of barium and strontium from the cathode's emissive coating tend to migrate on to the screen grid. Since this gets pretty hot when the valve is producing large quantities of RF, the material on the screen emits free electrons, resulting in yet more negative screen current. Old age and overheating at any period in the valve's history will make matters worse, and with incorrect loading a particularly tired specimen may produce more peak negative-going screen current than the screen supply can sink. The result will be a rapid increase in screen voltage, probably followed by a flashover between anode and screen. This is the most likely reason for small 4CX tetrodes to be pulled from professional transmitters and appear on the surplus market, so beware...

For that reason, 4CX250s of unknown vintage can be a bit of a handful unless you're familiar with the problems which can arise, and your power supply can cope with their nasty tendencies. In particular, you can't really set up a new amplifier properly unless the valves are behaving according to the manufacturer's data sheet. As we shall see, you'll be relying on the screen-current meter as an indicator of correct loading. If the readings are heavily affected by screen emission, the amplifier will end up seriously under-loaded, spreading plus and minus 30kHz and in grave danger of flashover on every speech peak.

COMMISSIONING A NEW AMPLIFIER

Let's now fast-forward through the detailed legal-limit power amplifier designs in Chapters 8 and 10, and assume that your new Thunderbox is absolutely complete. The HV supply and the main power-supply and control unit have been separately tested and, as far as you can see at this stage, all the right volts are appearing in the right places at the proper times. You can still smell the paint you used for the front panel – and your fingers still ache from terminating multiway screened cables with those horribly fiddly Cannon connectors recommended by that fool of an author in Chapter 11. All in all, you reckon you're ready for testing to commence. Where do you start?

Believe it or not, at the local photographic shop. Before you hook up any volts to a new amplifier, trot round to the nearest suitable emporium and invest in a couple of large cans of compressed air normally used for cleaning photographic apparatus. When you get back to the shack, open up the amplifier, take the valves out and stand the amp on end with the valve bases uppermost. Then do the Han Solo bit and blast everything in sight with a jet of air – especially every part of the valve bases, variable capacitors and anywhere else that swarf or metal particles might be hiding. Take special care around the valve bases, especially those in which the screen decoupling capacitor is not totally sealed, since it's quite possible to get a piece of metal swarf in just the wrong place. Then vacuum out any bits which may have fallen to the bottom of the amplifier compartment. This is important because metal fragments tend to emerge months later and cause havoc at the most inconvenient times, like two minutes into an Es opening.

The next job is to make sure that the anode and grid tuned circuits resonate in the band you want the amplifier to work on. This is easy to say but quite hard to do because at VHF and UHF the resonant frequency will inevitably be modified by the covers over the grid and anode compartments, and a conventional dip oscillator can't be coupled to tuned lines. One way to do the job is to power the blower and heaters, apply a small negative voltage to the grids via the metering (try –9V from a battery with a milliammeter in series if it isn't convenient to extract it from the PSU) and feed some RF in to the grid circuit. You'll see some indicated grid-current as you tune the grid-tuning capacitor through resonance, which will give you an idea of whether you've got it right, although this won't necessarily be the exact point at which the grid tuning will peak in practice. Another way to check for resonance is to apply some RF via a VSWR indicator; the actual VSWR readings won't be meaningful but the reflected power indication will flicker as you tune through resonance. You can also use this method to check the anode circuit for resonance. In a push-pull amplifier you may find some interaction between anode tuning and the grid-current meter, which indicates that the machine will need to be neutralized – don't worry, we'll come to that in a minute.

If you find to your dismay that a tuned-line grid circuit won't resonate, don't worry. If the lines are too short, you can add a low-value silver-mica across the grid-tuning capacitor or

COMMISSIONING A NEW AMPLIFIER

twist up a couple of inches of wire to achieve the same effect. If they're too long, remove a plate from each rotor of the split-stator capacitor. If the anode circuit won't resonate, however, you've got a bit of a problem and you'll need to do some fairly major surgery somewhere – but not before re-checking all the dimensions against the drawings!

Having checked roughly that the tuned circuits will resonate, connect a VSWR meter in series with the amplifier's input. With the heaters active (don't forget the blower) and a little grid bias, apply a whiff of RF drive and spend a happy half-hour adjusting the grid tuning and loading controls (and if necessary the position of the input coupling loop) to get a VSWR something like 1:1. If the amplifier is accurately built from a published design you should have no difficulty in finding the right settings. If it doesn't want to know, check your dimensions and wiring. It's amazingly easy to omit a crucial grounding strap or assume that a ground return exists when in fact it doesn't.

NEUTRALIZING

Eventually you should end up with a reasonable VSWR on the input and a grid circuit that seems to be tuning up. What's next? Probably neutralizing, if it's a push-pull amplifier.

★ ★ ★ DON'T PANIC ★ ★ ★

For some reason the prospect of neutralizing frightens amplifier builders to death; as soon as someone mentions the word, they start gibbering and muttering and finding excuses not to play with the new amplifier this evening. Actually, neutralizing a well-built amplifier is dead easy. You don't need any test gear whatsoever; all you need is a source of RF and the existing PSU metering and – if you've never done it before – a bit of patience.

Maybe it'll help if we work through what we're trying to do and why. Think of it like this. A 4CX tetrode amplifier at VHF or UHF has a fair bit of gain: a two-valve grid-driven 144MHz machine with reasonably high-Q input and output circuits should have a power gain in the region of 26dB, implying that if we put in 1W of drive we'll get 400W of output power. This means that we need at least that amount of isolation between the 'grid' and the 'anode' sides of the amplifier. Lacking this, the amplifier can turn into a very effective and powerful TPTG (tuned-plate tuned-grid) oscillator. Actually, Great Brains who know all about things like Bode plots and Nyquist criteria will tell you that you need at least 3dB more isolation than the power-gain figure, to meet the proper criteria for stability. All in all, let's say that your amplifier requires at least 30dB isolation between its input and output circuits.

From the constructional point of view it's easy to achieve 30dB isolation provided you take a little care – solid metalwork with lots of screws in the right places and plenty of screening and decoupling. Isolation really shouldn't be a problem unless you do a Walter and run the grid leads through the anode compartment, or run an unbypassed HV lead alongside the grid-tuning capacitor.

Unfortunately, no matter how well you did the metalwork, as soon as you plug the valves into their sockets the whole thing turns to worms. The reason is quite simple – it's quoted in the data sheet as C_{AG}, the capacitance between the control grid and anode of the valve, and it represents a feedback path between the input and output circuits. Because a tetrode has a second grid which screens the control grid from the anode, C_{AG} is pretty small – for a 4CX250B it's 0·04pF – but when multiplied by something known as the Miller effect, it provides enough feedback to turn your amplifier into an oscillator. It's this feedback path that you're trying to 'neutralize' so that, as far as the amplifier is concerned, it isn't there any more. For the purists, we ought to add that in theory any common screen or cathode inductance needs neutralizing out as well, but this shouldn't enter into it if you're using proper bases and solidly grounding each cathode pin directly to the metal body of the base.

You might think that a feedback capacitance of 0·04pF isn't a lot, however much it might be multiplied by Mr Miller. In point of fact, most single-valve amplifiers using a 4CX250B

TRANSMITTERS, POWER AMPS & EMC

Fig. 6.8. Above-chassis neutralizing flags for a W1SL amplifier (Chapter 8)

can manage without neutralizing if they're well made, using a proper base with effective screening between the grid spigot and the other pins, plus the usual built-in screen decoupling. However, most push-pull two-valve amplifiers **will** need the treatment, basically because they have a bit more gain and twice the effective feedback capacitance. You can sometimes just get away without neutralizing a two-valve amplifier on 144MHz, if you use damping resistors across the grid circuit to reduce the power gain, or if you use SK610A bases and 4CX250Bs and don't have more than about 1500V on the anodes, but it's a bit marginal.

When you neutralize an amplifier, what you'll actually be performing is an old country ritual known as 'balancing a bridge'. It's similar to what you do every time you switch your analogue multimeter to measure resistance and use the 'set zero' control to compensate for the fact that the battery's got a bit sleepy since you last used it. In the case of an amplifier, you null out each valve's anode-grid capacitance by sticking in what amounts to an equal and opposite capacitance. Hey presto *foof!* the feedback capacitance disappears in a puff of smoke and the isolation returns to its former high figure, making the amplifier stable again.

If the amplifier designer intended it to be neutralized, he'll have provided 'neutralizing capacitors' with which to do it. These will be connected between the grid of one valve and the anode of the other (see the description of the W1SL amplifier in Chapter 8). The odds are that they won't be capacitors of the sort you can buy in a component shop: they'll probably consist of little bits of copper sheet which are attached to the top of small and rigid pieces of wire. Each of these 'neutralizing flags', as some people call them, will sit on a feed-through insulator somewhere near one valve's anode, as shown in Fig. 6.8. On the grid side it will probably be wired to the grid spigot of the other valve base (Fig. 6.9). In other words, you've 'cross-coupled' the anode of each valve to the grid of the other, via the small capacitances of the neutralizing flags as shown in the W1SL circuit (Fig. 8.22).

All you need to do to neutralize the amplifier is to vary these capacitors so that they both equal the feedback capacitance of the valves.

There are actually quite a few ways of neutralizing grid-driven amplifiers. Here's the one which seems easiest and works very well. First of all, take the anode-compartment cover off so that you can get at the neutralizing flags, but leave the grid-compartment cover screwed firmly in place. Make sure that the anode and screen supplies are disconnected and that there's no DC return path for either – in other words, take the HV connector off the amplifier completely and take the screen changeover relay out of its socket (you did use a socket, didn't you?) so that the screens are left floating. The presence of the screen bleeder resistor and VDR won't matter, so you don't need to remove them. Power up only the blower, heaters and bias supply – if you're using a power supply with clever interlocking of various functions as described in Chapter 11, you'll need to override some of the interlocks – and then apply some RF drive from your prime mover. Tune the grids to resonance and adjust the RF drive level so that you see about half-scale deflection on the grid-current meter of each valve. If there's any adjustment to balance the RF drive between

COMMISSIONING A NEW AMPLIFIER

Fig. 6.9. Cross-connection of the neutralizing flags shown in Fig. 6.6 to the opposite grids

the two grids – *eg* a differential capacitor across the grid inductor – give that a tweak so that the same grid current is indicated on both meters.

Having done that, watch the meter needles carefully as you swish the anode tuning control gently through its range. As you pass through resonance you'll see the grid meters give a little flick backwards, which shows you that a bit of RF energy is finding its way into the anode circuit via the feedback capacitance of the valves. The idea is to stop that happening, so that you can swing the anode tuning through resonance without it having the slightest effect on the grid current. To do that you'll have to get the capacitance between one anode and the other grid equal and opposite to the feedback capacitance. You do this by poking at the neutralizing flags with an insulated tool and varying their positions, watching the effect on the grid meters. Don't be afraid to bend the tags all over the place and even to start removing pieces from them if you need to. But remember that you're dealing with very small changes in capacitance, so go carefully and try and keep the two flags roughly physically symmetrical.

After a bit of trial and error you should find that a particular position or size of the neutralizing flags has reduced the amount of 'twitch' of the grid currents by quite a lot. When

you've reached this stage, have a general re-tune and re-tweak of the grid tuning, loading and balance; then pause for breath or a nice cup of tea. Having done that, now is the time to start watching those grid-current meter needles *really* carefully. Start adjusting the neutralizing capacitors in extremely small stages until you reach the point where there doesn't seem to be the slightest flicker as you go through resonance. When you've got there, re-tune and re-balance the grids again and look very suspiciously indeed at the grid-current meters while you swish the anode tuning and loading controls through all their possible positions. If you see the slightest movement, have yet another go at the neutralizing capacitors. Don't be content until the grid-current meter needles both look as though they're stuck in position on their scales, whatever amount of knob-twisting you do.

That's it – neutralized. No problem, was it? If you haven't done it before, don't be surprised if it takes you an hour or six to get it exactly right. Don't be surprised either if you seem to end up with much less neutralizing capacitance than the original design had. Many good amplifier designs like the W1SL are quite old now and may well have originally used valve bases which had less isolation than modern ones. In updating the W1SL design in Chapter 8 I've tried to suggest something that's close to the correct final value.

Having got the neutralizing right, spend a little time with the VSWR bridge and recheck all the grid input tuning, loading and balance. You're looking for a 1:1 input VSWR and a nice sharp peak in grid current as you take the grid-tuning capacitor through resonance. In a push-pull amplifier you also want to see the same value of grid current in each valve, which you achieve by adjusting whatever grid-drive balancing arrangements the designer has provided. Again, spend some time checking these things carefully and getting the feel of how everything works.

When you're happy, check the neutralizing one last time and make absolutely sure that no amount of messing about with the anode tuning and loading controls affects the grid current in the slightest.

READY TO RUN

We're getting closer. Put the screen changeover relay back in its socket. Then take out some strategic fuses in the power supply so that the only voltage applied to the valves when you switch it on will be to the heaters. Double-check that the HV supply is unplugged and the HV connector is removed from the amplifier. Then switch on and measure the heater voltage, using a good DVM if at all possible. The figure you want to see depends on a number of factors including what frequency the amplifier is to be used for – it might be worth having a quick look at the section dealing with heater supplies in Chapter 11 to remind yourself of the requirements before you power up. In simple terms, if it's anything over 6·00V you ought to be pondering the wisdom of knocking it back a bit before you go any further. Bear in mind that when an air-cooled triode or tetrode is in its socket, it's physically impossible to measure the heater voltage (or any other voltage, come to that) and at the same time keep the proper amount of cooling air flowing through the valve, so don't hang around when making measurements.

You should be able to slide the bottom cover of the amplifier back over the grid compartment, leaving a gap for the meter lead to come out; seal this with a bit of Blu-Tak or similar. Another way to achieve the same end is to cover the entire grid compartment with a sheet of clingfilm, secured to the sides of the amplifier with Sellotape or insulating tape. You can then gently push the meter probe (you only need one since the other is an earth return and can go anywhere on the metalwork) through the clingfilm and clip it on to the heater pin. The film will 'balloon' like mad when you fire the blower up, but it'll do an excellent job of forcing some air to flow where it'll do most good. From bitter experience, it's fatally easy to crack the ceramic at the bottom of a 4CX if you apply heater volts and no cooling air, so do take a little time to work out how to keep the air going through it.

Having checked the heater voltage, switch everything off and remove the valves from the amplifier. Now you can replace the fuses and restore the other supplies to working order, but DON'T connect the HV supply up yet. The next check is that the screen voltage is switched to its Class-AB1 value and that the voltage on pin 1 of each valve base is zero on receive and the proper screen voltage on transmit. When you've done that, switch to a resistance range and check that pin 1 is solidly grounded on receive. Then check that the grid voltage switches nicely between the proper transmit and receive values when the PTT is operated. If you're using 4CX250Bs, set the grid voltage on transmit to approximately –65V. Switch off, replace the valves and put the cover back on the anode compartment with all the requisite screws – this is essential, both for safety and also because if you don't you can get some very odd effects (not to mention massive feedback) due to distortion in the RF circulating-current pattern. Finally, take away any little bodges you may have put in to fool the power-supply interlocking for the tests you've been doing.

Before you do *anything* else, you MUST now connect and double-check your safety ground braid between the amplifier chassis, the main power supply and the HV supply as discussed in Chapter 11. Use an ohmmeter to check for continuity between the amplifier chassis, the chassis of the PSCU and HV supply, the prime mover and ground. When you're happy, make *absolutely sure* that the HV supply is *not* plugged into the mains and then re-connect the HV cable to the amplifier. Connect the station dummy load to the amplifier output via a power meter set to a low range such as 10W. Connect the prime mover to the amplifier but make sure that (a) it's in receive and (b) the output power is backed right off to zero. Set the frequency to somewhere in the middle of the SSB sub-band.

STATIC TESTS

Now then – you're shortly going to run the

COMMISSIONING A NEW AMPLIFIER

amplifier for the first time and see the culmination of all your hard work. Double-check that the dummy load is connected and that the power meter is set to a low range. Switch on the blower, heater and all supplies but at this stage make sure that the HV cannot come on until you're good and ready for it to do so. Allow ten minutes or so for everything to warm up and settle down – then power up the HV supply and see what happens. Actually, nothing whatsoever should happen and the only meter needle which ought to move is the one on the HV PSU indicating anode volts. If there are any pyrotechnics or if fuses blow, shut down and find out what went wrong – it's probably something quite simple like inadequate clearance somewhere (look for carbonising across a gap) or a decoupling capacitor which didn't like 2kV very much. If you can't see anything, take the valves out and carefully inspect both them and the inside of the bases for any evidence of flashover. If you still can't see any reason for the pops and bangs, leave the valves out, replace the anode-compartment lid and disconnect everything except the HV supply and its ground return. Turn the HV on again. If there are now no flashes and bangs, change the valves; one of them obviously has an internal fault. It might be a good move to check that nothing in the PSCU has taken umbrage, incidentally, before you power up again – make sure you physically unplug the HV supply and then check as before that the screen and grid supplies are doing their thing.

Assuming that you've now got the amplifier happily sitting there in the receive condition and with HV on, what next? Check again that the drive power is still set to zero and then (great moment, this) switch to transmit!

Several things will happen when you do this, and it's good to have a clear idea what to expect. Assorted relays will go over, for a start – the antenna relays, the screen relay and any other PTT relays you may have – and the sound of these all operating at once can really startle you if the adrenalin's flowing a bit. The screen-current meters will swing up to a centre-zero position if you're using the scheme outlined in Chapter 11, and the anode-current meter will also show something. If you've set the grid bias correctly it should show about 100mA if the amplifier has two valves and about 50mA if it has one. Anywhere in that region is near enough for now. Having established that all these things happen and that there are no untoward events or clouds of noxious black smoke, let the PTT go and switch back to receive. Switch between transmit and receive a few times to get used to how placidly the amplifier behaves.

I've just described what *should* happen. What *could* happen when you switch to transmit is that the grid meters and the anode-current meter go hard over, the HV transformer emits a deep hum, the shack lights go dim and the RF output-power meter shows about thirty trillion watts going into the dummy load. If so, the amplifier has become bored with merely amplifying and has decided to make a career move into high-power oscillation. No problem; switch everything off, read the section on neutralizing again and this time do it properly! Actually, this little nasty usually strikes when you're doing your third or fourth amplifier and have become over-confident. First-time builders do their neutralization with care and dedication so, when they switch to transmit, the new Thunderbox just sits there quietly. Hardened amplifier-bashers don't, so it doesn't.

Assuming everything's looking good, go to transmit again and lock the PTT. Wave your hand over the valves' air outlet; the exhaust air should be pleasantly warm and should be at about the same temperature for both valves. You can then adjust the standing current to the requisite value for low intermods, which for Class AB1 is 100mA for a single valve and therefore 200mA for a pair. Those who have owned commercial amplifiers might think this sounds a bit high compared with the figures quoted in their manuals; you may draw your own conclusions. Switch back to receive and check that the standing current falls to zero. Go back to transmit and check that the anode current rises smoothly to the value you just set. Stay in transmit and check that no power

6 - 33

TRANSMITTERS, POWER AMPS & EMC

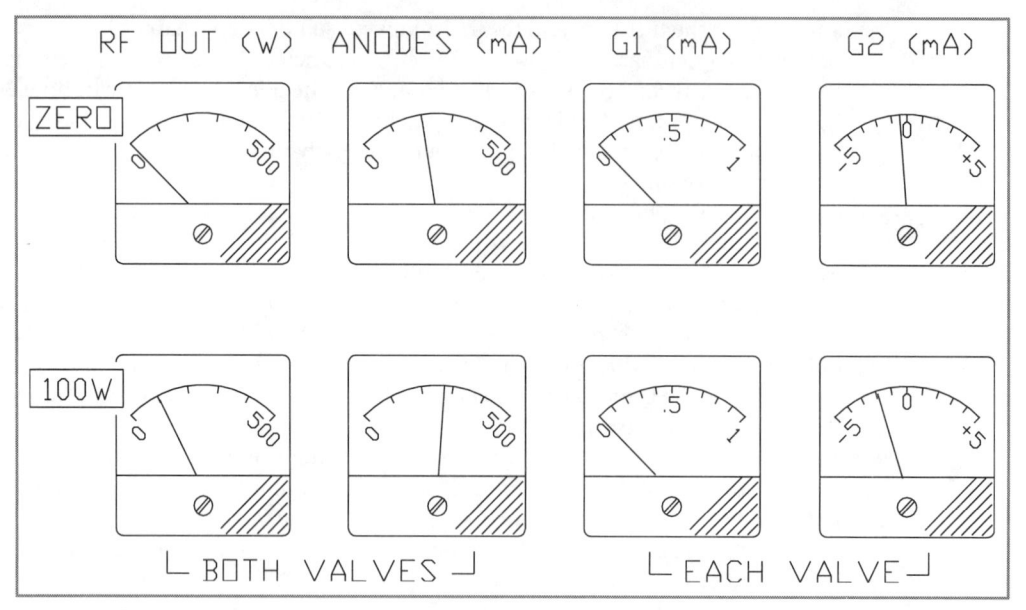

Fig. 6.10. Meter readings for a two-4CX250B amplifier, correctly set-up for standing current (zero RF output power) and at 100W RF output power. The RF output and anode current readings are for both valves combined, while the G1 and G2 currents are for each valve

is being indicated on the output wattmeter. There shouldn't be so much as a microwatt if the neutralizing was done properly.

RF TESTS

Now for some RF. Incidentally, you'll need to note some meter readings shortly, so have a writing implement and some paper somewhere in the vicinity. First of all, set the output loading capacitor to maximum loading – that's fully unmeshed for the W1SL design with loop-coupled output (Chapter 8), or with the string slack in the K2RIW (Chapter 10). Switch the output power meter to a slightly higher range, *eg* 50 or 100 watts.

Switch to transmit and lock the PTT. Gently begin to feed in some RF drive from the prime mover while watching the output meter. As soon as you see any indicated power on the meter, tune the grid and anode tuning controls to peak it up. **Don't** adjust the loading control at this stage. When grid and anode tuning are both peaked, increase drive to bring the output level up to about 100W and note the readings on the anode and screen current meters. You should find that the anode current is somewhat higher than the standing-current figure, but don't worry too much about its exact value for the moment. You're much more interested in the screen-current meters, which should be reading negative (Fig. 6.10). Adjust the output loading capacitor so that each screen meter shows about –2mA; the reading should be fairly equal but don't worry if one meter says –1 and the other says –3mA for example. Repeak the grid and anode tuning for maximum RF output power. Now turn the power down and back up to 100W or so a few times; make sure that all the meter needles move smoothly and that the output power goes up and down without any funny little jerks and jumps.

If you see any weird meter readings or the needles start twitching about, find out what's happening before you go any further. You're unlikely to have a problem with the amplifier itself if you've come this far successfully, though, and about the only thing which could be causing trouble is RF feedback into the prime mover (see later). If you're using a proper dummy load, that won't happen; if you're doing a Walter and trying to run up a

COMMISSIONING A NEW AMPLIFIER

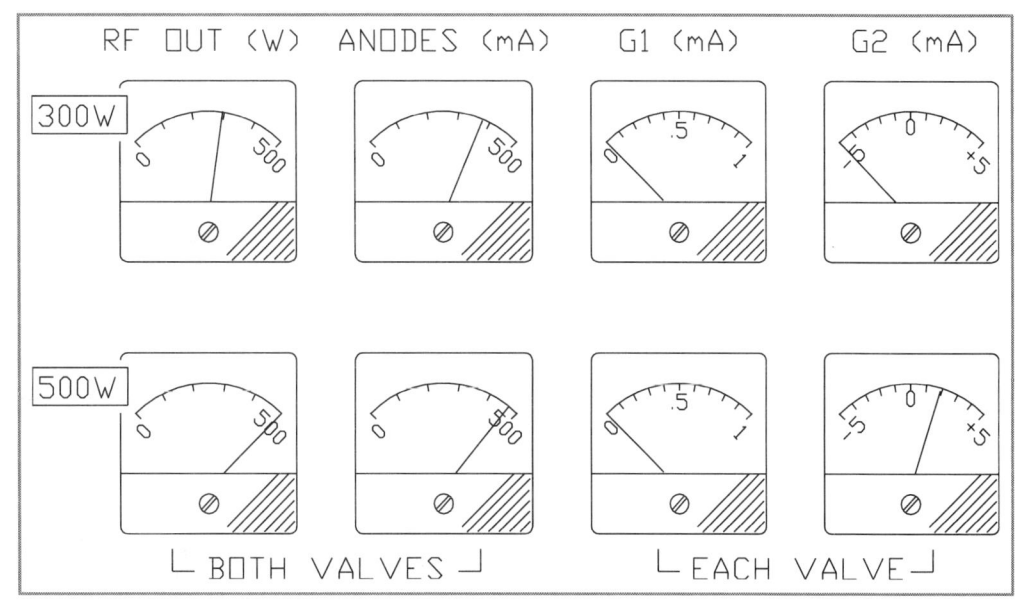

Fig. 6.11. Meter readings for a two-4CX250B amplifier, correctly set-up for 300W and 500W RF output power. The RF output and anode current readings are for both valves combined, while the G1 and G2 currents are for each valve

brand-new amplifier into your antenna, don't expect any sympathy from me.

Now change the power meter's range to 500 or 1000 watts. Go to transmit and establish the 100W power level again; peak the grid and anode tuning for the highest RF output. Then slowly bring the power up to about 300W, keeping a wary eye on the screen-current meters as you do. If they begin to move in the positive direction at any point, give the loading control a tweak in the direction that takes the screen currents negative again, so that at the 300W mark they're both around –5mA. If you have less than about 1800V on the anode, or the anode-supply regulation isn't all that hot, you might not be able to get the screen current quite so negative. If that's the case, simply adjust the loading control to make the screen currents as negative as possible at the 300W output point as shown in Fig. 6.11.

Still with 300W indicated, have a quick go at the anode and grid tuning to peak the RF output; you may be able to get a little more. Whatever you do, *please don't* try peaking the RF output by twiddling the loading control. You will totally wreck and ruin the amplifier's intermod performance if you do. The golden rule is to think of the loading control in a tetrode amplifier *only* as something to adjust to set the screen current to the figure you want, and only ever adjust it when you're looking at the screen-current meters. So *never* tweak the loading while gazing at the RF output-power meter. They say it sends you blind.

FULL POWER

Good, isn't it? Now go for broke and wind the drive up to the point at which 500mA is indicated on the anode current meter. You should see about 450-500W RF output and the screen current will have moved in the positive direction to +2mA or thereabouts (Fig. 6.11). If the screen current is somewhat more positive than that, try and get it back down by adjusting the loading. If you can't – because the loading is at maximum and the screen current is still much above a few mA – you haven't got enough loading range. Make a mental note that you'll have to increase the output coupling by bringing the coupling loop closer to

6 - 35

the anode line, or by fitting a bigger flapper if your amplifier uses capacitive loading. If the screen current is a lot higher than it should be, like 10 or 15mA, shut down and do this before you go any further. The coupling is much too light and you're in danger of a flashover.

Now take the input drive down to zero and back up to 500mA indicated anode current a few times. Again, all meters should move smoothly; in particular each screen-current meter should go from zero to about −5mA and then back towards zero, finally ending up at about +2 mA at full output. For VHF and UHF use, in which heavy loading is good both for valve life and intermod performance, that order of screen current is what you want to see in each 4CX250B. If you're using 4CX250Rs, note that the intermod performance of this valve is very good if the single-tone screen current is around zero but it deteriorates rapidly if you use lighter loading and let the screen current go very positive. Keep the screen current of '250Rs around zero at full anode current for best results. For the 4CX350A and FJ you should aim for no more than zero screen current (*ie* never positive) at 250mA anode current per valve.

Incidentally, all these screen-current figures are given for 4CXs with 350V on the screen and something like 2kV on the anode. If the anode voltage is lower than that, the screen currents will be more positive. I'm also assuming that the valves are in reasonable condition. Remember that if you have clapped-out valves in a new amplifier you're quite likely to see lots of negative screen current and so you can't possibly know whether your loading is correct or not.

By this stage you might be wondering why I haven't mentioned the grid-current meters – shouldn't they be showing something? In a word, no. If you see any indication at all on the grid meters at 500W out, you've either got very low anode voltage or some rather tired valves. In Class AB1 you can forget all about the grid meters; they should only indicate anything when you switch to Class C, and if the valves are any good it'll only be a few milliamps before you hit the legal limit.

For some 4CXs in Class AB1 the grid-current meter(s) may move **backwards** a small amount – normally only a matter of microamps – on speech peaks, just before the onset of normal grid current. This indicates that you won't be able to wind the drive up very much more before departing from Class AB1. Not all valves show reverse grid current, and some only start to do it after a few tens of hours' use. Reverse grid current is the sum of gas current, leakage current and control-grid emission current (bet you wish you'd never asked) and it will generally increase somewhat with life. If you see much more reverse grid current than usual, the odds are that the valve is running hot; check the blower and any ducts, filters, etc for blockages.

After a few years' hard use you may also begin to see signs of positive grid current at your accustomed full power level. This suggests that the valve is getting a bit past its prime, and you'll have to be a shade less exuberant with the drive control. Even when the valve is on its last legs, you still should *never* run a Class-AB1 linear amplifier into positive grid current.

GOING ON THE AIR

Having persuaded the Thunderbox to deliver 500-odd watts into a dummy load and satisfied yourself that it's stable and properly loaded, you can think about going on the air with it. Switch to SSB and adjust the drive level to produce about 10% less than the maximum achievable output, or the legal power limit at the antenna. Having worked through the commissioning procedure I've described, in theory you should have good intermodulation performance from the word go. Your next move is to phone a local who knows what he's doing and ask him to conduct some tests with you (see the panel on page 6-11). While he's firing up his equipment, replace the dummy load with the antenna and have another loading and tuning session. Remember the steps:

1 Adjust the grid and anode tuning for a peak in RF output at about 100W and then do the same at 300W.

2 Adjust the loading so that the screen current per valve is around –5mA at 300W output.

3 Check that at 500mA total anode current you have no more than about +2mA screen current per valve.

CONTROLLING THE DRIVE LEVEL

Earlier on I was saying "...vary the drive level so that the amplifier is producing 300W". This is all very well if your rig has some form of variable power control, but what if it hasn't?

First of all, here's how **not** to do it. As I've said already, you shouldn't use the microphone gain as a general-purpose RF output control. If you turn the microphone gain down a long way, hoping maybe to wind a 25W rig down to the 2W or so required by a pair of grid-driven 4CXs, your carrier suppression will be more than a little degraded – not to mention the fact that speech peaks may still emerge at a good deal more than 2W. Equally, don't try to reduce the drive by detuning the grid circuit of a valve linear amplifier, because the intermodulation performance of the prime mover will suffer owing to the bad VSWR you're presenting to it. Above all, you really mustn't turn down the drive power by reducing the DC supply voltage to the transceiver! (Until someone actually did that, for a contest if you please, I'd never have believed that even Walter would have tried it...)

Can we do a little better than that? Many rigs nowadays have an 'official' RF output control which operates through the ALC circuitry, and this is the best way to make moderate day-to-day adjustments in power levels. For rigs with no RF power control but which do have an external ALC input, the same effect can be achieved by feeding an adjustable negative voltage (typically 0 to –9V) into that socket. But don't expect too much of the ALC circuitry, because unless its dynamics are extremely good you may see a spike of full-power output at the beginning of every syllable or sentence. This will not be detectable in static tests with a tone generator but will cause severe splatter on speech, and can also cause damage to low-drive solid-state amplifiers.

TRANSMITTERS, POWER AMPS & EMC

In spite of everything I've said about the virtues of under-running the prime mover to obtain a clean drive signal, you may have no options. If the available power output is vastly more that your amplifier needs, your best approach overall is to run the prime mover at one-third to half its rated power and dispose of the rest of your excess drive in an attenuator connected permanently into the transmit path (see Chapter 12 for design information).

STAYING LEGAL

The next issue in day-to-day operation is how to measure the output power of your station, and how to set up your routine drive levels as opposed to those used for the initial tests. The terms of the UK licence restrict you to +26dBW PEP at the antenna, which is 400W PEP. When officialdom appears on the doorstep one day for a station inspection, the odds are that they're going to take one look at your gleaming Master Blaster and ask you to demonstrate how you measure its peak envelope output power and how you keep it within the licence limits.

The only way to measure VHF/UHF peak envelope power that's at all practical in an amateur station is with a good peak-reading wattmeter. Sure, the Bird 43 will measure carrier power to within 5% but it won't measure the peaks of the RF envelope, and neither will any other instrument based simply on an RF detector and an ordinary moving-coil DC voltmeter. If you tell the inspector that you set up a carrier level that's legal and then, er, assume that speech peaks equal carrier level, he'll probably reach for his notebook and ask you to repeat that for the court case. Your amplifier probably gives about 10% more PEP than it does sustained carrier power, because the HV doesn't sag on speech peaks in the same way that it does with a constant carrier or slow Morse. So build or buy a PEP-reading wattmeter, or modify your existing wattmeter using the module in Chapter 12. Commercial PEP wattmeters have been reviewed by a number of authors. Try and borrow Angus McKenzie's *Buyer's Guide to Amateur Radio*; alas it's now out of print.

The next question was: on SSB, how do you *keep* your PEP output within the licence limits? Answers involving esoteric mental disciplines or speaking in square waves will have the inspector reaching for his notebook again. Instead, how about a decent speech processor? You wouldn't find a professional SSB transmitting system that didn't have some form of peak limiting – it's regarded as vital for commercial SSB users not to overdrive and thus cause adjacent-channel interference. Also, it's well known that the human voice has a fairly poor ratio of peak-to-mean energy, so some compression of the dynamic range of speech is a Good Move to increase the intelligibility or 'talk power' of your signal.

Speech processing of some sort is included in almost all modern HF rigs, and increasingly in VHF/UHF rigs as well. Although clipping increases talk power, it's still basically a form of distortion and will generate a wide spectrum of intermodulation products. Any clipper must be followed by a filter to restore the clipped signal to its original bandwidth, but the presence of distortion products within the bandwidth of a heavily clipped signal will limit the attainable improvement in intelligibility. Modern rigs usually use RF (strictly IF) clipping, and pass the peak-limited SSB signal through a second IF filter to clean it up. The virtue of RF clipping is that most of the intermods fall far away from the wanted signal, permitting a heavier degree of clipping before the audible distortion becomes too noticeable. In contrast, many of the distortion products of an AF clipper fall within the wanted audio bandwidth and sound exceedingly unpleasant.

So the best option for a loud, clean SSB signal is an RF clipper, preferably integrated into the rig. Next best is an external RF clipper between the microphone and the rig. This device converts your audio signal to SSB, clips it, cleans it up and finally converts it back to audio. The Datong devices are highly recommended, especially the ASP Automatic Speech Processor which includes automatic level control of the voice signal presented to the peak limiter and also a tone generator to set

Going on the Air

the carrier level you need. Having set the levels using the tone, your PEP on speech cannot overshoot. Even the simple AF clipper in Chapter 12 is far, far better than no peak limiting at all.

SETTING UP A SPEECH PROCESSOR

All forms of speech processor need to be set up with great care. Otherwise you see no improvement in 'talk power', or more likely the processor ensures that you overdrive more consistently than before! Here is a general procedure for setting up a speech processor.

1 Connect the rig to a dummy load. *Please* don't connect it to an antenna because you're going to make some foul noises before you get everything behaving properly.

2 Identify a variable gain control *ahead* of the clipper input. Regardless of what the manufacturer called it, we're going to call this the CLIPPING LEVEL control.

3 Identify another gain control *after* the clipper output. We'll call this the DRIVE LEVEL control for the rest of the rig.

4 Turn CLIPPING LEVEL to maximum and emit a "Waaaaahlo" into the microphone. Adjust DRIVE LEVEL until the RF output reaches the maximum achievable.

5 Then back off the DRIVE LEVEL until the RF output drops by 10% (see panel on page 6-7). This step is the one which gives you a clean signal, and it's *vital*.

6 Listen to your signal on another receiver – use headphones to avoid feedback. Advance the CLIPPING LEVEL control from zero until your voice sounds loud and crisp but not distorted. Do not touch the DRIVE LEVEL. If the background noise in your shack is very noticeable, back off the CLIPPING LEVEL.

7 Check the width of the signal with a local station. If it's broader than it was without the speech clipper, turn down the DRIVE LEVEL until it's actually *narrower* than before – because the clipper is limiting your speech peaks below the level that overdrives the rest of your transmitter. Note the setting of the DRIVE LEVEL control *and don't touch it again*.

8 Any improvement in 'talk power' will be most noticeable when your signal is weak, so check the effectiveness of the clipper with DX stations, not the locals. Adjust the CLIPPING LEVEL as necessary.

The locals should appreciate your reduced bandwidth, but don't expect them to give glowing reports on your audio quality. Your signal will have more background noise than before and won't sound as smooth, so you may need to compromise on the CLIPPING LEVEL. This applies particularly to contest stations: although clipping is good for raising the DX, you need the points from the locals too, and they won't bother to call you if your audio sounds unpleasant. As a matter of fact, some people say they can't abide speech clippers. Maybe that's because they've never heard one that's been set up properly – or more likely they have, but haven't realised it because a properly-processed signal just sounds 'loud'.

If I've managed to persuade your conscience that some form of PEP limiting is necessary but you *still* won't use a speech clipper, your only other option is to use ALC. This involves generating a negative-going signal from an RF detector at the output of the linear amplifier and feeding it right back into the prime mover, thus enclosing the entire transmitting system within the level-control loop. This is emphatically *not* simple. The ALC signal must have the correct characteristics of voltage swing, attack time and decay time to suit not only your particular model of transceiver but also the power gain within the extended ALC loop. The transverters in Chapters 8–10 have internal ALC loops for protection against overdrive, and may provide some initial circuit ideas. But it isn't possible to be specific about the ALC requirements of your particular setup; you'd have to experiment. *Now* have I persuaded you to try a speech clipper?

FLASH-BANG-WALLOP!

So at last you're on the air with the atom-smasher and it all seems to be working well. When you transferred the output from the dummy load to the antenna, the amplifier tuned and loaded a little differently, but that's only to be expected because antennas don't look exactly like 50Ω resistive. Still, there's plenty of range in the loading control to match anything reasonable. This, of course, is one of the beauties of a valve amplifier as opposed to solid-state amplifiers which have no user-settable loading at all. But *never* try to run a big valve amplifier without a load of some sort. If you do, there will be the father-and-mother of a flashover as the peak voltage in the tank circuit winds itself up to a hundred kilovolts or more.

Where does a hundred kV come from, when the HV is only 2kV? The basic reason is the difference between the 'loaded Q' and 'unloaded Q' of the RF tank circuit. In normal operation the peak RF voltage in this circuit will be the DC anode voltage multiplied by the Q-factor, which is fairly low because of the loading effect of the antenna – so typically you'd expect peak RF voltages of 4–6kV depending on the specific design. However, the unloaded Q of the tank circuit is much higher than its loaded Q, by at least a factor of 10 and in many designs much more. So if you disconnect or lose the load from your Thunderbox and go "Waaaaahlo", the RF voltages developed in the tank circuit will be ferocious – at least 40–60kV, and probably a lot more. Something has to give, and what usually breaks down is the external path across the ceramic anode/screen insulator of the valve. The flashover from the anode has to go somewhere, and it's usually straight down to the metal screen-grid ring. This is why you need to take precautions, since the 1kV-working capacitor in the valve base won't handle a huge overvoltage, and the components associated with the screen supply probably won't like it either.

Flashovers are not something you want every day of the week, but they're not the catastrophes some people seem to think. As long as you take a few precautions, like having some suitable resistance in the output of the HV supply to limit the prospective short-circuit current and using voltage-dependent resistors (VDRs) to limit the voltage surge on the screen supply rail (Chapter 11), a flashover won't necessarily harm anything – although it'll usually wake you up.

The other celebrated cause of flashovers is letting the screen or control grid float relative to the cathode. The type of flashover doesn't involve RF at all, but the result is equally unpleasant. A floating screen grid takes about two-fifths of a nanosecond to assume the full anode voltage, so about 2kV or whatever you happen to have on the anode appears across the 1kV-working screen-bypass capacitor in the valve base. This type of flashover isn't even big and entertaining; it's usually a sharp *ffft!* followed by a small sizzle, and you need a new base. To prevent this happening to you, first of all *don't* adopt a suggestion in the German *UHF Compendium* – to wit, sticking a 100kΩ resistor between the screen of a 4CX and deck in the fond hope that it'll "...prevent static damage". You might as well hang some garlic in the grid compartment – it'd probably be more effective. A better defence against the floating-grid flashover syndrome involves mounting a suitably rated voltage-dependent resistor (VDR) directly on the valve base in addition to about 33kΩ of bleeder resistors, with back-up VDRs in the PSU. Even in the best-run households, wires to the screen can simply drop off; if a new base costs nearly £100 and a VDR costs 70p, how can anyone justify not fitting one or more? Try the V250LA40 for a 300V screen supply, or a V275LA40 for 350V.

RF FEEDBACK

If you've got RF feedback, the usual symptoms are that people complain about squawks and serious distortion on your speech peaks – or when you stop talking, the transmitter

GOING ON THE AIR

doesn't. The problem is usually due to stray RF getting back into the microphone circuitry and making the whole transmitter howl. Turning the microphone gain to zero will usually stop it, but that's not exactly a workable solution.

RF feedback is most likely to occur when you start to add accessories to the basic radio. So if you suffer from this annoying problem, the best approach is to strip everything down to the bare rig and re-assemble the system step-by-step. Begin testing with a dummy load and check that your microphone, leads and plug are properly screened and bypassed against stray RF, and that all the RF connectors and cables are properly screened. Also check that RF on the mains or power leads isn't driving a stabilized power supply crazy. Then connect the antenna. If the feedback now appears, check with a field-strength meter to ensure that you're not sitting right inside the RF field from the antenna, and that there isn't excessive RF on the outside of any screened cable. At each stage of your tests, increase both the microphone gain and the RF drive to maximum to see what margin of stability exists.

Adding a power amplifier may cause problems, either due to the increased RF field or because the additional signal, power and control wiring has created new paths for pickup of stray RF. Take particular care to avoid creating ground loops involving low-level audio signals; you may find that a speech processor fitted in the microphone lead requires an internal battery. And don't forget the PTT line which runs right alongside the microphone wire inside the curly cord.

The cures for RF feedback in transmitters are exactly the same as for audio breakthrough in consumer electronics – ferrite rings, ferrite beads, bypass capacitors and careful attention to innocent-looking leads which can carry stray RF into sensitive places.

EMC

There's one other tiresome possibility we ought to consider. This is that, having got your Thunderbox working and delivering lots of lovely potent power to your antenna, there's a Wagnerian hammering on the front door accompanied by a brisk *obbligato* for solo doorbell. You open up and there's your neighbour from three doors away. He's seven feet tall and six feet across, and principal of the local night-club bouncers' training college.

"WOT ARE YOU DOING TO MY TELEVISION, MATEY?"

"Er – er...."

"IS IT YOUR BLARSTED CB RIG AGAIN?"

"Er, well, no, not actually: you see it's...."

"WOTEVER IT IS YOU WANT TO STOP IT, 'COS OTHERWISE PEOPLE COULD GET A BIT UPSET, KNOW WHAT I MEAN?"

"Er...."

Your scenario might not be quite as extreme as this, but running high power in an urban environment may well attract the attention of your neighbours. You never can tell whether any kind of consumer electronic equipment is going to bomb out when hit with a little RF. Technically, such problems of electromagnetic compatibility (usually referred to as EMC) are usually very simple to solve, and I'll mutter a few words about the technical side later. But first I want to have a look at the 'people' aspects, which are often the **real** problem with EMC.

You'll have noticed that I've occasionally referred to a mythical radio amateur by the name of Walter, who is a kind of living embodiment of crassness and the uncrowned king of stupidity. In fact, about the only thing psychoanalysts agree on is that there's a little of Walter in every one of us – in your angry neighbour, in yourself, and in me too. Each of us knows that there are certain kinds of situation that make us react in a totally irrational way. Mercifully these situations are mostly different for each of us, but EMC – including EMC problems **within** the amateur bands – touches on a sensitive spot for most of the human race. The reason is that you're not dealing with a calm, civilized and rational adult. You're talking to an internal part of the same person that's a good deal more primitive and considerably less intellectual.

Here's an example of how it works within our own ranks. The rational response you might expect when you tell someone politely that there's something amiss with their transmission would be a reply of "Oh, sorry, thanks for telling me", and then some attempt to fix the problem. But what frequently happens is that people come back with a variety of daft replies along the lines of –

"You're the only one that's complained, old man"; or

"Oh, well I must be putting a very strong signal into your receiver, old man"; or

"Well, you're about 50kHz wide yourself, old man"; or

"No, I can't turn the mic gain down – it'll invalidate the guarantee"; or

"But I only bought it yesterday from the approved importer, so it must be all right".

What's **really** happening here is that the unconscious Walter in many of us thinks that the words "Your transmission is defective" are actually conveying the message "Something

EMC

about *you* is defective; you are an inadequate and pathetic human being". Hence the irrational reactions, especially since we've all been pretty systematically misled to suspect that the accusation might be true. Have another look at the responses above and you'll notice that most of them can be translated as "Even if what you say about my transmission is true, I don't know what to do about it". More reinforcement of those lurking feelings of inadequacy. In other words, what they really need is a bit of help – starting right here in this chapter.

It's interesting to note that some of the typical responses are also extremely defensive ("Well, you're 50 kHz wide yourself, old man"). Maybe our transmitters – or guns, cars, steam locomotives, football teams or whatever – serve as symbols for an unconscious part of ourselves, and when that symbol is criticized it's as though we ourselves are under attack, and we automatically start to fight back. Worse still, an unreasonably defensive reaction can send the whole situation to pot by provoking the Walter in the person – myself, for example – who was originally just trying to be rational and helpful but just CAN'T STAND UNREASONABLE PEOPLE!!! (Sorry. Deep breath. Carry on...)

By extension there's a Walter in each of your neighbours. If you start breaking through into someone's hi-fi or television, you're invading their privacy – which is a civilized term for that very primitive, un-evolved and unconscious part of ourselves which is concerned with safety and security. If you threaten that, don't for one moment expect a rational adult response. You'll get a response which is probably thirty or forty years out of date; it'll be exactly the same response as the individual made when they were very small and subjected to a similar invasive stress. This isn't a psychological textbook and I'm not Melanie Klein. Just don't be surprised when rational reasoning doesn't work and is met by anger, rage and other manifestations of hostility; and don't get drawn into it.

Whatever happens, *it's up to you* to remain rational, calm and positive; you can't always rely on the other person because frequently you're not dealing with the real rational human being at all, but with a rigid and unconscious set of behaviour patterns. Grasping that simple fact and working with it is the first step in doing something about EMC problems.

How do the 'people' factors interact with the rest of the domestic EMC situation? Well, you certainly don't need me to tell you that in 99% of cases it's the neighbours' equipment that's defective – it doesn't possess adequate immunity to RF signals. But if you tell your neighbours straight out that their beloved hi-fis, TVs, videos or telephones are defective, they'll all arrive in a deputation headed by Matey from three doors away! On some primitive level we all like to think that the things we choose to buy and own are perfect, and discovering that they are not is painful. It's the same pain that we felt as small children when we discovered that our parents weren't perfect either, as a matter of fact.

The staff of the Radio Investigation Service are no longer able to play the mediating role that they could when they worked for the old Post Office. That isn't necessarily to their personal liking, so don't blame them for it. It just means that you'll have to do the mediating yourself, as well as investigating and solving the technical problems. It's quite hard to convince people that because the problem only starts when you're transmitting, that doesn't make it the fault of the transmitter – to the non-technical that takes quite a leap of logic. Try explanation by analogy – if the builder leaves a slate off your roof you'll get water in, but only when it rains. The solution to the problem is not to complain about the rain, but to get the builder back to fix the slate.

In the UK, direct legal responsibility to do something about deficiencies in your neighbour's equipment rests with the supplier of the goods, especially if they're new and always if they're rented. Unfortunately, for much the same irrational reasons as described above – plus ignorance and sheer laziness in some cases – many dealers are quite prepared to lie

TRANSMITTERS, POWER AMPS & EMC

to your neighbour and tell him that of course it's your fault. When you and your neighbour have negotiated your way past that obstacle, the ultimate responsibility lies with the manufacturers of the equipment; some can be extremely helpful both to yourself and to their dealers, while others couldn't care less. It depends very much on getting past the administrators to find an engineer who knows enough about EMC problems to admit that they exist. Having reached that point, a technical solution is not far away.

This isn't the place to go into the technical minutiae of how to solve domestic EMC problems – thousands of words have been written about that and probably there are thousands more in the pipeline. It is worth stressing, though, is that two heads are better than one, and there's a lot to be said for asking the guys at the radio club if they can help. There are various comprehensive filter kits around, and it makes sense for a club to own one so that a variety of filters can be tried to find a cure. All you then have to do is to buy your own sample of the one that worked. You may also find that someone at the club is either the local EMC wizard, or that an outsider with the right combination of social and technical insight can play that vital mediating role between you and your neighbour.

If all else fails and you're a member of the RSGB, try asking their EMC Committee for help. In some cases they've done a very good job, but their resources are limited and their volunteers can't be everywhere at your convenience. And don't expect them to rescue a social situation that has already been allowed to degenerate into trench warfare.

Finally – whatever you do, don't give up. You may have to exercise the patience of Job, and you'll certainly receive some interesting insights into the nature of the human unconscious – but keep at it. The main thing is not to become embroiled in arguments which contain words like "rights" or "the source impedance at the base of TR2"; both are equally futile.

Good DX, and save some for me!

CHAPTER 7

BEAM ANTENNAS & FEEDLINES

BEAM ANTENNAS AND FEEDLINES

by Günter Hoch DL6WU

When we talk about beam antennas, gain is usually the most prominent topic. We sometimes forget that other features may be equally important, and that there are strong interrelations between them and gain. So I'll begin by describing all these antenna performance parameters and how they are related. This will introduce such important topics as capture area and the stacking of antennas. Then I'll look at the various types of beam antenna, which among other things will explain why the Yagi is the most popular beam antenna for VHF and UHF. Yagi antennas are one of the main topics in this chapter, and as well describing their functions in detail, I'll give details of the DL6WU long-Yagi designs and how to make Yagis work.

Finally I'll look at the feedline, which is almost as important as the antenna itself, and also discuss power dividers and baluns.

You can think of an antenna as a coupling device between your feedline and free space. It works the same either way, whether you're transmitting or receiving. This concept of *reciprocity* is an important and very useful idea, because it's easier sometimes to think of an antenna as a radiating device and at other times as a device for scooping up incoming radiation out of free space. Both ways are equally valid, and I'll switch freely between them in the interests of making things clear.

Another result of thinking of the antenna as a coupling device to a propagating wave is that the best antenna for a given type and path of propagation is not necessarily the one with the highest gain: it's the one which couples most efficiently with the propagating medium at the end of the path. The discussion of optimum antenna choice for various types of communication will provide some practical examples.

Indeed, this whole chapter will be a mixture of practical and theoretical topics because in order to understand antennas you need your feet firmly on the ground, as well as your head in the air!

GAIN

Gain is obviously one of the most important parameters of a beam antenna. It is also one of the least understood. In fact the very name 'gain' is false, because a passive device like an antenna cannot have true power-gain. The RF power that is radiated cannot be more than the RF power fed into the antenna. What we have come to call 'gain' is the increase in power radiated in one particular direction, at the expense of power radiated in other directions.

It is common practice to compare an antenna's gain – in its best direction, of course – against that of some standard antenna. The two most common standards are the isotropic radiator and the half-wave dipole. An isotropic radiator is a simple idea – it radiates equally well in every direction – but it doesn't exist in reality. On the other hand, a half-wave dipole radiates better in some directions than others, so we have to specify that we're comparing against *its* best direction too.

Antenna gain is usually expressed in decibels (dB) relative to the standard antenna. Decibels mean nothing unless the reference is named. It is good practice to write 'dBd' for gain over a

BEAM ANTENNAS & FEEDLINES

half-wave dipole, or 'dBi' for gain referenced to the isotropic radiator. Even in scientific publications, one is often forced to search the text for a clue to the reference used. In professional RF engineering there is a convention that antenna gains below 1GHz are given in dBd, and above 1GHz in dBi, so in this book we'll generally be working in dBd unless there's a good reason to use dBi. Either way, we'll always say which we're using.

PATTERN

The spatial radiation pattern of an antenna describes the distribution of power passing through the surface of an imaginary sphere surrounding it. This idea is difficult both to picture and to measure, so we normally plot the power in the horizontal and the vertical plane. The familiar 'butterfly' patterns (*eg* Fig. 7.1) are generated on a test range by plotting relative power on a circular chart which is rotated synchronously with the antenna. The advantage of this circular or *polar* diagram is that it's very easy to visualize what happens when you turn your antenna, because that's exactly how the measurement was made. An alternative method is to record the power level on a linear strip-chart as the antenna is rotated; although this shows up the low-level details more clearly, it makes the real-world behaviour of the antenna harder to visualize.

To compare antenna patterns, they have to be drawn to the same scale. Scientific and serious amateur publications almost always use charts with equally-spaced rings in dB steps, the outermost ring being the 0dB or

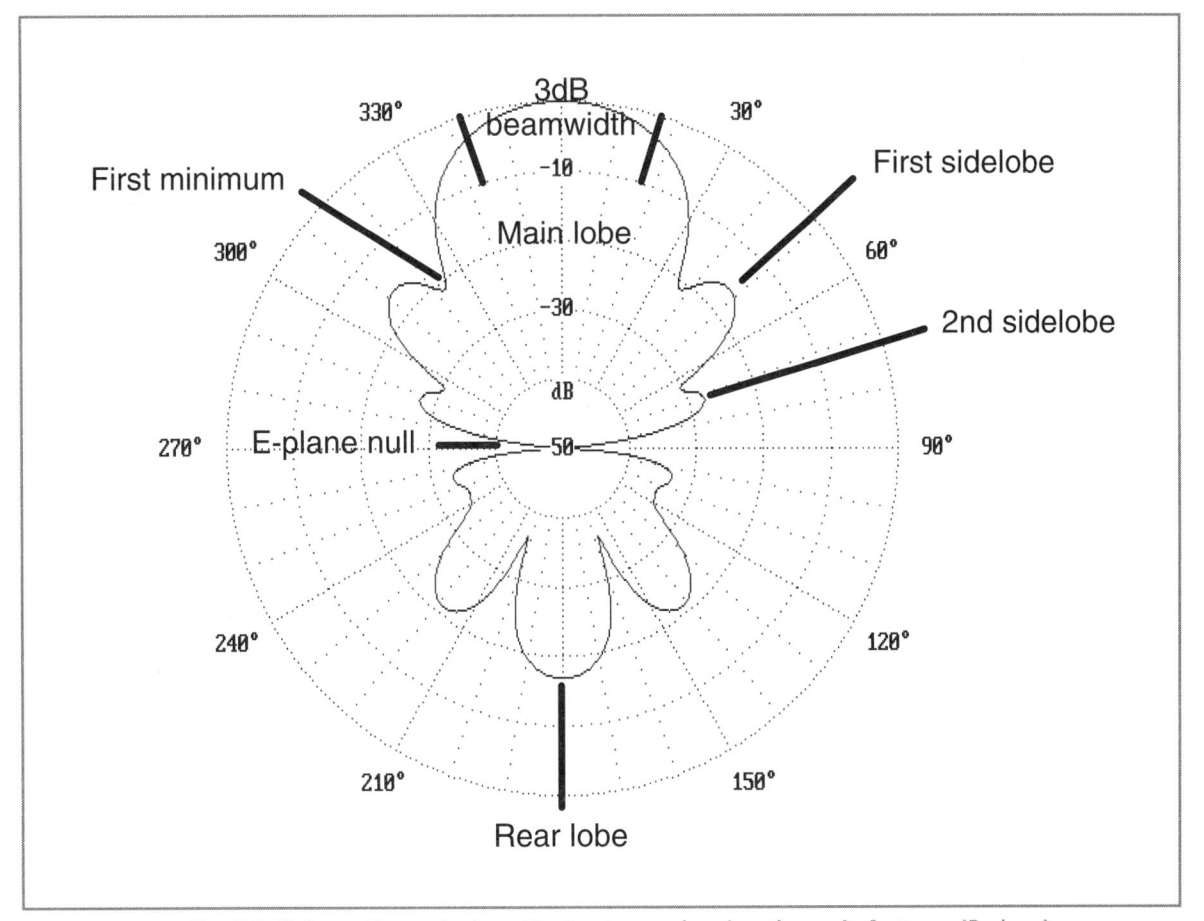

Fig. 7.1. Polar pattern of a long Yagi antenna showing the main features (E-plane)

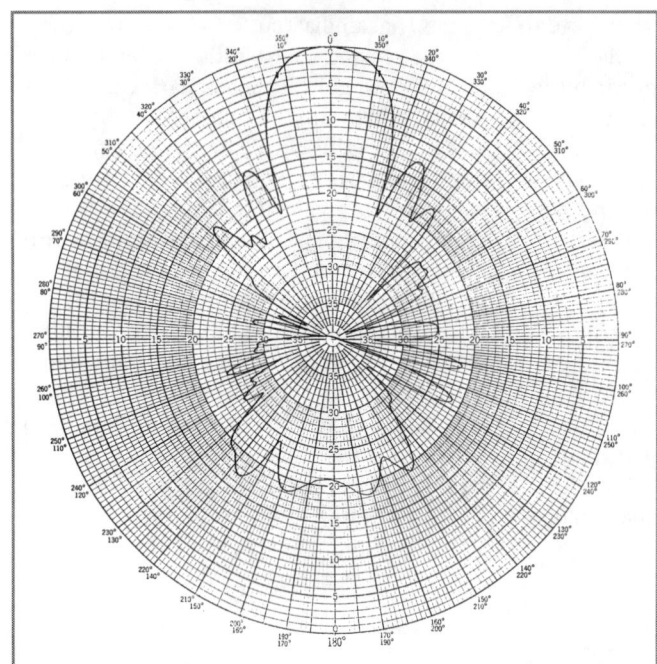

Fig. 7.2a. Polar pattern of a long-Yagi antenna, linear dB scale. Minor lobes are excessive, because the antenna is being operated below its design frequency

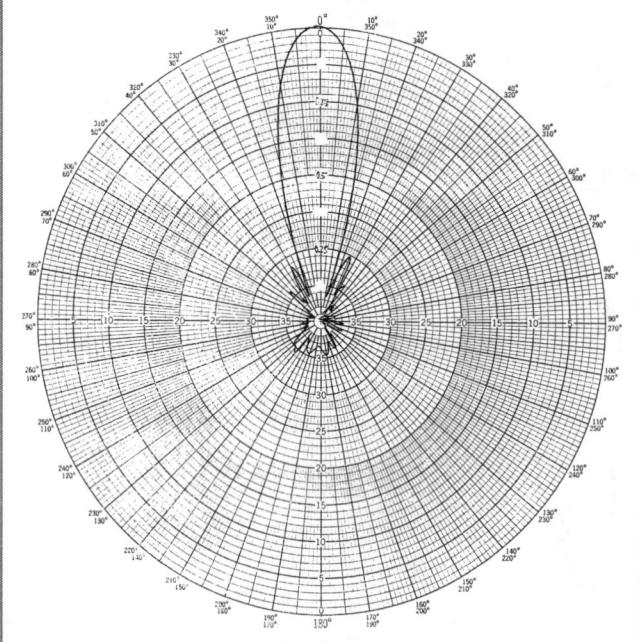

Fig. 7.2b. Same pattern as Fig. 7.1, but on a linear voltage scale which conceals the problems

reference level. A sensible scale goes down to −40dB or −50dB at the innermost ring or dot, or 0 to −40dB or −50dB across a strip-chart. This is adequate for good test-range measurements. Do not trust plots which are voltage-linear, or use shortened dB scales ending at −30dB or less, or even dB scales which become narrower towards the centre. These may hide important details (compare Figs 7.2a and 7.2b) and many amateur-band antennas *do* have something to hide! Even when you understand these pitfalls, it's still difficult to compare patterns plotted in two different ways; you generally need to re-plot both patterns on to a proper 40dB or 50dB polar scale.

The directional pattern of a gain antenna plotted in the normal way, with maximum radiation at the top of the 0dB ring, will look something like this:

(1) It will have a *main lobe* (also called the *front lobe* or *major lobe*) with a maximum in the desired direction. The main lobe is actually three-dimensional and shaped something like a baseball bat; the pattern shows a cross-section through it in the chosen plane of polarization.

(2) In almost all beam antennas, the whole pattern should be symmetrical about the direction of maximum radiation – though it often isn't quite.

(3) There will be several *sidelobes* of energy squirted out in other directions. The main features are likely to be the *rear lobe* at 180° and the pair of *first sidelobes* on either side of the major lobe. Each pair of sidelobes is actually a cross-section through a three-dimensional hollow *cone* of radiated power.

(4) Between the lobes are radiation *minima*. Antennas based on horizontal dipoles should have nulls at ±90° in the horizontal plane, just like an ordinary dipole would.

Fig. 7.1 shows all these features. Also marked on the diagram is the width of the main lobe, as measured between the half-power (−3dB) points on either side of the direction of maximum radiation. The angle between the −3dB points is called the *half-power beamwidth* or *aperture angle* and is usually given the symbol ϕ (the Greek letter *phi*).

The pattern of a horizontal dipole, measured in the horizontal plane, is called the *E-plane pattern* because it contains the electrical vector of the electromagnetic field. The pattern perpendicular to the dipole element(s) is called the *H-plane pattern* because H is the symbol for the magnetic vector. Except for very complicated antenna arrays, just these two plots will tell us all we need to know about the radiation behaviour. From the E- and H-plane plots the overall spatial distribution of radiated energy can be estimated, a procedure called *pattern integration*. Measuring the radiation patterns can therefore be an approximate method of measuring the gain.

Strictly speaking, pattern integration doesn't tell you the gain – it tells you the 'directivity', ignoring any non-radiative power losses. If you measure the E-plane and H-plane beamwidths ϕ_E and ϕ_H, you can estimate the directivity using a very simple formula derived by John Kraus W8JK:

$$D_I = 10\log_{10}\left\{\frac{42000}{(\phi_E \times \phi_H)}\right\} dBi$$

As I mentioned earlier, a dipole has some directivity and hence some gain over the imaginary isotropic radiator; 2·15 dB in fact. So we can amend the Kraus formula to give a result in dBd, thus:

$$D_D = 10\log_{10}\left\{\frac{42000}{(\phi_E \times \phi_H)}\right\} - 2\cdot15 dBd$$

Having said all that, you must be very careful with the Kraus formula – it involves a lot of assumptions which may not be correct. Firstly, if you're going to equate directivity with gain, you must obviously assume zero losses. This simplified formula also assumes a rather narrow front lobe (ϕ less than 40°), and negligible back and side lobes (at least –20dB). If there is significant energy loss to these minor lobes, the directivity and gain are reduced accordingly – which may mean *severely!* Although it is possible to refine the formula to take some account of sidelobe levels, you are just piling one assumption on another, and your final estimate of gain won't be worth much. What you *can* usefully do with the Kraus formula is to work it backwards, to check what beamwidths you can reasonably expect from a high-gain array; I'll say more about this later.

FEED IMPEDANCE

As seen from the feedpoint, an antenna presents a complex impedance with resistive and reactive components. The resistive component is composed of the radiation resistance, which 'consumes' power by radiating it, together with the loss resistance which really does consume power by heating up the antenna. The reactive part is caused by the antenna being made up of elements which behave like tuned circuits. Both the resistive and reactive components of the antenna feedpoint impedance will vary with frequency, the reactive component usually more rapidly.

An antenna should of course be impedance-matched to the source of RF power – usually the feedline. If it doesn't already match the feedline, you can obviously use some impedance-transforming device. But remember that transformers are not lossless, and loss always rises with the transformation ratio; so an antenna design which 'wants' to match the feedline is always desirable. The best gain and pattern do not always come at the frequency of a 'natural' impedance match, so it's always good to know what is happening to the feed impedance as the frequency is swept across the amateur band – and possibly beyond. Some preamps and transmitter final amplifiers have high gain outside the amateur bands, and may break into parasitic oscillation at frequencies where they see a particularly bad mismatch.

BANDWIDTH

The operating frequency range of an antenna is determined by two factors: gain bandwidth and impedance-matching (VSWR) bandwidth. These two factors are quite independent of each other, but they both limit the frequency range where operation is possible without retuning of some sort. For beam antennas, the most practical measure of gain bandwidth is probably the frequency span between the –1dB points. For Yagi antennas the gain drops much

BEAM ANTENNAS & FEEDLINES

more rapidly on the HF side of the peak than on the LF side, so it isn't always wise to design for maximum gain in the middle of the operating band.

VSWR bandwidth is generally defined as the frequency range within which the VSWR is less than 2. Different types of antennas behave very differently in this respect, depending largely on the Q of their matching networks.

CAPTURE AREA

As I mentioned earlier, antenna 'gain' is actually a misnomer. In the receiving case, an antenna's ability to gather energy out of the propagation medium is much better described by the term *capture area*. At a given field strength, the total power collected by the receiving antenna corresponds to a surface area which can be measured in square wavelengths. This is most obvious for a horn antenna, which really does capture almost all the radiation entering its open end. It is also easy to see how a large parabolic dish captures the energy striking the surface and focuses it onto a smaller antenna at the feedpoint.

Wire and rod antennas have next to no surface area, yet they still capture RF energy. It doesn't really matter how they do it, but the amount of energy corresponds to some *equivalent* capture area – which often has no obvious physical extent. We can discover the capture area of a wire antenna from its gain. Capture area is related to gain by a simple formula:

$$A = \frac{G_I}{4\pi}$$

where A is the capture area (in square wavelengths) and G_I is the gain over isotropic (as a ratio, not in dB). Thus a half-wave dipole with its gain of 2·15dB over isotropic has an effective capture area of 0·13 square wavelengths. A Yagi antenna with a gain of 13dBi (a factor of 20) has an capture area of about 1·6 square wavelengths which could be visualized as a circle of about 1·4λ in diameter. Although you can't see this shape, or guess its size without a little calculation, the capture area is obviously much greater than the sum of the physical element cross-sections – see the panel about **Antenna Areas**.

There is a complication with scatter modes such as aurora and troposcatter, because the incoming wavefront is not coherent (Chapter 2). This prevents the antenna from developing its full gain on receive, so the effective capture area is reduced. The shortfall in gain is called the *aperture to medium coupling loss*.

To sum up: capture area is not directly linked with physical antenna dimensions. Although it is sometimes useful to think of the capture area as an actual surface with definite boundaries, don't take this too literally – its real shape is not always obvious, and the edges are fuzzy. So capture area is really just an alternative to talking about gain.

ANTENNA NOISE

Chapter 4 has a plenty to say about antenna

ANTENNA AREAS

There are several different types of 'area' associated with antennas, so let's be clear what we mean in this chapter.

Capture area is a measure of the antenna's ability to collect incoming energy from free space. It is not a true geometrical quantity which you could measure with a ruler; in fact it's often very different from any area you could see when looking at the antenna.

Aperture area is a term I would only apply to antennas which possess a genuine, solid, physical radiation-collecting aperture, *ie* dishes and horns. It is the geometrical frontal area you could see when the antenna is beaming straight at you, and could measure with a ruler. Capture area is always smaller than aperture area; the ratio of the two is the *aperture efficiency*.

Frontal area is the term I'd use for other antennas apart from single Yagis, *eg* broadside arrays, or arrays of Yagis. It's the area enclosed by lines drawn around the outermost elements of the array, as viewed from the front. Capture area is usually somewhat greater than frontal area.

Physical area is what you (or the wind) would see when the antenna is beamed straight at you. For an aperture antenna, physical area and aperture area are generally almost identical. For antennas like Yagis and colinear stacks, physical area is generally much smaller than either frontal area or capture area.

noise-temperature, so I'll only mention it briefly here. Noise sources in antennas are either intrinsic (resistive losses) or external, *ie* pickup of noise radiation. Resistive losses can be kept small by using high-conductivity metal for radiating elements or reflector surfaces. Some designs are lossier than others and therefore more sensitive to element conductivity – see the next section on efficiency.

The greater part of the antenna noise is almost always picked up from outside. When pointed at the horizon, the antenna's field of view is normally half-filled by ground whose temperature is in the vicinity of 290K. The sky filling the other half may or may not be 'cold' (in the radio sense – see Chapter 4), depending on the frequency and on the celestial radio sources in that particular direction. The sun is the only strong source of extra-terrestrial radio noise on 432MHz, while on 144MHz and lower frequencies the Milky Way may produce a quite noticeable background hiss. On UHF and microwaves much of the sky is 'cold', so an antenna pointed away from earth can be very 'quiet' indeed. If the antenna is properly designed, resistive losses in the antenna itself add only an insignificant amount of noise. So the main source of antenna noise in EME and satellite operation on 432MHz and above may well be pickup of noise from the warm ground and other sources via the back and side lobes. It follows that a clean pattern is essential if full advantage is to be gained from ultra-low-noise preamplifiers. Your target should be at least 25dB suppression of lobes pointed at the ground or the sun. Poor balance in antennas and feedlines can also lead to noise pickup from the feedlines themselves.

To sum up: antenna noise temperature is not as clearly defined as some other system design parameters. It depends strongly on antenna orientation, and differs widely from band to band. Antenna noise temperatures are only significant in above-horizon work on UHF and microwaves. But that's only the antenna – don't forget that lossy feedlines and impedance transformers can severely degrade your system's noise performance.

EFFICIENCY

On VHF and UHF we don't hear much about efficiency of antennas. In contrast, on HF a lot of energy may be absorbed by lossy ground and loading coils, so efficiency is a common topic when talking about short-wave antennas. Although VHF and UHF antennas used to be considered lossless, we're having to change our views to some degree now that amateurs have access to computer programs that can actually calculate resistive losses and antenna efficiency.

As I mentioned earlier, a measure of gain known as 'directivity' can be computed from pattern data. Real gain is always somewhat less, and computer simulation shows the difference to be due mainly to skin-effect loss. So the quantity:

$$\frac{(gain)}{(directivity)}$$

can rightly be termed 'efficiency', and the lost power goes to heat up the antenna and its surroundings.

Skin-effect loss is caused by a concentration of current into the outer surface of the element, especially at higher frequencies. For example, on 144MHz over 99% of the current flows in the outer 40µm thickness of an aluminium element, rather than through the whole cross-section. This results in a higher RF resistance per unit length, and greater I^2R losses. Obviously the choice of conductor material has a strong influence on antenna efficiency, but there are other factors too. Similar radiation patterns can be generated by diverse configurations of element lengths and spacings, involving widely different geometries and numbers of elements. For a given type of element material, the antenna design which needs the fewest elements and the lowest currents to produce the same directivity will also have the highest efficiency, because the least energy is wasted in I^2R losses. A special warning about so-called "computer-optimized" designs: many simulation and optimization programs do not compute efficiency and may therefore produce results which cannot be duplicated using real-world materials. I'll return to computer optimization later in this chapter.

Before we leave the general topic of efficiency, I'll just mention a different kind of efficiency which relates to horns and parabolic dishes. This is the *aperture efficiency*: the ratio of the effective capture area to the physical aperture area (see box on **Areas**). Aperture efficiency is always less than 1 – not because of resistive losses, but because of such things as incorrect shape of a reflector, poor focusing, loss of energy by scattering and spillover or failure to capture all the energy because the holes in a mesh dish are too large. A warped and battered dish with a poor choice of feed antenna will have low aperture efficiency – and if it's covered in rusty mesh it will have high resistive losses too.

STACKING OF ANTENNAS

It is common practice to combine two or more directive antennas into groups, for additional gain. This immediately raises the question of optimum spacing.

What *is* 'optimum', anyway? It is obvious that two receiving antennas cannot pick up more than twice the power of one, so the limit of stacking gain is 3dB. To achieve this, the antennas must be separated far enough that their capture areas do not overlap, *ie* they don't compete for the available RF energy. Greater separation would not increase the gain but would give rise to increased sidelobe levels and unnecessary mechanical problems. So the optimum separation is the ***minimum*** spacing which achieves the full stacking gain. Experiments and calculations have shown that the optimum spacing (D_{OPT}) is a function of the half-power beamwidth (ϕ) in the stacking plane.

$$D_{OPT} = \frac{\lambda}{\left\{2\sin\left(\frac{\phi}{2}\right)\right\}}$$

where D_{OPT} is the spacing (in wavelengths) at which the stacking gain is maximum and ϕ is in degrees. In practice the maximum attainable stacking gain for two antennas is about 2·9dB. From this maximum you must subtract the losses of the power divider and the additional feedlines, so you'd be doing well to achieve 2·5dB net increase. Both theory and experiment show that D_{OPT} remains the same for more than two antennas stacked in a row.

Usually the beamwidths ϕ_E and ϕ_H in the E- and H-planes are different, so the optimum stacking distance must also be different in the two planes. A typical short Yagi antenna (about five elements) might have beamwidths of 50° in the E-plane and perhaps 70° in the H-plane. For a box configuration of four antennas with horizontal polarization, this would mean an optimum separation of 1·18λ horizontally and 0·87λ in the vertical plane (centre-to-centre). If the four antennas were simply stacked vertically up the mast, a total height of 3 times 0·87λ would be required. These values are theoretical and may need some correction, especially if there is evidence of severe interaction between antennas – as indicated perhaps by a bad impedance mismatch.

Slightly closer stacking is often recommended as a precaution against excessive sidelobes in the final array. Sidelobes arising from stacking rather than from the individual antennas are sometimes called *stacking sidelobes* or *grating sidelobes*. Two antennas separated by D_{OPT} will have a first stacking sidelobe at –11 to –12dB in the stacking plane, and this can be used as a test for correct spacing.

Reducing D to 80% of D_{OPT} costs about 0·5dB in gain but reduces these sidelobes to about –20dB; this is valid for stacks of **two** antennas only! If four antennas are stacked side by side or one above another, only the ***third*** stacking sidelobe is affected by variations of spacing; and closer stacking reduces the gain without lowering the overall sidelobe level significantly.

The above rules apply to any kind of antenna, so groups of groups (*eg* 4 x 4) may be designed by using the same basic formula. All you need to know are the half-power beamwidths in both planes. Finally, the most important generalization of all about stacking is this: antennas which have clean individual patterns will also behave the best in stacks.

TYPES OF DIRECTIVE ANTENNAS

Having covered the range of general antenna concepts, let's see how they apply to some real antennas.

This is a book on VHF/UHF DX communication, so we are not concerned with low-gain omnidirectional antennas. To produce gain and directivity on metre and decimetre waves, Yagi antennas are now used almost exclusively. But this hasn't always been so. Gain can be generated by other mechanisms and it seems appropriate to take at least a glance at them, if only to explain why Yagis are so popular.

ARRAY ANTENNAS

If RF energy from a transmitter is split up and radiated from two or more points, the emanating waves interact to form pattern of 'dark' and 'bright' zones. The directions in which addition or cancellation occurs are determined by the phases, amplitudes and locations of the source points. The same applies to real RF sources like dipoles, even though each dipole has a radiation pattern of its own; the familiar 'doughnut' shape. If all dipoles in an array are pointing the same way, the pattern of the whole array is made up of the dipole pattern superimposed on the pattern which point sources alone would generate. Where either pattern has a null, the whole array pattern must have a null too. That's why the patterns of complex arrays become broken into sidelobes. It is also the reason why the pattern of any array of parallel dipoles (including of course the Yagi antenna) must have nulls at ±90° in the E-plane, just as a single dipole does.

This principle – the *superposition principle* – holds for any array of elements. The only reservations are that the elements must be parallel and of a similar type, and that interactions between individual elements do not distort the current distributions. The superposition principle is equally true for groups of grouped antennas, up to the largest conceivable arrays.

BROADSIDE ARRAYS

The oldest application of the array principle consists of half-wave dipoles arranged in a plane at regular intervals and fed in phase. Maximum radiation is broadside-on to the plane of the array (in and out of the page in Fig. 7.3A), hence the name *broadside array*. The basic 'unit' of any broadside array is nearly always an 'H' arrangement of four dipoles spaced a half-wavelength centre-to-centre. Each pair of colinear half-wave dipoles has a high-impedance feedpoint at the centre, which matches well to the high impedance of the open-wire line connecting the two halves of the array, and the two quarter-wavelength sections of open wire transform this to a conveniently low impedance at the centre of the 'H'. For horizontal polarization the 'H' is laid over on its side as shown, giving the basic unit a colinear broadside array its familiar Wild-West name of 'Lazy-H'.

To make a larger array with more gain, two or more of these lazy-H cells are fed in phase, preferably by connecting their midpoints through equal-length feedlines to a central feedpoint (Fig. 7.3B). The radiation pattern of this array is bi-directional front and back. If one-sided radiation is desired, a reflector is placed behind the array. For VHF and UHF this reflector can either be a solid or mesh plane or an array of Yagi-style reflectors. The power gain of eight dipoles spaced 0·5 wavelength is hardly more than 7dBd, though this gain is achieved in two opposite directions – which can sometimes be very useful for routine net or relay operation over fixed paths. Adding a reflecting plane raises the gain by close to 6dB if the plane is very large. It is more common to use a screen measuring 1·5λ x 2λ to add about 5dB of gain, bringing the total to about 12dBd. This 3λ² frontal area corresponds to a theoretical gain of about 13·6dBd, in fairly good agreement with practical results. Amateurs sometimes use dipole reflectors instead of the reflecting plane; the reflectors are tuned to be self-resonant at 5-10% below the operating frequency, and are spaced about 0·2λ behind the radiators (Fig. 7.3C). The gain of this 16-element colinear broadside array is about 11·5dBd.

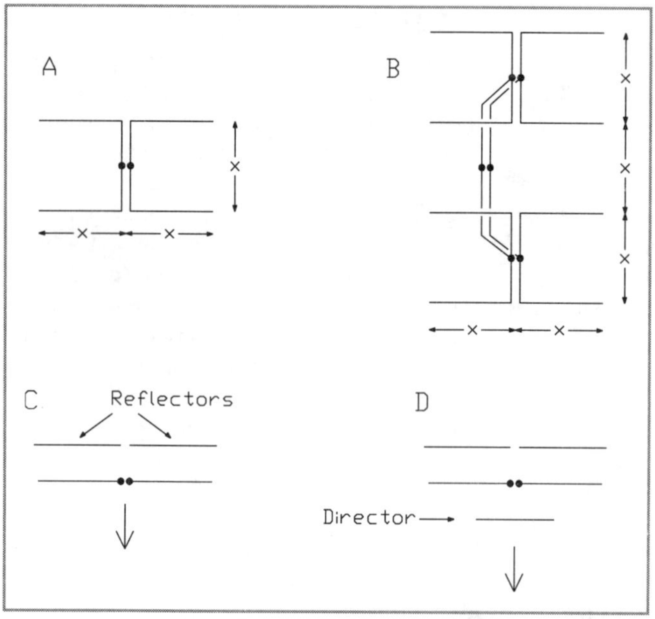

For many years the '16-element stack' was the dream antenna of 144MHz operators. Fairly simple mechanically, almost fool-proof electrically, with a clean pattern and over 10% bandwidth – this seemed like real performance in the early days of VHF. Enthusiasts were claiming gains of up to 16dB over a dipole! Some giant arrays were built – especially in the USA, of course – grouping four or even more of these 16-element antennas. Another significant development was the addition of an extra parasitic director to each colinear bay (Fig. 7.3D), giving a total of 20 elements and about 1·4dB extra gain. This type of array is still marketed by Cushcraft, and was instrumental in the 144MHz EME revival in the early 'seventies.

Theories (or rather myths) developed to explain the necessity of very large frontal areas for successful DX work – for surely such good results *had* to involve something more than mere mortal gain? Our antenna folklore is still coloured by this period in the 1950s and 1960s; and in retrospect I can't help wondering why, because some outstanding Yagi designs were already beginning to emerge by then. Perhaps one of the reasons was that big colinears are extremely tolerant of gross constructional mistakes, which Yagis are not.

Whatever happened to all those colinear broadside arrays? Well, other factors are important besides gain – for example wind-load, weight, rotating volume and moment of inertia. Broadside arrays are poor in all these respects; so if gain is your prime target, you've got big mechanical problems. A 64-element broadside array for 144MHz consisting of four 16-element groups is at least 8m high and 5m wide. Yet the gain is no more than 17dBd, which is easily obtained with four medium-length Yagis requiring a much smaller rotating volume and only four feedpoints instead of 32.

What basically happened to all those colinear broadside arrays is that they were blown down and replaced by Yagis. With today's antenna technology at 144MHz and

Fig. 7.3.
A 'Lazy-H' antenna – dimensions 'x' are λ/2
B Eight-element colinear broadside array
C Top view of a colinear broadside array with Yagi reflectors
D Top view of a 20-element colinear broadside array with an extra director at each level

TYPES OF DIRECTIVE ANTENNAS

below, broadside arrays are best regarded as a relatively easy way to reach moderate amount of gain but little more than that. However, where the objective is a broad, flat beam (*eg* for beacon transmitters) they may still be the best choice.

The situation for 432MHz is essentially the same. The mechanical problems of a colinear broadside array are not quite as great because of the smaller dimensions, but more gain is required for effective DX operation than on 144MHz so you're quickly into astronomical numbers of elements and feedpoints. Once again, the Yagi is a clear winner.

ENDFIRE ARRAYS

Any departure from equal-phase excitation of a row of parallel dipoles will make the pattern squint. And if the phase difference between successive elements in a long row becomes large enough, both main lobes will shift from broadside to 'endfire', **along** the direction of the row. Unlike the broadside array, the pattern of an endfire array is basically **uni**directional, because both broadside beams are shifted to the same side by the de-phasing.

The proper phase relationship between the elements can be generated in two ways: by driving each element via a feedline of appropriate length or by radiation coupling. In the latter case the non-driven elements are called 'parasitic' since they feed off the radiated energy, and re-radiate it with appropriate phase and amplitude to create the desired pattern. This is the Yagi principle and will be treated in a separate section.

Practically the only application of an all-driven endfire array is the *logarithmic-periodic* (LP) antenna. Fig. 7.4 shows two of the several possible ways of providing the required phase shift (close to 180°) between adjacent ele-

ments. The LP principle calls for a constant ratio of length and spacing of successive elements. The operating frequency range is determined by the longest and shortest elements, and the gain by the taper rate. Only those elements within about ±10% of half-wave resonance draw sufficient current from the feedline to be involved in the radiating process. So an LP antenna designed to work over a wide frequency range is actually a succession of several limited-band antennas on one boom. Usually only three or four elements are operative at the same time. If the taper is very gradual, many more elements can become operative and considerable gain can be developed, but only over a narrow frequency range. Although theory predicts a very high gain per unit length for antennas of this type, they are hardly ever used in practice – a Yagi is simpler.

The main use of the log-periodic antenna in VHF/UHF DX-chasing is to listen for propagation outside the amateur bands. One manufacturer (KLM Electronics) has taken advantage of its broad VSWR bandwidth by using a small log-periodic cell instead of a dipole driven element in wideband Yagis.

APERTURE ANTENNAS

Instead of building up large radiating surfaces from a multitude of elements, solid reflectors can be used to shape a beam and achieve gain. If the reflector dimensions are large with respect to wavelength, the laws of geometrical optics will apply and parabolic or other shapes of reflector can be used to concentrate the radiation from a dipole or some other small feed antenna. Nearly 100% aperture efficiency can be reached if the radiation comes from 'within' the antenna, as in the case of a horn. One heroic radio amateur has actually built a huge steerable horn for 432MHz and above, but they are pretty impractical for the rest of us.

To function properly, the diameter of a parabolic dish must be at least five or six times the wavelength and is better at ten times – so a 6m dish for 432MHz is electrically quite 'small'! Even for a dish of this size, the aper-

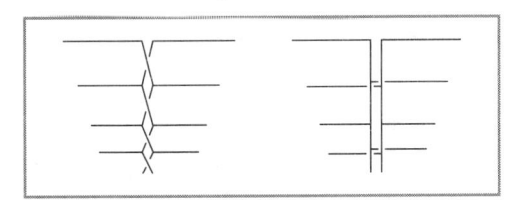

Fig. 7.4. Feed arrangements for log-periodics

ture efficiency will probably not exceed 50%. The problem is to illuminate the dish evenly from the primary feed antenna at the focus. For maximum gain the primary feed has to radiate a sharp cone of RF into the dish. This is not feasible at lower frequencies since the dimensions of the feed itself must be small to avoid blocking the beam. So a compromise must be made between 'spillover' (radiation past the reflector's rim) or poor illumination of the outer regions. A dish with spillover on transmit will also pick up noise from the ground on receive, especially when elevated, so small parabolic antennas are not very quiet; 50K is about the achievable minimum antenna noise temperature.

Despite these problems – not to mention the weight and windload – a number of amateurs are using parabolic dishes, mostly for EME work on 432MHz and above. In this high-gain bracket there are several reasons for choosing a reflector antenna. Firstly, it can be used on more than one band with a simple change of feed, or even a multi-band feed. And however high you go in size and gain there is still only one feedpoint, allowing a simple connection to the low-noise preamp without lossy line and transformer sections. Also, the polarization can be switched, or even continuously rotated, very simply at the feed.

The 432MHz band is about the lower frequency limit for amateur-sized dishes, though larger professional dishes around the world do appear occasionally on the moon-bounce scene – including sometimes the 300m monster at Arecibo, Puerto Rico, which I have to admit is big enough even for 144MHz!

SPECIAL ANTENNA TYPES

There are a few more types of antennas occasionally used in VHF DX work which I should at least mention.

Helical antennas consist of a corkscrew-like spiral conductor extending perpendicularly from a reflector plate. The pitch of the helix is around $\lambda/4$ and the circumference near to 1λ, and the gain increases with length as you'd expect. A minimum of about five turns are necessary for reasonably 'round' circular polarization. The sense of the polarization (left-hand or right-hand circular) follows the thread of the spiral. Gain is lower than with Yagis of the same length, although their simplicity makes helicals very useful for satellite work. However, several studies have shown the inferiority of circular compared with horizontal polarization in tropospheric propagation. The 3dB loss associated with contacts to ordinary plane-polarized antennas further reduces the merit of helical antennas in all-round use.

Backfire antennas gained a lot of publicity in the 'seventies but never achieved widespread use apart from a few military applications. There are actually two types: short-backfire (SBF) antennas and long-backfire (LBF) antennas. The SBF type consists of a 2λ diameter low-rimmed open cavity, a dipole in front of it, and a 1λ reflector plate behind the dipole. This arrangement exhibits about 13dBd gain over a relatively wide frequency range. The LBF type has some features of a Yagi antenna. A rod waveguide structure extends from a large reflector plate or cavity towards a smaller auxiliary reflector. With growing length of the waveguide section, the main reflector must become larger. Functionally it is analogous to a Fabry-Pérot resonator (better known today as a laser cavity) in which a narrow beam is formed by multiple reflections. The Yagi rod structure is used to hold the energy together between the reflectors. The smaller disk corresponds to the semi-permeable laser mirror which allows radiation to 'leak' out. Unfortunately the main reflector must be made rather large for high gain, and for best results must approach a parabolic profile in quarter-wave steps. LBF antennas with large reflectors have the same gain as equally large parabolic antennas, so there seems to be an advantage only in the transition region below about 5λ diameter where true reflector antennas cease to operate effectively. Little successful amateur use has been reported to date.

Rhombic antennas with high gain are often used in short-wave point-to-point work. So why not use them on VHF? Well, remem-

TYPES OF DIRECTIVE ANTENNAS

ber that ground reflection plays an important part in the build-up of gain from a rhombic, so rhombics always squint upwards. A single rhombic in free space would have a vertically split main lobe with a *null* at 0°. Stacking rhombics may lower the wave angle but won't bring it down to zero, and there's always a problem with ground-noise pickup. A complete redesign and a stack of four would be necessary to get the same gain from a 'groundless' rhombic array high enough for grazing-angle radiation. One entertaining feature of a very long rhombic is that you don't need a terminating resistor to make it unidirectional; by the time the RF arrives at the far end, most of it has been radiated and the rest has forgotten the way back!

Only one successful VHF DX application of rhombics is known, which is 50MHz and 144MHz EME. VK5MC and a few other big-time antenna farmers use extremely long stacked rhombics on 144MHz for the few minutes each month when the moon passes through the narrow fixed beam. In fact the first 144MHz EME QSO from the UK between strictly amateur stations was accomplished with a rhombic 100λ long (yes, 200 metres – 600ft!), erected in a farmer's field of course.

YAGI ANTENNAS

Although the Yagi is really just a special type of endfire array, its importance has become so immense that Yagis deserve a special section of their own. The antenna itself was more or less a by-product of VHF propagation studies and it's interesting to note that the first description in a 1928 IRE paper by Yagi and Uda referred to a 'long' Yagi array, very similar to what we use today. In that paper, Yagi and Uda described their brainchild as a "disrupted-rod waveguide structure" and that viewpoint explains some features not easily described by the usual element-by-element method.

A row of rods, electrically shorter than ½ wavelength and also spaced closer than ½ wavelength, will conduct RF energy just like a tubular waveguide. Like any other waveguide, it has a characteristic impedance and a phase velocity, and a passband within which it can be made to convey energy from a source to a terminated load with low loss and virtually no radiation. These properties are determined by the lengths and spacings of the rod elements. When used as an antenna, the rod waveguide structure itself may radiate very little energy; almost all the radiation comes from the impedance discontinuities at the two unterminated ends. And that's basically how short Yagis work – their patterns are formed essentially by two sources of energy at opposite ends of the boom.

If the Yagi structure is long enough – several wavelengths – the impedance discontinuity can be distributed to make the whole structure 'leak' radiation, rather like a tubular waveguide antenna with multiple slots in the wall. The lengths and spacings of the rods can be altered to cause local changes of phase velocity and characteristic impedance, resulting in radiation. Tapering towards the open end makes the structure increasingly 'leaky', and at the same time raises the characteristic impedance, thereby improving the match to free space; only a very small part of the energy is reflected at the last element. So, unlike the electrically short HF versions, the VHF/UHF long-Yagi is a *travelling-wave* antenna.

Beam formation by a distributed radiation source is immensely complicated, but fortunately the Yagi itself isn't aware of this fact. It's really very easy to build a Yagi that works after a fashion. But it has taken decades of research to explore the *full* potential of long Yagi antennas. All of this work was done by experiment, with great time and effort; and much of it was done by radio amateurs, I'm proud to say.

Until relatively recently, the mathematical treatment of realistic Yagi arrays lagged far behind experiment. After all, a large parasitic (*ie* radiation-coupled) array is quite a problem because each element influences all the others, and all the others affect it too. The accuracy of results depends largely on the mathematical representation of the conducting elements, and also on the numerical methods chosen to solve the complex integral equations. So the early theoretical work was limited to small,

GROUND REFLECTIONS

Ground reflections can play just as important a role at VHF and UHF as they do at HF. Although our antennas are at least a few wavelengths above ground, that does not automatically guarantee low-angle radiation.

For any horizontally-polarized antenna above ground, there is a *null* in the radiation pattern at zero wave-angle (horizontal takeoff), and the main lobe is tilted upwards. This is equally true of single antennas or arrays, and still applies even if you tilt the antennas themselves.

The bottom curve in the graph below shows the wave angle of the main lobe produced by an antenna and its reflection in a flat, perfectly-reflecting groundplane. Even a height of 10λ, the main lobe is tilted upwards by a degree or so. But antennas for 50MHz and 70MHz are usually much lower in terms of wavelengths, so the upward tilt can be quite marked.

Above the main lobe there is a pattern of alternating minima and maxima in the vertical radiation pattern. The wave angles for the 1st minimum, 2nd maximum, 2nd minimum and 3rd maximum are also shown on the graph, and there are still more at higher angles. At the best angles, you can enjoy an increase in gain of 6dB from ground reflections; at worst, you get severe attenuation. The effect of imperfect ground is to reduce the maximum increase and to fill in the nulls.

These maxima and minima explain why an antenna without elevation capability can produce some puzzling sun-noise measurements, and why you cannot always rely on ground gain around moonrise and moonset. They also show that a small, low-mounted antenna can sometimes be better for high-angle short-skip propagation than the monster at the top of the tower!

On 50 and 70MHz you can lower the wave angle by mounting the antenna on a long slope, or at the top of a suitable hill [4, 22]. The antenna height h required to produce zero wave angle on a downslope of angle Φ to the horizontal is given by:

$$h = \frac{1}{(2 \sin 2\Phi)}$$

Note that the slope must be many wavelengths long in order to produce the desired effect [4].

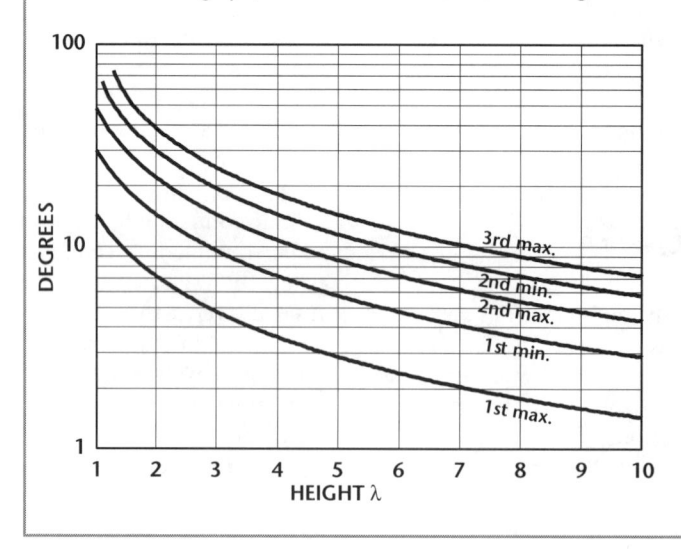

Wave angles of maximum and minimum radiation, as a function of antenna height above perfect ground. A logarithmic scale has been used to show the low angles more clearly

simplified arrays and a few special cases of larger arrays, while amateurs and professional engineers forged ahead developing *practical* long Yagis. Although modern computers and numerical modelling methods have now caught up with experiments, very little has changed as a result because the experimenters had already almost arrived at fully-optimized designs.

It takes just as much know-how and common sense to get sensible results out of a mathematical model of a Yagi as it does to develop and measure the real thing. Unless you thoroughly understand what Yagi antennas are about, either approach leaves plenty of room for mistakes. The first breakthrough of computer modelling into amateur Yagi design was the publication of a classic series of articles by the late Jim Lawson (W2PV) in 1980 issues of *Ham Radio*, later published as a book [1]. Unfortunately Lawson was followed by others who did not share his depth of understanding coupled with practical experience, and a number of perfectly good VHF antennas were sawn up and sacrificed as a result of blind computer-worship.

The next step forward was the availability of the MININEC program, a simplified version of the well-established Numerical Electromagnetics Code (NEC) for mainframe computers. The arrival of MININEC on the amateur radio scene coincided with rapid growth in the power of personal computers, so MININEC was soon running on a variety of PCs and minicomputers. Although the W2PV method is fast and surprisingly accurate, it doesn't predict the fine sidelobe structure of Yagis to the same accuracy as good-quality range measurements or MININEC [2]. More recently it has become possible to check the MININEC calculations on long Yagis using NEC itself, and although MININEC is a definite improvement over the W2PV program, some puzzling discrepancies remain [3].

To sum up the present status of computational methods: in skilled hands, a program such as NEC can give results which are as reliable – in general terms – as good-quality range measurements. But neither computation nor measurement can be trusted blindly, so the way forward lies through a combination of computation and practical experiment. Steve Powlishen K1FO and Rainer Bertelsmeier DJ9BV have been particularly active in developing computer-optimized Yagis for 432MHz and then testing them successfully on the EME path.

SHORT YAGIS

Yagis with small numbers of elements are now extremely well understood. Lawson's articles and book [1] give a very thorough explanation of what we VHFers would call electrically 'short' Yagis for HF and 50/70MHz, so I will just recall a few basic facts.

Currents induced by an electromagnetic field in wire or rod elements close to $\lambda/2$ long will depend in phase and amplitude on the relative tuning of these elements. The element is an electrical half-wavelength long when it resonates, which creates the maximum coupling with the electromagnetic field surrounding it. Longer elements behave inductively – the phases of the currents lag behind the exciting field. Elements shorter than the resonant length show capacitive behaviour – the currents lead the field. Tuning also depends on element diameter; thicker elements resonate at lower frequencies than thinner ones. Within reasonable limits, elements exhibiting the same reactance are interchangeable.

An inductive (longer than $\lambda/2$) element placed close to a dipole radiator builds up currents tending to offset the field in its direction, so it is said to act as a *reflector*. A capacitive (shorter) element enhances the field – and thus the energy flow – in its direction, and is therefore called a *director*.

There are mountains of literature on the optimum dimensions of Yagis with small numbers of elements, for the reasons I've already mentioned. Except perhaps for use on 50MHz and 70MHz, such short arrays are of little use to the VHF DXer because of their relatively low gain. If you're not bothered about a few tenths of a dB, scaling down a proven HF Yagi should serve the purpose if a small antenna is all you desire. Although no more than about 8dBd can be obtained from a full-size 4-element Yagi, this gain is a bargain compared to the small area required. For short-range ionospheric and meteor-scatter working, where the wave arrives at a significant angle above the horizon, up to 6dB more gain can be added by ground reflection if the antenna is at the right height above an unobstructed foreground [4] – see the panel on *Ground Reflections* opposite. So short Yagis can be useful for medium-range DXing on 50MHz in particular, for exactly the same reasons that similar Yagi antennas are widely used for HF communications.

But most VHF/UHF propagation modes require high gain and low-angle radiation. Short Yagis have very broad vertical (H-plane) patterns which mean a lot of useless high-angle radiation when polarized horizontally. To overcome this problem, and to fulfil our ever-increasing demands for gain, two solutions have emerged: vertical stacking of simple antennas, which I've already begun to discuss, and the development of very long Yagi arrays.

LONG YAGIS

The history of long-Yagi development is a fascinating tale, marked equally by experimentation on a heroic scale and by a catalogue of unnoticed mistakes, proliferating errors and almost total lack of co-ordination. Some of the best work of the 'sixties remained unpublished for twenty years, by which time it was already becoming obsolete; I refer of course to the NBS Yagis [5].

It is only a slight over-simplification to say that the breakthrough for very long Yagi antennas was reached by returning to Yagi and Uda's original concept – looking at a Yagi as a waveguide structure. As I said earlier, a chain of director elements forms a kind of waveguide. As the frequency increases, the directors become electrically longer and eventually become resonant half-waves. When this happens, or when the element spacing becomes greater than a half-wavelength, the structure will no longer carry a travelling wave; the velocity of propagation becomes zero. That is why the performance of a long Yagi falls catastrophically above its high-frequency cutoff point. Conversely, making the directors shorter increases the velocity of the travelling wave towards the velocity of radio waves in free space.

The same wave velocities, and hence the same overall antenna performance, can be achieved by an infinite variety of element lengths and spacings. But instead of experimenting with an infinite pile of aluminium rods (or these days, an infinite number of computer optimization loops) it's much simpler to look at the velocity **profile** of the travelling wave. By proper distribution of the wave velocity along the directors, you can optimize the gain, sidelobe attenuation and bandwidth all at the same time, and almost without compromise. The proper velocity distribution for this purpose is close to logarithmic – changing by the same **ratio** in each unit of boom length – and is achieved in practice by element length and spacing tapering which is typical of most modern long-Yagi designs. In other words, modern designs **are** close to optimum.

What kind of performance can you expect? Well, the gain of a good long-Yagi antenna should be pretty close to:

$$G = 8 \frac{L}{\lambda}^{0.78}$$

expressed as a ratio relative to a dipole; in decibels that would be:

$$G = 7.8 \log_{10} \frac{L}{\lambda} + 9 \text{dBd}$$

For example a Yagi 5λ long should have a gain of 28 times, or 14·5dB over a dipole. The presence of the exponent 0·78 means that gain is not quite proportional to length. So a Yagi 10λ long should have a forward gain of 48·2

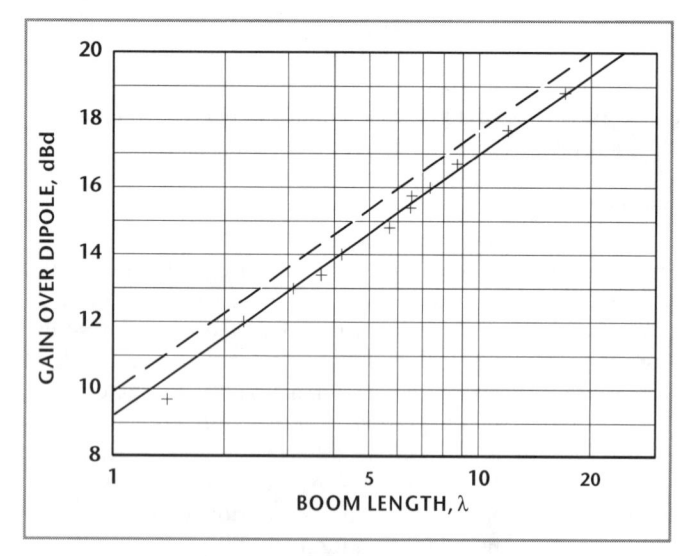

Fig. 7.5. Yagi antenna gain versus boom length.
Solid line and crosses: gain of Yagi antennas developed and measured by DL6WU at 432MHz and 1296MHz.
Dashed line: maximum gain obtained by computer optimization (theoretical).

TYPES OF DIRECTIVE ANTENNAS

times, or 16·8dBd, so doubling the length has increased the gain by only 2·4dB rather than 3dB as we might have hoped.

Fig. 7.5 shows that the gain relationship works over a very wide range of antenna lengths, from about 2λ to over 50λ! Even so, it is not a rigid limit, and some experimenters claim to have surpassed it by a few tenths of a dB. Unfortunately the high gain-per-length values of Yagis shorter than 2λ are not to anyone's present knowledge obtainable with long antennas, and there also seem to be good theoretical reasons not to expect that particular kind of miracle [4].

The difference between E-plane and H-plane patterns of short Yagis diminishes with increasing length, and beyond about 5λ boom length the beam cross-section approaches a circle. Even so, the first H-plane sidelobes remain consistently around 2dB stronger than those of the E-plane. The first sidelobes should always be attenuated to –15dB or better, and all higher sidelobes should be at least 20dB down.

Front/back ratio should exceed gain over a dipole by at least 3dB; in other words, whatever the forward gain, the rear lobe should register –3dBd or lower. The attainable front/back ratio appears to depend on the overall boom length in a periodic fashion, and certain lengths seem 'naturally' better than others. Even so, a particularly good front/back ratio only represents a dimple in a cone of energy radiated in the general backwards direction, and the *total* amount of backward radiation is more important than the spot figure at 180°. Multiple reflectors can reduce the frequency-consciousness of the front/back ratio, particularly if they approximate a solid reflecting screen, but it also seems that a well-optimized Yagi leaves less scope for improvement by using exotic reflector structures [6].

The final aspect of performance in a good long Yagi is that the gain and VSWR bandwidths should both be at least 5% of the operating frequency. This can be achieved with a perfectly ordinary dipole feed. In other words, a well-optimized long Yagi is also very frequency-tolerant, and an excessively narrow bandwidth indicates that the antenna is *not* optimized.

All these performance features are readily obtainable with modern long Yagis. ***Do not settle for anything less!***

Although many existing Yagis fall short of what can be achieved today, there are plenty of other reasons why the Yagi has become almost the universal VHF antenna. Construction is relatively simple; electrical connections are restricted to the driven element; and the main extension is horizontal rather than vertical, making Yagis easy to mount and rotate. Recognition of all these advantages by industry has paved the way to mass-production of Yagis for TV reception. Many TV antenna manufacturers have produced amateur Yagis as well, and of course some still do.

STACKING YAGI ANTENNAS

Like any other antennas, Yagis can be stacked to produce more gain. However, unlike some other antenna types, these stacked arrays make use of all three dimensions – depth as well as height and width – to produce a mechanically balanced structure.

Among the most frequently asked questions about Yagi antennas is that of correct stacking spacing. Although Yagi antennas obey the general rules given in the earlier section on **Stacking**, a few additional remarks are in order. Optimum stacking rules are based on the assumption of negligible mutual influence, *ie* no detuning of one antenna by another. It is hard to predict which antennas are critical in this respect, though we do have a few clues. Practice has shown that antennas with low overall sidelobe levels are less sensitive to proximity effects than those with a more 'ragged' pattern. Antennas operating at the high-frequency limit of their bandwidth are more likely to be pushed over the edge – this can usually be recognized from the pattern by the absence of a pronounced null between the main lobe and the first sidelobe.

The decision whether to stack a few very long Yagis or a greater number of shorter ones depends to some extent on the desired shape

BEAM ANTENNAS & FEEDLINES

of the overall pattern – see the following section on the choice of antennas for various propagation modes. From a standpoint of volume economy, the shape of the array should approach a cube, which would mean arranging four medium-length antennas in a box configuration, or 16 very long Yagis in a 4·4 stack. Anything more ambitious than that looks like a case for a parabolic dish.

Another problem for the all-round VHF/UHF DXer is how to stack antennas for different bands. This is an area where computer studies can cast some light, and the general conclusions are just what you'd expect anyway. Basically, the antenna for the lower band doesn't mind how close you put the higher-frequency antenna. From the viewpoint of a 144MHz Yagi, a 432MHz antenna is only an electrically small object, even though it's inside the capture area of the 144MHz antenna. But from the viewpoint of the 432MHz array, any part of the 144MHz Yagi within its capture area will cause a major disturbance. So the minimum stacking distance should be the radius of the higher-frequency antenna's capture area, *ie* somewhat more than half the recommended stacking distance for a pair of those antennas.

Stacking single antennas for several bands is therefore pretty simple. You start the 'Christmas tree' with the largest and lowest-frequency antenna at the bottom and work upwards with gradually decreasing vertical spacing, until you reach the highest band of interest – or until the whole thing gets blown over. In principle it would improve the low-angle radiation on the lower VHF bands if you put the longer-wavelength antennas at the top and work downwards, but practical mechanical considerations usually forbid this.

Problems may arise when stacking pairs of long Yagis for different bands such as 144MHz and 432MHz, or nesting boxes of four Yagis, because the total separation between the 144MHz Yagis may not be sufficient to fit in the 432MHz Yagis without a clash of capture areas. Also, the antennas for the higher band are still within the capture area of those for the lower band, and it's something of an exaggeration to say that you can ignore their presence completely. The performance of a highly-optimized 144MHz antenna is likely to be noticeably affected by 432MHz antennas within its capture area.

A few other rules of caution will also apply, which could be summed up like this:

● Keep anything which might resonate out of the high field-strength regions.
● The higher-frequency antenna will rarely do much harm, but cables and mounting frames may.
● The sensitive space around a Yagi is something like a tube along its length, spreading out into a cone beyond the end; so watch out for very long lower-frequency Yagis protruding forwards into that cone, regardless of the vertical separation.

CHOICE OF ANTENNAS

A German saying has it that "He who has the choice has the pain". Most of us have to live with limitations which keep us from putting up all the antennas we might wish to install, and we therefore have to settle for a compromise between our ambitions and our resources.

In this section I will take a look at the antenna requirements for the different propagation modes on each band, so that you can decide what combination serves your own purposes best.

TROPOSPHERIC PROPAGATION

'Tropo' – the ability of the atmosphere to deflect radio waves beyond the optical horizon – is the most-used propagation phenomenon for VHF and UHF DX. As described in Chapter 2, it is present on all bands. Actually there are two tropospheric modes: ducting and scatter. Tropo ducting occurs in the lower atmosphere, so it requires low-angle radiation. Only radiation below 1 or 2° elevation can couple into the refractive layers. Although a favourable location may be the most decisive factor for this, the antenna must be high enough to allow a low-angle take-off; 10λ above ground is not too much.

The tropo-ducting specialist would want an antenna with a flat, wide beam to pick up weak signals slightly off-heading when searching the band. Such an antenna would get most of its gain from vertical stacking but, contrary to earlier beliefs about the magical powers of broadside arrays, it makes no difference what kind of antennas are stacked as long as it's done correctly. There is no minimum gain for tropo-ducting DX, and there's never too much either! So vertically-stacked Yagi antennas – **lots** of them – are recommended for tropo ducting on all the VHF and UHF bands.

Troposcatter is a mode which relies on atmospheric turbulences. In contrast to ducting, troposcatter can sometimes make use of higher take-off angles, so height above ground and a good site may not be quite so important.

The path losses for medium- to long-range troposcatter communication are much higher than for duct contacts, so high-gain antennas are mandatory. Though extended troposcatter is of course possible on 50 and 70MHz, its amateur use is precluded in most countries by power limits. Furthermore, other lower-loss DX propagation modes exist on these 'long' metre-wave bands. On 144MHz and above, power levels of 50kW EIRP are not excessive for serious long-distance experiments, *eg* 1kW RF and 20dBi gain. Long Yagis in stacks and quad arrangements are what's needed for troposcatter.

Contest operation relies mostly on tropo: refraction and forward-scatter for the local and middle-distance contacts, and ducting and extended forward-scatter for the real DX with similarly-equipped stations. A big contest station needs a reasonably broad beam to avoid missing calls from off the sides, yet it also needs high gain for a big signal and some DX potential.

To get the best of both worlds when effort is no object, many top contest groups use extremely tall stacks of single Yagis.

7 - 19

IONOSPHERIC PROPAGATION

There are several ionospheric propagation modes occurring on one or more of the VHF bands, namely Es, F2, FAI, meteor-scatter and aurora. I'll deal with the last two separately, although common to all of them is their dependence on high ion or electron concentration in regions at least 50km above ground. Except at extreme ranges, the take-off angle can be quite high. Random polarization is another characteristic of many ionospheric propagation modes.

In single-hop sporadic-E openings, path attenuation can often be so low that 'anything goes' and a highly directive array can even be a hindrance. A short- to medium-length Yagi is really all that's needed. It might also be a good idea to install a log-periodic antenna covering about 40-150MHz for band-scanning purposes.

F2 propagation sometimes reaches the 50MHz band and may perhaps reach 70MHz. Signals are weaker than by Es, so a little more gain wouldn't hurt. At least a full-sized 4- or 5-element 50MHz Yagi should be considered. At these longer wavelengths, insufficient height above ground can be a real limitation on lowering the take-off angle, so it may be better to concentrate on getting a smaller antenna to a good height rather than building some monster which you daren't mount above the rooftops.

TEP (transequatorial propagation) and FAI (field-aligned irregularities) were discovered by radio amateurs only quite recently. There still is a lot of discussion about these propagation mechanisms (Chapter 2), which sometimes seem to be similar or even the same and are difficult to separate from other better-known propagation effects. This is especially true for 50MHz, where other ionospheric propagation modes may exist at the same time. So any recommendations on antennas must at present be tentative and aimed towards use in experimentation.

Antennas having enough gain for F-layer contacts should suffice for transequatorial 50MHz DX – relatively little is known about FAI propagation on this band, and even less on 70MHz. Antenna requirements for 144MHz TEP depend on the stations' location with respect to the geomagnetic equator. The borderlines for this mode appear to be around 4000km to the north and south of it, suggesting equipment of at least troposcatter grade. Stations closer to the geomagnetic equator report stronger TEP signals, but may need elevation capability. FAI on 144MHz is another weak-signal mode. Although there seem to be preferred paths in central and southern Europe, antennas for FAI work need to have virtually EME standards of performance and elevation capability. Little information is available on the existence of these phenomena on 432MHz.

METEOR-SCATTER

Meteor-scatter is particularly popular on the 144MHz band, although it is also used a good deal on both 50MHz and 70MHz. The 70MHz band is something of a special case, being available in so few countries; even dealing more generally with 50MHz in Europe, the possible meteor-scatter paths are defined more by licensing than by propagation or antennas. For the short-haul paths on these lower bands, *eg* G–EI or G–GM, a fairly low antenna is required because of the high take-off angles. But we should also recognise from experience in the USA that 50MHz MS is in fact an extremely reliable mode for distances of the order of 1000km, especially when using some hundreds of watts and at least a 1·5λ Yagi.

On the 144MHz band, MS has been called "the poor man's DX". One can bridge 1500km with ease by using just 100W and a single medium-length Yagi antenna – given a major shower, the right operating techniques (Chapter 3) and some patience. The true MS specialist is more demanding. To make DX contacts using meteor trails barely above the horizon, high-gain arrays (and more power) are needed. However, the chances for medium-range contacts are definitely lowered because many useful trails are outside the 'window' of a highly directive antenna.

Meteor-scatter on 432MHz is definitely not for everyone; the echoes are rare, weak and

CHOICE OF ANTENNAS

short. High power and high-gain antennas are mandatory. But once again, the narrow pattern of high-gain antennas will reduce the chances to 'see' the same meteor trail from both locations. Precision planning is needed to maximize the chances of a successful contact, and this may include elevating the antennas except on extreme horizon paths. The general system requirements for 432MHz MS are not far below those for EME, together with the high-speed keying and decoding equipment as used on 144MHz, which is essential with the short bursts typical of 432MHz.

AURORA

Radio aurora occurs on all VHF bands, and also on 432MHz. Antenna requirements are similar to those for meteor scatter, but they also depend on your geomagnetic latitude and the typical location of auroras in your part of the world. A station situated in the north will be able to make many contacts using relatively small antennas, though more gain (and elevation control) would greatly extend the range. Stations further south can do without elevation capability, but need as much gain as possible. This is particularly true for 432MHz where signal levels are 15 to 20dB lower than on 144MHz. Adjustable polarization could be useful, but is difficult to arrange, especially in a steerable and tiltable Yagi array.

SATELLITES

Satellite communication might not be generally accepted as a variety of VHF DX operation, but that is a matter of definition. There is a certain technical challenge in exploring the extreme conditions which still allow contacts – low power, horizon passes *etc*. Specialists tend to use antennas with circular polarization, preferably switchable from left-hand to right-hand. Under most operating conditions there is a definite advantage over linear polarization; the reduced spin modulation plus the (theoretical) 3dB gain over an equal-length linear antenna can be quite noticeable. However, many operators throw away the advantage by improperly mounting or feeding their antennas and thereby severely distorting the polarization's circularity. Close to the horizon, linear polarization is often advantageous, horizontal being favoured in practically all cases over vertical. So all in all, amateurs who don't specialize in satellite work are not sacrificing much by using linear polarization. A good tropo combination array for 144MHz and 432MHz, interspaced with due care and providing full steerability, has all the performance necessary for satellite operation on these two bands.

MOONBOUNCE

EME (earth-moon-earth) operation is perhaps the greatest technical challenge to the VHF/UHF amateur. Success depends on a number of prerequisites, each of which are highly demanding. There are physical limits to receiver sensitivity, and legal limits to transmitter power. Operating skill can be learned, but the decisive role is played by the antenna system. It must be large – there's no way around that – and it must be *quiet*. It must also be accurately steerable, and should preferably be strong enough to survive the next winter. Even though many parts today can be bought off-the-shelf, it still takes plenty of engineering work to turn them into a functioning system. There is a good deal of special literature on EME antenna techniques, and only a general idea can be given here. But anybody setting out to erect a large VHF/UHF array for *any* purpose should also at least consider the EME requirements before making the final decisions.

The 50MHz band is not very favourable for moonbounce operation. In Europe, power and licence limitations preclude operation on this band or 70MHz in any event; the hurdles are high even with the power levels permitted in the USA. Theoretically the round-trip path loss is around 178dB, so antenna arrays providing about 15dBd should be sufficient. This would call for a quad array of four medium-sized Yagis spaced some 10m apart. For example, WA4NJP has had good success with four eight-element Yagis of the W1JR design [7,8], and has worked stations using as few as two long-Yagis – although I really do mean *long!*

The 144MHz band carries most of the EME activity, closely followed by 432MHz. Newcomers often ask where they should start. If you have plenty of space and are not afraid of building a big antenna, 144MHz is easier. Four 13dBd Yagis spaced about two wavelengths apart will do the job. There are several stations already using arrays of this size for terrestrial DX, so if you're contemplating an array of this size, consider adding an elevation drive for EME. The main lobe of a four-Yagi array as described is about 15° wide in both planes, making steering fairly uncritical although heavy-duty mechanisms are of course mandatory. A station equipped along these lines is not top-notch for 144MHz EME but is certainly capable of contacting at least 50% of the other active EME stations around the world – enough to broaden anyone's operating horizons!

Moonbounce operation on 432MHz places higher demands on equipment performance. The path loss is 9dB higher than on 144MHz, so the antenna system must make up for half the difference on each end. Still, the size of a sufficiently high-gain array is much more handy than on 144MHz. If you have limited space or intolerant neighbours, you'd do better to try this higher band. Four really long Yagis (eg 10λ boom length) at approximately 3λ spacing can produce the necessary gain of about 23dBd. The antennas must be of modern low-sidelobe design and have a good front/back ratio to make full use of the low sky-noise levels at 432MHz. Though contacts have been achieved with even smaller arrays, the recommendation given here should be rated as a minimum to be able to hear your own echoes frequently and therefore to work stations with similar or better equipment.

You can't cut corners with such a small array for 432MHz EME. You need short, high-grade cable connections between the feedpoints and a low-loss power divider mounted out in the rear of the array, and a good preamplifier connected straight on to the divider. Better still (if you do it correctly) would be open-wire feeders using no power divider. The next step up from four Yagis is an array of six, which also allows relatively short cable runs to a single 6-way divider. Compared to any 144MHz EME array this is still a rather

RECOMMENDED MINIMUM ANTENNA EQUIPMENT FOR SERIOUS VHF/UHF DX

MODE	50/70MHz	144MHz	432MHz
Tropospheric refraction	8dBd, 1λ Y	10dBd, 1·5λ Y	16dBd, 8λ Y
Tropospheric scatter	16dBd, 2 x 3λ Y =	19dBd, 2 x 8λ	Y =
Sporadic-E	Y, LP	10dBd, 1·5λ Y	
F2-layer	8dBd, 1λ Y		
Meteor-scatter	8dBd, 1λ Y	10dBd, 1·5λ Y	19dBd, 2 x 8λ Y = E
Aurora	8dBd, 1λ Y	10dBd, 1·5λ Y	19dBd, 2 x 8λ Y = E
FAI, TEP	8dBd, 1λ Y	19dBd, 1·5λ Y	23dBd, 4 x 10λ Y B E
Satellite		10dBc, 1·5λ XY F (13dBd, 2 x 1·5λ Y)	15dBc, 5λ XY F (18dBd, 2 x 5λ Y)
EME	15dBd, 4 x 1λ Y B F	19dBd, 4 x 3λ Y B F	23dBd, 4 x 10λ Y B F (7m parabolic)

KEY:
Y	Yagi antenna	XY	Crossed yagis	LP	Log-periodic
=	Stacked vertically	B	Box arrangement		
dBd	dB over a dipole	dBc	dB over crossed dipoles (circular polarization)		
E	Elevation steering advantageous			F	Full steerability required

CHOICE OF ANTENNAS

seldom used for all-round operation, and may therefore be put on a low separate mast to minimize wind loading, visual impact and sleepless nights.

DESIGN GUIDELINES

Before plunging into an antenna construction project – **think twice**. Make up your mind which bands and propagation modes are most important to you. Consider the gain and pattern requirements and try to combine them in as few arrays as possible. The chart opposite may help you make these decisions. Meanwhile I'll describe a few Yagi designs you may wish to consider.

Fig. 7.6. Photo of the six-yagi 432MHz array at DL6WU, with four short 144MHz Yagis stacked between

compact antenna, and even accommodates an interspaced medium-gain 144MHz array without undue interference (Fig. 7.6). Aiming must be accurate to within 2° because the –3dB beamwidth is of the order of 6–10° in azimuth and elevation. Stacking three bays vertically is strongly recommended to reduce the sidelobes towards the ground, and also to make the antenna more useful for terrestrial operating.

Larger EME arrays have of course been built for both bands, consisting of eight, twelve, sixteen or more individual Yagis. These are

DL6WU YAGI DESIGNS

The outcome of my own investigations into Yagi design and performance has been a 'family' of Yagis which can be designed to meet your own size and performance requirements using just a few charts and tables. These are developed from material originally published in *VHF Communications* [9, 10], and more recently in a Dutch translation by PA0MS [11].

The design guidelines include optimized director spacings and lengths which give a combination of high gain and good suppression of minor lobes over a broad bandwidth. Using these dimensions, longer or shorter Yagis can be designed by simply adding or subtracting elements, down to *a minimum of ten*. Another design objective was a feed-point impedance close to 50Ω, to avoid the need for a high-Q impedance transformer which would restrict the VSWR bandwidth.

ELEMENT SPACINGS AND LENGTHS

Element spacings are given in the table on this page, both individually and measured cumulatively from the reflector. As you will see, the minimum number of ten elements implies a minimum boom length of 2·85λ, *ie* almost 6·0m (20ft) on 144MHz and *pro rata* on the other bands. The spacing increases gradually towards the constant value of 0·40λ for the 13th director and beyond and further directors can be added at this spacing, without limit as far as anyone can see. Yagis of more than 50 elements have been built for 1·3GHz and perform exactly as expected.

To select a suitable boom length, you can start with some idea of the physical size you want, or use the estimated gain figures in the table of element spacings. You must then choose the Yagi which most closely fits your requirements, and accept that particular boom length. You can**not** squeeze or stretch the spacings to fit some stock length of tubing!

Although any DL6WU Yagi will perform well, some are better than others by virtue of having a 'naturally' good front/back ratio in excess of 20dB, which may also lead to slightly

ELEMENT SPACINGS FOR DL6WU LONG-YAGIS

Element	Distance along boom (λ)		Approx gain dBd
	Relative	Total	
Reflector	0	0	
Driven	0·200	0·200	
Director 1	0·075	0·275	
2	0·180	0·455	
3	0·215	0·670	
4	0·250	0·920	
5	0·280	1·200	
6	0·300	1·500	
7	0·315	1·815	
8	0·330	2·145	11·9
9	0·345	2·490	12·4
10	0·360	2·850	12·8
11	0·375	3·225	13·2
12	0·390	3·615	13·6
13	0·400	4·015	14·0
14	0·400	4·415	14·3
15	0·400	4·815	14·6
(constant)			
20		6·815	15·8
25		8·815	16·6
30		10·815	17·3
...and so on			

DL6WU YAGI DESIGNS

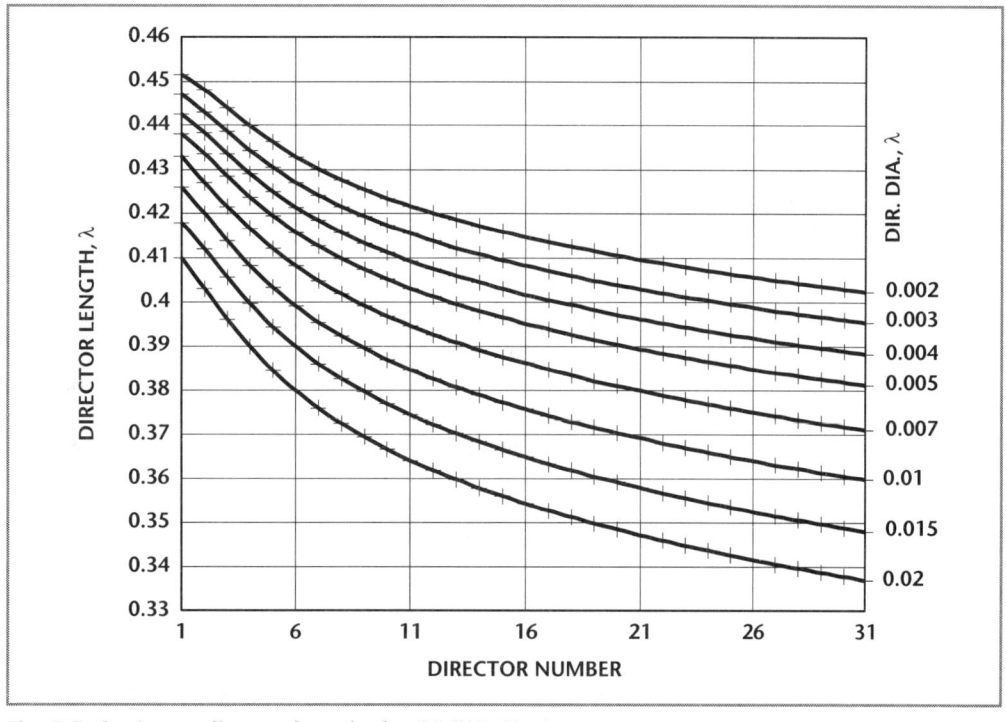

Fig. 7.7. Optimum director lengths for DL6WU Yagis

improved forward gain. These especially favourable lengths occur in the Yagis with ten elements, 14 or 15 elements, 19 or 20 elements, and so on at intervals of approximately five directors or 2λ in boom length. On the other hand, if you are mostly interested in gain, just go for the longest boom you can manage.

As already stated, the performance of a parasitic director is closely related to the reactive component of its impedance. Elements with equal reactances are essentially interchangeable. Reactance depends on both electrical length and diameter, so a chart of optimum element lengths and diameters will look very much like a reactance chart. Fig. 7.7 shows the optimum length for the first, second, third... director as a function of element diameter in wavelengths. This chart is valid for Yagis with eight or more directors, *ie* ten or more elements in total.

As you might expect, the element lengths 'taper' along the boom. After the first few elements which launch the travelling wave, the director lengths reduce in a logarithmic manner, each being a constant fraction of the length of the one before. Also, thicker elements need to be shorter to give the same reactance as thinner ones.

To use Fig. 7.7, first select an appropriate element diameter. If it doesn't correspond to one of those plotted, sketch in your own curve using the plotted curves as guidelines. Then simply read off the length of each director, in wavelengths, and convert to the frequency of your choice.

The driven element and reflector lengths should both be a little less than 0.5λ, though both may need to be adjusted. The driven element length should be adjusted for the best match to 50Ω, and the reflector length for best front/back ratio. Neither length is particularly critical as to element diameter except for very thick elements, and the following table gives suitable starting values (in wavelengths) for experiment:

Element diameter	Driven element length		Reflector length
	Simple dipole	Folded dipole	
0·0001	0·483	0·488	0·494
0·001	0·477	0·482	0·489
0·01	0·456	0·470	0·480
0·02	0·445	0·464	0·474

In fact it is seldom practical to match a simple dipole driven-element directly to 50Ω; you should plan to use a folded dipole and a half-wave balun, especially on the higher bands – see later. The length of the folded dipole is surprisingly independent of the spacing between the two parallel limbs.

Instead of working with charts and tables of dimensions, you can use a home-computer program to automate the design procedure. This has the advantage of allowing you to explore a range of possibilities quickly and easily, without a lot of tedious arithmetic. Jerry Haigwood (KY4Z) has developed a simple and accurate numerical representation of the chart shown in Fig. 7.7. An extended version of his DL6WU Yagi design program is publicly available as a BASIC listing or on disk; further details of availability of computer programs are given elsewhere in this book.

The DL6WU designs are really intended for *long* Yagis, and are mechanically feasible only at 144MHz and above. There are many suitable designs for shorter Yagis for the lower VHF bands, some of which are recommended in Chapter 9, and also of course there are plenty of other long-Yagi designs. I recommend that, in making your choice, you use only modern designs backed up by some form of accurate evaluation – either by measurements on a test range or at the very least by extensive tests on the EME path – a hard test which tells no lies.

I don't want to decry computer-designed Yagis, but I wouldn't recommend any design that has not also been proven by practical testing in the real world.

ELEMENT LENGTH CORRECTIONS

The element lengths in Fig. 7.7 refer to elements somehow suspended in free space. Lengths of practical elements generally need to be corrected for the effect of the boom and mounting clamps. Also you may want to adapt other Yagi designs besides my own, and this may involve changing the element diameter. If so, you'll need to correct all the element lengths, so I'll show you how to do that too.

BOOM EFFECTS

Unless you use only insulating materials to support the elements, the presence of a nearby metal boom and other metal fittings will partially 'short-circuit' the centre of the element and reduce its electrical length. So you will always have to *lengthen* the elements to compensate. Ignorance of these boom effects led to Yagis gaining a reputation for being difficult to reproduce, and later to some curious attempts to avoid the need for corrections. However, boom-effect corrections are now quite well understood and you can approach them with some confidence.

The required correction depends on the

Fig. 7.8. A selection of element mounting methods

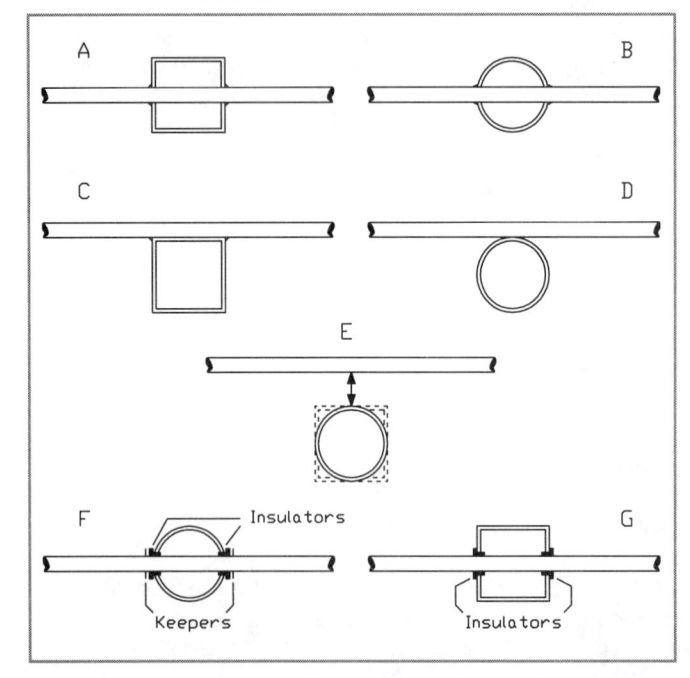

DL6WU YAGI DESIGNS

element mounting method, and on the thickness of the boom. It also depends to a smaller extent on the length and diameter of the element itself. But your first option is whether to use a metal boom at all or something made of an insulator such as wood or fibreglass. The attraction of insulating booms is that they require no element length corrections. Wooden booms are often touted as being low in cost, but in fact the costs of buying good-quality, straight-grained, knot-free timber and of weatherproofing it so that it *remains* an insulator are not insignificant. Fibreglass is a better insulator than wood, but is less rigid and much more expensive.

If you are using a metal boom, your basic options are to connect the elements securely to the boom or to insulate them in a fixed mechanical relationship to the boom. Of the two, the all-metal method has been more thoroughly explored but is open to the effects of corrosion. Insulated mounting schemes to side-step corrosion problems are more recent and less well characterized, but fortunately the corrections are smaller. Fig. 7.8 shows some of the options for element mounting.

Corrections for elements mounted through the centre of a square or round metal boom *and securely connected to it* (Fig. 7.8A, B) are shown in Fig. 7.9 as a function of boom diameter in wavelengths. The correction to be added to all element lengths is approximately one-third of the boom diameter for a boom diameter of 0·01λ, and a somewhat larger fraction of the boom diameter for thicker booms. A diameter of 0·01λ corresponds to 20mm at 144MHz, so you'd need to add about 6mm or 0·25" to all element lengths if you chose this mounting method. For elements sitting on top of a square metal boom (Fig. 7.8C), the correction is about the same as that for mounting through the middle; but once again you'd need to ensure a solid long-term electrical connection. The correction for elements either touching or almost touching the top of a round boom (Fig. 7.8D) is approximately halved. Elements mounted on insulators further away from the top of the boom (Fig. 7.8E) require very little adjustment, the correction dwindling rapidly from 50% of the through-boom value to 5% or less for elements mounted one boom-radius away from the boom surface.

Another mounting method which has recently gained popularity is to mount the element through the centre of the boom on insulating shoulder washers. The correction required depends on the type of washer, and also whether metal 'keepers' or pure friction are used to hold the element in place (Fig. 7.8F, G). K1FO has published corrections for elements of $3/_{16}$" diameter at 432MHz, for a specific mounting method using shoulder washers and metal keepers [6, 12]. A correction of 5mm has been determined at 432MHz for 4mm diameter elements mounted in 6·3mm diameter holes through a 20mm square boom, as in Fig. 7.8G. This is about 50% of the through-boom value in Fig. 7.9, and I would estimate a similar fraction of the through-boom value for 144MHz also.

It is difficult to predict the effects of the many element-clamping arrangements used in

Fig. 7.9. Boom-effect correction to be *added* for elements passing through the centre of a metal boom, with a metal-to-metal connection at each side (as in Fig. 7.8A and 7.8B), or connected across the top of a square boom (as in Fig. 7.8C)

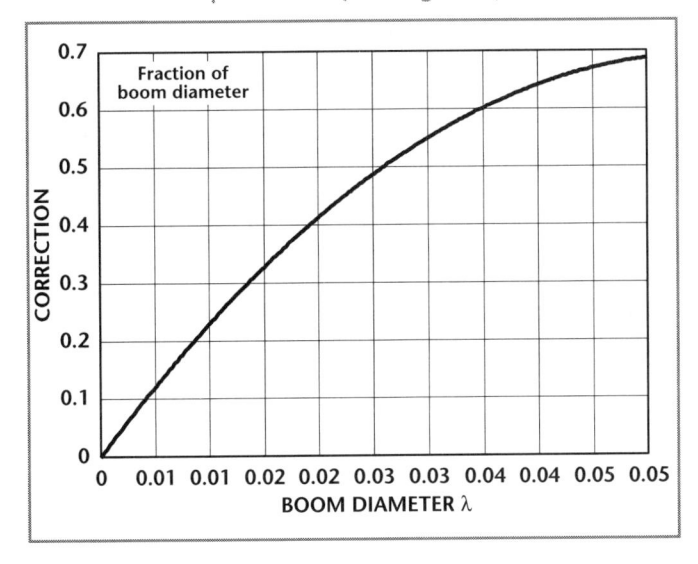

commercial antennas from which you might salvage the fittings. Fortunately it's quite easy to measure the corrections using a two- or three-element Yagi with an insulating boom: just replace the original director with a test element mounted in your chosen manner and change its length to restore the original VSWR conditions [13, 14].

ELEMENT DIAMETER

In presenting the design tables for DL6WU Yagis, I've already mentioned the effects of element diameter (thickness). The same effects apply if you want to build some other published Yagi design using elements of a different diameter. This often happens because continental Europe works in metric sizes, the USA works in inches, and the UK is caught between the two – which means that whatever size you need, it will always be out of stock!

The way to tackle this problem is to remember that elements having equal reactance are interchangeable. So you first calculate the reactance of the original element, and then calculate the length of the new element having a different diameter but the same reactance.

A formula for element reactance is:

$$X = 430\left[3\log_{10}\frac{2\lambda}{D} - 320\right]\left[\frac{2L}{\lambda} - 1\right] + 40W$$

where D is the element diameter and L its length. This is a bit of a mouthful, but is easily handled by a programmable calculator or a home computer.

Let's assume that we start with a 144MHz director element of $^3/_8$" diameter, length 35", and we want to convert this to use 4mm-diameter aluminium welding-rod. Inserting the numbers into the formula (don't forget to use λ in inches, as with the other dimensions) gives a reactance of –79·1Ω. Now work the formula backwards to calculate a new value for L, given X and the new D=4mm.

$$L = \left[\frac{(X-40)}{430\left[3\log_{10}\frac{2\lambda}{D} - 320\right]} + 1\right]\frac{\lambda}{2}$$

The new value for L is 914·2mm, which is 25mm longer than the original 35", as you'd expect for a thinner element. The program disk mentioned earlier has a simple BASIC program called ELE to automate this entire process.

Whatever element diameters you select, the best material is generally aluminium alloy – preferably a corrosion-resistant grade, especially if you live in an industrial area or near the sea. Welding rod of 4mm diameter is an excellent choice for both 432MHz and even 144MHz, although it may not be available in the metre lengths required for the latter band. Hard-drawn enamelled copper wire of about 3mm diameter can be used for 432MHz elements. Brass is not a good material for antenna elements, because it is prone to fatigue cracking. Although stainless steel is springy and highly resistant to corrosion, it is also lossy; its use may cost up to 0·5dB of gain at 144MHz and even more at 432MHz.

Element diameters for 144MHz and below depend critically on whether your winter climate includes storms which build up thick and heavy layers of ice. If so, the elements must be strong enough to support the ice loading, which rules out thin aluminium rods and tubes. Ice storms are fortunately rare in most of the UK, so element diameters as thin as 9mm ($^3/_8$") can be used down to 50MHz. Elements for 50 and 70MHz made from thick-wall 9mm tubing can withstand practically any gale, if reinforced at the centre with a 12mm sleeve (no length correction is necessary if the sleeve is less than about 200mm long).

By the way, if you are modifying someone else's practical design, you'll need to remove his boom-effect correction before applying the above formulae and then apply a new correction for your own element mounting method. The hardest part of this is to estimate someone else's boom-effect correction, though fortunately more and more designers are mentioning this in their published articles.

MAKING THE ANTENNA WORK

Adjustment of a DL6WU or any other type of

DL6WU YAGI DESIGNS

Yagi should be straightforward, especially if you know what to expect.

First, check the VSWR. All the DL6WU Yagis give a fairly good match to 50Ω, although this can be improved in any particular case by minor adjustments to the lengths of the folded dipole and the first director, which functions mainly as a matching element. Note that I said *minor* changes in those two element lengths, to improve a match that's already reasonably good. If the antenna shows absolutely no sign of matching properly, there's a fault somewhere in its construction and you will need to find it and fix it. Other types of antennas which require matching adjustments are best dealt with by pointing the antenna towards the sky and adjusting the driven element while it's comfortably within reach (Fig. 9.43).

Any major faults in antenna construction will show up in the radiation pattern. As a preliminary check, look for a symmetrical E-plane pattern with deep nulls at ±90° and reasonably well-suppressed minor lobes. You can't tell too much from the rear lobe, although it should be well-suppressed, but the first sidelobes can be very revealing. The following remarks apply only to long multi-element Yagis; Yagis with six or fewer elements tend not to have first sidelobes at all, or else they are completely hidden by the ±90° nulls.

All the DL6WU long Yagis have first E-plane sidelobes about 17dB down, which should be separated from the main lobe by a distinct null. If the nulls are too deep and the first sidelobes are too well-suppressed (–19dB or more), the antenna is operating on too low a frequency, *ie* its elements are too short. This can easily happen if you've underestimated the electrical shortening effect of the boom. On the other hand, if the first sidelobes of a long Yagi are poorly suppressed (–15dB or less) or merge into the main lobe, the antenna is operating on too high a frequency, *ie* its elements are too long. Either case – all elements too long or all too short – will produce less than the maximum attainable forward gain.

The gain reduction is less rapid if the elements are all too short, so it's better to err on that side than to fall over the steep 'cliff' on the HF side of the gain-frequency curve. Rain will also move the antenna HF on its gain-frequency response, so all the DL6WU Yagis contain a 1-2% element length allowance to guard against this possibility. K1FO has found that stacking his own highly-optimized Yagis tends to detune them in the same HF direction; he believes that antennas that are already close to the gain cliff can be pushed over when stacked in arrays, with disappointing results [6, 12]. If part of the process of optimization has been to move the Yagi to the top of its gain curve, any change due to interactions or rain will obviously be even worse.

WHICH COMMERCIAL YAGI?

It can be very difficult to choose a commercial Yagi on the basis of the manufacturer's claimed performance, because many claims are notoriously inflated. To give you some standard for comparison, overleaf are two tables of Yagi performance predicted by Rainer Bertelsmeier DJ9BV using the NEC3 computer program [15, 16]. The apparent accuracy of NEC in these applications is about ±0·2dB on gain and ±0·5dB on sidelobe levels down to −20dB; this is at least as good as any commercial antenna range at VHF or UHF. Moreover, NEC provides a constant yardstick against which we can judge antennas, regardless of where and when they were manufactured.

The antennas featured in these tables are a mixture of commercial products and published amateur designs. Most of the designs are for long Yagis, although the 144MHz table includes some shorter Yagis which could be suitable for 50MHz or 70MHz. The tables show the boom length, number of elements, and the calculated and claimed forward gains. As an additional measure of performance, the gain calculated by NEC is compared with the gain to be expected from a Yagi of the same boom length using the equation I gave earlier; this column is labelled 'Gain/exp'. Data on beamwidths and front/back ratio are also given.

As you might expect, some antennas are good for their length, and one or two are very bad! This is shown most plainly by the deviations from the expected relationship of gain versus length. 50MHz and 70MHz enthusiasts will already have noticed that the five- and six-element NBS designs are good performers for their size. A few of the longer Yagis also seem to give a little more gain than expected, especially if they have a favourable boom length like the DL6WU 30-element, or have been highly optimized like the K1FO 22-element. Within reason, the number of elements is not relevant, especially when some designs use multiple reflectors and others do not.

Unlike simpler types of gain calculation, including those using the W2PV program and MININEC, the calculations by DJ9BV using NEC take account of skin-effect losses, which affect some designs worse than others. Typical efficiencies for antennas using aluminium elements are 98% or greater, equivalent to resistive losses of less than 0·1dB. However, these losses are proportional to the squares of the currents flowing in the elements, so the effect of undue concentrations of current into a few elements can be quite marked. Computer analysis of well-optimized Yagis shows that the magnitudes of the currents in all elements are quite similar and vary in a progressive manner. However, optimization using computer programs which take no account of resistive losses can easily go 'off the rails', leading to designs which do not perform well in reality.

Considerably greater loss of efficiency arises from the use of elements made from poorer conductors than aluminium. For example, although the HAG range of antennas are DL6WU designs and should in principle fall close to the standard gain projection, the use of thin stainless-steel elements leads to low efficiency and loss of gain.

WHICH COMMERCIAL YAGI?

YAGI PERFORMANCE

144MHz

Design/type	Boom (λ)	Ele	Gain (dBd) NEC	Gain (dBd) Claim	Gain/exp (dB)	φE (°)	φH (°)	1sl (dB)	F/B (dB)	Notes and references
CC DX-120	0.7	20	12.1	14.0	–	47.0	27.5	–	22.8	Colinear (Fig. 7.3D)
NBS 5-el	0.8	5	9.0	9.2	+0.5	48.0	58.0	25.0	14.1	[5]
JB Q6/2M	1.2	6	9.85	10.9	0.0	46.5	49.0	21.5	11.0	Quad
NBS 6-el	1.2	7	10.2	10.2	+0.3	41.5	47.5	19.2	17.4	[5]
Tonna 9-el	1.65	9	11.0	10.95	0.0	40.0	45.4	20.4	16.7	Newer version only
W1JR 8-el	1.75	8	11.0	>10.85	–0.2	37.2	41.0	17.0	16.7	[8]
Tonna 13-el	2.13	13	11.35	11.85	–0.5	36.3	40.0	17.0	17.5	
HAG FX-224	2.35	11	11.8	12.4	–0.4	35.2	38.5	16.8	17.1	DL6WU (steel eles)
JB PBM14	2.8	14	12.6	13.7	–0.15	30.7	33.5	13.1	13.6	Parabeam
Tonna 16-el	3.1	16	12.65	15.65	–0.4	34.0	37.0	18.8	21.0	
CueDee	3.1	15	12.9	14.0	–0.2	33.0	35.5	17.5	21.5	
Tonna 17-el	3.2	17	12.9	13.15	–0.3	33.0	35.7	19.0	30.0	
CC 3219	3.2	19	13.1	16.2	–0.1	29.4	31.2	13.5	30.5	
DL6WU	3.6	15	13.6	13.6	0.0	30.5	32.5	18.2	36.0	[17]
HG 215B	4.1	15	12.85	15.65	–1.2	25.0	26.0	10.4	15.6	
KLM LBX-16	4.1	16	14.05	14.5	0.0	28.0	29.5	15.6	22.6	Mod DL6WU
CC 4218	4.2	18	14.1	17.2	0.0	27.0	28.5	15.4	18.5	Mod NBS
KLM LBX-17	4.5	17	14.4		0.0	27.0	28.5	15.8	20.0	Mod DL6WU
M2-5WL	4.8		14.4	15.0	–0.2	26.3	27.5	16.5	20.0	

432MHz

Design/type	Boom (λ)	Ele	Gain (dBd) NEC	Gain (dBd) Claim	Gain/exp (dB)	φE (°)	φH (°)	1sl (dB)	F/B (dB)	Notes and references
KLM-16	5.2	16	14.15	–	–0.7	25.0	26.0	11.5	19.8	
K2RIW 19-el	5.6	19	14.95	15.1	–0.2	27.0	28.3	17.0	20.4	[18]
K1FO-22	6.1	22	15.7	15.7	+0.3	24.7	25.7	17.2	21.2	[12]
Tonna 21-el	6.7	21	15.75	16.05	–0.2	24.0	25.0	15.7	40.0	With balun
HAG 723	7.3	23	15.4	15.8	–0.6	23.0	24.0	17.5	18.5	DL6WU (steel eles)
CC 424B	7.6	24	15.9	18.2	–0.2	20.3	20.9	12.7	29.2	
DL6WU 30-el	8.8	30	16.8	16.6	+0.2	22.1	22.9	16.1	40.0	
W1JR 31-el	10.4	31	17.55	17.25	+0.35	20.8	21.2	18.8	28.1	Mod DL6WU [19]
K1FO-32	10.5	32	17.55	17.5	+0.35	20.0	20.5	17.2	23.7	[6]
M2-13WL	13.3		17.95		–0.05	19.3	19.8	15.4	24.4	

Manufacturer abbreviations: CC = Cushcraft, JB = Jaybeam, HG = Hy-Gain, M2 = M^2 (K6MYC)

You'll also surely have noticed that there is scarcely a single instance of over-modest claimed gain! Some claims are grossly inflated, especially for the older designs and for certain manufacturers. It is ironic that many of the antennas with highly exaggerated claimed gains are actually quite good, so the advertising 'hype' was unnecessary. Some manufacturers, for instance Antennes Tonna, are now taking great trouble to make realistic and verifiable performance claims for their current production antennas. This is a highly creditable policy, especially since new antennas may have *lower* claimed gains than the products they replace. A manufacturer who has had the courage to take this step should ***not*** be rewarded by 'clever' people asking embarrassing questions about their old catalogues!

FEEDLINES, TRANSFORMERS AND BALUNS

A chapter on antennas would be incomplete without consideration of the feedlines connecting them to your station. After all, a 3dB cable loss means converting half your precious power into heat, and to make up for that loss you'd have to double the size of your antenna at least. Impedance transformers are important because they are used as power dividers in stacked arrays. I'll also have something to say about the much-misunderstood subject of baluns.

FEEDLINES

In 99% of all cases, 'feedline' today means coaxial cable. But it wasn't always so; VHF amateur radio began with home-made open-wire feedline, and open wire has recently made a comeback in the highly demanding area of moonbounce antennas. So let's begin by taking a critical look at both types of feeder.

Open-wire feeders – or parallel lines – are characterized by their spacing and wire diameter. These parameters determine the characteristic impedance (Z_0) by the simple relationship:

$$Z_0 = 276 \log_{10} \frac{D}{r}$$

where D is the centre-to-centre spacing and r is the wire radius.

In principle, open-wire feeders can have very low losses. Apart from the insulating spacers, the dielectric between the lines is air and only a vacuum has lower dielectric losses. Radiation losses are surprisingly low provided that the wire spacing is a small fraction of a wavelength, leaving only the ohmic resistance of the conductors themselves as a source of loss. As the power dissipated is proportional to the square of the current, higher values of Z_0 are more favourable. For example 2mm wires spaced 100mm apart have a characteristic impedance of 552Ω. Assuming a λ/D ratio of 25 to be acceptable, this line could be used on HF and up to 70MHz with excellent results. For 144MHz, the spacing must be reduced to avoid radiation losses, and rather than use thinner wire it's generally better to accept the slight increase in resistive losses due to the lower value of Z_0. Such an open-wire line would still outperform half-inch coax as far as loss is concerned.

The situation on 432MHz is a little worse, though open wire still can have lower losses than coaxial cable. A further reduction in spacing is dictated by the wavelength, but it also becomes reasonable to increase the wire diameter to make the feeder semi-rigid and almost self-supporting. With a spacing of 25mm and 2mm wire, Z_0 is reduced to 386Ω. Other problems then emerge, like keeping the spacing constant, preventing dirt from forming current bridges on the spacers, and the field distortion by the spacers which can upset your calculations of impedance and velocity factor. Even so, there is still a case for using open-wire feeders for interconnections within an array of several Yagis for 432MHz EME, where 0·1dB may make a noticeable difference. 432MHz marks the upper-frequency limit of parallel-wire feeders. I won't even discuss TV ribbon, which has no place in serious VHF/UHF work.

Turning to coaxial cable, the advantages are

obvious: the dielectric and current-carrying surfaces are inside, free from mechanical, electrical and climatic influences. Ground potential on connectors and the cable sheath reduces hazards and the danger of short-circuits. Unwanted radiation, signal pickup or coupling to other circuits are almost non-existent. Cables are (mostly) flexible and easy to install, *etc etc*. The main drawbacks of coaxial cable are weight, cost and particularly loss, which can be a real headache for VHF and UHF amateurs.

The main sources of loss in coaxial cables are – in approximate order of importance – resistive (skin-effect) loss in the centre conductor, RF leakage through a braided outer conductor, dielectric losses in the insulating material, and resistive losses in the outer conductor. The reason that losses in the centre conductor are so important is that the current density on its surface is much greater than on the outer conductor, and this is made worse by the low impedance of coaxial cables. A 50Ω cable carries ten times the current of a 500Ω line at the same power level. Dissipation in the centre conductor alone is 50 times greater than in open-wire line with the same conductor diameter!

This example shows how unfortunate the choice of 50Ω as standard impedance was. It owes its widespread acceptance to its abundant use by US armed forces during World War II and afterwards – as also does the horrible PL259/SO239 'UHF' combination. The 60Ω standard used in European broadcast applications was chosen for maximum breakdown voltage in a given outer diameter, and 75Ω TV cable was designed for lowest loss per unit weight. In principle, either would be preferable to 50Ω in applications where loss is critical, but cost and compatibility usually make it more economical to use a heavier 50Ω cable instead.

For a fixed cable diameter, loss is not dictated by the characteristic impedance alone because the diameter of the inner conductor also depends on the dielectric. In coaxial lines the characteristic impedance is given by approximately:

$$Z_0 = \frac{138}{\sqrt{\varepsilon}} \log_{10} \frac{D}{d}$$

where ε is the dielectric constant of the insulator, D the **inner** diameter of the outer conductor and d the diameter of the inner conductor.

The higher the dielectric constant of the insulating material, the thinner the centre conductor needs to be, which may be useful if low cost is more important than low losses. Our priorities are different: we are always looking for lower losses in our coaxial cables and are generally prepared to spend a little more. Moving upmarket, foam insulation has a lower dielectric constant than a solid. This means that the diameter of the inner conductor must be increased to maintain a 50Ω impedance, so the cable has lower losses but costs more. Taking this one step further, air-dielectric cable is widely used in professional installations and sometimes appears on the surplus market. It is recommended only if it has always been pressurized to exclude moisture, and if you're prepared to continue to look after it in the same way (not too difficult: use a fish-tank pump). Semi-air-spaced cables like H100, Westflex 103 or Belden 9913 are just as vulnerable to moisture ingress, and can be a load of trouble unless you seal the joints with extreme care [20].

A small part of the attenuation in coaxial cables is caused by leakage and resistive losses in the outer conductor. A smooth, solid tube would be best, of course, but some flexibility is almost always necessary. The first steps up from ordinary single-braid coax (*eg* RG213 or URM67) are double-braid (RG214) or cables with braid over copper foil (H100, W103 and 9913), both of which have good flexibility. Better still is solid copper or aluminium, ribbed for flexibility. Despite the lower conductivity of aluminium compared with copper, it is acceptable as an outer conductor because of the lower current density and also allows considerable savings in cost and weight over solid copper.

Further cost savings are possible by taking advantage of the fact that the RF current on the inner conductor flows only on the surface, so the entire conductor doesn't need to be solid copper. A solid copper conductor is preferable in small cables because it has lower losses than a stranded conductor, but in larger cables the centre conductor is often either a copper coating on an aluminium core or a hollow copper tube.

The table of cable data is intended mainly to illustrate the differences between various cable types rather than as a complete reference. It also serves to show that RG213 (URM67), the old standby, should really be used only where it cannot be avoided. The other 50Ω cables are manufactured and sold under a variety of names (Flexwell, Cellflex, Heliax *etc*). Except for the low-cost H100/W103/9913 the best cables are of rugged construction with solid corrugated copper shields. For amateur installations the polyethylene and PTFE foam types are the easiest to use.

Special highly flexible versions of these cables (*eg* Kabelmetall HCF) are ideal where frequent bending is necessary, although the connectors are heavy and costly.

Summing up the practical implications of all these various kinds of coaxial cables, it is advisable to make the long cable run from stiff, low-loss cable and use a short piece of flexible line to bridge the rotator. If the bridge is made using good connectors (a pair of N-types has less than 0·05dB loss at 432MHz) that is a tolerable sacrifice.

Be prepared to spend good money on the main cable run, and look after it well, because it's one of the most important parts of your station.

IMPEDANCE TRANSFORMERS AND POWER DIVIDERS

Transmission lines find widespread usage as matching devices in VHF/UHF equipment. Although in principle they can accomplish any kind of complex impedance transformation, we amateurs rarely have instruments suitable for measuring complex impedances at VHF and UHF under outdoor conditions so we try to make our antenna impedances straightforward and non-reactive. We can therefore manage very well with simple quarter-wave transformers for matching to the feedline.

COAXIAL TRANSFORMERS

Transformation between 'real' (*ie* non-reactive) impedances is easily checked by VSWR measurement and substitution of loads by terminating resistors – see Chapter 12 for a full explanation. To transform between two real impedances Z_1 and Z_2, all one needs is a quarter-wave line with the characteristic impedance $Z_0 = \sqrt{Z_1 Z_2}$. Coaxial transformers for outdoor use are most easily made from square tubing, with cylindrical wire or tubing for the centre conductor (Fig. 7.10). The characteristic impedance of such a line is:

$$Z_0 = 138 \log_{10} 1 \cdot 08 \frac{D}{d}$$

MATCHED ATTENUATION OF 50Ω COAXIAL CABLES

Type	Atten (dB/100m)				OD (mm)	CC (mm)	Dielectric	Notes
	50MHz	70MHz	144MHz	432MHz				
UR76	12	14	19	32	5·0	7/0·32	PE	Near equiv. RG58
UR67	4·6	5·6	8·4	15·5	10·3	7/0·77	PE	Near equiv. RG213
H100	3·2	3·8	5·5	9·8	9·7	2·5	PE/air	
Westflex 103	2·0	2·5	4·5	7·5	10·3	2·7	PE/air	
¼-inch	3·8	4·6	6·9	12·6	various		Foam PTFE	Corrugated hardline
½-inch	1·5	1·8	2·8	5·1	various		Foam PTFE	Corrugated hardline
⅞-inch	0·9	1·1	1·5	2·8	various		Foam PTFE	Corrugated hardline

Values for amateur bands are estimated from data for other frequencies. Values for hardline are typical of that size.
OD   outside diameter over plastic sheath   PE   polyethylene
CC   centre conductor diameter   PTFE   polytetrafluoroethylene

FEEDLINES, TRANSFORMERS & BALUNS

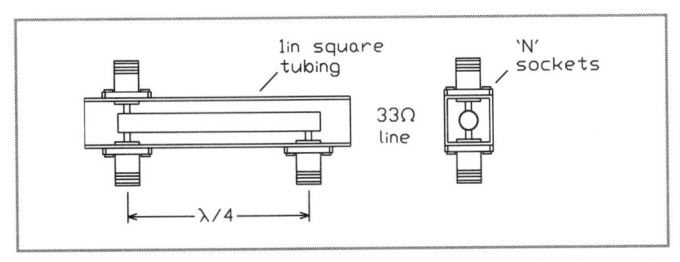

Fig. 7.10. A 50Ω two-way power divider made from square-section tube

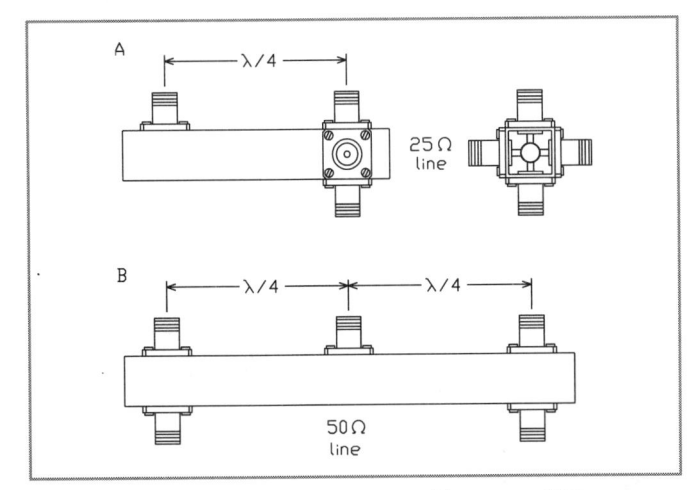

Fig. 7.11. Two types of four-way 50Ω power divider: quarter-wave (A) and half-wave (B)

Fig. 7.12. Two possible ways of feeding four 200Ω Yagis using 200Ω open-wire line. Dimension 'x' is the optimum vertical stacking distance for the Yagis in use. A and B show alternative methods, depending on the horizontal stacking distance required

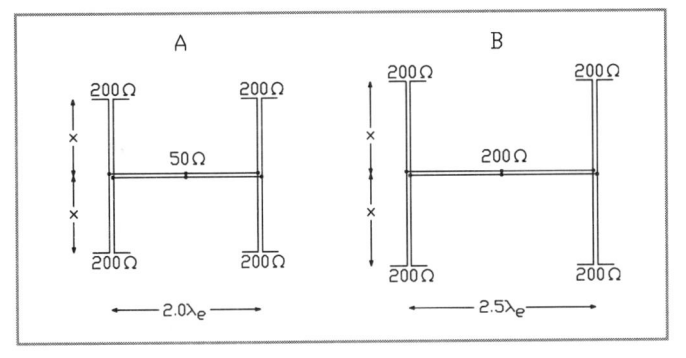

where D is the *inner* width of the square tubing and d the *outer* diameter of the centre conductor. Air is used as the dielectric so there is no velocity-factor correction; the end-to-end length of the inner line must be ¼ of a free-space wavelength.

When line transformers are used as combiners (also known as power dividers) in stacked arrays, some thought should be given to low-loss design. As pointed out before, high currents should be avoided; this means keeping impedance as high and VSWR as low as possible. Take for instance the case of four 50Ω antennas to be connected to a common 50Ω line. This could be done by paralleling all four and then transforming the resulting 12·5Ω with a 25Ω quarter-wave section (Fig. 7.11A). Or it could be done in two steps (Fig. 7.11B): combine each pair of antennas to give 25Ω, and use a 50Ω quarter-wave transformer to give 100Ω; when the two 100Ω points are connected in parallel at the centre, the result is 50Ω. This latter method, called a half-wave combiner, is far superior to the simple quarter-wave combiner in both bandwidth and low-loss performance.

As a general rule when working with coaxial lines, and especially on the higher bands, treat interconnection and impedance transformation as two separate functions. Transforming should be done in low-loss air-dielectric line sections, and interconnecting by (short) runs of matched cable. In other words, don't use ordinary coaxial cable for impedance transformation if you are at all concerned about losses.

Construction of 'square-line' impedance transformers is very simple. Begin by finding some 25mm or 1" square aluminium tubing, which will conveniently accept N-type connectors on its flat faces (Fig. 7.10). Measure the inside dimension and work out what diameter of copper or brass tubing you need for the inner line. The only difficult bit may be to find this tubing! If in doubt, you can use the next smaller size and bring the characteristic impedance down to the exact value you require. You can do this either by mounting the tube slightly off-centre, or by lining the tube with a thin flat sheet of aluminium. If

7 - 35

you use these tricks, make sure the effects apply equally to both sides of the power division; you can only do this for two-way or half-wave four-way dividers.

You can leave the ends of the inner tube open. The ends of the square tubes are also field-free if they extend a few centimetres beyond the connectors, so they only need to be sealed against water ingress. It would be nice to make one end-cap from something transparent, so you can see at a glance if any water has crept in.

The following table shows the impedances to be used for quarter-wave and half-wave 50Ω combiners, and the corresponding D/d ratios. Since we all seem to have access to differing sizes of tubing, you probably need to take a pocket calculator to the metal shop and work out what can be done with the materials available.

		Z_0	D/d
Two-way	λ/4	35·4	1·67
Three-way	λ/4	28·9	1·50
Four-way	λ/4	25·0	1·41
Four-way	λ/2	50·0	2·13
Six-way	λ/2	40·8	1·83

The largest number of Yagis to be joined in a single combiner is six. For an eight-Yagi system, make two groups of four and then connect the two groups to a two-way combiner at the central feedpoint. A sixteen-Yagi system will require a total of five four-way combiners – plus at least 25 N-type plugs and sockets... quite a strong argument for open-wire feedline!

OPEN-WIRE FEEDLINE

The inherent balance of parallel open-wire lines makes them well suited for both interconnections and impedance transformation. Yagi antennas with 200Ω folded-dipole driven elements are ideal for feeding with 200Ω open wire. Fig. 7.12 shows two schemes for combining four Yagis in this way, with very simple matching to 50Ω coaxial feeder. Since the Yagis and the open-wire feedline have the same characteristic impedance, any vertical stacking distance can be used without affecting the impedance matching. The length of the horizontal cross-piece of feeder depends on the horizontal stacking distance for the particular Yagis in use. Shorter Yagis which require horizontal spacings in the range 1·9-2·2λ can use a 2·0λ (electrical) cross-piece as shown in Fig. 7.12A. This transforms the feedpoint impedance of the array of four Yagis to 50Ω, which can be connected to the coaxial feeder via a λ/4 sleeve balun. For longer Yagis requiring horizontal stacking distances of 2·2-2·5λ, the cross-piece should be 2·5λ (electrical) which gives a feedpoint impedance of 200Ω, the same as an individual Yagi.

The velocity factor relating the physical to the electrical length can be very high for open-wire feeders – as high as 0·975 if the dielectric spacers are few in number and located away from voltage maxima. Measurement of the velocity factor is very simple. Connect a length of the open-wire line to a low-power RF source (*eg* by a 4:1 coaxial balun), terminate the far end and monitor the VSWR in the coax line from the RF source. If you lay an insulated screwdriver across the open wire and slide it along, you will see a disturbance in the VSWR which repeats itself every electrical half-wavelength.

Doing away with baluns, cables, transformers and connectors saves a lot of weight, loss and money. But there is a price for everything; open-wire feeders are quite critical in dimensions and installation and they can also be weather-sensitive and mechanically unstable unless you're very careful. And open-wire lines cannot be lengthened or shortened very easily, so you'd better be sure about your velocity factor and stacking distances before you commit yourself.

When working with open-wire feeders, small impedance errors tend to accumulate and the bandwidth of satisfactory performance is usually very narrow; fortunately, this doesn't matter for DX antennas. Be prepared to do a lot of experimenting, but also be prepared for some benefits: you can trace the voltage maxima and minima by just waving your hands around the feedline and watching the VSWR, and you can also fine-tune the impedance match by bending the wires in the

FEEDLINES, TRANSFORMERS & BALUNS

right places! Naturally, these hands-on adjustments should be made using extremely low power (Chapter 12). Most open-wire feed systems require an adjustable impedance transformer somewhere in the system, probably at the central feedpoint. This can be combined with the symmetry transformer (balun) required for the transition to coax.

The rules for installing open-wire feeders have been set down by DL9KR and G3SEK in K2UYH's *432MHz Moonbounce Newsletter*.

1 Use heavy-gauge enamelled copper wire, *eg* 3mm, to make the feedline largely self-supporting. Avoid clear polyurethane enamel; it doesn't stand up to sunlight. The best grade of enamel is dark brown, and very difficult to scrape off for soldering.

2 Buy soft-drawn wire, and stretch it out straight immediately before use. For heavy wire, you'll need a stout post or tree and a towing hitch on the car!

3 Use only straight runs of feeder and make all joints at right-angles. This is *vital*.

4 Pre-stretch the individual wires, and then construct each section of feeder to the correct length *before* installation. Keep the two sides of each section of feeder *exactly* the same length.

5 Use a minimum of spacers made from solid PTFE rod, and avoid placing them at voltage maxima.

6 To make a straight joint between two wires, use a small sleeve made from brass or copper tubing. To make a T-joint, drill a hole across the end of a piece of tubing. Slip one of the wires through the hole and the other wire into the open end of the tube. Crimp the wires in place and then use an enormous soldering iron. Waterproof all joints with polyurethane spray; although susceptible to sunlight, polyurethane varnish has the great advantage that it doesn't need to be scraped off before soldering if repairs should be necessary at a later date.

7 Keep the feeder well away from other conducting objects.

8 Avoid using mechanical supports for the feeder, other than the antennas and the junction to the main feedline.

9 Avoid high VSWR in any section of feeder, to minimize the effects of rain and frost at high-impedance points. The two arrangements in Fig. 7.12 involve a VSWR nowhere greater than 2.

The overall result may be well worth the effort. EME operators have reported up to 3dB improvement in noise performance over crude coax feed in large 432MHz arrays, although that is extreme. More often you could expect an open-wire feed system for a large 432MHz EME array to be at least as good as a coaxial system made from heavy hard-line – and probably cheaper, lighter and easier to construct. But don't expect miraculous results in horizon modes and on lower frequencies. If you want an all-purpose array for both ends of the band and in any kind of weather, open-wire feed is probably not for you.

BALUNS

Most antennas are symmetrical, *ie* balanced with respect to ground, so a transition to unsymmetrical coaxial cable must be provided somewhere in the system. A balance-to-unbalance transformer or *balun* is usually fitted at each individual antenna feedpoint, or at the central feedpoint if open-wire line is used within the array. There are many types of balun; some of them involve impedance transformation as well, although that is not essential to the balun function.

There are a lot of misconceptions about baluns and whether you really need them, so

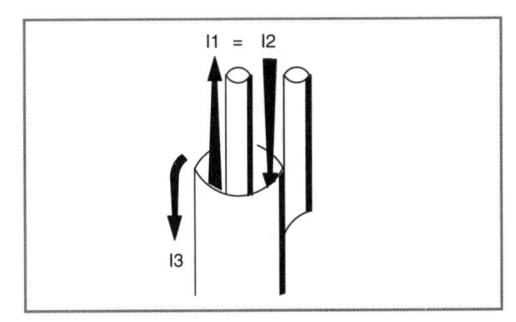

Fig. 7.13. Currents flowing on the inside and outside of a coaxial cable. Currents I1 and I2 on the inside of the cable are always equal. If the system is unbalanced, a current I3 will flow on the outside of the outer sheath

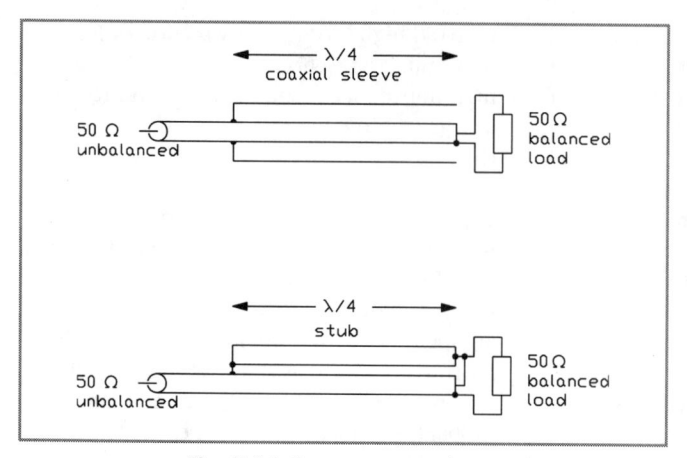

Fig. 7.14. Quarter-wave sleeve and stub baluns, which give no impedance transformation

let's establish what a balun needs to do. Beginning at the coaxial side of the junction, the currents I1 on the inner conductor and I2 on the *inside* of the outer conductor are always equal (Fig. 7.13), because the inner and outer conductors of the coax are very closely coupled. However, as a result of the 'skin effect' which confines RF currents to the surfaces of conductors, there is another possible path for a current I3 on the *outside* surface of the outer conductor. If the antenna feedpoint is truly balanced with respect to ground, a voltage minimum occurs at that precise point along the driven element. When you connect the coaxial feeder, there would be no EMF to drive the current I3 along the outside surface and all would be well.

In real life, however, the antenna feedpoint is never exactly balanced so an outer-surface current **will** flow unless something is done to stop it. Otherwise, stray RF currents can flow along the boom and the boom braces, down the mast, into the rotator cables, down the main coax feeder and into the shack. These in turn give rise to unwanted radiation, pattern distortion, RF feedback, TVI and all kinds of funny effects. So you *do* in fact need a balun, to force the system into balance and actively prevent currents from flowing down the outer of the coax.

To see whether you have problems with feedline radiation, check for energy on the outside of the feedline and for radiation off the sides of the beam where the ±90° nulls should be. Another fairly reliable test for RF currents on the outside of the coaxial cable is to see whether different lengths of cable affect the VSWR. In principle this shouldn't happen, because the VSWR is an indicator of conditions *inside* the cable and an extra length of the same type of cable should have no effect. But if there are currents on the outside of the cable, changing the length will affect the current distribution all the way up to the feedpoint, and will therefore affect the VSWR. If merely grasping the outside of the cable changes your VSWR, something is *very* wrong!

Low-impedance symmetrical feedpoints should *never* be connected to coaxial cables without some sort of symmetry transformation. The very least you can do is to ground the cable shield to the metal boom ¼ wavelength (in air) from the feedpoint. This creates a high impedance against outer-surface currents at the feedpoint and hopefully discourages current flow. That technique was recommended for many years by Antennes Tonna and is better than nothing, but it's still rather a weak method of forcing balance upon the system. If you want a 50Ω balun without an impedance transformation, it's much better to use a proper quarter-wavelength sleeve balun or a quarter-wave stub (Fig. 7.14). On 50MHz and 70MHz you can also use a trifilar-wound balun, either air-cored or wound on a ferrite rod [20], but take great care to keep the wires closely coupled and all leads short [4]. One-sided matching devices – gamma matches and the like – are not very satisfactory because they don't actually make a proper balance-to-unbalance conversion and you may have to take additional precautions to eliminate the last traces of current on the outer surface of the feedline and along the boom.

The most successful VHF/UHF balun is the half-wave coax loop transformer which converts 200Ω symmetrical to 50Ω unsymmetrical (Fig. 7.15). This method is state-of-the-art and to be preferred to most others, because the close coupling between the

FEEDLINES, TRANSFORMERS & BALUNS

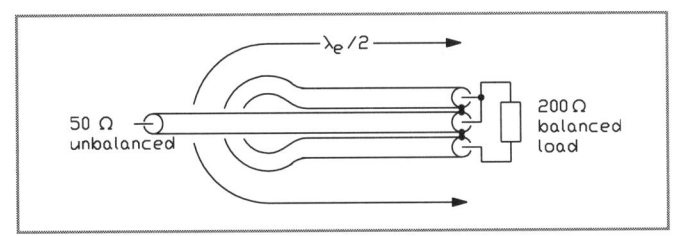

Fig. 7.15. Half-wave balun, which also transforms 50Ω unbalanced to 200Ω balanced. The loop is an electrical half-wavelength long

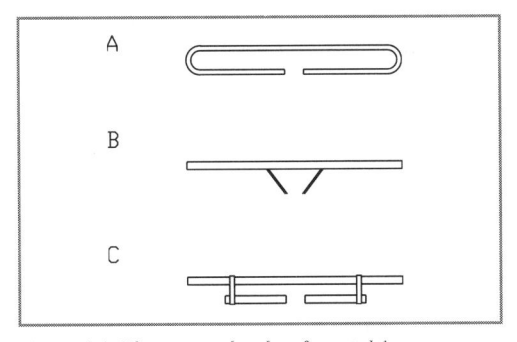

Fig. 7.16. Three methods of matching a driven-element to 200Ω balanced, for use with a half-wave balun (Fig. 7.15). If the driven element impedance would be 50Ω, a simple folded dipole (A) can be used. The Delta-match (B) and T-match (C) are adjustable

inner and outer conductors of the coaxial cable forces the system towards balance. The DL6WU Yagis have a natural feedpoint impedance close to 50Ω and could therefore be used with a simple sleeve or stub balun. However, it is far better in practice to go up to

Fig. 7.17. Constructional details of a folded dipole and half-wave balun for 432MHz, using a commercial plastic moulding and an N connector. Fill with closed-cell aerosol foam for waterpoofing

200Ω by means of a folded dipole (Fig. 7.16) and then to use a half-wave balun for the transition to 50Ω coaxial. Fig. 7.16 also shows the delta match (B) and the T-match (C), the latter being the symmetrical form of the gamma match. Both of these can be adjusted to accommodate 'odd' and reactive feedpoint impedances; the width between the tapping points mainly affects the transformation ratio, and the element length can be shortened for reactance compensation.

Fig. 7.17 shows constructional details of a half-wave balun. The length of the loop should be an *exact* electrical half-wavelength, so don't forget to allow for the velocity factor. Do not use thin low-grade cable for the balun line; doing so would cause unnecessary extra loss and might result in breakdown at high power levels. The characteristic impedance of the cable is unimportant; if available, low-loss 60Ω or 75Ω cable can sometimes be used to advantage in 50Ω systems. If the cable ends cannot be enclosed as shown in Fig. 7.17, use semi-rigid line which doesn't suck up water as readily as braided cable, and use sealant [21].

The loss in a good coax balun is much less than 0·2dB on 432MHz, and less than 0·1dB on 144MHz and below. By the way, if you're not sure of the loss in some dubious device, put power through it. A loss of 0·2dB at 500W means about 25W dissipated. If the balun catches fire, it was too lossy anyway!

Having fitted a balun, don't imagine that you can then forget about symmetry. Unless the coaxial cable is taken symmetrically along the boom (preferably on the opposite side from the elements), and then runs away at right-angles to the plane of polarization, you'll be back in trouble again. This applies particularly to crossed Yagis: the *only* satisfactory way to mount these is on an insulating mast, and to take the cables symmetrically out of the *rear* of the Yagis. Final traces of current on the outer surface of coax feedlines can be eliminated by threading the cable through giant ferrite beads, or by winding it on a ferrite rod or toroid, but this should be tried only as a last resort after a thorough search for some underlying problem.

CONCLUSION

Yagi antenna technology has stabilized in recent years. The application of computers has shown that decades of experimentation have left but a very narrow margin for improvement. Therefore anybody who expected sensational revelations from this chapter must have been disappointed.

On the other hand, design of antennas and especially of large arrays is one of the last domains of amateur ingenuity. It is not difficult to design and build an antenna that works – but it's just as easy to commit a series of minor errors that will keep it from performing optimally. I hope I've alerted you to most of them.

REFERENCES

[1] James L. Lawson W2PV, *Yagi Antenna Design*. ARRL (1986). ISBN 0-87259-041-0. Published after his untimely death, this book is an edited, revised and extended version of his original series of articles which appeared in *Ham Radio* through most of 1980.

[2] J.C. Logan and J.W. Rockway, The new MININEC (Version 3): A Mini- Numerical Electromagnetics Code. *NOSC Technical Document 938*, September 1986. (Both the report and the associated program disks are publicly available, but can be difficult to obtain outside the USA.)

[3] Peter Beyer PA3AEF, Antenna simulation software. Third International EME Conference, Thorn (NL), 1988.

[4] L. A. Moxon G6XN, *HF Antennas for All Locations*. RSGB (1982). ISBN 0-900612-57-6.

[5] P.P. Viezbicke, Yagi Antenna Design. *NBS Technical Note 688*, 1976.

Joe Reisert W1JR, How to design Yagi antennas. *Ham Radio*, August 1977.

[6] Steve Powlishen K1FO, High-performance Yagis for 432MHz. *Ham Radio*, July 1987.

[7] Ray Rector WA4NJP, Update on 6 meter EME. *Proceedings of the 22nd Conference of the Central States VHF Society*. ARRL (1988). ISBN 0-87259-209-X.

[8] Joe Reisert W1JR, VHF/UHF World: Optimized 2- and 6-meter Yagis. *Ham Radio*, May 1987.

[9] Günter Hoch DL6WU, Yagi antennas. *VHF Communications* 3/1977.

Günter Hoch DL6WU, More gain from Yagi antennas. *VHF Communications* 4/1977.

[10] Günter Hoch DL6WU, Extremely long Yagi antennas. VHF Communications 14, 1/1982.

[11] Günter Hoch DL6WU and Peter Maartense, PA0MS, *Zelf Ontwerpen en Bouwen van VHF en UHF Antennes*. VERON (NL). ISBN 90-70756-49-8. (In Dutch.)

[12] Steve Powlishen K1FO, An optimum design for 432-MHz Yagis. *QST*, December 1987 and January 1988.

[13] J. Edward Pearson KF4JU, Element length disturbances due to end-chamfering and insulated metal booms. *Proceedings of the 22nd Conference of the Central States VHF Society* [7].

[14] Günter Hoch DL6WU, Yagi Antennas for UHF/SHF. *The ARRL UHF/Microwave Experimenter's Manual*. ARRL (1990). ISBN 0-87259-312-6.

[15] Rainer Bertelsmeier DJ9BV, Effective noise temperatures of 4-Yagi-Arrays for 432MHz. *DUBUS-Magazin* 4/87.

[16] Rainer Bertelsmeier DJ9BV, Gain and performance data of 144MHz antennas. *DUBUS-Magazin* 3/88.

[17] See Chapter 10.

[18] Steve Powlishen K1FO, On 432 No. 5: The RIW-19. *VHF/UHF and Above*, Vol 2 No 12, December 1985 (publ. KA0HPK).

[19] Joe Reisert, W1JR, VHF/UHF World: A high-gain 70-cm Yagi. *Ham Radio*, December 1986.

[20] Ian White G3SEK, Balanced to unbalanced transformers. *Radio Communication*, December 1989.

[21] John Nelson GW4FRX, In Practice: Waterproofing. *Radio Communication*, January 1989.

[22] Mike Gibbings G3FDW, 'Moxon slopes' at VHF and other thoughts. *Radio Communication*, May 1988.

CHAPTER 8

144MHz

THE SUFFOLK 144MHz TRANSVERTER

by Sam Jewell G4DDK

The *Suffolk* transverter is designed to meet our stringent requirements for a transverter fit for use on today's crowded 144MHz amateur band. It is the result of many hours of design, building and laboratory testing, followed by evaluation of further prototypes built from the description which follows. The transverter consists of separate high-performance transmit and receive converters, each of which has been designed to be easy to construct, align and use. Emphasis has been placed on ease of construction, even to the extent of providing full PCB layouts.

Signals received in the range 144–146MHz are converted to 28–30MHz, where they may be received on any radio covering that frequency range. Similarly, any mode capable of being transmitted by the HF transceiver in the range 28–30MHz is converted by the transmit converter to 144–146MHz. If you have an HF receiver covering 28–30MHz and want to listen to signals in the 144MHz amateur band, you can build the *Suffolk* receive converter alone. If you like what you hear, you can then add the transmit converter and join in the fun.

The transmit side consists of a low-level (100mW output) converter and a 10W PEP linear amplifier. An automatic level control (ALC) loop is incorporated within the transverter to maintain linearity, and an external ALC output can extend the control loop to include the HF driver as well. A true DXer won't want to stop at 10W output, and the *Suffolk* can be used to drive any tetrode power amplifier such as the W1SL described later in this chapter.

The *Suffolk* transverter was designed to be easy to build, but that does not mean performance has been compromised. It provides a level of performance which is equal to that of most current and expensive commercial designs, while still retaining a 'home-built' flavour. Good design doesn't just come from the ability to obtain the latest components, or access to a laboratory full of test equipment. You need to know what level of performance is required, and – just as importantly – what can be tolerated under various practical operating conditions. After that, you need to spend a lot of time planning the block diagram. A PC running TCALC (Chapter 5) will crunch the numbers for the receive converter, allowing you more time to consider the implications. Eventually you will have a paper design which could work.

Construction can then begin, although each individual stage still needs to be checked for correct operation according to the design parameters. If you got it right, the completed transverter will perform according to your calculations; if not, more work is needed to find out why.

RECEIVE CONVERTER

More than any other part of the *Suffolk* design, the receive converter reflects a personal approach to obtaining the required performance from a 144MHz transverter. The overall block diagram of the receive converter is shown in Fig. 8.1. A good deal of time was spent in getting the design 'right' using TCALC, and in measuring the performance of individual circuits to meet the requirements of

SUFFOLK 144MHz TRANSVERTER

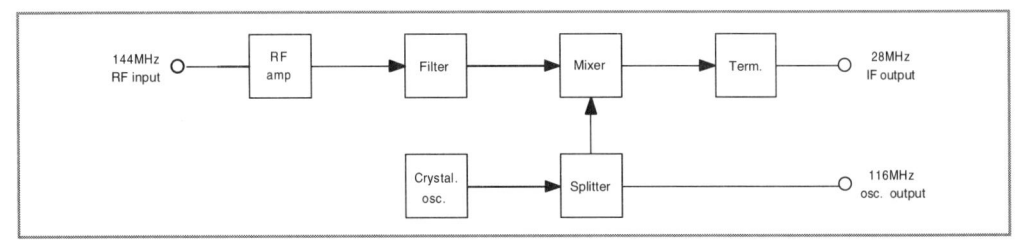

Fig. 8.1. Block diagram of the *Suffolk* receive converter

the optimized block diagram.

The TCALC analysis of the *Suffolk* receive converter is explained at length in Chapter 5. The converter was designed to be used with a modern high-performance HF transceiver. If you use it with an older HF rig, the performance of your receiving system may well fall short of that obtainable with modern transceivers. With the *Suffolk*, it will almost always be the HF radio which limits the overall receive performance, not the transverter itself. After close examination of many HF transceiver reviews, I decided that it was reasonable to expect an input third-order intercept of +20dBm at 28MHz, together with a noise figure of 10dB. If your HF radio has worse performance than this, you will not achieve the best possible overall performance using the *Suffolk* receive converter.

Given the above level of HF receiver performance, what can we reasonably expect from the receive converter? We learned in Chapter 4 that a receive noise figure of approximately 2·2dB is adequate for most terrestrial operation at 144MHz. With this noise figure, a degradation of less than 3dB in signal-to-noise ratio may be expected as compared with an idealized noiseless receiver.

Dynamic range is important when the 144MHz band is crowded with stations, as during an opening or contest. What dynamic range can we expect – and more importantly, achieve? An input third-order intercept (IP3) of 0dBm is an acceptable target for the receive converter, but is itself misleading unless the IP3 of the following HF radio is also taken into account. With an IP3 of +20dBm for the HF receiver, 0dBm for the receive converter and a gain of 20dB in the receive converter, the system IP3 will be –3dBm. The spurious-free dynamic range (SFDR – see Chapter 5) will depend on the receive system's noise figure; if this is 2dB, the SFDR will be about 93dB. Obviously a poorer intermodulation intercept in the HF receiver will lead to a smaller dynamic range for your whole 144MHz receiving system. All that I could do was to make the *Suffolk* front-end sufficiently robust that it won't let your HF receiver down.

The quest for an even better signal-to-noise ratio – a smaller degradation of S/N, that is – will inevitably lead to poorer system dynamic range. The balance between sensitivity and dynamic range achieved in the *Suffolk* receive converter is close to optimum with today's band conditions and electronic devices. So how is this balance achieved?

The key lies in the use of a modern dual-gate MOSFET for the RF amplifier. These devices are capable of low noise figures and high dynamic range simultaneously. You may be surprised to discover that I did not use GaAsFETs. Although they are capable of very low noise figures, GaAsFETs require low-loss input circuits which are not very suitable for printed-circuit construction; equally, the dynamic range of GaAsFETs is not very good unless negative feedback is used. At 144MHz, the best overall performance comes from dual-gate MOSFETs such as the BF981 from Philips Components.

To assess the repeatability of the MOSFET amplifier design for the *Suffolk*, I built and tested six units. A Hewlett Packard HP8790 noise-figure meter was used to measure noise figure and insertion gain, and an HP 8568B spectrum analyser to check stability. Only one amplifier gave a noise figure in excess of 1dB,

144MHz

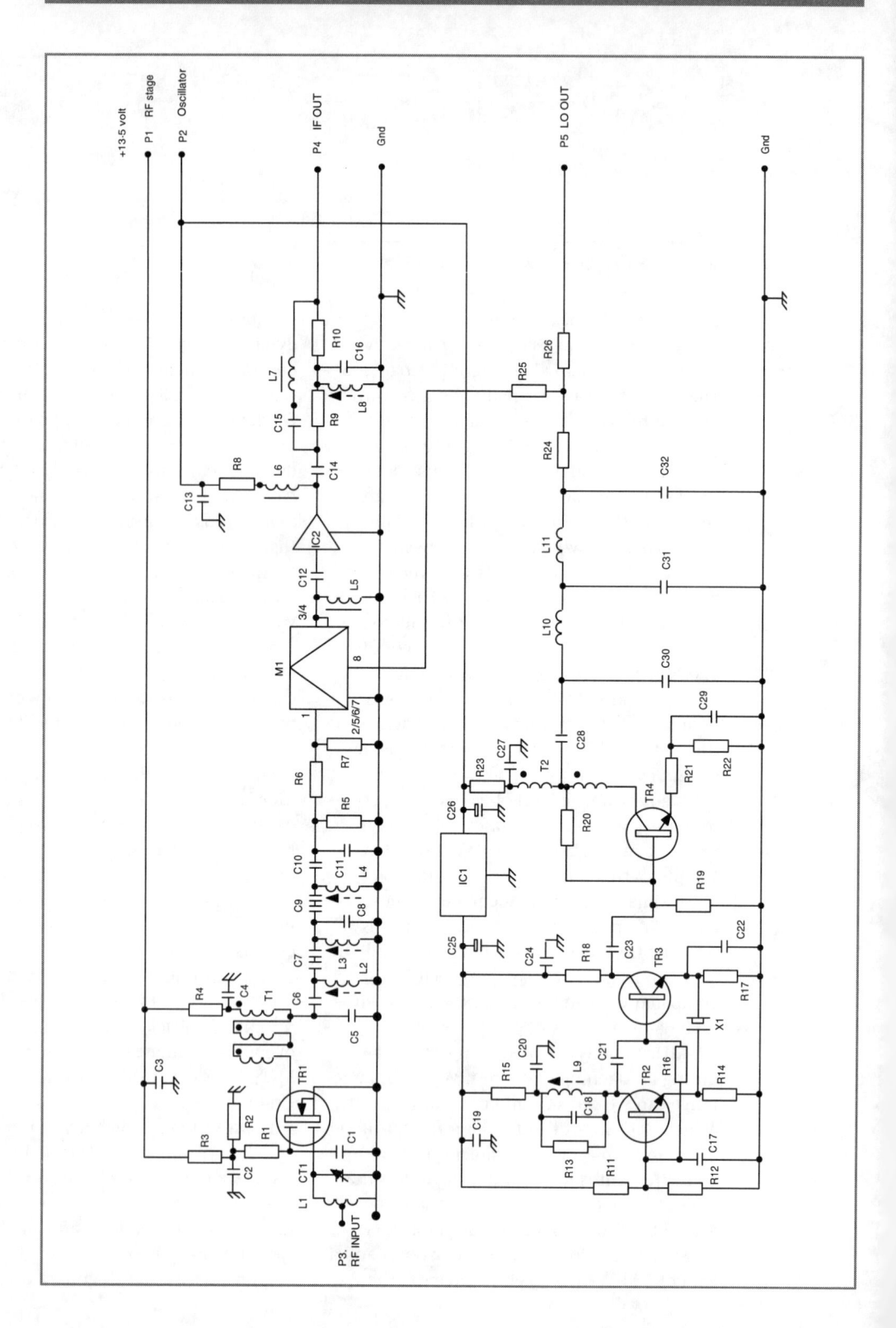

Fig. 8.2. Circuit diagram of the *Suffolk* receive converter

SUFFOLK 144MHz TRANSVERTER

all the others giving a value between 0·6 and 0·9dB. Insertion gain measured 26dB ±1dB on all units, apart from the one that was giving a high value of NF; when the MOSFET was changed in the suspect amplifier, its performance fell into line with the other units. Dynamic-range measurements on the same amplifiers indicated an input third-order intercept in the range 0 to +4dBm. At no time was I able to obtain an IP3 as high as this with any of the dual-gate GaAsFET RF stages I tried.

Optimum dynamic range is achieved in the MOSFET amplifier with a drain load impedance of about 500Ω. Matching this to the filter's 50Ω characteristic impedance is achieved with a trifilar-wound broadband transformer. A 3:1 winding ratio achieves a 9:1 impedance transformation, which is close enough. With this drain load the insertion gain measured 26dB when the BF981 amplifiers were adjusted for lowest noise figure.

Fig. 8.2 shows the circuit of the *Suffolk* receive converter and local oscillator. As a 144–146MHz bandpass filter (image rejection filter) to follow the MOSFET RF amplifier TR1, I initially intended to use a commercial helical filter but was disappointed with its performance, particularly the insertion loss of 8dB for the three-stage filter. I also found that the return loss (Chapter 12) increased dramatically outside the passband, leading to a poor termination for the preceding RF stage, and problems in maintaining its dynamic range. As a result I designed a three-stage bandpass filter with top-capacitor coupling which exhibits almost ideal characteristics for the intended purpose. Input and output impedance were designed to be 50Ω so that the filter can be easily aligned and tested. The filter is isolated from the mixer by a 50Ω attenuator which also serves to optimize the overall RF gain (Chapter 5).

For consistent high performance the diode-ring mixer takes some beating. The biggest disadvantage seems to be the high cost of the better devices. For this reason the mixer specified has been deliberately limited to one of the medium dynamic-range types. If you have worked through the dynamic-range calculations for the *Suffolk* in Chapter 5, you'll realise that there is little to be gained in this design by using a higher dynamic-range mixer. Having said that, I recommend the full-specification SRA-1 rather than the budget SBL-1. The reason for this is the assured performance of the SRA-1, the cheaper device having a slightly wider tolerance. If cost is all-important, use the SBL-1; you may never notice the difference in performance. The mixer is slightly overdriven by the local oscillator to obtain a small but significant increase in dynamic range. The mixer is specified for +7dBm LO injection, giving an IP3 of approximately +12dBm. With +10dBm of LO input, IP3 rises to about +15dBm.

If the mixer is to operate correctly, all its ports must be terminated in a non-reactive 50Ω load at *all* frequencies generated in the mixing process. This can be extremely difficult to arrange, particularly when you consider the wide range of frequencies involved. If the termination is reactive, mixer conversion loss will increase and intermodulation performance will deteriorate. The IF port is particularly critical in this respect.

The simplest broadband termination is an attenuator, but to be effective it should have a value of at least 10dB and preferably 20dB. At the LO port I compromised at 6dB to gain the advantage of the higher LO power level. But even 6dB of attenuation at the IF port would have caused major difficulties in achieving the desired overall noise figure and dynamic range. Diplexers are commonly used to overcome this problem but most types of diplexer are difficult to align without access to a network analyser, so they cannot be recommended for home construction.

To overcome these problems, the active IF termination has become popular. These devices are chosen for their broadband 50Ω input impedance and high dynamic range. A popular choice is a grounded-gate RF power FET, for instance one of the Texas Instruments P8000 series. These devices perform well but can be difficult to obtain. My own choice was a monolithic microwave integrated circuit (MMIC), as used extensively in other parts of

144MHz

the transverter. The MMIC chosen for use in the receive converter at IC2 has high dynamic range, low noise and a constant 50Ω input impedance from DC to well over 1GHz. Just as important as their electrical characteristics, MMICs are simple to use, low in cost and easy to obtain.

What is the input impedance of your HF receiver at 28MHz? It is almost certainly not 50Ω, so in order to present a reliable 50Ω load impedance to the MMIC I have included a simple diplexer at the output of the transverter [1]. The demands upon this diplexer are far less than those at the mixer IF output, so alignment is no problem. In addition to providing a proper impedance termination for the MMIC, the diplexer provides a degree of bandpass filtering and a well-defined source impedance for the input filters of the HF receiver.

Frequency-translation stability in the converter is important, especially for moon-bounce and meteor-scatter where being on the right frequency can make the difference between success and failure. It is very important to use a high-stability crystal in the local oscillator. A cheap crystal will almost certainly lead to frequency drift. A fifth-overtone 116MHz crystal is used in a Butler oscillator (TR2-TR3). This circuit is better known as a harmonic generator, but experience has shown it to be the best choice for frequency setting and reliable starting as well as low noise and high stability.

Local-oscillator drive for the transmit converter is also supplied from the receive converter. Because of this, a higher output level is required than can be supplied from the oscillator alone. An amplifier (TR4) increases the power output, which is then divided to the required level for each mixer. The feedback amplifier design used here is based on a design published in the excellent *Solid State Design for the Radio Amateur* [2]. Both shunt and series negative feedback are used to achieve the required stable gain with good input return loss. A lowpass filter follows, to reduce the harmonic level, after which the output level at 116MHz is +16dBm. The power splitter has a loss of 6dB, giving two outputs of +10dBm. If the transmit converter is not used, it is necessary to terminate the unused output in 50Ω. Failure to do so will lead to excessive oscillator drive to the receive mixer, which will increase the conversion loss.

A full TCALC design analysis of the *Suffolk* front-end is presented in Chapter 5. The calculated overall gain is 18·5dB; the noise figure is 2·25dB and IP3 is −5·2dBm. The measured values for the real-life transverter are given later.

The system noise figure of 2·25dB is obtained at the receive converter input, and therefore does not take into account the contribution from the feeder between antenna and converter. Cable loss should be kept to the minimum possible, to minimise the additional noise and to reduce power loss on transmit. If your HF radio has a lower noise figure than 10dB, the overall system noise figure can be significantly lower than 2dB – but remember that this will probably be at the cost of dynamic range.

Because 50Ω interstage impedance matching has been used throughout, it is possible to change various stages to optimize the performance to meet your own particular requirements. But please remember that the *Suffolk* design as published is already optimized for terrestrial DXing. Any 'improvements' that you attempt will inevitably result in a reduction in performance somewhere else.

TRANSMIT CONVERTER

The transmit converter consists of two separate units – the frequency converter and the power amplifier. The frequency converter provides an output of approximately 100mW with extremely good intermodulation-distortion (IMD) performance. This may be enough for some purposes such as driving a second transverter to some other band such as 1296MHz, and in fact the original design work was inspired by just that need. The transmit converter is not designed as a stand-alone unit; it requires local-oscillator drive from the receive converter.

For most amateurs, 100mW will not be

SUFFOLK 144MHz TRANSVERTER

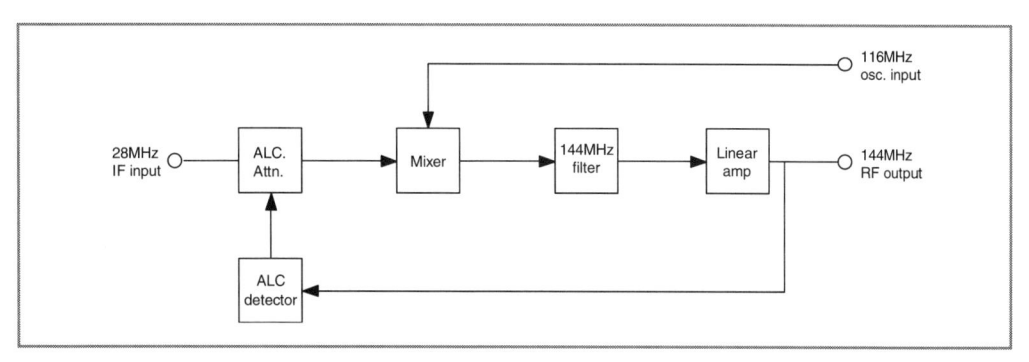

Fig. 8.3. Block diagram of the *Suffolk* transmit converter

sufficient output power – it's hard to make a dent in a pile-up with this level of QRP. The second unit, which I will describe shortly, is therefore a linear amplifier designed to take the low power up to 10W.

The block diagram of the transmit converter is shown in Fig. 8.3, and the full circuit in Fig. 8.4. Starting at the transmit converter input, many modern HF radios have extremely low transverter-output drive levels, while others have no transverter output but can only deliver their full HF power. The *Suffolk* can accept HF drive levels up to a few hundred milliwatts (+24dBm), or down to about one hundredth of a milliwatt (–20dBm). It is relatively easy to cope with the excess drive – all you need is an attenuator. Very low levels of drive need an amplifier at IF to increase the level to something more usable. Greatest versatility is therefore achieved by using a wide-range variable attenuator at the transmit IF input, followed by an amplifier to establish the correct drive level for the mixer.

It is good practice to incorporate automatic level control (ALC) in a transverter. ALC allows the transverter to be set up for a specific output level at which the transverter operates linearly. This is considerably less than the flat-out maximum power – see Chapter 6. Your voice could drive the transverter beyond its linear power range, resulting in distortion, and ALC prevents this from happening. Once ALC is incorporated in the transmitter chain, it becomes simple to add some additional circuitry to protect the output transistor in the event of a poor match at the antenna connec-

tion. ALC is achieved by a PIN-diode attenuator (D1) at the IF input, which can reduce the IF drive to the transmit mixer by at least 20dB if necessary. A separate variable attenuator, RV1, is used to set the initial IF drive to a reasonable level so that the ALC doesn't have to do all the work.

Following the PIN attenuator, a broadband MMIC amplifier (IC1) provides 8dB of gain to increase the IF level into the transmit mixer. If your HF rig has an even lower output than the transmit converter is designed to accept, the MMIC can be changed for a higher-gain version. Details of this alternative are given later.

A diode-ring mixer (M1) converts the 28–30MHz IF input to 144–146MHz, using local oscillator drive at a level of +10dBm from the receive converter board. Mixing from 28–30 to 144–146MHz requires care in the choice of mixer, since an in-band product will be generated at 5 x 29MHz = 145MHz. This fifth-order product is only inside the amateur band when the mixer is driven between 28·8 and 29·2MHz, but at other IF drive frequencies it is still present. The 144–146MHz bandpass filters can do little to remove this product, so you are relying on the mixer to avoid generating it at excessive levels. Diode-ring mixers provide the highest level of fifth-order rejection of any type of mixer I tested, but in order to achieve this the IF drive level *must* be kept low.

Third-order distortion must also be kept low to ensure a clean-sounding signal, and higher orders must be suppressed to avoid splatter

144MHz

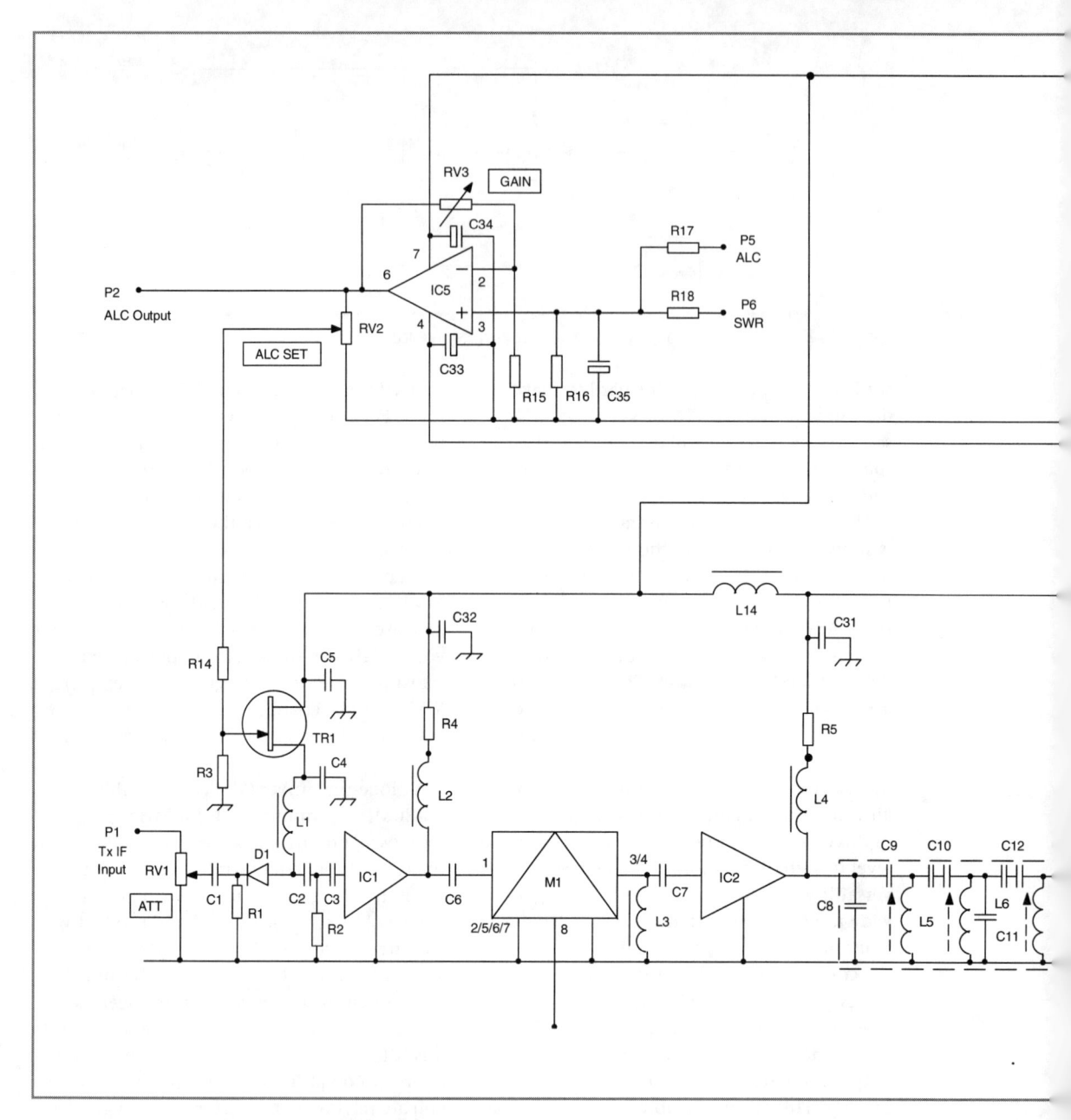

(see Chapter 6). The target for the mixer output was set at better than −50dB for third-order products, and once again a low IF drive level ensures the best results. This creates a problem of its own. The mixer has typically 6dB conversion loss, so assuming that the IF drive is kept to −13dBm maximum, the wanted mixer output at 144MHz will be −19dBm. To reach a final output of 10W (+40dBm) thus requires a gain of very nearly 60dB, and such high gain in the same frequency range can be the cause of instability. The use of MMIC amplifiers, negative-feedback amplifiers and intentional loss between stages allows the required gain to be achieved with stability.

SUFFOLK 144MHz TRANSVERTER

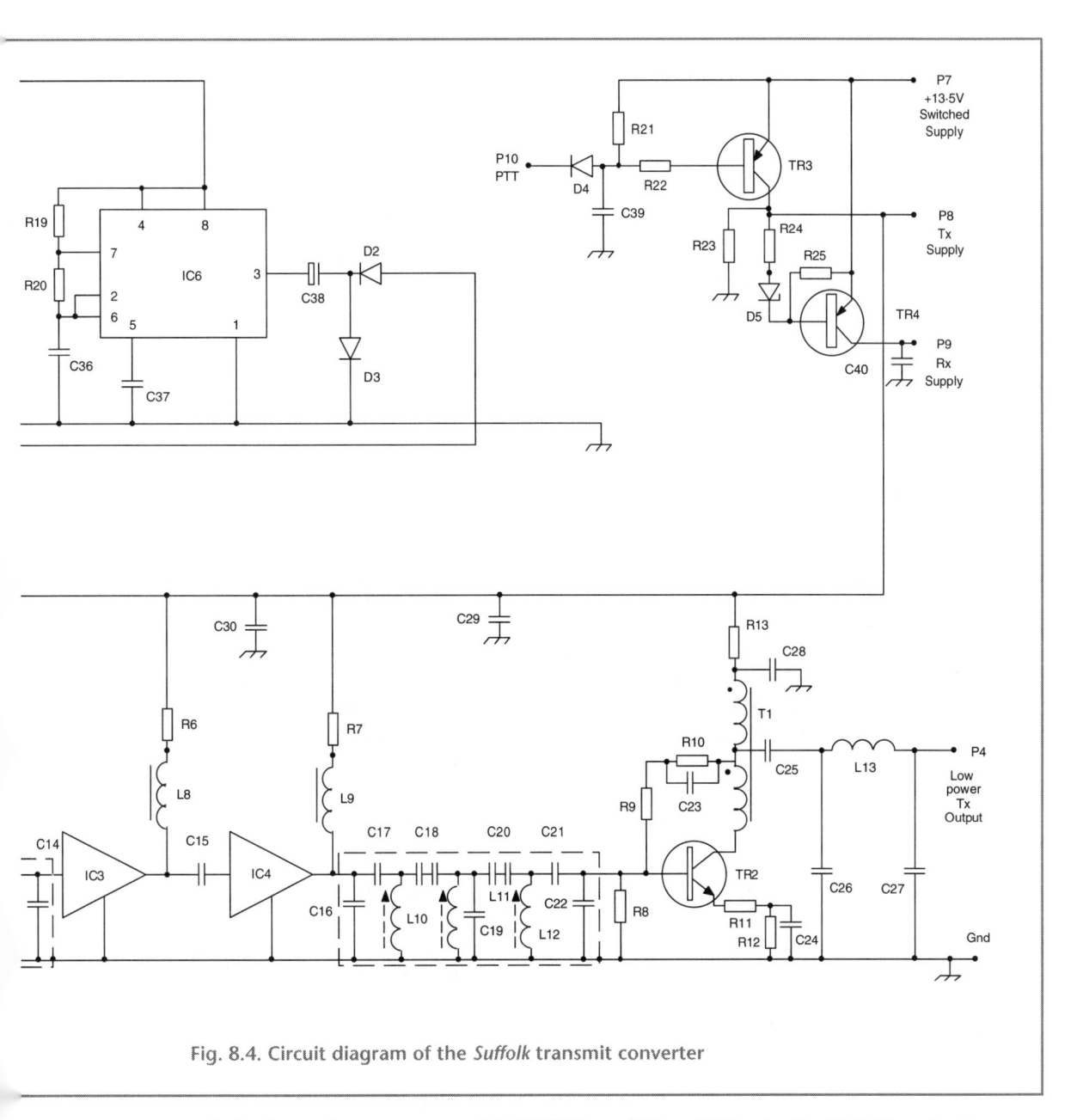

Fig. 8.4. Circuit diagram of the *Suffolk* transmit converter

As in the receive converter, an MMIC (IC2) is used to terminate the transmit mixer output. This provides a high return loss at the output, ensuring excellent mixer IMD performance. The MMIC is followed by a three-pole bandpass filter (BPF) to remove the image frequency of 88MHz and LO leakage at 116MHz. Two further MMIC amplifier stages, IC3 and IC4, raise the 144MHz output to a higher level, and a second BPF ensures that any last remnants of spurious outputs are effectively removed.

A BFR96 amplifier stage with negative feedback [2] provides over 100mW PEP output with third-order IMD products typically 46dB below PEP (–40dB below either of two tones).

144MHz

The output stage is followed by a lowpass filter with a cutoff frequency of 150MHz.

Transmit/receive power switching is provided by the additional circuitry associated with TR3 and TR4. On receive, TR4 conducts as a result of forward bias through D5, R24 and R23. When a press-to-talk (PTT) ground is applied to P10 to initiate a changeover to transmit, TR3 is forward-biased, removing the forward bias from TR4 and hence switching off the receive-converter RF stage. TR3 also provides power to the transmit converter and linear amplifier connected to P8.

The diode D4 at the PTT input provides isolation from positive voltages that might appear on the PTT lead, and C39 decouples any stray RF. In extreme cases it may prove necessary to add more decoupling here, to prevent unwanted switching to transmit in the presence of strong RF fields from other transmitters.

Amplification of the low-level ALC and VSWR detector output voltage is provided by the FET-input operational amplifier IC5. Gain is set by the ratio of RV3 to R15. By making RV3 variable the gain can be adjusted to deal with detector output from either the 10W amplifier or the 'barefoot' transmit converter.

ALC operation calls for controlled attack and decay of the ALC attenuation. These time constants are determined by the values of R16, R17, R18 and C35. Rising power at the ALC detector, produced by a peak of modulation, causes a negative-going detector output voltage. Initially C35 is discharged and therefore appears as a short-circuit to the detector voltage. As C35 charges through R17, the voltage at the non-inverting input of IC5 goes negative. The amplified negative-going voltage from the output of IC5 biases TR1 towards cutoff, reducing the bias current for the PIN diode D1. This in turn reduces the mixer drive, with a consequent reduction in transmit-converter output to the value determined by the setting of the ALC SET potentiometer RV2. When the modulation peak is over, C35 slowly discharges through R16 to restore full transmitter gain gradually. The values chosen for R16, R17, R18 and C35 give what I believe are optimum ALC attack and decay times. If you want to change them, the section on alignment tells you how.

Optimum control of modulation peaks is achieved by feeding ALC voltage all the way back into the HF transmitter itself. If the power and attenuation levels are set up correctly, all the ALC is then provided by the HF transmitter, leaving the transverter's own internal ALC loop with very little to do. Most HF transmitters require a negative-going ALC voltage for gain control, and the *Suffolk* provides this facility. To generate the required polarity from a positive operating voltage requires the use of a polarity inverter. IC6 generates a high-frequency pulse stream which is connected to a diode pump consisting of C36, D2 and D3. This circuit produces a rectified and smoothed voltage of approximately –10V to the operational amplifier IC5. The output of IC5 is made available at P2 for connection to the ALC input of the HF radio.

LINEAR AMPLIFIER

This section describes the linear amplifier producing 10W PEP output for less than 100mW input. A Mitsubishi M57713 amplifier module is used for convenience, reliability and its small size. These devices are available from several distributors and the cost is competitive with a discrete-transistor design.

For the constructor, the main advantage of the module is the low risk of transmitting a signal with severe intermodulation distortion or parasitic oscillations. Most people who have tackled transistor VHF power amplifiers have experienced some problems with stability, linearity and (all too often) the destruction of an expensive transistor. I experimented extensively with a 10W PA using discrete transistors, and its performance was truly excellent – except when it oscillated! After a lot of effort, I reluctantly decided that I could not guarantee good performance under all circumstances and changed over to the hybrid PA module. These devices give little trouble provided the correct supply and bias voltages are used and the RF drive level is not exceeded; the power amplifier is extremely easy to build

SUFFOLK 144MHz TRANSVERTER

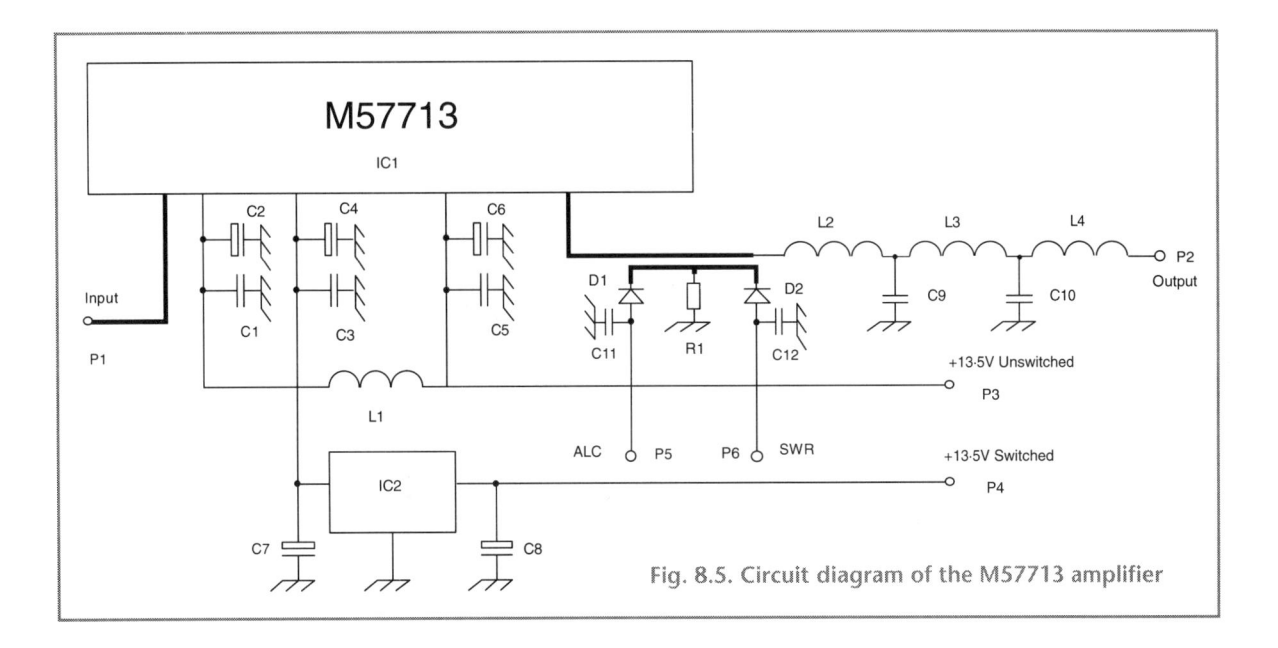

Fig. 8.5. Circuit diagram of the M57713 amplifier

and requires no alignment. The disadvantage lies in the intermodulation performance. As with any 'linear' amplifier, modules display high levels of IMD if overdriven, and at maximum rated output power the level of intermodulation products is just about tolerable. But unlike most amplifiers, you cannot produce a super-clean signal from a module merely by under-driving it. When the drive is reduced, the levels of the odd-order intermodulation products with respect to the main signal do not decrease significantly until the output has fallen below about 1W PEP. I suspect that this unusual intermodulation performance may be due to deliberate design compromises in the interstage and output matching.

So the hybrid module approach is a compromise. There should be no problems caused by instability, and the intermodulation performance will be adequate, though it will not reach the very high standards we recommended in Chapter 6. Having said that, I should also point out that many commercial transceivers use just this kind of PA, and when modules with better intermodulation performance come along in the future – as they surely ought to – the *Suffolk* will accept them. And don't forget that the transverter's overall intermodulation performance is also influenced by earlier stages in the transmit converter. If the bandpass filters in the transmit converter are not aligned properly, intermodulation in the poorly terminated MMICs can be a problem. Follow the alignment instructions carefully!

The amplifier module is bolted to a heatsink on the transverter rear panel, with its connecting leads soldered to the amplifier printed-circuit board. The circuit (Fig. 8.5) is almost ridiculously simple. An integrated-circuit 2A regulator (IC2) provides a 9V bias supply to the module (IC1). The linear amplifier is switched to transmit mode by turning on the bias regulator *via* TR3, which is located on the transmit-converter board. To avoid having to switch high currents, collector voltage remains connected to the module during both transmit and receive.

The module's RF output is connected via a directional coupler to a 150MHz lowpass filter, designed to reduce the level of all harmonics to less than –60dBc. The directional coupler provides negative voltage outputs from D1 and D2, corresponding to forward and reflected power. These voltages are amplified on

144MHz

Fig. 8.6. General layout inside the case

the transmit-converter board, and are used for ALC control as already described. Protection against output mismatches is achieved by feeding the 'VSWR' signal from D2 back into the ALC loop to reduce the drive power.

CONSTRUCTION

Almost any modern construction technique can be used to assemble the transverter, but for guaranteed results I recommend the use of printed-circuit boards made according to the designs presented in the following section. These boards fit into the recommended case to produce a professional-looking piece of equipment, as shown in Figs 8.6, 8.7 and 8.8.

In line with the modular approach used for each of the transverter units, three separate printed circuit boards are used in the construction of the transverter. These boards are:

1 – Receive converter
2 – Transmit converter
3 – Linear amplifier

Board size was determined by the case used to house the complete transverter. If you want to build the receive converter only, this will fit into a popular size of die-cast box (see component list).

Double-clad epoxy-glass board 1·6mm thick is used for all three PCBs (Figs 8.9, 8.10 and 8.11). The top of each board is left fully copper-clad to act as an RF groundplane, and the interconnection tracks are etched on the reverse side. Components are generally mounted on the groundplane side with their

Fig. 8.7. Front panel of the *Suffolk* transverter

Fig. 8.8. Rear panel of the *Suffolk* transverter

SUFFOLK 144MHz TRANSVERTER

leads passing through the board and soldered on the track side. Where component leads pass through the groundplane without need for grounding, the copper is removed around each hole on the groundplane side to prevent short circuits.

A second groundplane is used to interconnect all grounded leads on the track side of each board. Component leads which need to be grounded are soldered on both sides of the board, thus bonding the top and bottom groundplanes together electrically. This form of construction ensures stability in VHF/UHF applications without the expense of a board with plated-through holes.

The linear-amplifier board differs from the converter boards in that all components are mounted on the track side, because the

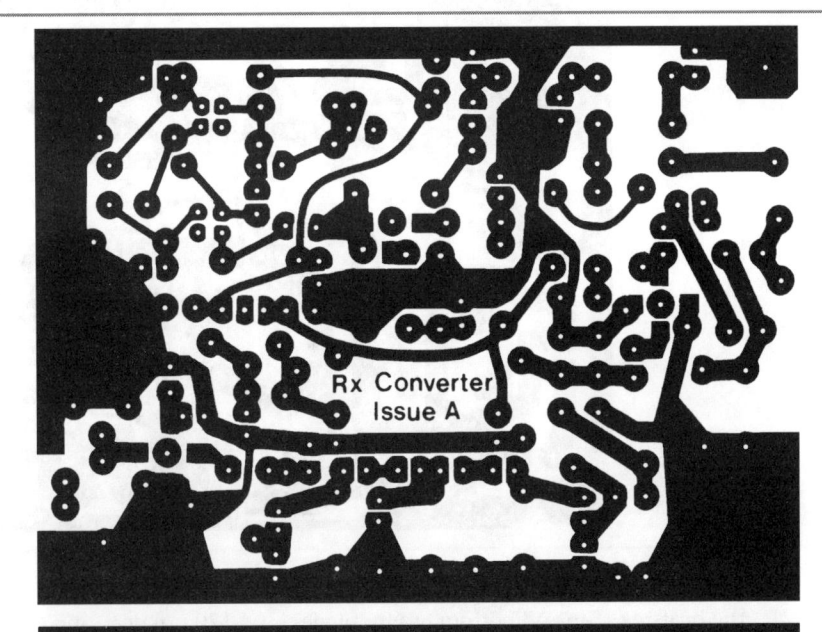

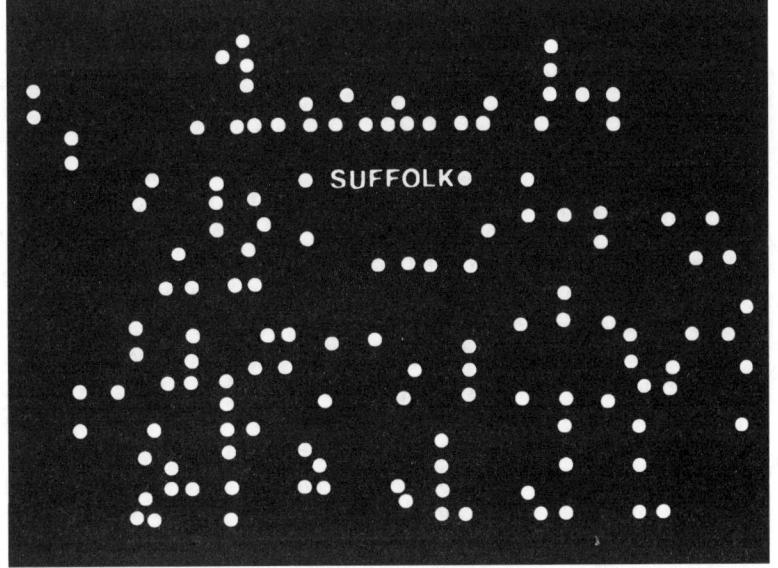

Fig. 8.9(a). Full-size PCB mask for the receive converter (underside)

Fig. 8.9(b). Full-size PCB mask for the receive converter (topside)

144MHz

Fig. 8.10(a). Full-size PCB mask for the transmit converter (underside)

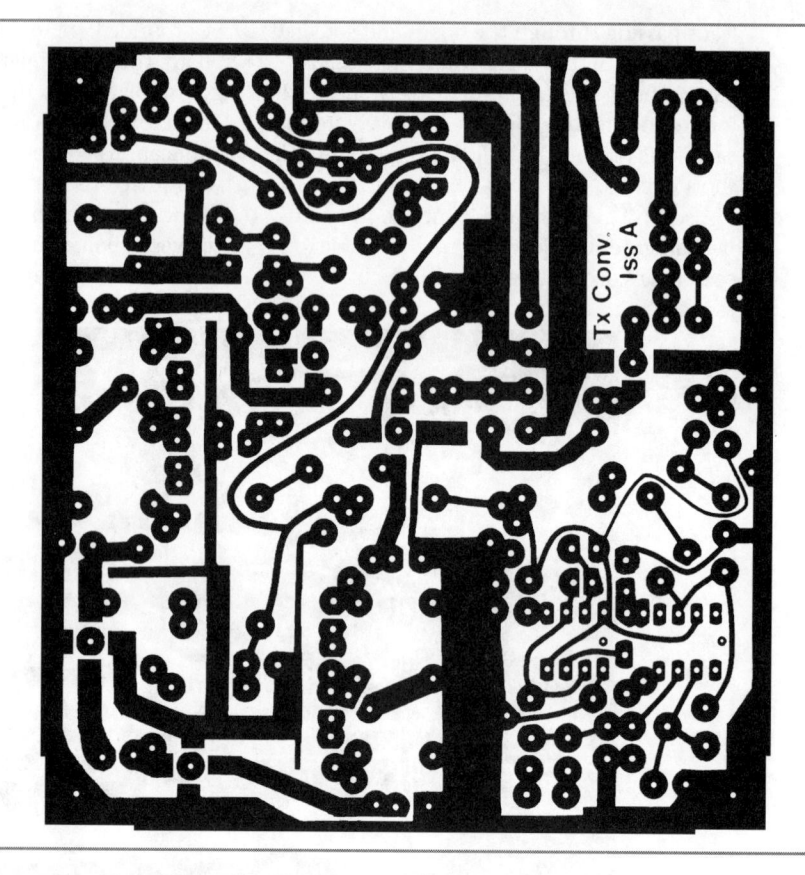

groundplane side of the board must be close to the heatsink. Top and bottom groundplane bonding is provided by short wire links through the board. The positions of these through-board links are shown as black dots on the component layout.

Figs 8.7 and 8.12 show the general layout inside the case. A 'U'-shaped aluminium screen separates the amplifier from the rest of the transverter. Feedthrough capacitors decouple the supply leads and ALC/VSWR detector outputs where they enter or leave the screened amplifier compartment.

Threaded nickel-plated brass pillars 12·5mm high are used to support and space the two converter boards from the bottom of the transverter case. The linear amplifier is secured to the rear panel and heatsink by four M3 screws which pass through the board and case, and are tapped into the heatsink as shown.

The recommended case for the *Suffolk* is the Centurion AE3, which is 280mm wide, 120mm deep and 65mm high. It consists of two pieces; a plain aluminium 'U'-shaped base and a durable brown epoxy paint-finished steel top. The two parts of the case are held together by four screws. Some problems were encountered in finding a suitable heatsink for the amplifier. Eventually I decided to use a readily obtainable heatsink which is larger than required, and to cut this down to the right size for the intended case. I recommend that you get a local engineering workshop to do the cutting for you, unless you really enjoy hard work with a hacksaw and file! The Siefert KL-139 75SW heatsink (black anodised finish) needs to be cut down to 64 x 105mm to fit on the rear panel.

For a professional appearance, you need to paint the bare aluminium parts of the case. I chose Coral Beige paint, which provides a very pleasing effect when lettered with black or

SUFFOLK 144MHz TRANSVERTER

Fig. 8.10(b). Full-size PCB mask for the transmit converter (topside)

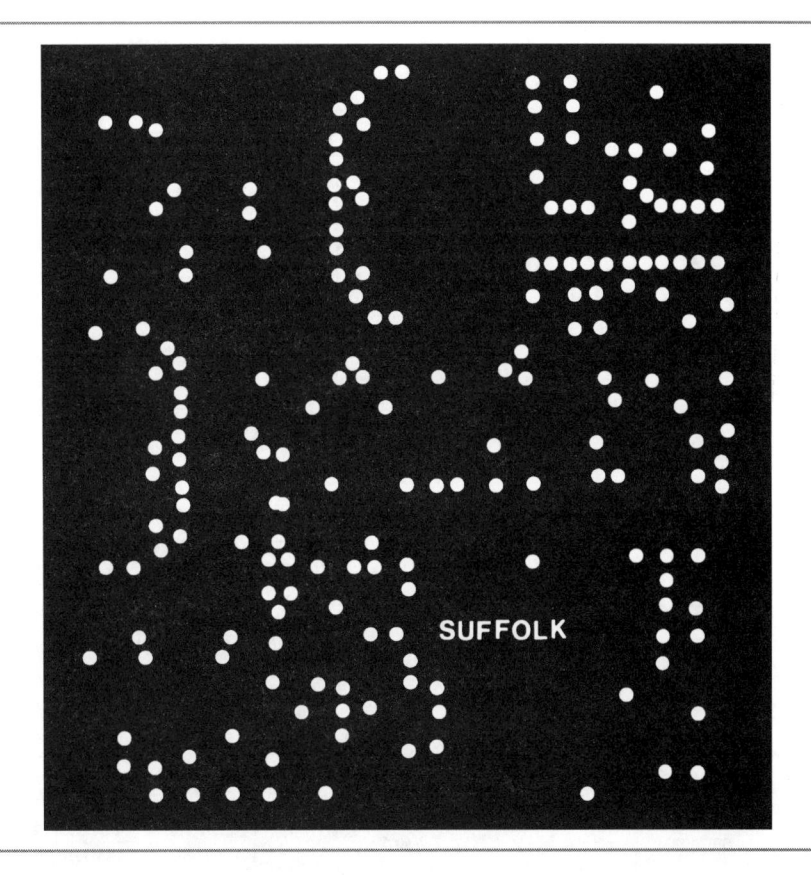

white rub-down transfers. Holes need to be drilled in the front and rear panels of the case to mount input and output coaxial connectors, power switch, power connectors, indicators, mounting pillars *etc*. To avoid scratches on the front panels, cover the panels with masking tape to mark and drill the holes, and paint the panels *after* drilling.

Fig. 8.11. Full-size PCB mask for the M57713 linear amplifier (topside; underside is fully copper-clad)

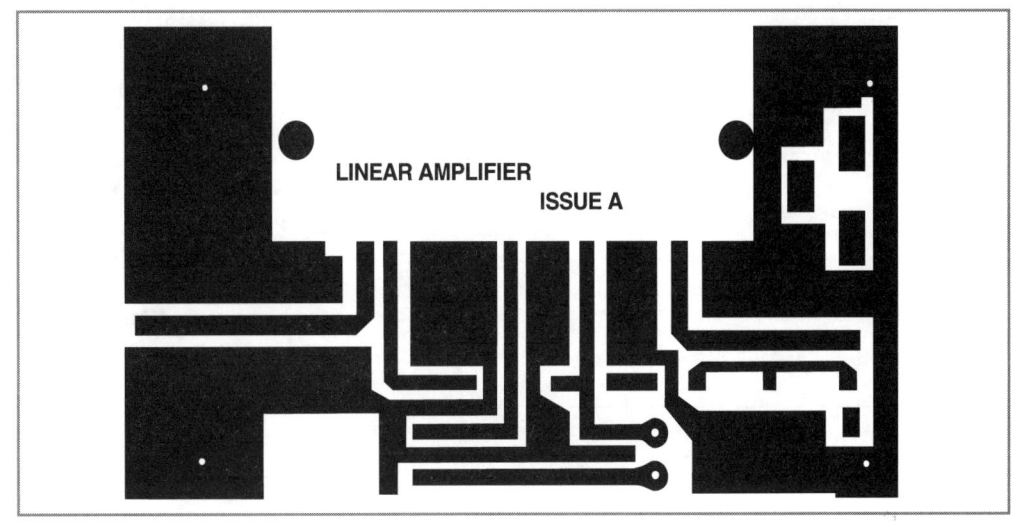

144MHz

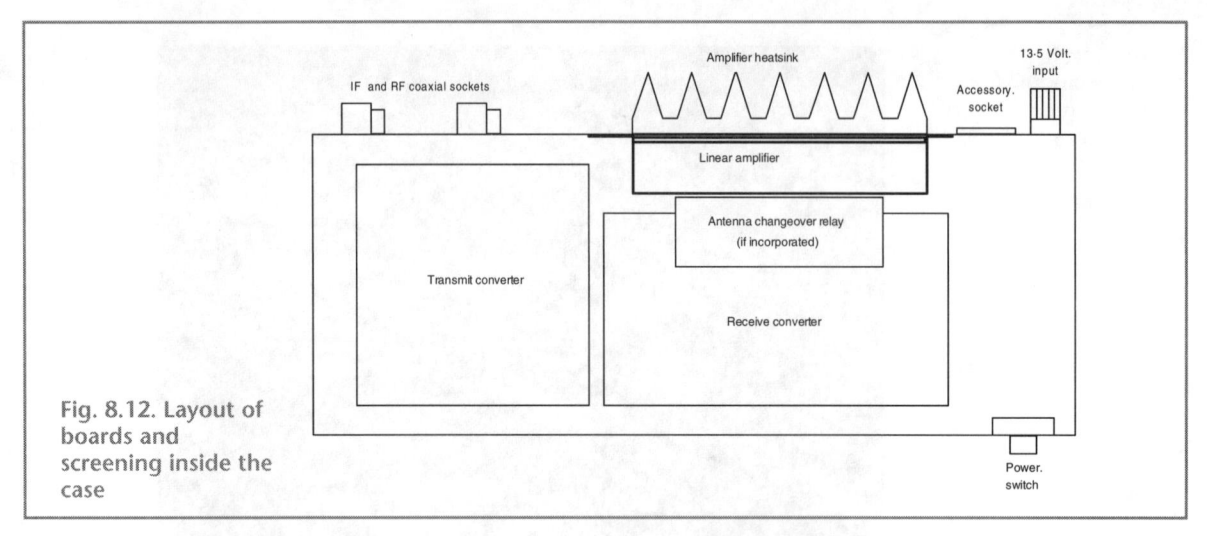

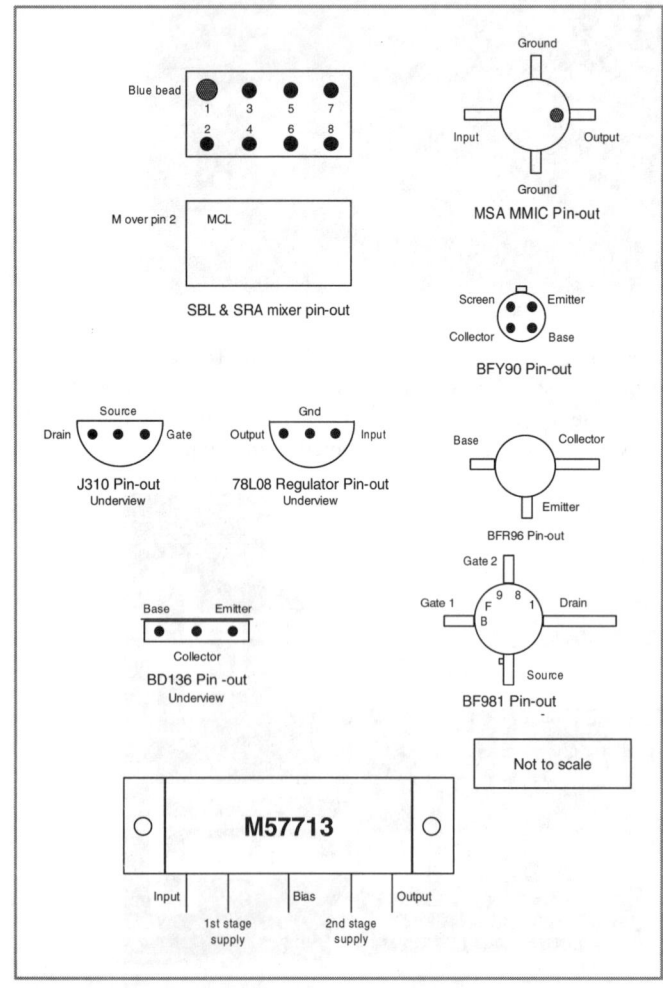

Fig. 8.12. Layout of boards and screening inside the case

ASSEMBLY

Assembly of the printed-circuit boards should begin with the receive converter, since this provides the local-oscillator signal for the transmit converter and can also be used in the preliminary alignment of the transmit converter.

Component layouts have been provided for all three printed-circuit boards making up the transverter, and Fig. 8.13 shows the pin-out diagrams of the semiconductors. The component designations (R1, C2, *etc*) are marked on the layouts, and the value of each component is in the list associated with each board. Please use only the recommended type of components – it really does matter! Where a component lead is connected to the groundplane, this is marked on the layout by a black dot on the lead.

RECEIVE CONVERTER

Fig. 8.14 is the component layout for the receive converter. All resistors and RF chokes in the receive converter are mounted flat to the board with 10mm lead spacing. Where a lead has to be soldered to ground (denoted by a black dot in Fig. 8.14), remember to solder it

Fig. 8.13. Pin-out diagrams of semiconductors used in the *Suffolk* transverter

SUFFOLK 144MHz TRANSVERTER

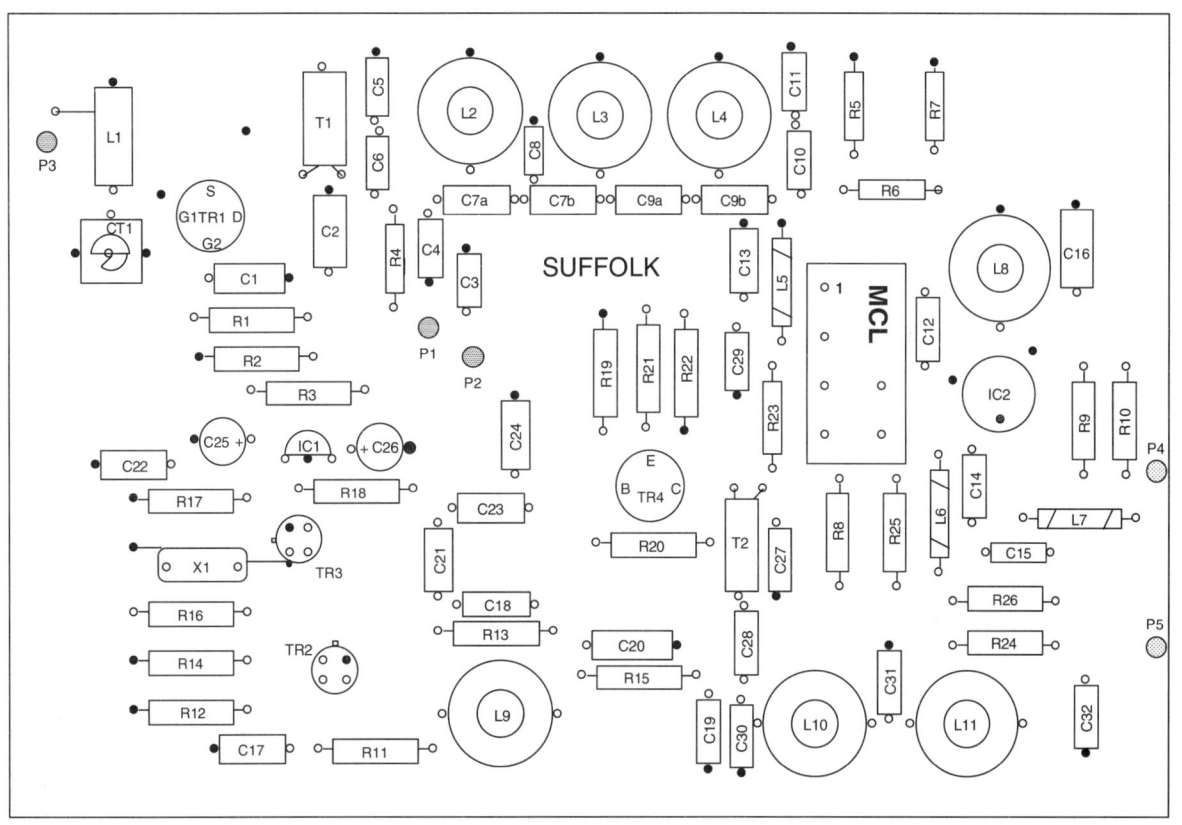

Fig. 8.14. Component layout for the receive converter

• Denotes groundplane connection

to both top and bottom of the board. Capacitors are mounted vertically with short leads; once again, solder grounded leads on both sides of the board, taking care not to overheat or fracture the body of the capacitor. The top-coupling capacitors in both the receive and transmit converter bandpass filters are each made from two 1·8pF ceramic capacitors connected in series. These may be replaced by a single tubular capacitor of 0·9pF if you can find one of suitable size.

The coils in the bandpass filters are mounted side by side as shown. Cut off the small lugs on the sides of the coil formers, so that the coils seat properly on the board with the correct spacing. Take care to mount all the coils in the bandpass filter the same way round, with the ungrounded ends of the windings closest to the board. Solder the grounded end of each coil to both top and bottom groundplanes, having first removed a small amount of plastic from the base of each coil to allow access for soldering to the top groundplane.

The input coil L1 should be wound carefully as shown in the component list. Do not use tinned copper wire; only silver-plated or enamelled copper wire of the correct size will give the lowest noise performance.

Care is needed when winding the toroidal transformers T1 and T2. For practical hints on winding toroidal transformers, see the panel on page 9-10. Be sure to spread the windings around the whole core, keep the interconnec-

8 - 17

144MHz

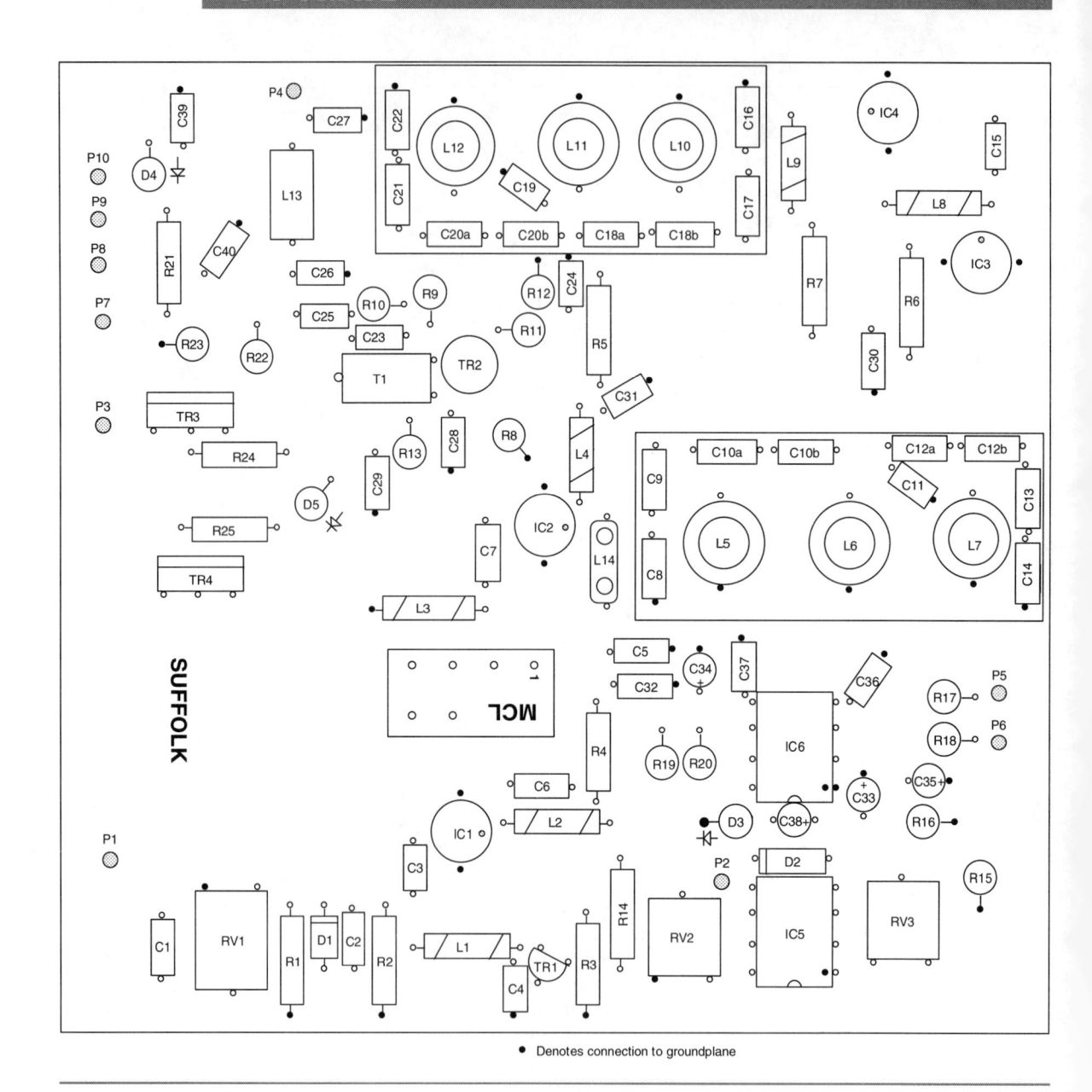

Fig. 8.15. Component layout for the transmit converter

tions short and double-check that they are correct. The start of transformer T1 (wire 1) goes to the drain of TR1, and the finish goes to the junction of R4 and C4. The junction of wire 2 and wire 3 is the tap, and connects to C5 and C6. The junction of wires 1 and 2 is not used; it should be carefully positioned away from the PCB groundplane. Transformer T2 is wound in the same way as T1, except that only two (bifilar) windings are required.

Solder the mixer, voltage regulator, crystal, transistors and MMIC into place, taking care to get the leads the right way round. This applies particularly to the mixer, which can be difficult to remove once soldered into place. The correct orientation of the mixer is shown

SUFFOLK 144MHz TRANSVERTER

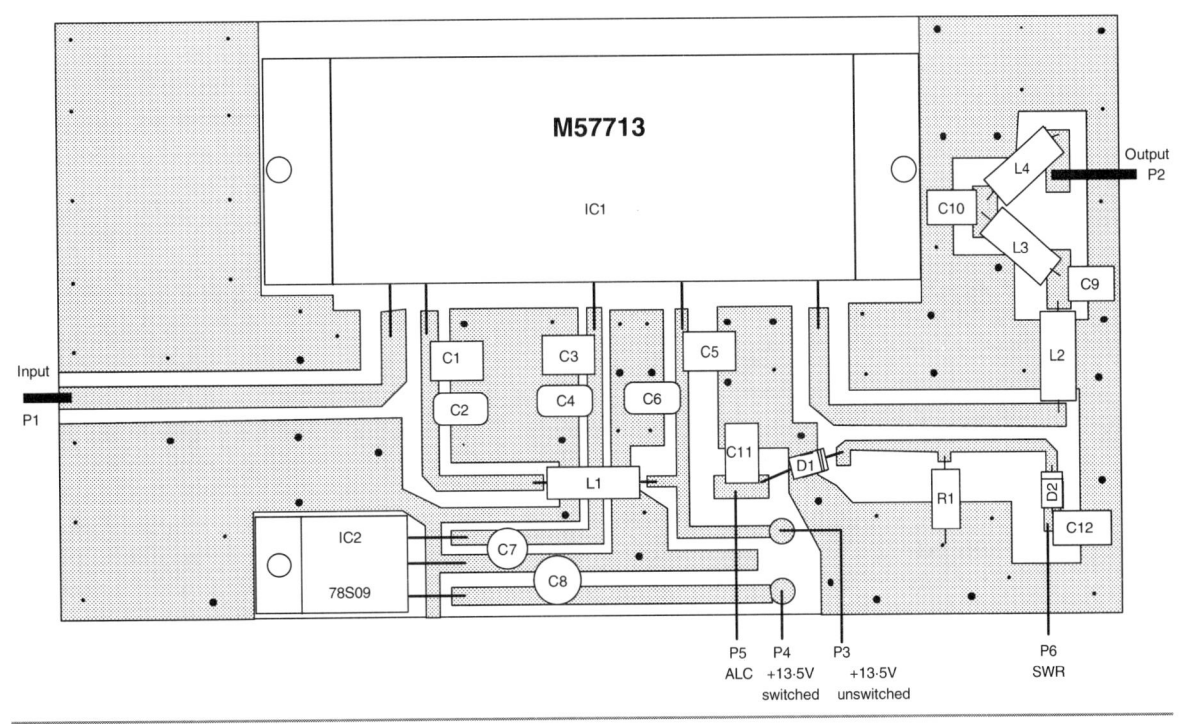

Fig. 8.16. Component layout for the M57713 linear amplifier

in the component layout. The name 'MCL' should appear in the corner as shown. The metal can of the crystal is grounded by two short wires, as shown in Fig. 8.14. Solder lightly and **very quickly** to the rim of the crystal can.

This completes assembly of the receive converter. Crop all component leads short on the etched side of the board, and carefully re-check all connections and component placings before applying power. Setting-up the receive converter is covered later in the section on alignment. I recommend that you do this before assembling the transmit converter, since the receive converter can be used to align the bandpass filters on the partially-constructed transmit converter.

TRANSMIT CONVERTER

Assembly of the transmit converter is similar to the receive converter and Fig. 8.15 shows the component layout. Begin with the two bandpass filters. These are assembled exactly as in the receive converter; once again, make sure that the ungrounded ends of all the coils are closest to the board. When the filter coils and capacitors are mounted on the board, tinplate screens can be soldered into place around them as shown. These screens will need to be made from off-cuts of sheet since they are not available commercially. Each screened enclosure is 22mm x 38mm x 18mm high, and two are required. Align the filters as described later, *before assembling the rest of the converter*.

It was found necessary to mount a few resistors vertically on this board as a result of limited space in some areas of the PCB. IC5 and IC6 can be mounted in sockets if required.

POWER AMPLIFIER

Fig. 8.16 shows the simple component layout. Start construction of the amplifier by soldering short lengths of tinned wire through the holes marked in the PCB. These through-board links are marked by black dots in Fig. 8.16, and they bond the top and bottom groundplanes together and ensure amplifier stability. File the blobs of solder almost flat on the reverse side

144MHz

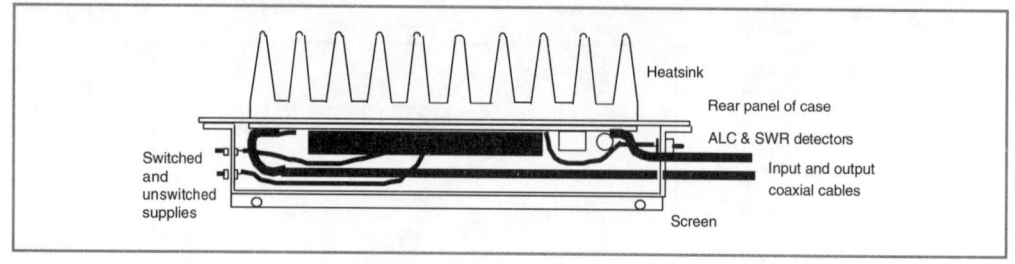

Fig. 8.17. Mounting the M57713 amplifier to the rear panel of the transverter case. Note the 'U' shaped metal screen, and that the PA module is mounted directly on the heatsink through a cut-out in the rear panel

of the board. Solder all remaining components into place as shown, except for the amplifier module itself.

When all the other components are in place, connect +13·5V to the switched (P4) pin and check that +9V appears at the regulator output. Any more than 9V could damage the amplifier module, so rectify any problems *before* soldering it into place.

Carefully fit the amplifier board to the heatsink, and position the module into the cut-out in the board. Check that it fits properly. Now remove the module, smear a thin, uniform layer of heatsink compound on to the mating face of its metal flange, and screw it back to the heatsink. The amplifier module can now be soldered into place. Fig. 8.17 shows how the amplifier is wired up behind its screen, and mounted through a cut-out in the rear panel of the transverter case.

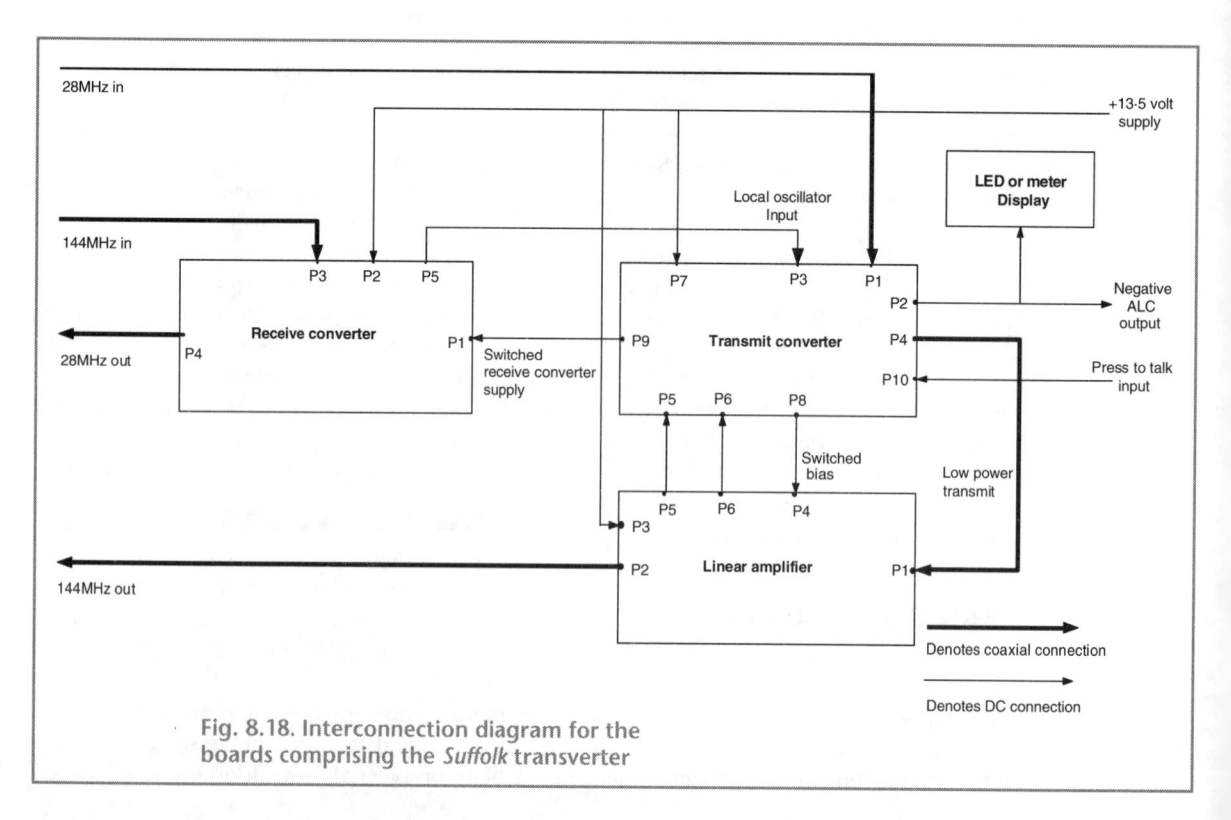

Fig. 8.18. Interconnection diagram for the boards comprising the *Suffolk* transverter

SUFFOLK 144MHz TRANSVERTER

WIRING UP

Each board has its input and output connection points indicated by a 'P' on both the circuit and layout diagrams. Fig. 8.18 shows how these points should be connected together to produce a complete transverter – but please read the instructions on alignment before you wire the units together and tie everything down into the case! All DC connections should be made with plastic-covered stranded wire. All RF interconnections should be made with thin 50Ω coaxial cable such as RG174.

I chose small PCB-mounting coaxial sockets for IF and RF inputs and outputs. It is not essential that you use these, but they make life easier if you want to make changes to the design or check performance. SMB or SMC are the preferred socket types, since they are physically small and available at low cost from several sources.

ALIGNMENT

Some basic items of test equipment are needed to align the transverter properly. You should have most of them in your shack already, not only to comply with the licence requirements but also because they are invaluable if you want to build your own equipment! Chapter 12 contains plenty of good ideas for specialised VHF/UHF test equipment.

You will need:
- Multimeter
- Sensitive absorption wavemeter covering 100 to 150MHz
- 50Ω RF power meter(s), 10mW to 20W
- 0·5W 50Ω attenuators of 6dB and 3dB
- Two low-power 50Ω terminations
- Three 100-200mm thin coaxial leads with suitable sockets

The receive and transmit converters are best aligned before they are finally fitted into the transverter case. If a fault is found, it's much easier to sort it out while you have easy access to both sides of the board.

RECEIVE CONVERTER

Begin alignment with the receive converter. Connect the oscillator and RF stage supply pins together (P1 and P2). Connect the short coaxial leads to the RF input, IF output and local-oscillator output connection points on the PCB. Connect 50Ω terminations to the oscillator output and antenna sockets.

Connect +13·5V to the converter and check that the current drawn is no more than about 130mA. If it is significantly higher, check for short circuits, wrongly placed components, faulty components *etc*. When you are satisfied that all is well, proceed to the alignment of the converter.

Tune your wavemeter to 116MHz and place its pick-up coil close to L9. Rotate the core of L9 until the meter indicates oscillation at 116MHz. Switch the converter off and on, and check the oscillator restarts reliably. If it fails to restart, turn the core slightly and recheck. Final setting of the core to set the oscillator exactly on 116·0000MHz can be done later.

Now place the wavemeter coil close to L10 or L11 and check that the meter indicates the presence of a strong 116MHz signal. If you have a power meter capable of reading 10 to 20mW full-scale, connect this in place of the 50Ω local oscillator termination and confirm that the output is approximately 10mW; then replace the termination.

Connect a receiver to the converter IF output. Tune the receiver to 29MHz, and you should detect at least a small increase in noise output when the converter is switched on.

Set the cores of L2, L3 and L4 level with the tops of their respective formers. A further small increase in noise output will be noticed if the filter and the RF stage are both working. Slightly screw in the core of L3 by up to two turns. The noise should show a further peak. Slightly retune the cores of L2 and L4 for maximum noise.

Adjust the core of L8 to increase the noise output. This peak is very broad, and may be hard to identify for certain.

Connect a 144MHz antenna to the converter RF input. Tune CT1 to increase the noise output from the receiver. The proper setting for CT1 is the most difficult adjustment in the whole transverter, since it needs to be set for optimum signal-to-noise ratio.

144MHz

This setting *does not* coincide with maximum noise. CT1 is best adjusted with the aid of an automatic noise-figure meter. However, not everyone has access to one, so some alternative methods are needed. The best 'amateur' method is to use an alignment aid such as the G4COM design mentioned on Chapter 12. If you have one, you've probably already read the instructions in the article, so I won't waste space by repeating them here. Failing that, you'll have to use a weak off-air signal. Judging small differences in signal-to-noise ratio by ear is very difficult, so I recommend that you use a weak FM signal as described in Chapter 5 for setting up the gain of preamplifiers.

The adjustment procedure described above is likely to result in the converter being peaked in a relatively narrow frequency range, somewhere between 144 and 146MHz. If you have access to a sweep generator, a network analyser or a spectrum analyser with tracking generator, the bandpass filter can be adjusted to give a flat response within 1dB across the 144-146MHz band. Otherwise, just re-peak the converter to cover the segment of the band of interest. In any case, the noise figure should finally be optimized at the DX end of the band. No preamplifier should be required for terrestrial DXing unless the feeder loss is very high – in which case, change the feeder! Although optimization at the LF end of the band may leave the receiver slightly deaf for satellite working between 145·8 and 146·0MHz, you will probably be using a preamp for this application anyway.

TRANSMIT CONVERTER

Alignment of the transmit converter is easy, especially if you align the two bandpass filters before completing the rest of the construction. When the two filters have been built, connect short coaxial leads to the input and output of one filter, and connect an attenuator to each lead. If you have access to the necessary test equipment, this can be used to align the filters for a nice symmetrical response across 144 to 146MHz. It should be possible to obtain a response flat within 1dB. Attenuation at 116MHz should be about 50dB, and greater than 70dB at the image frequency of 88MHz.

If you don't have access to sophisticated test equipment, you could align the filters using a 1000Hz tone-modulated AM signal generator and the detector and selective AF level meter described in Chapter 12. Otherwise you will have to align the transmit-converter filters by listening to off-air signals using the receive converter. Whichever way you do it, use the highest possible values of 50Ω attenuators on either side of the filter. Insufficient attenuation will lead to gross errors in filter alignment, and intermodulation in the transmit converter.

Once the two filters are aligned, finish building the transmit converter while trying not to disturb the filters you've just so carefully adjusted!

Connect the short coaxial test leads to the IF input, RF output and local oscillator inputs. Adjust RV2 so that its rotor is at the ground end of its travel. The setting of RV3 is not too important at this stage.

Connect +13·5V to P7. Check with your multimeter that +13V appears at P9, the receive-converter switched supply. No voltage should appear at P8, the switched supply for the transmit converter and linear amplifier bias. Temporarily connect P10 (PTT) to ground, and check that +13V now appears at P8 but not P9. While still switched to transmit, check that the output connections of IC1, IC2, IC3 and IC4 are all at +4·5–5·0V. Also check for –8 to –10V at pin 4 of IC5. If all the correct voltages are not present, look for wrong connections, faulty components *etc*.

Wire P2 of the receive converter to P7 of the transmit converter, and P1 of the receive converter to P9 of the transmit converter. Connect the receive converter local oscillator output to the transmit converter local oscillator input (RX P5 to TX P3) using thin coaxial cable. Connect a power meter capable of measuring 100 to 500mW full-scale to the low-power transmit output (P4). Adjust the input attenuator RV1 for maximum attenuation, with the rotor at the grounded end of its travel.

SUFFOLK 144MHz TRANSVERTER

Switch on, and check that the power meter at the output indicates zero output with no 28MHz drive. Connect a low-level 28MHz transmit signal to the TX IF input (P1). The level of this signal should not exceed 250mW. Gradually rotate RV1 to increase IF drive to the transmit converter. The power meter should now start to indicate some output at 144MHz. Saturated output at maximum drive should be in excess of 100mW. If you have a calibrated IF drive signal, from a signal generator for example, set the input attenuator to give 100mW (+20dBm) output at 144MHz for –17dBm input at 28MHz.

The alignment of the transmit converter is now complete except for setting the ALC level, which must wait for the linear amplifier to be commissioned. A spectrum analyser connected at the transmit converter output should show a very clean signal, although with harmonics at higher levels than are normally considered acceptable. If the transmit converter is to be used without the linear amplifier, you should still use a lowpass filter.

LINEAR AMPLIFIER

Connect short coaxial leads to the amplifier RF input (P1) and RF output (P2). Wire P3 to P4. Connect a 10W 50Ω power meter (or an in-line power meter and 50Ω dummy load) to the output from P2. Connect +13·5V to P3/4 and check again that the voltage at the regulator output is 9V, plus or minus a fraction of a volt; if not, fix it.

The transmit converter is capable of overdriving the amplifier and this could cause difficulties in adjustment, so temporarily place the 6dB attenuator between the converter output and amplifier input. Drive the transmit converter to deliver about 100mW output to the attenuator, which will give 20–30mW at the PA input. The power meter at the amplifier output should now indicate between 3 and 10W.

Now switch off and reduce the 6dB attenuator to 3dB. Re-apply drive and note that the output has increased. I strongly recommend that the 3dB attenuator is left permanently connected between the transmit converter and PA input. The presence of the attenuator prevents damage due to overdrive, and is a further aid to stability.

This completes alignment of the individual transverter modules, which can now be mounted in the case and neatly wired together.

ALC ADJUSTMENT

With the transverter boxed and almost ready to operate on-air, the final step is to set the ALC. Connect the RF power meter to the transverter output. With ALC and VSWR protection connected to P5 and P6 of the transmit converter, drive the transverter to full output. Measure the voltage at P2 of the transmit converter (ALC output). Adjust RV3 (GAIN) to confirm that the maximum obtainable negative voltage is approximately –8V, and then back off RV3 until the voltage is about 200mV below the maximum.

Now adjust RV2 (ALC SET), and note that the output power starts to fall as TR1 reduces the PIN-diode attenuator current. Adjust RV2 for the required RF output, which should be *no more than 10W*.

If the ALC attack and decay times do not suit your voice characteristics you may alter them by changing the value of R16 and R17. R16 sets the decay time D according to:

$$D = C35 \times R16 \text{ seconds}$$

while R17 sets the attack time A according to:

$$A = C35 \times R17 \text{ seconds}$$

With the values given, the attack time is 220ms and the decay time 330ms. These are fairly gentle times, because I have tended to err on the safe side so as not to cause rough-sounding audio. However, if you do modify the time constants, don't be tempted to use the ALC of the *Suffolk* to increase your 'talk power' because that's not what it is intended for.

R18 sets the VSWR protection reaction time P as:

$$P = C35 \times R18 \text{ seconds}$$

This time constant is best kept short, for

144MHz

obvious reasons, though if the value of R18 is too small it can affect ALC operation.

External ALC to the HF transceiver can be taken from P2 on the transmit converter. Almost all HF radios require a negative-going voltage for ALC control, and the voltage provided by the *Suffolk* should be adequate. A 20kΩ potentiometer may be connected between P2 and ground to provide an adjustable external ALC voltage. To set the external ALC, begin by adjusting the transverter as previously described, with the external ALC disabled. Next connect the external ALC to the HF radio and adjust the 20kΩ potentiometer so that the 144MHz output just starts to fall from the previously set output power. This ensures that the transverter is not being overdriven, and that the ALC circuit of the HF radio is in overall charge.

Experience has shown that many HF transmitters are capable of much more transverter output drive than specified by the manufacturer. However, this is often at the expense of spectrum cleanliness, so you can use the external ALC control to reduce the HF drive to more acceptable levels. You can generally afford to do this, because the *Suffolk* is designed to accept very low levels of transverter drive. If you cannot check the 28–30MHz drive signal with a spectrum analyser, an output level of 10 to 15dB below the maximum available will normally ensure a clean source signal.

If low-level spurious signals are noticeable at 5–7kHz either side of the transmitted carrier, these are probably due to the negative-voltage generator on the transmit-converter board. Extra decoupling capacitors are probably required in addition to C33 and C34.

MODIFICATIONS

Several modifications are possible to the basic *Suffolk* design. These are easy to do, thanks to the modular nature of the transverter.

The transmit converter was designed for use with modern transceivers which often feature very low-level transverter drive outputs. The Suffolk will accept 28–30MHz input levels between –20 and +24dBms (10µW to 250mW) for full power output at 144MHz. Lower levels can be catered for by changing IC1 (the transmit IF amplifier) from the specified MSA 0404 to an MSA 0304, at the same time changing the bias resistor R4 to 270Ω. This will extend the range of HF drive levels down to –24dBm (4µW).

During the time the receive converter was being developed, a new MMIC was announced. The MSA1120 (soon to be followed by the lower-cost MSA1104) appears to have been aimed specifically at mixer terminations. It features a gain of 12·5dB, a noise figure of 3·5dB and an IP3 of +18dB at 28MHz, making it a worthwhile replacement for the MSA0404 (IC2) in the receive converter. However, the increased gain will reduce the system dynamic range unless it is compensated by increasing the attenuation between the BPF and mixer. TCALC shows that an increase to 9dB of attenuation will retain the same system noise figure, with an improved overall system IP3 of –2dBm. Alternatively an attenuation of 8dB will reduce the system noise figure to 2dB with the same IP3 as the original converter. Notice how sensitive the overall performance is to very small changes in gain. As this and other new MMICs become available, the choice is yours.

If more than 10W PEP output is required, the M57727 can be used in place of the M57713. The 100mW transmit converter will drive this module to over 25W PEP output. Unfortunately the higher-power module is not a passport to a cleaner signal at the 3-10W level required to drive a big valve amplifier. Details of these modular amplifiers and many others can be found in the *Mitsubishi RF Power Devices* catalogue, available from Mitsubishi semiconductor distributors.

RESULTS

The following table of results was taken from measurements on the prototype transverter. TCALC predictions are given in square brackets.

RECEIVE CONVERTER
Gain +18dB [+18·5dB]
Noise figure 1·5 – 1·7dB (144 – 146MHz, excluding IF noise)

SUFFOLK 144MHz TRANSVERTER

Fig. 8.19. Spectrum of the *Suffolk* transverter transmitter output, showing as far as the third harmonic. All harmonics are more than 60dB down on 8W output, and all higher harmonics are below the analyser noise floor

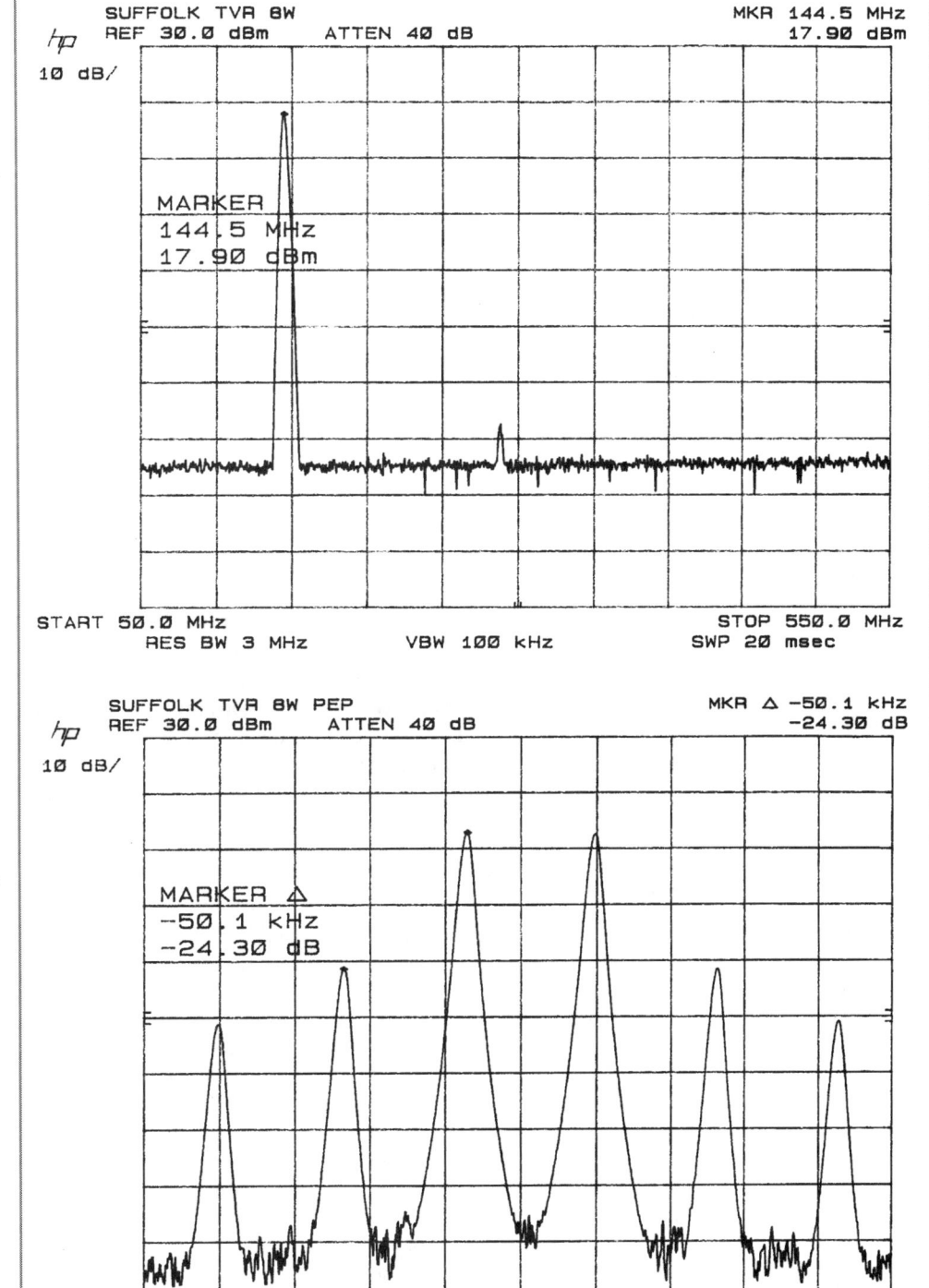

Fig. 8.20. Output spectrum of the *Suffolk* transverter showing two-tone intermodulation performance at 8W PEP output. Third-order products are more than 30dB down (–24dB down from either tone). Fifth-order products are more than 40dB below PEP

144MHz

	1·9 – 2·1dB (including IF receiver NF of 10dB) [2·25dB]
Input 3rd-order intercept	–6dBm (including IF intermod at IPI3 = +10dBm) [–5·2dBm]
Image rejection (88MHz)	Better than 80dB
TRANSMIT CONVERTER	
Output power	>10 W PEP
Third-order intermodulation	Better than 30dB below 8W PEP
IF input level for 10W output	+23dBm to –20dBm (depends on setting of input attenuator)
Harmonics	Better than –60dBc
88MHz image	Better than –80dBc
Local oscillator	Better than –80dBc
ALC output voltage range	0 to –8V
Current consumption	1·75A at 8W output

The above figures were taken from my own prototype and Figs 8.19 and 8.20 show typical spectrum-analyser displays. Although I have attempted to make the design as reproducible as possible, some figures will inevitably vary from transverter to transverter. If there are any significant differences in performance between your model and the measurements above, you have a problem – most likely in the alignment. Please re-check carefully before blaming the designer...

INTERFACING TO THE HF TRANSCEIVER

Fig. 8.21 shows the simplest possible arrangement for interfacing the *Suffolk* to an HF transceiver and the rest of the station. A large part of Chapter 11 is devoted to the single topic of station control and transmit/receive switching, so these comments are confined to transceiver-transverter interfacing.

It is not always obvious how best to connect a transverter to your HF transceiver. Many different transceivers could be used with the *Suffolk*, and it would be impossible to give full details of how to interface them all. Therefore this section will deal with the general principles involved, and I'll have to leave you to sort out the details.

A word of caution at this point: incorrect connections could lead to expensive damage, and I can't be responsible for *your* errors! So be careful to check at every stage of the interfacing. If your questions about interfacing are of the most basic kind, you should ask yourself whether you're really ready to build your own transverter.

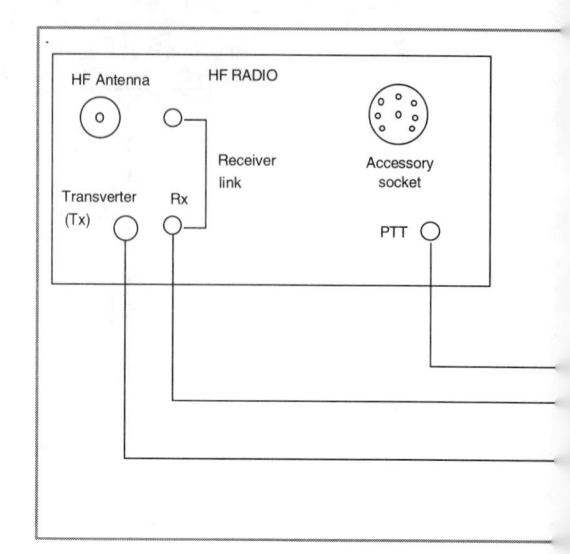

Perhaps the most common interfacing problem is how to separate the HF transmit and receive paths. If you are fortunate, or chose your radio well, the manufacturer will have done this for you, and you can follow Fig. 8.21. Otherwise you will need to make your own provisions.

Transceivers with a single HF antenna socket are generally the most difficult to deal with, since the transmit and receive paths need to be separated, and also a large amount of unwanted transmitter power needs to be lost in an attenuator before connection to the transverter. Separating the receive and transmit paths requires a relay switching box with appropriate TX/RX changeover sequencing. Even a short spike of HF power into the transverter input will do it no good at all! A changeover relay will be required to bypass the attenuator on receive, and for fail-safe operation that relay must be energised only on *receive*.

The *Suffolk* transverter can accept a maximum of 250mW input at 28–30MHz. Low-

SUFFOLK 144MHz TRANSVERTER

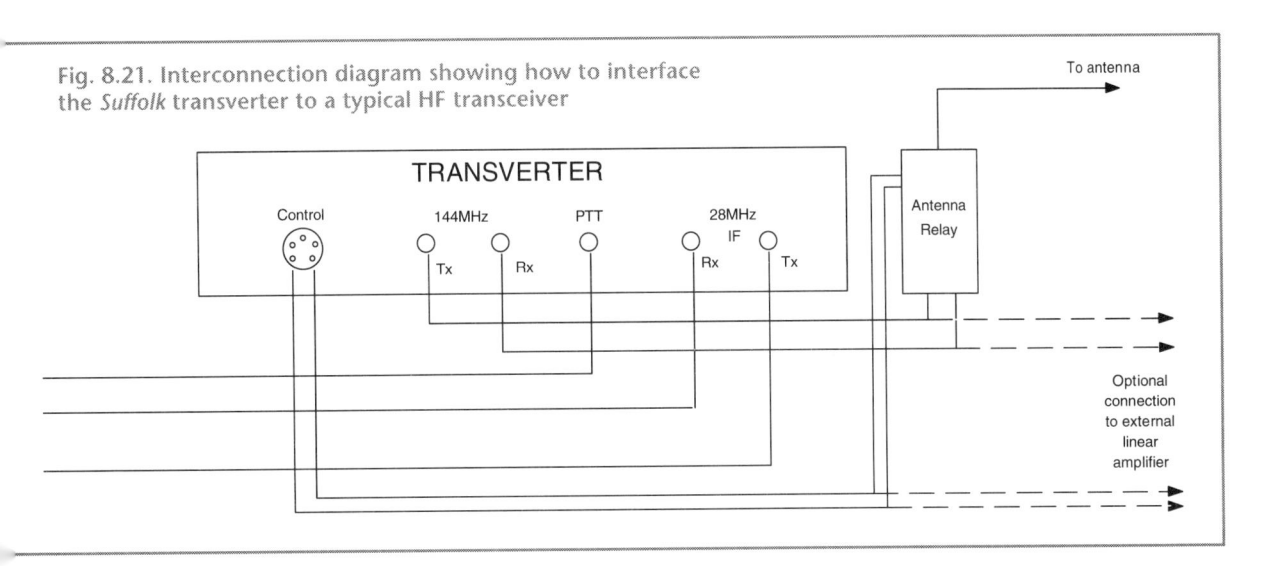

Fig. 8.21. Interconnection diagram showing how to interface the *Suffolk* transverter to a typical HF transceiver

power HF transceivers such as the Kenwood TS120V are fairly straightforward, and you simply need a 20dB attenuator (10W in, 100mW out; see Chapter 12). If the HF transmitter is capable of high power but also has an RF power-output control, it can be turned down to 10W or less in order to drive the transverter – although you also need to think what would happen if you forget to turn the power down from 100W! The resistors in the attenuator need to be capable of dissipating almost all the available HF power, and care must be taken to ensure low leakage around the attenuator and its relay switching.

HF transceivers with independent transverter sockets are much easier to interface. Usually a transmitter IF output of between 10μW (–20dBm) and 250mW (+24dBm) is provided, with the lower levels being typical of many modern designs, and the *Suffolk* is designed to cope with this entire range of levels. Simply connect the HF radio transverter output to the *Suffolk* 28–30MHz TX IF input (P1), using a suitable length of miniature coaxial cable. Try not to make this lead too long, or you may get breakthrough of signals on the 28MHz band as a result of cable leakage. An early European '144MHz DX record' has since been identified as a direct 28MHz contact between two leaky transverters!

HF transceivers with transverter outputs generally require some form of switching to disable the main HF PA. This can be done through the *Suffolk*'s accessory socket; consult the transceiver handbook for details.

The receiver connection should preferably be via a separate transverter receiver socket. If the receive transverter input is via the same transverter socket as on transmit, you will need to provide independent TX/RX switching, which is fairly straightforward. But if the receiver input is via the main HF antenna socket, you really ought to fit a relay box or modify the HF transceiver. Otherwise, sooner or later, you are going to stuff 100W of HF power into the receiver output of the *Suffolk*. It is often quite easy to identify a redundant socket on the back of the HF transceiver (the 'phone-patch' output, for example) which can be connected by coaxial cable directly to the receiver input, bypassing the antenna changeover relay.

A more sophisticated arrangement catering for several transverters connected to the same HF radio would use a switching box to connect the TX and RX IF to the wanted transverter via miniature screened relays or a rotary selector switch. The switching box also needs to disable the HF PA, and to route the PTT and ALC lines to the appropriate transverter.

If you can, arrange the switching so that all

144MHz

the transverters can remain powered while on standby, so that their local oscillators are already stabilized on frequency. A problem can arise here, due to strong-signal pickup in one of the unused transverters feeding through the limited isolation of the IF switch and being heard on the HF receiver. The solution in the case of the *Suffolk* is to switch off the receive converter during standby while retaining power to the local oscillator. Separate supply connections and unused pins on the accessory socket allow for this option, which I have implemented in my own station.

SUMMARY

I have presented a detailed description of a 28 to 144MHz transverter which is easy to build, but has the high performance necessary for today's crowded band conditions. The design and this description have both been thoroughly tested by a number of volunteer constructors, to whom we should all be very grateful. As a result, I hope that the *Suffolk* design will be a success with home constructors.

For anyone who has never used a well-designed transverter or transceiver at VHF before, the *Suffolk* can take a little getting used to. There is no excess gain, so don't expect your HF transceiver's S-meter to hover at S9 on noise. If you are easily impressed by lots of noise, you're probably not getting the best out of your DX system. You'd do well to re-read Chapters 4 and 5, and get used to the fact that weak signals can be perfectly copiable without moving the S-meter.

REFERENCES

[1] The diplexer circuit was devised by H. Paul Shuch, N6TX. It first appeared in *Ham Radio*, February 1977 and is used by kind permission of the author and magazine editor.

[2] The broadband Class-A negative-feedback amplifier used in both the receive converter and transmit converter appeared in Chapter 8 of *Solid State Design for the Radio Amateur* by Doug DeMaw W1FB and Wes Hayward W7ZOI, and is used by kind permission of the publishers, the American Radio Relay League.

SUFFOLK 144MHz TRANSVERTER

COMPONENT LIST FOR THE *SUFFOLK* RECEIVE CONVERTER

RESISTORS
0·25W 2% carbon film

Value	Qty	Ref
4R7	1	R21
18R	3	R24, 25, 26
33R	1	R23
39R	1	R6
47R	1	R22
51R	2	R9, 10
100R	2	R1, 18
150R	2	R5, 7
180R	1	R4
220R	1	R8
390R	1	R17
470R	1	R13
510R	1	R19
560R	1	R15
820R	1	R12
1k	3	R11, 14, 16
1k5	1	R20
47k	1	R2
82k	1	R3

CAPACITORS
Sub-miniature ceramic plate, Philips 682 and 629 series

Value	Qty	Ref
1p8	4	C7a, 7b, 9a, 9b
10p	1	C18
15p	1	C8
18p	1	C21
22p	4	C6, 10, 15, 22
27p	2	C30, 32
47p	2	C5, 11
56p	1	C31
1n	12	C1, 2, 3, 4, 17, 19, 20, 23, 24, 27, 28, 29
10n	3	C12, 13, 14

Polystyrene

Value	Qty	Ref
560p	1	C16

Tantalum bead, 16 to 35V wkg

Value	Qty	Ref
0·1μ	1	C25
1μ	1	C26

PTFE Foil trimmer Philips 809 05002

Value	Qty	Ref
10p	1	CT1

INDUCTORS
Toko S18 series (Cirkit or Bonex)

Ref	Qty	Description
L8	1	Red (2·5 turn) coil with aluminium core
L2,3,4	3	Orange (3·5 turn) coil with aluminium core
L10,11	2	Orange (3·5 turn) coil without core
L9	1	Green (5·5 turn) coil with aluminium core

RF CHOKES
Miniature moulded chokes with axial leads (Cirkit or Bonex)

Value	Qty	Ref
1·5μH	1	L7
10μH	1	L6
100μH	1	L5

L1 — 8 turns of 1mm diameter silver-plated wire, inside diameter 6mm, turns spaced one wire diameter. Tap at 2·5 turns from ground. Mount 2mm above the board.

TRANSFORMERS

T1 — 4+4+4 turns trifilar wound on Amidon T50-12 toroid (Cirkit or Bonex). Connect as a 3:1 auto-transformer (see text).

T2 — 6+6 turns bifilar wound on Amidon T50-12 toroid. Connect as a 2:1 auto-transformer.

TRANSISTORS

Type	Qty	Ref
BF981	1	TR1 (Philips Semiconductors, *etc*)
BFY90	2	TR2, 3
BFR96	1	TR4

INTEGRATED CIRCUITS

Type	Qty	Ref
78L08	1	IC1
MSA0404	1	IC2 Avantek (Bonex)

MISCELLANEOUS

X1 — 116MHz fifth-overtone crystal, high stability, HC18/U holder

M1 — SRA-1 or SBL-1 ring mixer (see text) (Cirkit, Bonex)

(Diecast box for receive-only converter. STC part number 46R CS00 043 A00)

COMPONENT LIST FOR THE *SUFFOLK* POWER AMPLIFIER

RESISTORS
0·25W carbon film 2%

Value	Qty	Ref
27R	1	R1

CAPACITORS
miniature ceramic plate

Value	Qty	Ref
27pF	2	C9, 10
1nF	2	C11, 12
3n3	3	C1, 3, 5

Tantalum bead 35V working

Value	Qty	Ref
0·1μF	1	C7
1μF	1	C8

Solid aluminium bead

Value	Qty	Ref
4·7μF	3	C2, 4, 6

DIODES

Type	Qty	Ref
1N4148	2	D1, 2

INTEGRATED CIRCUITS

Type	Qty	Ref
M57713	1	IC1: Mitsubishi 10W hybrid amplifier module
78S09	1	IC2: 2A 9V regulator

INDUCTORS

L1 — 10 turns of 24SWG enamelled copper wire, 6mm inside diameter, close wound. Self-supporting, mounted close to board.

L2, L4 — 3 turns of 24SWG enamelled copper wire, 3mm inside diameter, turns spaced one wire diameter. Self-supporting, spaced 3mm above the board.

L3 — 3 turns of 22SWG enamelled copper wire, 5mm inside diameter, close wound. Self-supporting, spaced 5mm above the board.

MISCELLANEOUS

Heatsink — Siefert KL-139 75SW, black anodized finish. Machined to fit the recommended case (see text). (Schaffner EMC, Micromark or GCA Electronics)

Screen to separate the amplifier from the transmit converter – see Fig. 8.12 and 8.17.

Four bolt-in feedthrough capacitors, 1-10nF, for connections to the screened amplifier compartment.

144MHz

COMPONENT LIST FOR THE *SUFFOLK* TRANSMIT CONVERTER

RESISTORS
0·25W 2% carbon or metal film

Value	Qty	Ref
4R7	1	R11
15R	1	R13
39R	1	R12
120R	2	R1, 2
180R	3	R4, 5, 7
330R	1	R6
560R	1	R9
680R	1	R8
1k	3	R10, 22, 23
2k2	1	R24
2k7	1	R14
5k6	1	R21
8k2	1	R25
10k	3	R15, 19, 20
33k	1	R18
180k	1	R3
220k	1	R17
330k	1	R16

Presets, carbon or cermet

Value	Qty	Ref
470R	1	RV1: 0·2" x 0·4" pin spacing
2k	1	RV2: 0·2" x 0·2" pin spacing
20k	1	ALC output adjustment (optional – see text)
1M	1	RV3: 0·2" x 0·2" pin spacing

CAPACITORS
Sub-miniature ceramic plate, Philips 682 and 629 series

Value	Qty	Ref
1p8	8	C10a, b, 12a, b, 18a, b, 20a, b
15p	2	C11, 19
22p	4	C9, 13, 17, 21
39p	2	C26, 27
47p	4	C8, 14, 16, 22
1n	3	C5, 25, 39
4n7	1	C37
10n	15	C1, 2, 3, 4, 6, 7, 15, 23, 24, 28, 29, 30, 31, 32, 36

Multilayer ceramic

Value	Qty	Ref
100n	1	C40

Tantalum bead, 16V wkg

Value	Qty	Ref
1µF	1	C35
10µF	3	C33, 34, 38

TRANSISTORS

Type	Qty	Ref
J310, 2SK55	1	TR1
BFR96	1	TR2
BD136	2	TR3, 4

INTEGRATED CIRCUITS

Type	Qty	Ref
MSA0404	3	IC1, 2, 4: Avantek (Bonex)
MSA0204	1	IC3: Avantek (Bonex)
CA3140	1	IC5
NE555	1	IC6

DIODES

Type	Qty	Ref
BA479	1	D1: PIN diode
1N4148	3	D2, 3, 4
5V6	1	D5: 5·6V 400mW Zener diode

INDUCTORS
Toko S18 series (Cirkit or Bonex)

Ref	Qty	Description
L5, 6, 7, 10, 11, 12	6	Orange (3·5 turn) coil with aluminium core
L13		3·5 turns of 1mm diameter tinned copper wire, 6mm inside diameter, turns spaced one wire diameter. Mount 2mm above board.
L14		2 turns of plastic insulated wire through a 'binocular' (2 hole) ferrite bead. Type not critical, *eg* Siemens type B62152 code A0007X060 (Electrovalue)

RF CHOKES
Miniature moulded type with axial leads (Cirkit or Bonex)

Value	Qty	Ref
10µH	3	L4, 8, 9
150µH	3	L1, 2, 3

TRANSFORMER

Ref	Description
T1	6+6 bifilar turns of 24SWG insulated wire on Amidon T50-12 toroid (Cirkit or Bonex). Start to TR2 collector, finish to R13, tap to R10.

MISCELLANEOUS

Ref	Description
M1	SBL-1 ring mixer (Cirkit or Bonex)

Tin-plate screens, 22mm wide x 38mm long x 18mm high – two needed.

THE W1SL 144MHz POWER AMPLIFIER

by John Nelson GW4FRX

This well-known amplifier design first saw the light of day in the February 1971 edition of *QST*, the magazine of the American Radio Relay League. Modestly billed as "New Ideas for the 2-Meter Kilowatt" by Thomas F. McMullen Jr, W1QVF and Edward P. Tilton, W1HDQ, the device is nowadays usually referred to as the "W1SL" since that was the callsign later obtained by McMullen. Both authors were well-known and greatly respected for their involvement in VHF and UHF activities, and with that pedigree it's not surprising that the W1SL is by far the best 144MHz twin-tetrode amplifier design ever – it's an embodiment of the Mies van der Rohe principle of "less is more". When you first look at a W1SL you can't help suspecting that it's too simple to be true. Don't be fooled: a lot of very subtle thinking went into this amplifier, which is why it's very difficult indeed to make one that works badly.

The W1SL – sometimes referred to by affectionate owners as the switched supply for the "Weasel" – is a two-valve push-pull amplifier with linear grid and anode circuits. The grid lines are formed from standard copper microbore central heating pipe and the anode line is cut from brass and bent to shape. The anode tuning arrangement lays to rest a prevalent problem with two-valve VHF amplifiers. Most of its 'plumber's special' predecessors used circular discs running on screw-threads or some similar arrangement, but for various reasons these can be a severe pain. One is that the effective tuning range you can achieve with them tends to be a bit limited; another is that you inevitably run into arcing problems sooner or later because the screw threads make poor RF contact (remember that with a loaded Q of 10 or so, you've got about 20A peak RF current in the anode tank circuit components of a twin-4CX amplifier). Yet another problem is that it's almost impossible to persuade the anode circuit to behave in a balanced way, so the loading can be very unsymmetrical and one valve ends up doing 75% of the work. The W1SL solves all these problems by a masterstroke. It uses a 'flapper', which has no contacts and doesn't rely on an RF ground return, to form what amounts to a split-stator capacitor.

Having said all these nice things, there are a few alterations and modifications which can be made to improve the performance of the device. For example, like most 144MHz amplifiers of its vintage the W1SL was designed primarily with Class-C use in mind and the output coupling loop in the original design gives insufficient loading for good linearity in Class AB1. The screen-grid DC circuitry also needs some updating.

The part you will undoubtedly have to modify in the anode compartment is the output coupling loop, because in the original it's much too small and in the wrong place. If you duplicate the dimensions in the original circuit, you'll probably discover the following: a fair amount of power out in Class C but not a lot in AB1; an inability to get the screen current to go negative (implying that the loading isn't heavy enough – see Chapter 6); and generally rather poor efficiency. If it really isn't your day there'll also be a flashover as

144MHz

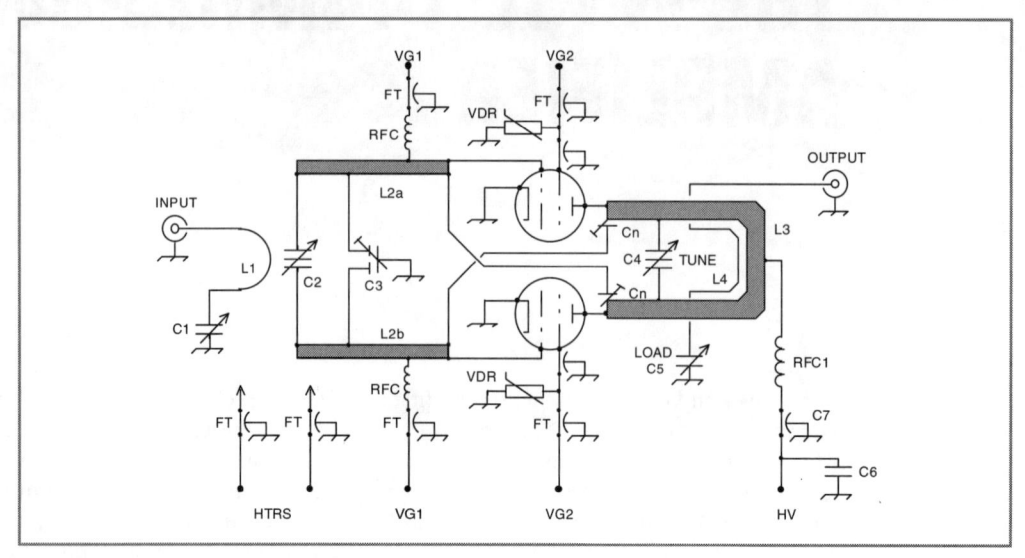

Fig. 8.22. Circuit of the W1SL power amplifier. See Fig. 11.5 for additional screen-grid protection components

you try to wind the power up. People have tried all sorts of clever ways to get round this loading problem, with Pawsey stubs and all sorts of fearsome devices being pressed into service. In fact, the original design seems to work perfectly well provided that you enlarge the coupling loop considerably, as shown in the diagrams and construction notes which follow.

As with almost every 4CX amplifier design in the history of the world, the original W1SL included a low-value series resistor in each screen feed – in this case 150Ω. For Class-C work, screen regulation doesn't matter so much and the resistors would help to isolate each screen from the common power supply. But if you include these resistors in a modern W1SL intended for Class-AB1 use, you instantly negate the effort which you put into building a screen supply which could source and sink vast amounts of current with practically zero variation in output voltage. Such resistors are also bad news if you have a flashover. If 150Ω is the first thing the current surge sees on its way out, the voltage on the screen grid will streak up and the screen decoupling capacitor will blow faster than you can say 'tetrode'. I'd recommend that you protect each screen using VDRs, parallel resistors and inductors as shown in Fig. 11.5. The W1SL circuit shown in Fig. 8.22 includes only the VDRs for clarity, but no doubt you can reason the rest out for yourself.

CONSTRUCTION

The amplifier is built in two 17" x 8" enclosures, following the general layout of Figs 8.23 and 8.24. The anode box is 3¾" deep; although the original grid box was only 3" deep, you need closer to 4" depth to accept the mouth of a decent-size blower as recommended in Chapter 6. The assembly is completed by top and bottom covers and a 19" rack panel, ideally 5U or 6U tall. All this is standard metal-bashing and you should be able to make that, or get it made locally, without too much bother. If you're a dab hand in the workshop you could make a W1SL chassis entirely from sheet, bar and angle stock.

For the non-specialist metal-basher, the anode line is probably going to be the most difficult part of the whole amplifier to make. Don't let that stop you, though – my first two 144MHz amplifiers were W1SLs, and in both cases the nearest I came to 'workshop facilities' was possession of one junior hacksaw and some extremely old and blunt files. The 'workshop' was the floor! All very character-

W1SL 144MHz POWER AMPLIFIER

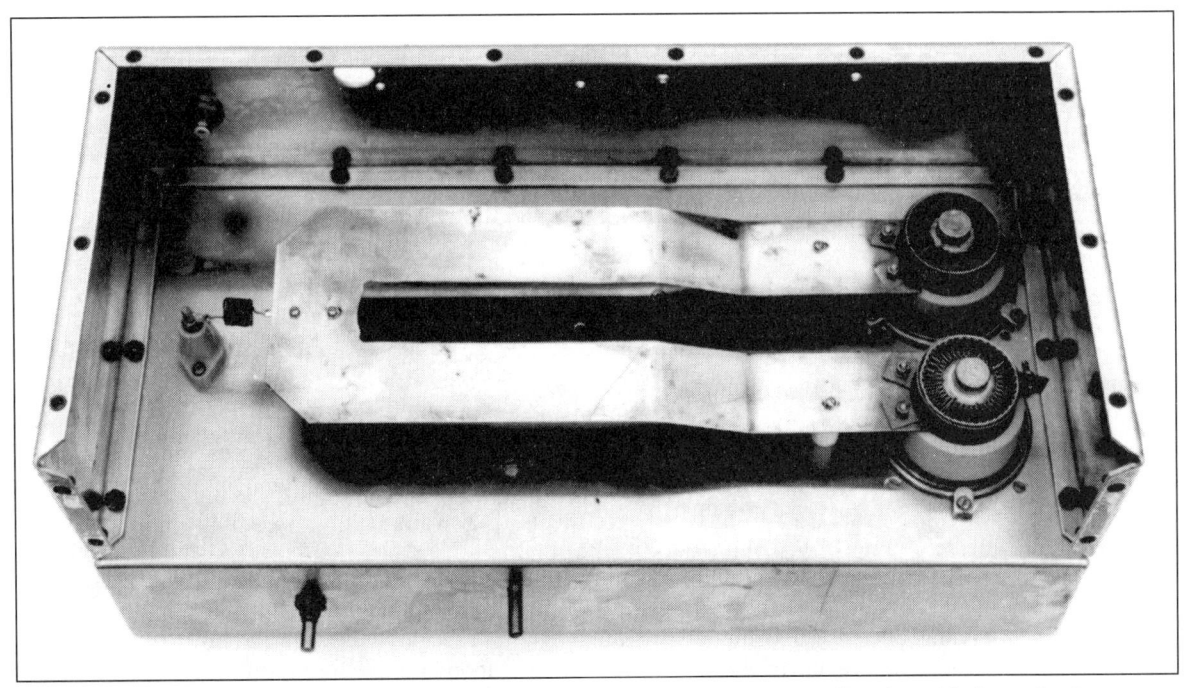

Fig. 8.23. Top view of a partly-constructed W1SL amplifier. Still to be added are the front panel, anode-tuning capacitor, output coupling and top cover

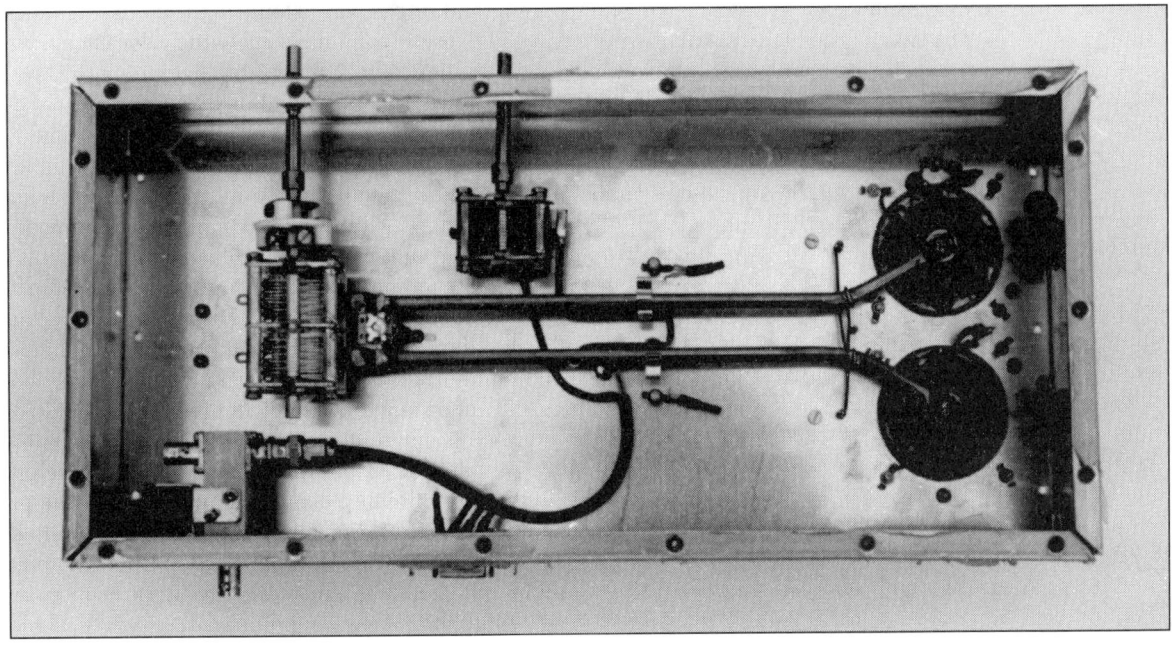

Fig. 8.24. Underside view of a partly-constructed W1SL amplifier, minus its DC wiring

8 - 33

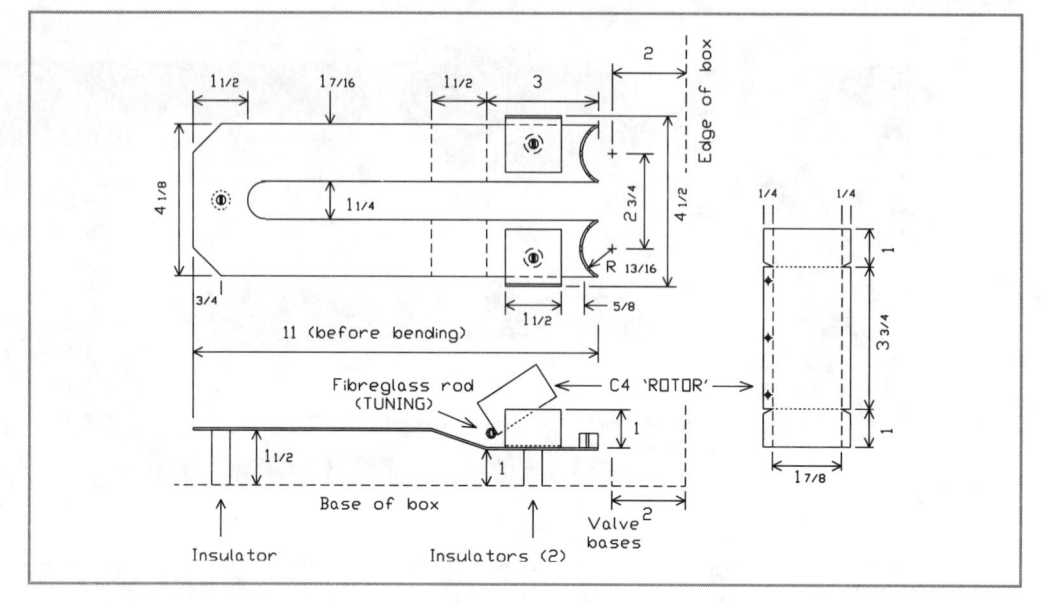

Fig. 8.25. Leading mechanical dimensions of the anode line and tuning arrangement (dimensions in inches)

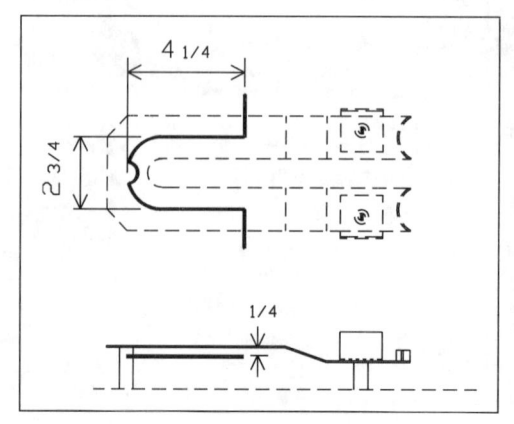

Fig. 8.26. Output coupling loop for the W1SL amplifier (dimensions in inches)

building, of course, and another argument to defeat those who moan about the impossibility of home-brewing anything, let alone big amplifiers. If you or yours draw the line at metal filings in the carpet, you should still be able to find someone who can run you up a set of W1SL metalwork in the lunch hour. Think of it as an initiative test in the true amateur spirit.

The layout is quite uncritical. Having drawn a centre-line along the top of the lower chassis, mark the centres for the two valve bases 2" from the right-hand edge, and 2¾" apart. Drill pilot holes and use these to mark the top cover. All the other major components – the anode line, the grid lines, the coupling loops and the tuning and loading capacitors – follow in sequence towards the left-hand end of the chassis. Fig. 8.25 shows the anode line and the vanes which form the stators of the anode tuning capacitor. The line is supported on stand-off insulators as shown, and is connected to the anodes of the valves by tabs and ring clamps. Elongate the mounting holes in the anode line to allow the valve clamps to be drawn up firmly before tightening the line down on to the stand-offs.

The flapper which forms the 'rotor' of the tuning capacitor is a box-shaped arrangement mounted on a quarter-inch fibreglass rod (Fig. 8.25). Vertical fins on the anode lines provide extra capacitance to the ends of the flapper. The rod is supported by a slow-motion drive on the front panel and a bearing at the rear of the anode enclosure. It's best to mark the positions of the two holes for this rod after mounting the anode line.

The output coupling loop is shown separately in Fig. 8.26. It starts half-way between the 'U'-shaped closed end of the anode line and the tuning flapper, follows the centre-line of the anode line around the 'U' (dodging past the stand-off insulator) and exits at the opposite half-way point. In other words, the loop sits directly under half of the anode line at a constant distance of ¼". There's no particular need to make the loop out of copper strip as per the original article; something like 12 or 14SWG copper wire does the trick nicely. With a loop of this type and a 35pF capacitor as specified, you shouldn't have any trouble at all in getting the loading spot-on.

The two grid lines are brought to half-wave resonance by a 25+25pF split-stator capacitor C2, whose rotor is isolated from ground. Balance and a ground return for RF are provided by a differential capacitor C3 across the grid lines. The grid tuned circuit is shown in Fig. 8.24, and in close-up in Fig. 8.27; C3 is the small component mounted where the grid lines connect to C2. This arrangement works very well on one condition: that C2 is a truly symmetrical split-stator capacitor with ceramic end-plates and very little supporting metalwork. Unfortunately, capacitors of this sort are anything but common. Most of us end up using the two separate sections of a 25pF two-gang with a metal frame, such as the Jackson U102 type shown in the photographs. These are often found surplus (actually you can still buy them new) and seem to be the ones which usually find their way into W1SLs. This is fine – but only if you make a few changes to suit. Many people who have attempted to use a two-gang instead of a true split-stator have found that the grid currents of the two valves don't quite balance; they peak at slightly different frequencies instead of keeping in step. What's worse is that the grid balancing adjustment (see below) is impossibly twitchy. The solution is to mount the capacitor on stand-off insulators, and to ground the centre of the frame with a nice wide copper strap joined also to C3 (Fig. 8.27 clearly shows the grounding strap between the grid lines). Hey presto! the grid circuit will now behave beautifully, and people will stop you in the street to congratulate you and ask for the secret of your success in life.

The grid balancing capacitor C3 is one of those awfully-hard-to-find low-capacitance differential things (not a butterfly – a differential capacitor is the kind that increases on one

Fig. 8.27. Close-up of the W1SL grid lines and input coupling loop. The two-gang capacitor C2 is mounted on stand-off insulators, and is grounded only by the vertical copper strip from the centre of its frame. The small differential balancing capacitor C3 is connected across the grid lines in front of C2, and also to the grounding strip

side and decreases on the other as you turn it). If you really can't find a suitable capacitor, you could try using two 5pF tubular ceramic trimmers, one on each grid line, and balance the grids by adjusting one clockwise and the other anticlockwise. The grid balance should be adjustable by means of an insulated trimming tool through a hole in the bottom cover plate of the grid compartment. You can make this hole quite large and then blank it with a snap-in metal grommet of the sort Halfords sells for use on cars.

The grid lines are made from $^3/_8$" copper pipe. For most of their length they run parallel and about 1" apart centre-to-centre. The ends

144MHz

which connect to the valves are bent apart, and each is flattened and drilled to accept a screw (Fig. 8.24). The opposite ends are soldered to C2 and C3. The original W1SL was presumably designed with the American 144–148MHz band in mind, whereas the UK and Europe have only 144–146MHz. Because of this, and also because of the fact that you'll probably be using the amplifier right at the bottom end of the band for most of the time, I'd recommend that you lengthen the grid lines from the original W1SL dimensions by at least ½". If you use a Jackson U102 25+25pF for the grid tuning capacitor and make the grid lines 11½" long before bending, you'll certainly be OK. 144MHz should tune at slightly more than half mesh, and 146MHz at slightly less – just what the doctor ordered.

From this latter point you will gather that the grid tuning of a W1SL is exceedingly sharp. The original write-up didn't mention a slow-motion drive and there wasn't one in the pictures, so one perforce has to assume that Messrs McMullen and Tilton both had muscular control and reflexes about a million times better than mine. Unless you're either a Tornado pilot or a concert harpsichordist, fit a 10:1 drive to the grid tuning capacitor; you should then find the grid tuning just about manageable. The anode tuning is quite sharp as well, and a slow-motion drive also helps to lock the mechanically unbalanced flapper in place. Don't forget that the bearing for the grid tuning shaft must be airtight since the grid compartment is pressurized, so choose a drive such as the fully enclosed Jackson 10:1 job which doesn't leak.

You could use a similar component for the input loading control as well, for the same reason.

The input coupling loop isn't at all critical in the W1SL. You should easily achieve a 1:1 VSWR with the arrangement shown in Figs 8.23 and 8.27. The coupling loop is about 2" long and ½" wide, and sits about ¼" below the grid lines, half-way along the straight part. The original design used copper strip, but here again 14SWG enamelled copper wire or thereabouts works perfectly well.

The DC connections to the grid lines are made at the points of minimum RF voltage, about half-way along. If you want to make a proper job of it, temporarily attach the grid feeds at roughly the right place, tune the lines to resonance, and slide an insulated screwdriver along one line to find the point of minimum disturbance. Then move the DC grid connections permanently to that location on both lines. This is not at all critical, so long as both sides are the same. Don't spend too long on the job, though, because not much air is going to go through the valves while you've got the bottom off the grid compartment – try and jury-rig something to ensure that they get at least a whiff of air over the seals and grid spigots whilst you're twiddling.

Because modern SK610 and SK620 bases have better isolation than those which were around in 1971, you will probably need a much smaller neutralizing capacitance than was shown in the original article. Try just ½" of the braid from some UR43 coax, soldered to the tops of the feed-through insulators which come out underneath the anode line (Figs 6.8 and 6.9). Chapter 6 explains how to adjust their position to neutralize the amplifier.

Incidentally, the wiring from the various electrodes of the valves to the multiway connector was done in UR43-type coax in the original. If you enjoy a challenge and don't mind spending hours and hours getting the bending radii right, you can do a supremely beautiful job by using semi-rigid coax – especially if you polish the outer screen with Dura-Glit or Brasso.

With a few demon tweaks along these lines, you should find that the W1SL is supremely easy to get going and a joy to use. If your blower is big enough, you shouldn't have to touch the anode or grid tuning for months on end – and once you've set the input VSWR and balance you can forget about them until you have to change the valves. Incidentally, I've also used the W1SL design as a basis for several amplifiers using 4CX600JAs and 4CX1500As, and they all worked a treat.

ANTENNAS FOR 144MHz DX

There are many commercial antennas for 144MHz and, if you want to buy rather than build, the table on page 7-31 provides the information you need to make a well-informed choice. In this chapter we present two well-proven 144MHz Yagi designs which are not available from the major antenna manufacturers.

The 10-element DL6WU Yagi in Fig. 8.28 is known as the 'PA0MS Yagi' in the Netherlands, where it has been popular and well-proven for several years. It is a compact design giving a gain of about 11·8dBd from a 4·5m (15ft) boom length. The pattern is very clean, as shown in Fig. 8.29. On its own this Yagi is suitable for situations where a longer boom might not be acceptable, and the clean pattern means that it can also be successfully stacked and bayed. According to the guidelines given in Chapter 7, suitable stacking distances for

Fig. 8.28. The 10-element DL6WU Yagi for 144MHz. Above it are two other DL6WU Yagis, a 23-element for 432MHz and a 49-element for 1·3GHz. All three antennas are fed with folded dipoles and half-wave coaxial baluns

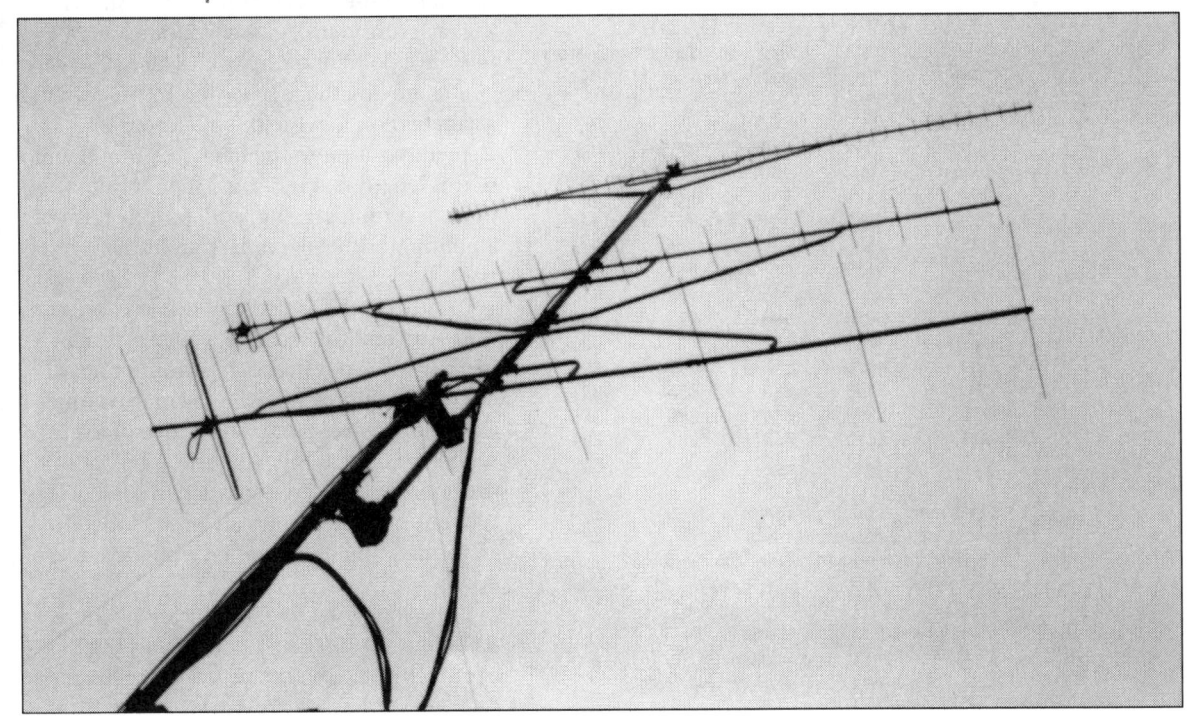

144MHz

10-ELEMENT DL6WU/ PA0MS YAGI

Element diameter 4mm, boom 25mm or 25·4mm square

ELEMENT POSITION		LENGTH		
		Free space	Full contact [1]	Insulators [2]
Reflector	0	1026	1034	1031
Driven [3]	390	980	980	980
D1	555	932	940	935
D2	930	924	930	927
D3	1380	919	927	924
D4	1905	914	922	918
D5	2490	904	912	908
D6	3120	894	902	898
D7	3780	884	892	888
D8	4470	874	882	878

[1] Elements mounted through or on top of square boom, in full electrical contact
[2] Elements mounted on insulating bushings through boom. Bushings: Heyco part No 61PR8000
[3] Driven element 5–8mm diameter, preferably folded dipole with 4:1 coaxial half-wave balun

15-ELEMENT DL6WU YAGI

Element diameter 4mm, boom 25mm or 25·4mm square

ELEMENT POSITION		LENGTH		
		Free space	Full contact [1]	Insulators [2]
Reflector 1	0 [3]	1028	1035	1031
Reflector 2	0 [4]	1028	1035	1031
Driven [5]	360	980	980	980
D1	505	938	945	941
D2	880	934	941	937
D3	1330	929	936	932
D4	1855	925	932	928
D5	2440	915	922	918
D6	3070	906	913	909
D7	3730	897	904	900
D8	4420	892	899	895
D9	5140	888	895	891
D10	5890	883	890	886
D11	6670	878	885	881
D12	7420	865	872	868

[1] Elements mounted through or on top of square boom, in full electrical contact
[2] Elements mounted on insulating bushings through boom. Bushings: Heyco part No 61PR8000
[3] 300mm above other elements
[4] 300mm below other elements
[5] Driven element 5–8mm diameter, preferably a folded dipole with 4:1 coaxial half-wave balun

horizontal polarization would be about 2·9m vertically and 3·2m horizontally. The dimensions of the antenna are given in the table. The element lengths are given in 'free space', *ie* without correction for boom effects, and also for the two different mounting methods shown in Fig. 8.30. The boom-effect corrections are made according to the information given in Chapter 7.

As with all the DL6WU designs, the 10-element Yagi can be fed with a simple folded dipole and a half-wave coaxial balun (*eg* a 685mm loop of URM76 or RG58). The resulting VSWR should be better than 1·5 without any adjustments, but if necessary it can be further improved by making small adjustments to the length of the first director, or by altering the diameter of one half of the folded dipole as shown in Fig. 8.28.

Construction is very straightforward. You should have no difficulty in cutting elements to length with far greater accuracy than is really necessary for 144MHz. When marking the positions of the elements along the boom, reference all your measurements from the position of the reflector. When your measuring tape runs out, accurately locate a new reference point and carry on as far as you can before moving the tape again; this is far more accurate than moving the ruler from one element position to the next. A 25mm boom of this length requires some support to prevent it from sagging, either in the form of aluminium struts or cords from further up the mast.

To support the folded dipole and connect the balun and the feedline, you can either make your own junction box from standard electrical components [1] or adapt a commercial dipole centre-piece. The folded dipole is supported only at the connection points, and the boom passes between the two parallel legs; the centre of the dipole can be supported by an insulator but must **not** be grounded to the boom.

The second Yagi design is distinctly more ambitious – a 15-element on a 7·4m (25ft) boom! This design goes beyond the cluster of NBS and commercial Yagis with boom lengths

ANTENNAS FOR 144MHz DX

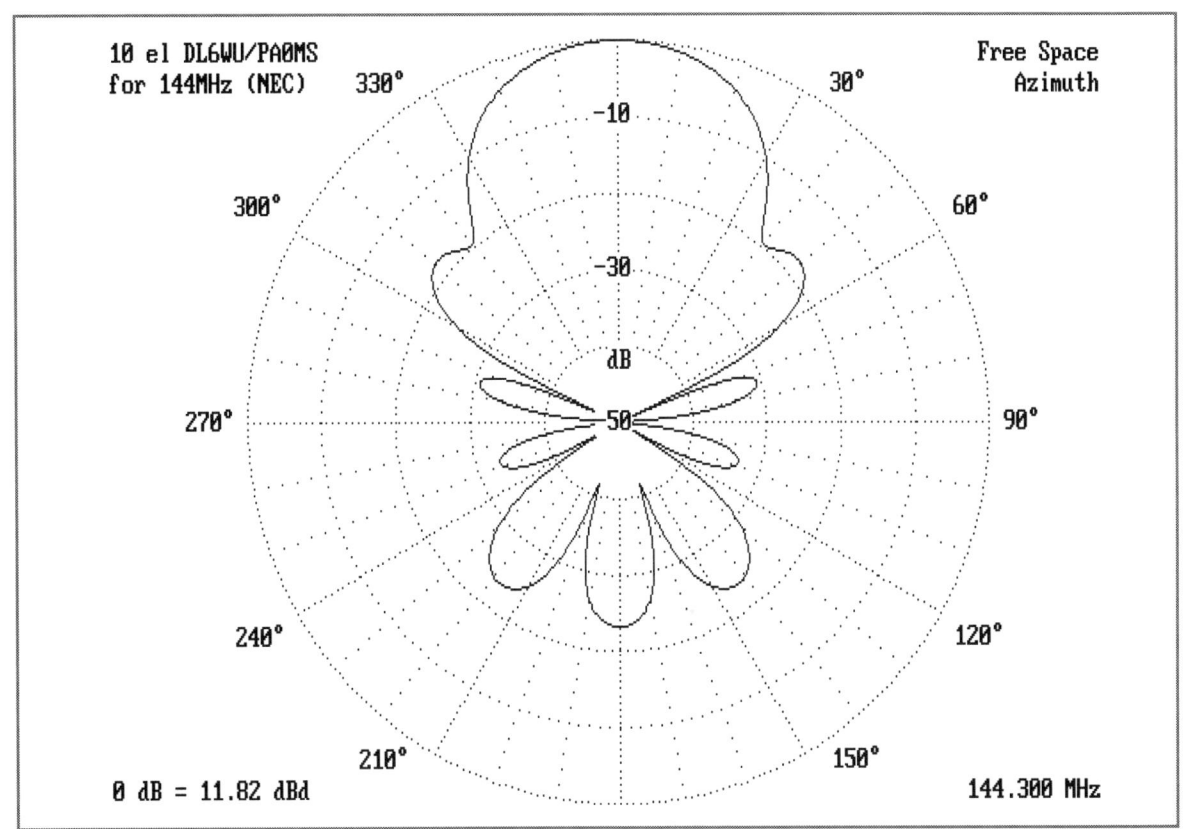

Fig. 8.29. E-plane pattern of the 10-element DL6WU/PA0MS Yagi, calculated by the NEC program. The front/back ratio should be well over 20dB, and the first sidelobes at about −18dB relative to the main lobe

Fig. 8.30. Two different mounting methods for elements on a 25mm (or 1") square boom

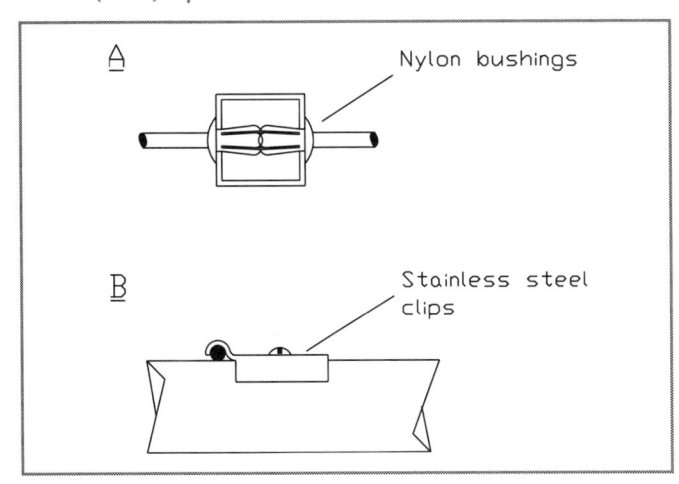

of around 6m, and is intended for really serious 144MHz DXers and contesters who have the masts, towers and rotators to handle such a monster. With a gain of about 13·6dB, even one of these Yagis makes a highly effective tropo antenna, and four will put you in the moonbounce league. Although it is actually quite an old design, the table on page 7-31 shows that this Yagi still out-performs many others of a similar boom length.

This Yagi departs from my general design philosophy that one reflector should be sufficient, by having dual reflectors above and below the plane of the rest of the elements. The overall boom length is chosen to coincide with a 'natural' high front/back ratio, and at the optimum frequency this can be in excess of 30dB. Fig. 8.31 shows the approximate E-plane pattern. According to the guidelines in Chapter 7, suitable stacking distances for horizontal polarization would be 3·7m vertically and 3·95m horizontally. The

144MHz

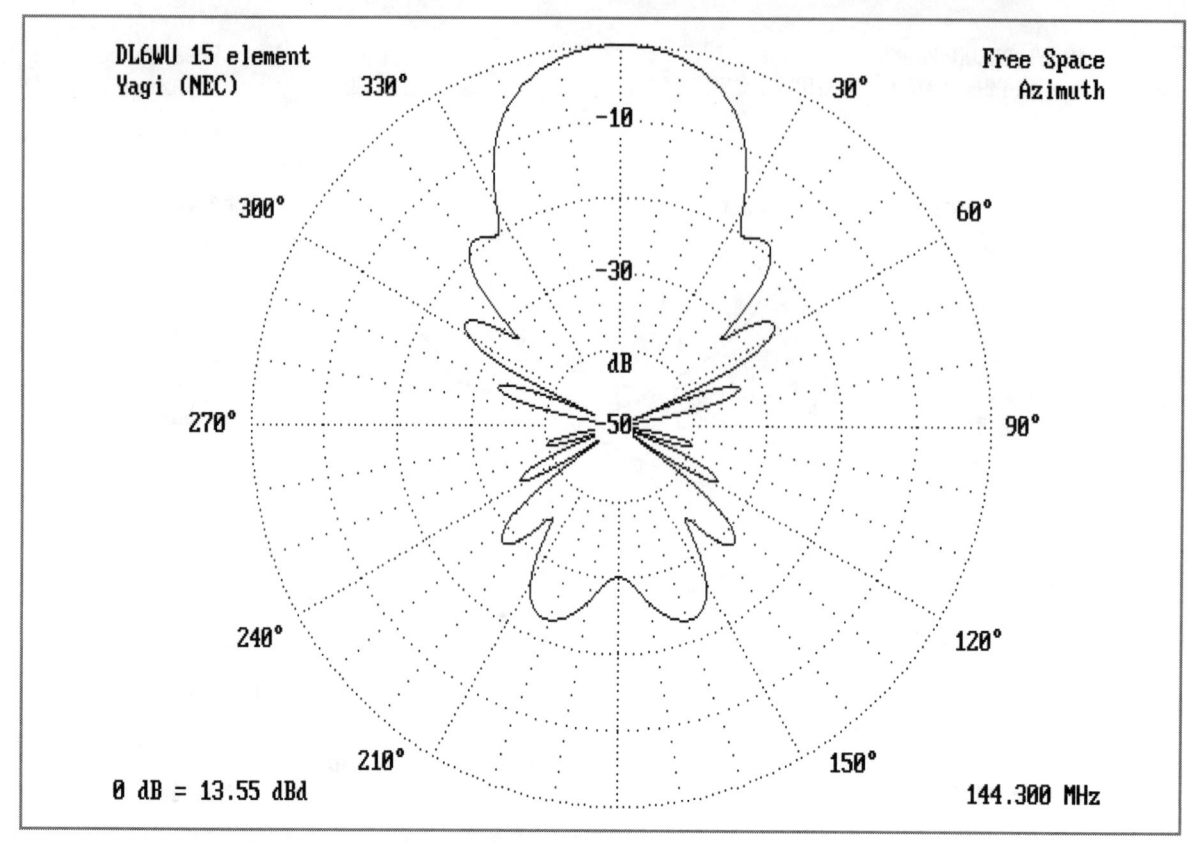

Fig. 8.31. E-plane pattern of the 15-element DL6WU Yagi, calculated by the NEC program. The front/back ratio should be about 30dB, and the first sidelobes should be about −18dB relative to the main lobe

dimensions of the 15-element Yagi are given in the table on page 8-38.

Construction follows the same principles as the 10-element Yagi, although the much greater boom length calls for additional reinforcement of the boom against sagging and the effects of side-winds. Aluminium struts can prevent sagging, but they contribute very little sideways strength. To prevent the boom from bending or snapping in a strong side-wind, it is important to make the middle section from one continuous length of tubing with no splices. For greater sideways strength at the mounting point you could telescope a short length of smaller thick-wall tubing or even hardwood inside the boom, and possibly add a short sub-boom or a 'trombone' brace as shown in Fig. 8.28.

OTHER DESIGNS

In addition to these standard DL6WU designs a range of 'enhanced DL6WU' designs for 144MHz has been published in DUBUS [2]. These have been optimized by DJ9BV using the NEC-2 computer code and range in length from 1·8λ to 4·8λ (10 metres) although the changes are relatively minor. Compared with the standard designs, the enhanced DL6WU Yagis have somewhat better sidelobe suppression but only 0·1–0·2dB more gain.

REFERENCES

[1] F. Schulze PE1DAB, Construction of folded dipole for 2m Yagis. *DUBUS* 4/90, p. 31.

[2] R. Bertelsmeier DJ9BV, Yagi Antennas for 144MHz. *DUBUS* 1/90, p. 19.

CHAPTER 9

50 & 70MHz

A HIGH-PERFORMANCE 50MHz TRANSVERTER

by Dave Powis G4HUP

The prime considerations in this design for a 28–50MHz transverter were that its performance should match the best currently available commercial HF rigs, and that it should also be suitable for home construction.

DESIGN OBJECTIVES

This transverter shares many of its design objectives with the *Suffolk* 144MHz transverter described in Chapter 8. For the combined receive converter and HF receiver, I aimed for a 0dBm third-order input intercept (IP3), with a suitable noise figure as discussed in Chapter 4. An output of 100mW (+20dBm) PEP was required from the transmit-converter board, with third-order intermodulation products at least 40dB down from the wanted signal. The transmit converter was to be followed by a 10W PEP power amplifier using a hybrid module.

This performance has been achieved using generally obtainable components, and at a cost which was reasonable in view of the level of performance. Marginally higher performance can be achieved in the receive converter, in terms of IP3 and dynamic range, but the extra cost involved was not felt to be justified. Moreover, the performance achieved in the present design is already limited by that of all but the most recent HF transceivers. On the transmit side, better IMD performance could be achieved but only by using Class-A amplifiers throughout, or a discrete-transistor PA capable of probably 50W saturated output.

Unfortunately it is practically impossible to cater for every combination of antenna, feeder and HF transceiver. Some assumptions have to be made: for example that separate 28MHz transmit and receive connections are available to the transverter. Chapter 8 considers the options for interfacing to other types of HF radio.

For the transmit converter, the primary assumption concerned the 28MHz drive level. From a survey of current equipment, a level of –20dBm (10μW) was taken as the likely minimum input signal, and the transverter will accept up to about +23dBm (200–250mW) from the transverter outputs of valved transceivers. Some older rigs have even higher transverter output levels which may need attenuation, although it seems that the modern trend is towards lower-level outputs. The output of some HF rigs is not particularly 'clean' in terms of intermod products and spurious signals; considerable improvement can often be obtained by incorporating a filter at 28MHz (Chapter 12), and also by backing off the drive level. For the latter purpose, ALC provisions are included in this design. For the control and switching, it has been assumed that all likely IF drivers will provide a ground on transmit from the press-to-talk (PTT) accessory output.

Formulating design assumptions for the antenna end of the system is rather more difficult. The parameters of importance here are the antenna noise temperature and the feeder loss. I measured the antenna noise temperature over a period of time at my home, which is in a rural location free from man-made noise and (regrettably) well-screened in most directions. This was done by comparing the antenna noise against thermal

50MHz TRANSVERTER

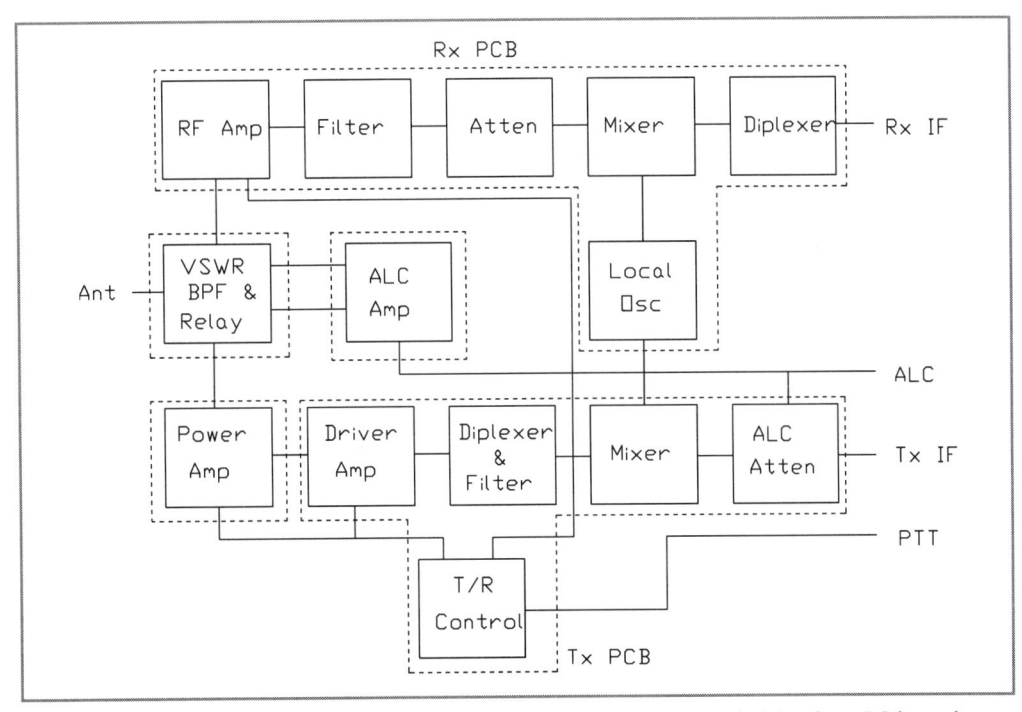

Fig. 9.1. Block diagram of the G4HUP 50MHz transverter, showing division into PC boards

noise from a resistive load, with due attention to accurate impedance matching. The lowest antenna noise temperature measured was 650K, and 1000K was more typical. This is lower than would be expected from Chapter 4, but I would not expect to find many quieter locations than mine at 50MHz; suburban or more exposed rural environments would be much noisier. Although I have used 600K as the target for the system noise temperature, you are free to optimize the receiver sensitivity for noisier environments, and this will bring corresponding improvements in the strong-signal handling. A noise temperature of 600K equates to a noise figure of about 5dB at the input of the receive converter (the loss and noise contribution of the feeder can be ignored in this context).

Turning to the HF receiver, current rigs have typical noise figures of about 10dB at 28MHz, with an input IP3 of about +20dBm. Working into such a receiver, the 50–28MHz receive converter therefore has to achieve an overall NF of 5dB and a system IP3 of 0dB as already specified. Thus the receiver block diagram begins to take shape.

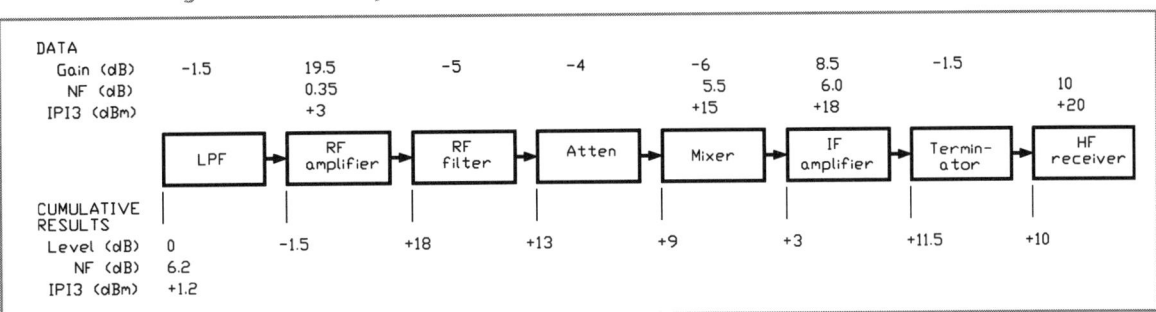

Fig. 9.2. TCALC analysis of the 50MHz receive converter and HF receiver

DATA									
	LPF	RF amplifier	RF filter	Atten	Mixer	IF amplifier	Terminator		HF receiver
Gain (dB)	-1.5	19.5		-5	-4	-6	8.5	-1.5	10
NF (dB)		0.35				5.5	6.0		10
IPI3 (dBm)		+3				+15	+18		+20
CUMULATIVE RESULTS									
Level (dB)	0	-1.5	+18	+13	+9	+3	+11.5		+10
NF (dB)	6.2								
IPI3 (dBm)	+1.2								

9 - 3

50 & 70MHz

An overall block diagram of the 50MHz transverter system is given in Fig. 9.1. As well as showing the major blocks, it also indicates the way I have partitioned the design into a number of modules, each on its own PC board. You can build whichever units you need: a receive converter, a complete low-level transverter, or even a stand-alone 10W PA.

RECEIVE CONVERTER AND LOCAL OSCILLATOR

The design of the receive section of the transverter was carried out using the concepts explained in Chapters 4 and 5, and the TCALC analysis is shown in Fig. 9.2. The 'paper' design was backed up by many measurements on individual stages, to make sure that the expected performance can be achieved in practice.

Working through the signal path of the receive converter (Fig. 9.3), the input circuit for the BF981 RF stage (TR1) uses a toroidal transformer. A lot of time was spent investigating low-noise matching into the BF981, using different arrangements of tuned circuit aimed at minimising losses, but the noise figures measured were far from satisfactory. In desperation, and against my intuition, I tried a toroidal transformer – and the first sample gave a noise figure better than 0·4dB at 50MHz! Having been educated to regard magnetic cores as inherently lossy, and therefore to be avoided wherever a low noise figure is important, this was something of a shock. However, thorough investigation failed to turn up any problems with the measurement technique, and certainly the performance of the receiver seems to agree with the calculations using this value of noise figure.

The output load of the BF981 is another toroidal transformer L3 (by now I was thoroughly converted) which is connected to the capacitive input matching of the filter. Two identical two-pole sections are used. Within each section, a common inductance forms the coupling element; this configuration is not often seen in amateur filter designs, but at these frequencies it is easier to control than the more usual top-coupling capacitance. It is sometimes possible to print the bottom-coupling inductors on the PCB, but the necessary inductances would have taken too much space so I settled for single-turn loops for L6 and L9. The insertion loss of the filter varies between 5 and 6dB over the frequency range 49·5 to 52·5MHz.

A 4dB resistive attenuator is used to get rid of excess gain before the mixer. It also serves to define the impedance into the mixer correctly, since the return loss of the filter is only around 15dB over the bandwidth quoted above. If you wish to adjust the overall gain of the receive converter, to optimize the gain to your particular RF noise environment, do it by

Fig. 9.3. 50MHz receive converter circuit

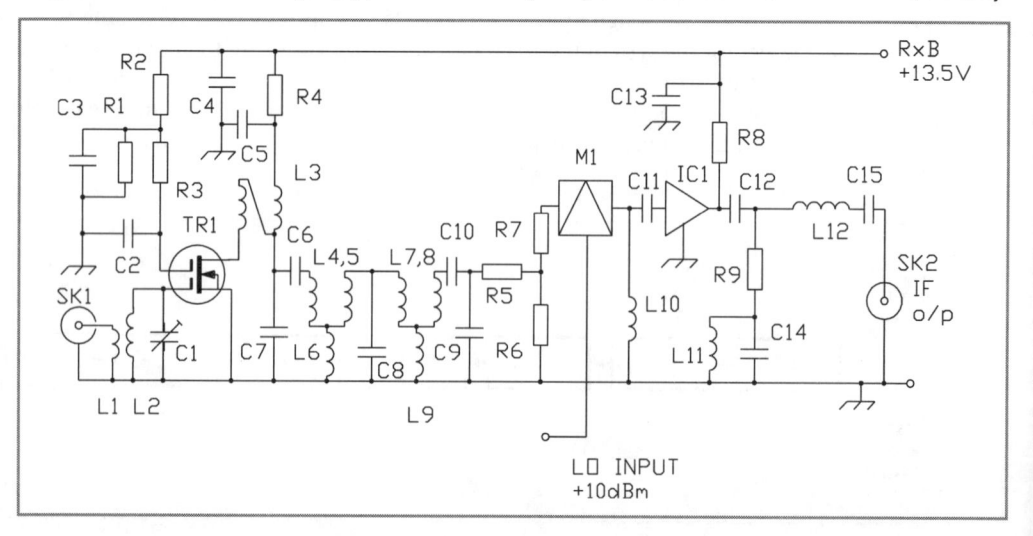

50MHz TRANSVERTER

Fig. 9.4. 22MHz local oscillator for the 50MHz transverter (on same board as receive converter)

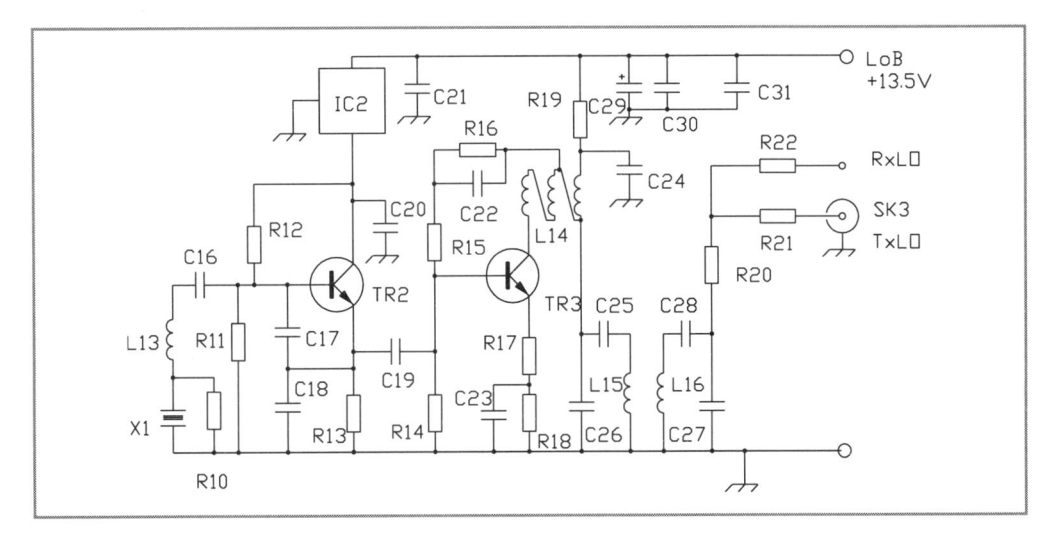

altering the value of this attenuator. This will ensure that the strong-signal handling attributes of the receive converter are not compromised. The mixer is an SRA-1 diode ring (or the less expensive SBL-1), driven at an LO level of +10dBm for greater dynamic range. This is followed by an MSA0404 MMIC as an active terminator, and then by a diplexer to provide a broadband load to the MMIC – see the *Suffolk* in Chapter 8 for further details.

The local oscillator is on the same board as the receive converter, and its circuit diagram is shown in Fig. 9.4. The oscillator is a Colpitts type using a bipolar transistor, and a voltage regulator (IC2) is used to maintain immunity against supply-voltage variations. The amplifier following the oscillator is a broadband stage with 16dB of gain. A simple two-pole filter reduces all unwanted signals to better than –50dBc, and a resistive power divider is used to provide a separate +10dBm output for the transmit converter.

TRANSMIT CONVERTER AND CONTROL SWITCHING

The circuit of the transmit converter is shown in Fig. 9.5. RV1 gives some preset control over the input signal level, and the PIN diode attenuator (D1) forms the control element of the ALC loop. The ALC range is in excess of

Fig. 9.5. 28 to 50MHz transmit mixer circuit

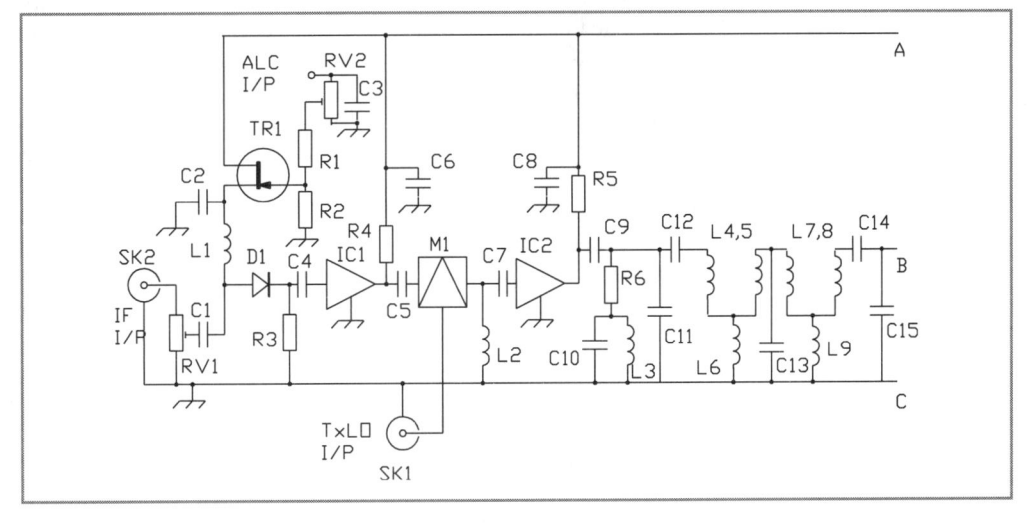

50 & 70MHz

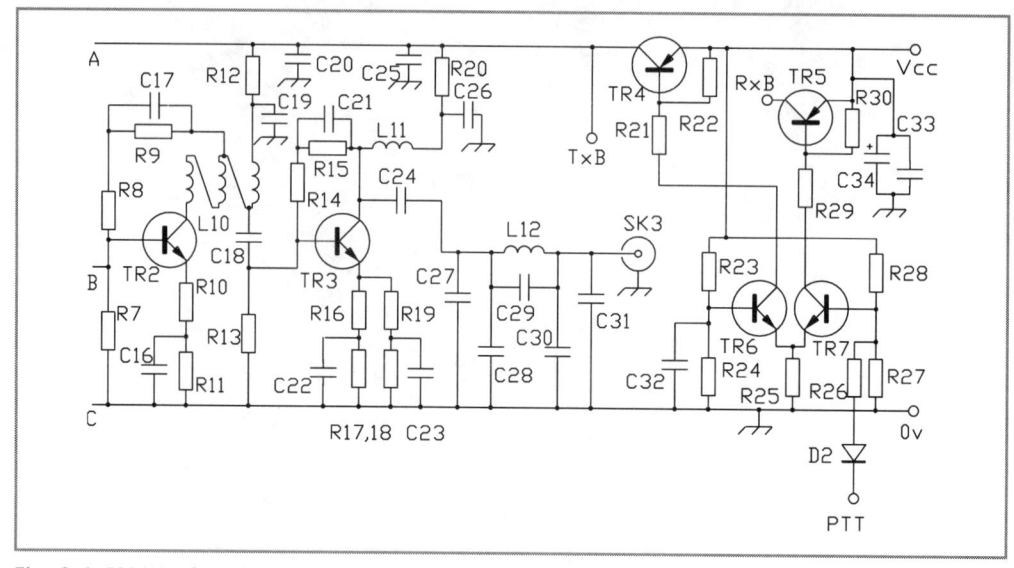

Fig. 9.6. 50MHz driver amplifier and TX/RX switching (on same board as transmit converter)

30dB, and is more than enough to protect the power stages from overdrive or damage. An MMIC (IC1) is then used to bring the signal level up to around –15dBm to drive the mixer, which is another SRA-1 or SBL-1. Although it is possible to drive the mixer harder, the IMD performance deteriorates. The following terminator, diplexer and filter are similar to those in the receive converter, though the diplexer has been simplified because the filter itself provides a reasonable termination over the wanted band of frequencies.

Two broadband Class-A transistor stages are used next (Fig. 9.6). The first has a BFR96 (TR2), running at a collector current of 40mA, and the second a BFG34 (TR3) at an I_c of 100mA. These relatively high currents are necessary to maintain the IMD performance to specification. The use of feedback in these stages enables the gain to be controlled, and in fact would allow the design to function at almost any frequency from around 10MHz up to about 180MHz, simply by changing the driving frequencies and the filters. The addition of the trapped low-pass filter ensures a very clean output, with no unwanted signals greater than –55dBc for a –15dBm input signal at 28MHz. Under these conditions the output is 200mW PEP with third-order intermodulation products better than –40dBc. The 44MHz second harmonic of the LO is better than –55dBc, and the 100MHz second harmonic of the wanted signal is better than –65dBc. Indeed, the transmit converter on its own would provide an excellent drive source for a second transverter to 1·3GHz or above.

Transmit-receive switching is mainly performed electronically, using PNP power transistors (TR4 and TR5) as the switching elements. These are driven by two NPN devices (TR6 and TR7) arranged as a long-tailed pair, to ensure that transmitter and the receiver cannot both be on at the same time.

PA MODULE

The availability of the Mitsubishi M57735 linear RF power module in the UK makes possible the simple construction of a reliable and robust PA. I have inadvertently operated mine into an open circuit with no damage although I don't recommend this as a regular operating practice! The major drawback is the high price in the UK, but a few mistakes with power devices in a discrete-transistor PA would soon add up to the same overall cost. The other drawback of these modules is that their IMD performance is not outstandingly good, as explained in Chapter 8.

50MHz TRANSVERTER

Fig. 9.7. 50MHz modular power amplifier circuit

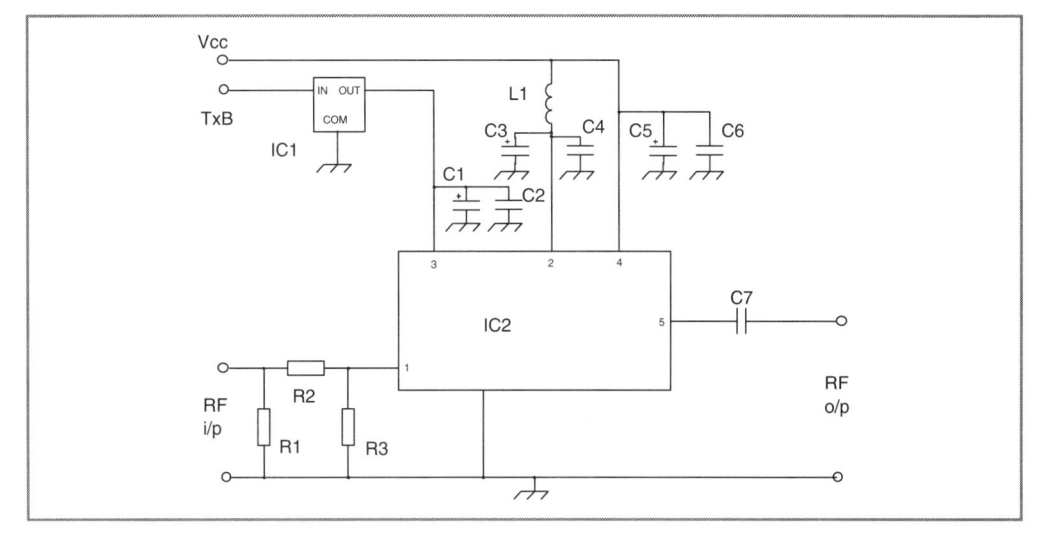

Fig. 9.7 shows the circuit diagram for the PA, including the bias control circuit. Switching between transmit and receive is by applying the bias; the supply to the collectors of the transistors is permanently connected. The TXB (switched 12–13·5V transmit) line used to power the transmit converter also switches the supply to the 78S09 bias regulator IC1.

Looking at the RF path, provision is made for a resistive attenuator at the input to the module, giving a further adjustment point to ensure optimum performance. The output signal has a DC blocking capacitor, and then will be cabled off to the VSWR and lowpass filter board. The DC supplies to the two transistors inside the PA module are brought out to separate pins (2 and 4) and each must be adequately decoupled at audio and RF. A choke is used to make the DC connection between the two. The DC bias input is similarly decoupled.

VSWR AND LOWPASS FILTER BOARD

This unit (Fig. 9.8) contains the VSWR detector, the main antenna changeover relay and the lowpass filter. The VSWR bridge provides 'forward' and 'reflected' DC outputs simultaneously, and has been tested up to 100W of RF in this configuration. The values of resistors R3 and R4 are specified to give a suitable signal to the ALC amplifier at the 10W output of this transverter. A CX-120P PCB-mounted coaxial relay is used as for antenna changeover; it is driven by the PTT line.

The lowpass filter is an implementation of a

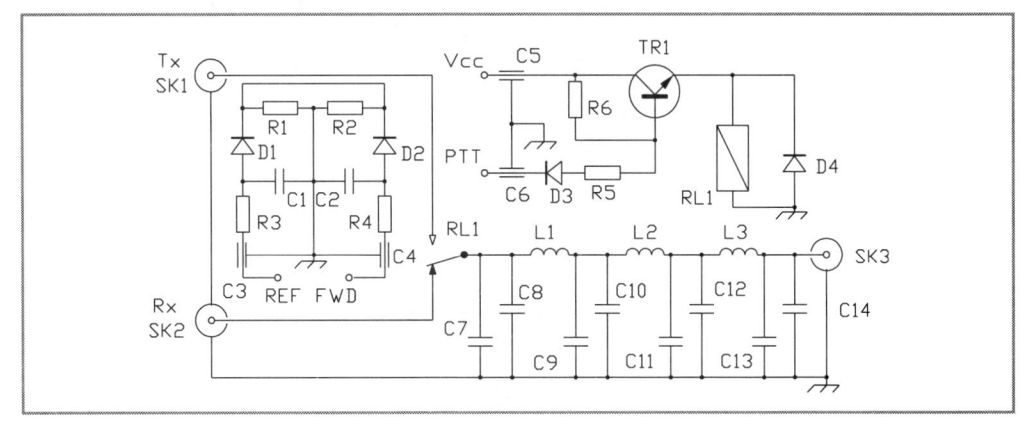

Fig. 9.8. VSWR and lowpass filter circuit for the 50MHz transverter

50 & 70MHz

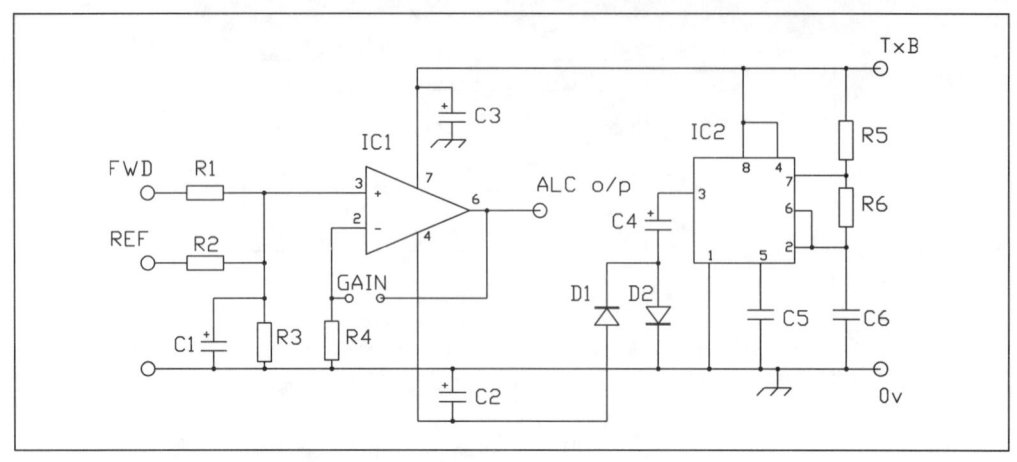

Fig. 9.9. ALC amplifier circuit for the 50MHz transverter

design by G4CXT. Like the VSWR bridge, the LPF has been tested up to 100W RF and can be used in other designs. The specified ceramic-plate capacitors are satisfactory for use with the 10W PA module, but silver-mica capacitors must be used at the 100W level. The insertion loss of the LPF on receive is about 0·2dB.

The entire unit is mounted inside a tin-plate box to maintain the integrity of the filtering.

ALC AMPLIFIER BOARD

The function of this board is to provide the interfacing between the output of the VSWR bridge, the ALC input on the transmit converter board, and the HF driver. Fig. 9.9 shows the circuit diagram. The forward and reverse VSWR detector outputs are the inputs to the op-amp IC1, and its feedback resistor is adjustable from the front panel to provide a power-output control. The output signal from IC1 goes to the PIN diode attenuator in the transmit converter, and also to a rear-panel connector to provide ALC control to the HF rig. IC2 and its associated components form a standard circuit to generate a negative supply voltage for IC1.

Fig. 9.10. Front view of the G4HUP 50MHz transverter

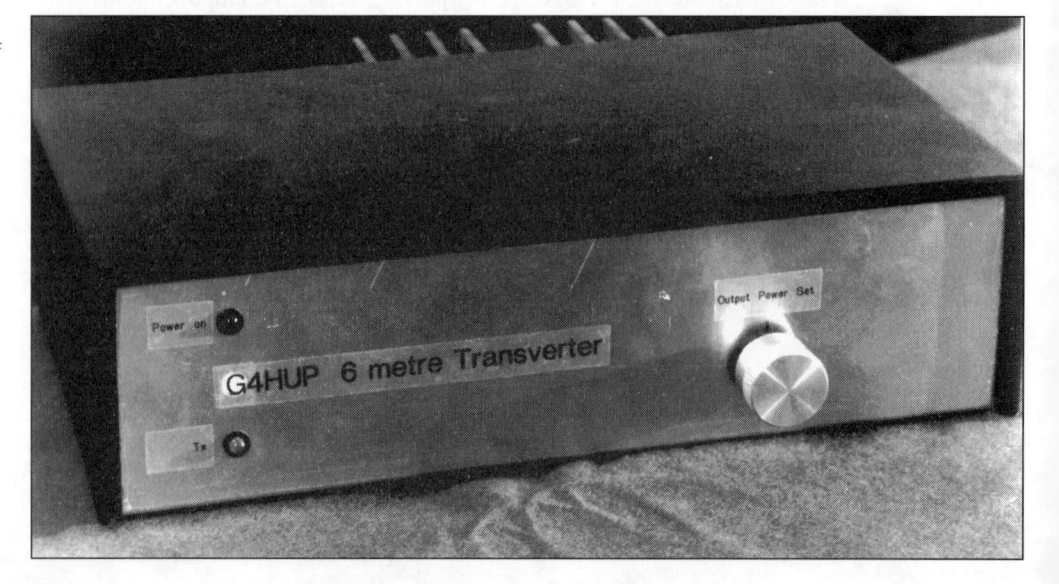

50MHz TRANSVERTER

CONSTRUCTION AND ALIGNMENT

I will take you through the construction and alignment of each board in turn, but first a few general remarks.

No specific outer case is recommended for this design – it's up to you to decide whether you build it into a die-cast box or a more stylish case. A large number of suitable cases are available commercially and other chapters in this part of the book offer several suggestions. The photographs (Figs 9.10 to 9.12) indicate the type of layout I used. The two main PCBs are mounted horizontally on 6–10mm metal spacers at the bottom of the case, with the RX-LO board towards the front, and the PA is mounted on the rear wall. I put a partition across from the back of the case, at the '50MHz' end, to support the VSWR-LPF

Fig. 9.11. Inside view of a prototype G4HUP 50MHz transverter. The modular PA is slightly different from that described in the text.

Fig. 9.12. Rear view of a prototype G4HUP 50MHz transverter. The recommended heatsink is larger than the one shown here.

board in its tin-plate screening box on one side, and the ALC amplifier board on the other. (Note that the transverter in the photographs was a prototype with slightly different circuits and board layouts from those given here.)

Connectors are another area where the choice is yours. As shown in the rear view (Fig. 9.12), I used an N-type for the antenna connector, BNCs for the IF connections, a multi-way D-type for the PTT and ALC, and 4mm sockets for the power. Within the transverter I used miniature SMB connectors for coax. All of these fit in with my own shack

WINDING TOROIDAL TRANSFORMERS

There is no problem in winding toroidal transformers – once you know how!

The gauge of wire depends largely on the size of the core. For low-power work, around 28 to 32SWG (0·4 – 0·3mm) is adequate. 'Self-fluxing' enamel which disappears as you solder the wire is very useful, and is generally coloured pink. However, if you can manage to find some different colours of enamel or insulation, it will make sorting out the connections much easier. Wire intended for rapid-wiring boards is available in a number of colours.

The two or three wires for each winding must be twisted together before winding them onto the core. Cut generous lengths of wire (250-300mm per winding), bare the ends and solder the wires in parallel. Make sure that the lengths are equal and none of the wires is slack. Next make a '?'-shaped hook from heavy-gauge wire and grip it in a hand drill. Hold one end of your parallel wires firmly in a bench vice, place the hook into the other end of the wires, and gently pull the wires taut. Slowly turn the drill handle to twist the wires together, making sure that you keep a constant tension on the wires to get an even twist. Aim for about 1 to 1·5 turns per cm (about three turns per inch). The ready-twisted wires can now be wound onto the core as if they were a single wire.

The number of turns on a toroid is the number of times that the wire passes *through the hole*. You cannot have half-turns on a toroid; even a straight wire through the core (*eg* the coaxial lead in Fig. 12.11) counts as one full turn. Wind the appropriate number of turns and distribute them evenly around the core. At these frequencies self-capacitance is important, so the spacing of turns can make a large difference to the effective inductance. Spacing the turns all the way around also means that the interconnections between the opposite ends can be very short. Secure the turns to the toroid with a few spots of super-glue, cut the twisted wires to leave an inch or two free, and untwist them back to the core.

You now need to connect the windings correctly. Circuit diagrams sometimes show these transformers with all the windings in a line, though they may also be shown in a 'folded' arrangement. The corresponding ends of each winding may also be indicated by a dot to show the correct phasing. The diagram shows all these possibilities, together with the interconnections on the toroids themselves.

Choose one end of your winding and call it the 'start'. Identify the 'start' and 'finish' ends of each wire; use an ohmmeter if the wires are all the same colour. Selecting one wire as wire number 1, its finish (F1) must connect to the start of wire 2 (S2) to make a tapping point. Twist these two wires together, but don't solder them yet. If the transformer is bifilar, that's all there is to it; the finish of wire 2 (F2) is a free end. For a trifilar transformer, F2 must connect to the start of wire 3 as shown in the diagram, to make another tapping point, and the free end is F3.

Double-check your windings using an ohmmeter, and when you're absolutely sure, solder the twisted joints right back to the core. Use a magnifying glass to make sure that both wires are fully tinned – faults caused by dry joints in toroidal transformers can take hours to find!

Until the leads are cut to their final lengths, it is important to label the various connections somehow – with small pieces of masking or insulation tape perhaps. Take particular care not to confuse the windings and taps when you finally solder the transformer into place with short leads.

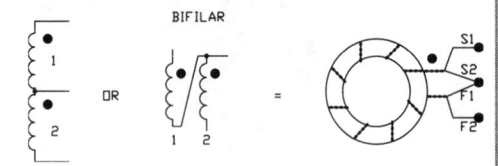

Circuit diagrams and winding details for bifilar and trifilar toroidal transformers.

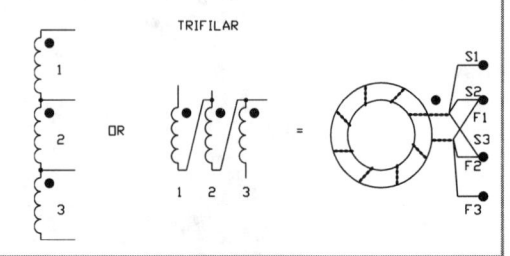

50MHz TRANSVERTER

standards, and I expect you'll want to do the same for yourself.

There are no real problems in construction of the boards. The only point of difficulty may be in winding the toroidal transformers – the panel opposite gives a few hints.

Alignment is a fairly painless process, thanks to the use of standard coils wherever possible. It is obviously easier if you have access to some test equipment, though you can get by with just a multimeter and some means of measuring RF power at levels down to about

Fig. 9.13 Interconnection diagram for the G4HUP 50MHz transverter

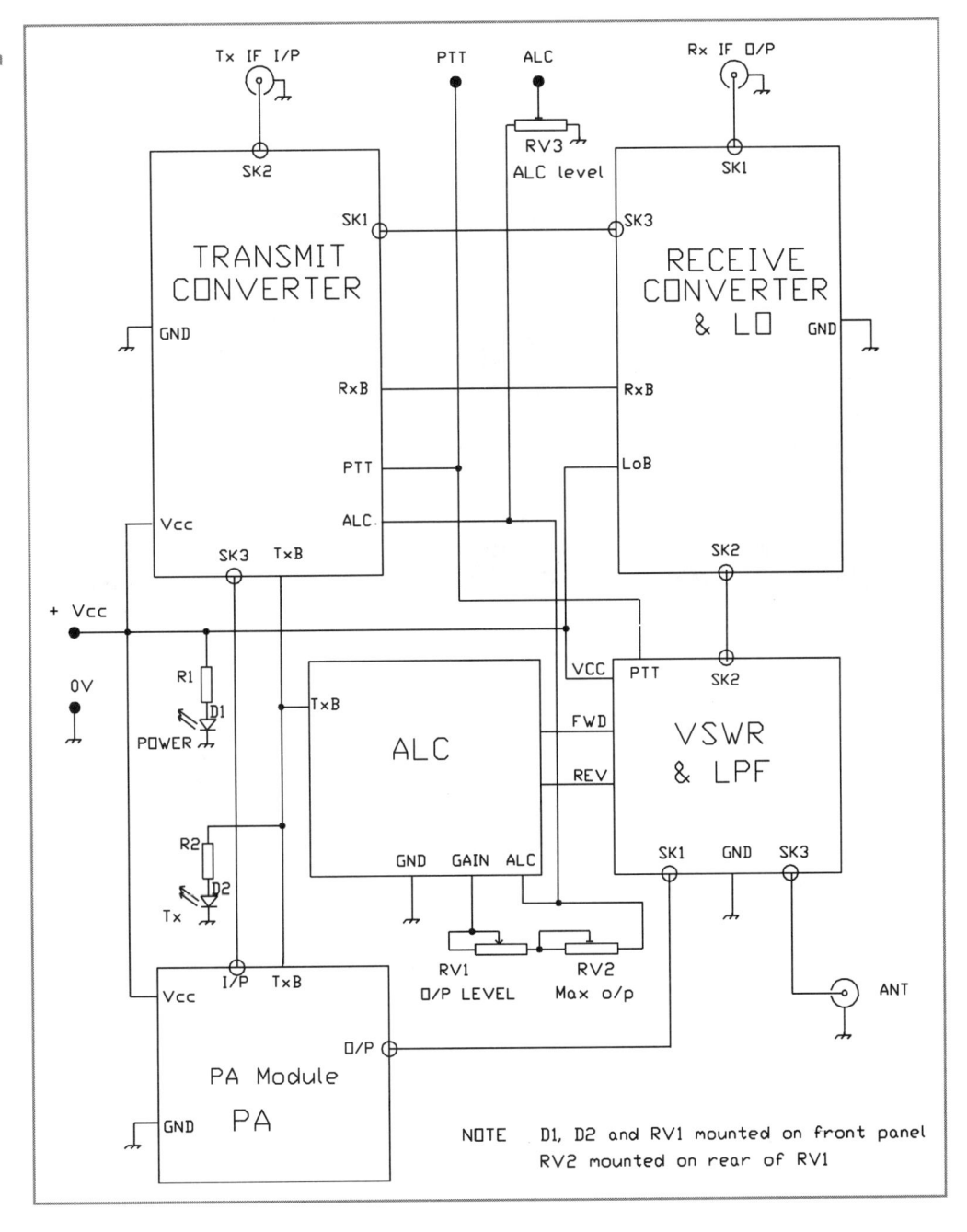

50 & 70MHz

10mW, such as the simple milliwattmeter described in Chapter 12.

Basic DC checks must be carried out first. Connect each board in turn to a 12·5-13·5V regulated supply (following the connection information given in Fig. 9.13) and check for correct DC conditions before proceeding on to the RF alignment.

RECEIVE CONVERTER AND LO

The circuits for this board were shown in Figs 9.3 and 9.4. Fig. 9.14 shows the PCB mask for the underside of the board. The top of the board is an un-etched copper groundplane, and no top-side mask is required; simply remove the copper from around the non-grounded holes using a 3-4mm drill held in your fingers. There is also a partial groundplane on the underside of this board, so some ground connections need to be soldered on both sides. Fig. 9.15 shows the component layout. All components are generally available,

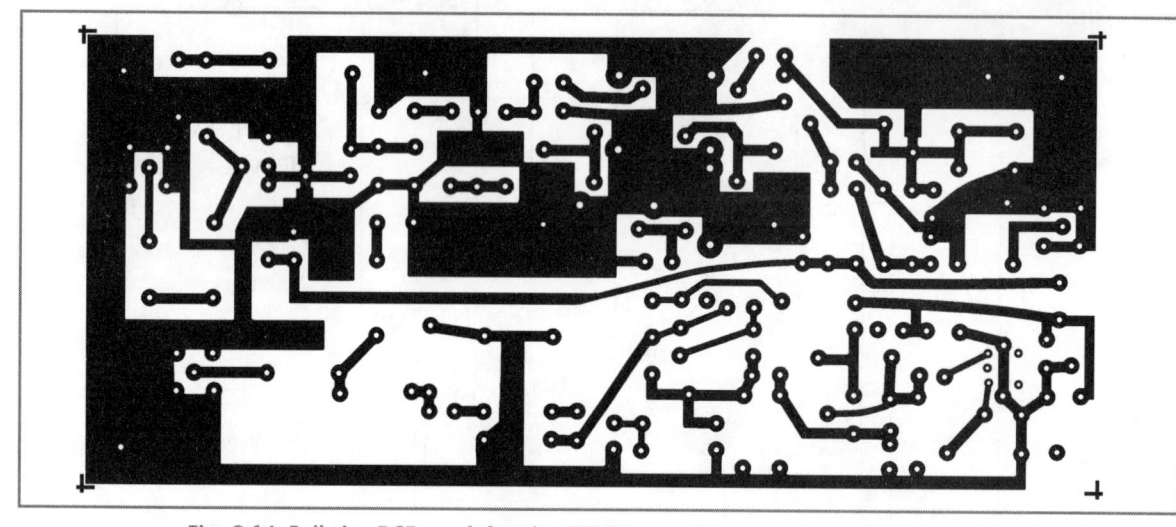

Fig. 9.14. Full-size PCB mask for the 50MHz receive converter and LO board (underside view; top side is un-etched copper groundplane)

Fig. 9.15. Component layout for the 50MHz receive converter and LO board

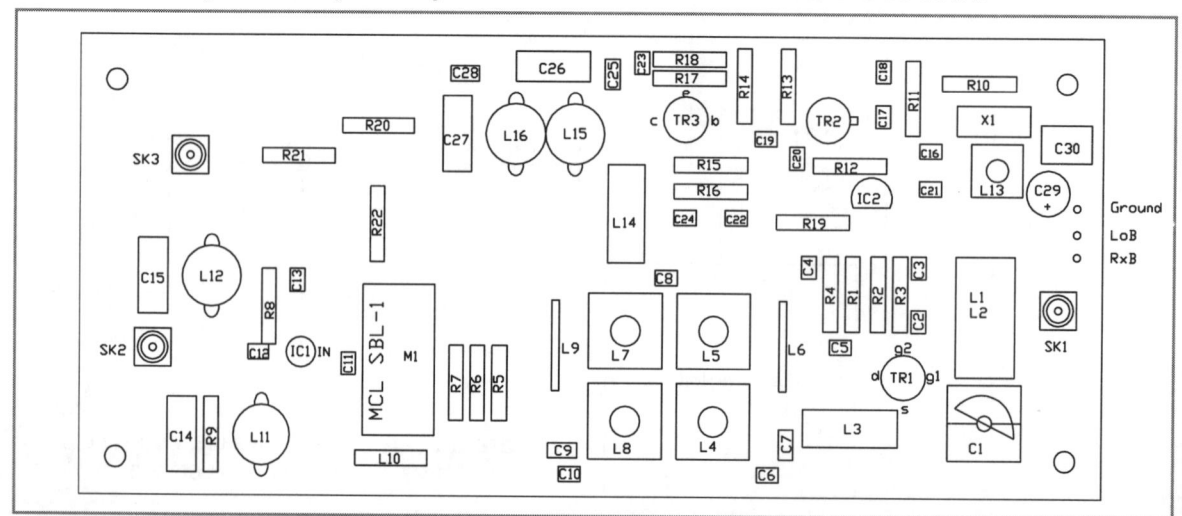

50MHz TRANSVERTER

and suggested sources are given in the components list.

Build the LO section first, including the amplifier and filter. You will have to remove the side lugs from the two S18 coils for the filter, so that they can be mounted close together on the PCB. The side of the coil with the lower 'shoulder' is the ungrounded end. If you have a receiver which covers 22MHz you can check that the oscillator is functioning. Use a milliwattmeter to check the local oscillator output level and peak the LO filter (start off with the two cores just proud of the top of the plastic coil body). You should find that the LO output to the transmit converter is +10dBm, and if this is correct (± about 1dB max) it is reasonable to assume that the RX LO drive is also correct. If you have a frequency counter, adjust L13 to give exactly 22·000MHz; if not, you'll have to wait until you can receive a beacon transmission.

Now build the rest of the receive converter, taking care to insert TR1, the mixer and IC1 the correct way round. The two coupling coils, L6 and L9, are each 20mm of wire in a 'U'-shaped loop.

Connect your HF receiver to the IF output and a 50MHz signal generator to the input of the board, with the TX LO output terminated in 50Ω. Set all the cores of the filter to be just proud of the can tops, and the cores of the mixer termination coils also just proud of the coil formers. Set the signal generator to 50·1MHz, CW, and tune the receiver until you find a signal at about 28·1MHz. Having heard your first signal, peak it using C1. If no signal appears, even at the highest setting of the signal generator, *do not adjust the filter cores* but look instead for a fault. Once you have located the signal and peaked C1, you can then attempt to set up the filter.

Since the transmit and receive filters are identical, the same procedure can be used to align them both, although in practice they hardly need alignment at all. A number of filters have been built to this design; in all cases an entirely acceptable performance between 50MHz and 52MHz has been obtained by setting the cores of the outer pair of coils (L4, L8) level with the tops of the cans, and those of the inner pair (L5, L7) a half-turn further out. Just make sure that the components you have used are exactly as specified. In practice, the benefits obtained by more accurate adjustment of the filter are minimal. If you have access to the necessary test equipment and can't resist the urge to tweak, go ahead and enjoy optimizing all four interacting adjustments.

TRANSMIT CONVERTER AND CONTROL

Figs 9.5 and 9.6 showed the circuit for this board. The style of construction is very similar to that of the receive converter; Fig. 9.16 shows the PCB mask and Fig. 9.17 the component layout. Begin construction by mounting the components associated with the TX/RX switching and the ALC circuit, as these are the most densely packed. Then work logically through the mixer, filter and amplifiers to the output, taking care to insert the mixer, ICs and transistors the correct way round. The two loop coupling coils in the filter (L6 and L9) are identical to those in the receive filter.

As in all construction projects, ensure there are no short circuits, particularly around the TX/RX switch since some of the components there are very close to one another. Connect +12·5–13·V and ground and check the DC voltages. Check the action of the TX/RX switch in response to a DC ground applied to the PTT pin of the board.

The RF alignment is quite simple. Set the cores of the diplexer just proud of the top of the coil body, and set the filter cores as detailed for the receiver. Set the core of the output trap filter just below the top of the coil body. To confirm the operation of the transmit converter, connect a source of 28MHz drive (up to about 200mW) to the input with RV1 set to maximum resistance. Connect the LO input from the receive converter, and a suitable load to the output.

Adjust RV1 to obtain a suitable output indication at 50MHz. Using a diode probe or milliwattmeter, you should be able to register at least 100mW at the output of the transmit converter board. Prototypes have typically given 200 to 250mW output at 50MHz.

POWER AMPLIFIER

The power-amplifier board is designed so that the entire assembly of components, rear panel and heatsink is held together by the screws which hold the board in place, along with those securing IC1 and IC2. In the prototype, the heatsink was drilled and tapped for M3 bolts using the PC board as a template. The same heatsink can be used as in the other transverters in Chapters 8 and 10, but I chose not to cut down the heatsink to fit behind a low-profile case. Make sure that there are no burrs round any of the holes, and do not overtighten the bolts or tighten them unevenly. Try out the assembly *before* you solder any components to the board, to make sure that everything fits all right without stressing the leads of the module. Two ways to destroy these modules are poor assembly practices and inadequate heatsinking, so be careful.

Fig. 9.7 showed the very simple circuit diagram – all the complication is inside the module. Figure 9.18 shows the PCB mask for the power amplifier board, and Fig. 9.19 the component layout. Note that all components are mounted on the top side of the board, which is single-sided and effectively uses the

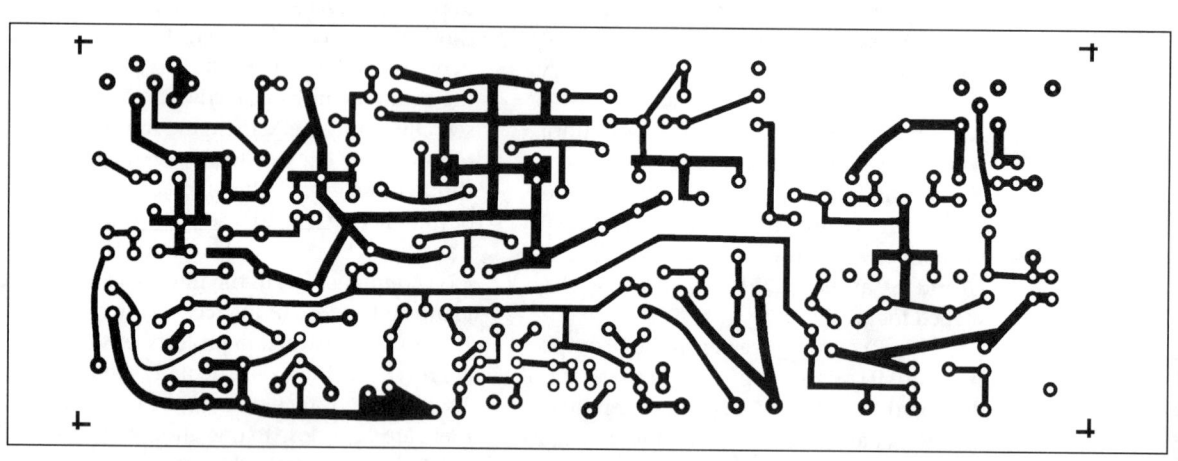

Fig. 9.16. PCB mask for the 50MHz transmit converter, driver and switching board (underside view; top side is un-etched copper groundplane)

Fig. 9.17. Component layout for the 50MHz transmit converter, driver and switching board

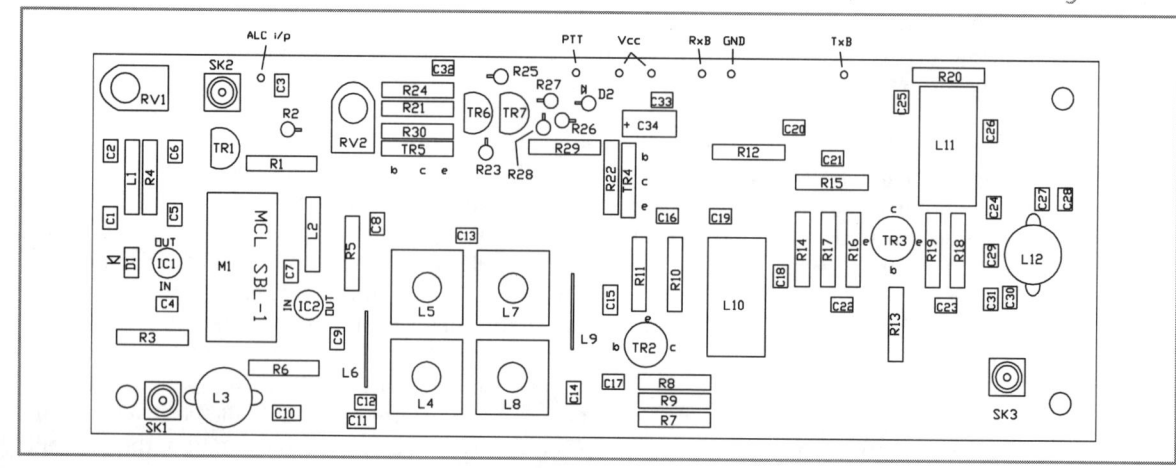

50MHz TRANSVERTER

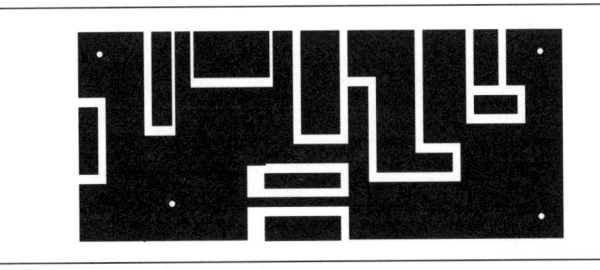

Fig. 9.18. Full-size PCB mask for the 50MHz PA board (top view; single-sided board)

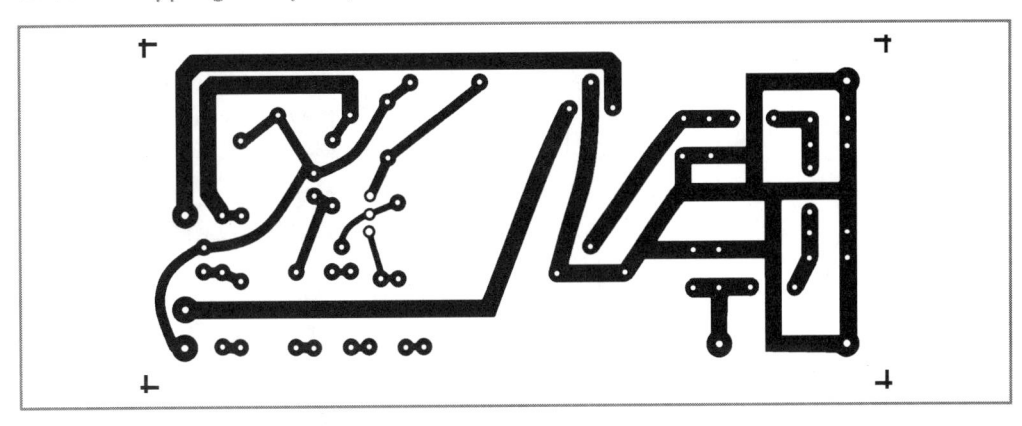

Fig. 9.19. Component layout for the 50MHz PA board

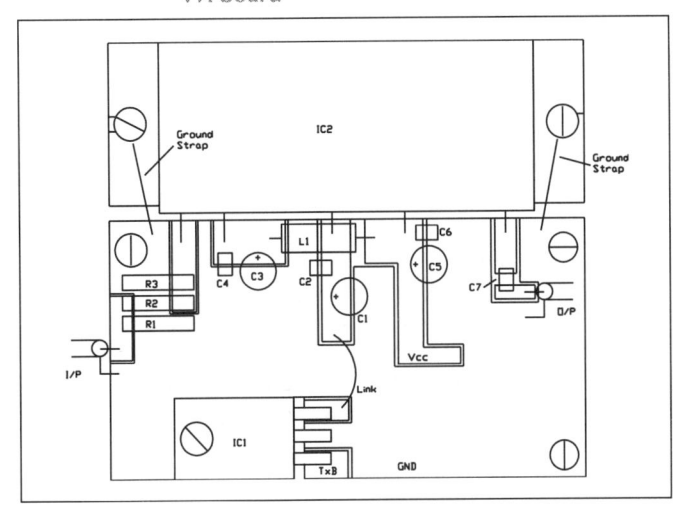

Fig. 9.20. Full-size PCB mask for the 50MHz VSWR-LPF board (underside view; topside is un-etched copper groundplane)

heatsink as an RF groundplane. There are two particular points to note in the construction of this stage. First, the RF ground connection of IC2 is via its mounting lugs. To give a good RF ground, two pieces of coax braid are used, one at each end of the module. Solder the braid to a solder tag, mounted under the bolt securing IC2 in place. The other end of the braid is then soldered to the copper groundplane of the PCB. The second point is that the link from the output of the regulator IC1 must be capable of carrying up to about 2A.

Apart from DC checks, be certain to check the 9·0V bias voltage *before* you solder in the expensive PA module. This stage requires no separate alignment.

VSWR AND LPF UNIT

The VSWR bridge is built in a screened box, along with the antenna changeover relay and the lowpass filter, to ensure that the good performance of the filter is not wasted by poor screening. The lowpass filter has deliberately been put on the antenna side of the relay, to give some extra protection against out-of-band strong signals on receive. The PC board is designed to fit the Teko 373 box, which is made of tin-plate with spring-off covers. Fig. 9.8 showed the circuit diagram, and the PCB mask and component placement are shown in Figs 9.20 and 9.21. The PC board is mounted on three metal spacers, and the coaxial sockets are mounted on the walls of the box, not on the board itself. Recommended drilling dimensions for the box are given in Fig. 9.22.

50 & 70MHz

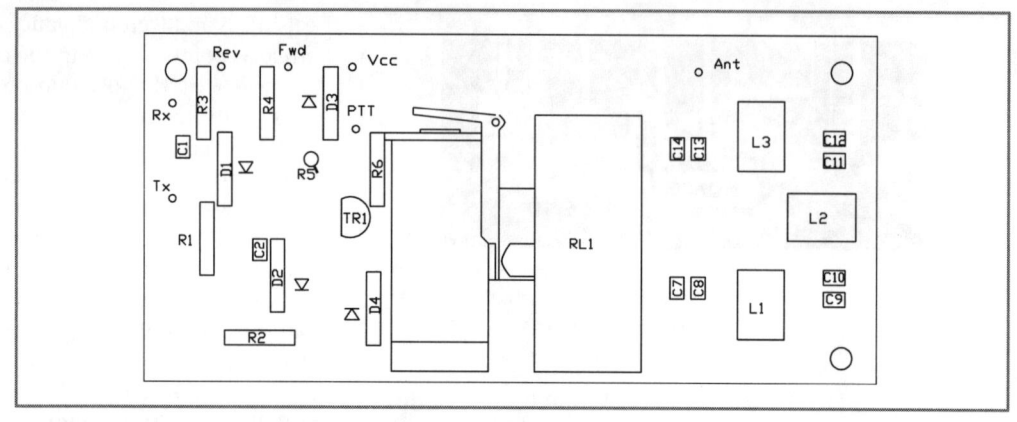

Fig. 9.21. Component layout for the 50MHz VSWR-LPF board

The diameters of the holes depend on the components available, *eg* the group of four holes on the side of the box are for the feedthrough capacitors.

Provided that the coil dimensions have been followed closely there is no alignment or testing to be done in this module, except to confirm the operation of the relay.

Fig. 9.22. Drilling diagram for the case housing the 50MHz VSWR-LPF board

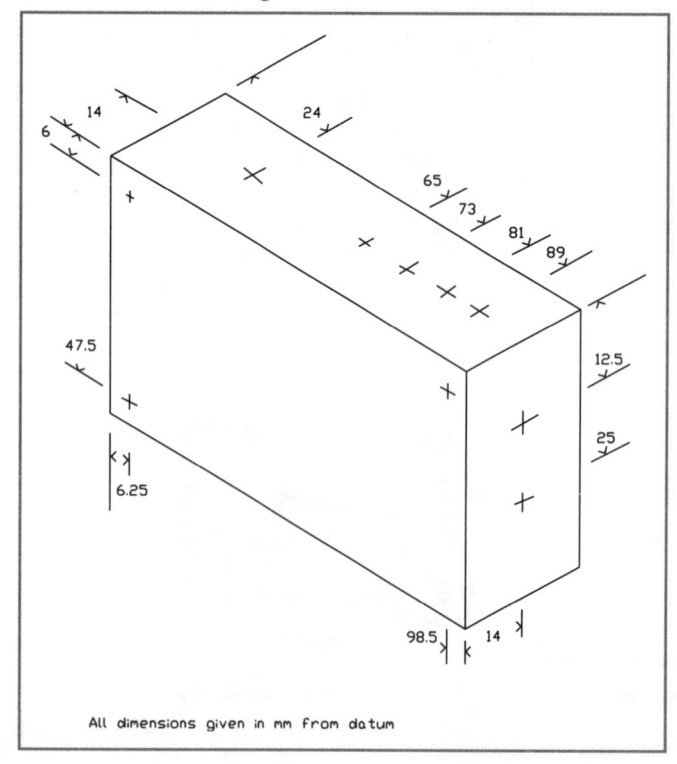

ALC AMPLIFIER

Fig. 9.9 showed the circuit diagram for the ALC amplifier board, and the PCB mask and component placement are shown in Figs 9.23 and 9.24. There is little that needs comment here, except to note the polarities of the electrolytic capacitors and the diodes of the negative-rail generator.

The panel-mounted OUTPUT LEVEL control shown on the diagrams and drawings is in fact two potentiometers – RV1 mounted on the panel and the preset RV2 mounted 'piggy-back' fashion on to the tags of RV1 (Fig. 9.13). This allows the transverter to be set up so that maximum on RV1 gives the full output from the unit without running into saturation. These two components should be wired so that minimum resistance occurs at the clockwise position, corresponding to maximum output.

To test the ALC board, simply confirm that the negative voltage generator IC2 is giving around –11·5V, measured at pin 4 of IC1.

OVERALL TRANSMITTER ALIGNMENT

Assuming that the PA is working correctly, the only aspect of the transmitter requiring further alignment is the setting up of the ALC levels. To do this, it is necessary to disable the ALC loop by setting RV2 of the transmit

50MHz TRANSVERTER

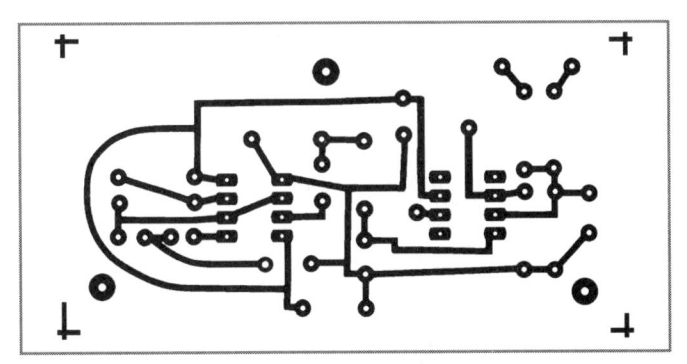

Fig. 9.23. PCB mask for the 50MHz ALC amplifier board (underside view; single-sided board)

Fig. 9.24. Component layout for the 50MHz ALC amplifier board

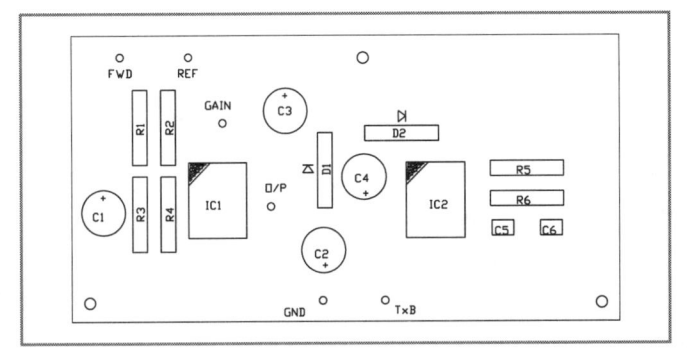

converter board to minimum. Also set RV1, 2 and 3 (Fig. 9.13) to maximum resistance. Connect a dummy load/wattmeter to the RF output, and a 200mW 28MHz drive source to the TX IF input. Ground the PTT line. Adjust RV1 (on the TX converter) to give maximum RF output, and measure the output voltage of the ALC amplifier IC1; typically this will be about –8V. Adjust RV2 (panel-mounted – Fig. 9.13) to reduce this by about 0·2V, to approximately –7·8V. Now adjust RV2 (TX converter) to give an RF output of 10W. This has now established the ALC levels for the transverter.

If the ALC loop is to be extended to include the HF driver, connect the ALC output of the transverter to the HF transceiver's ALC input. Now adjust RV3 (ALC, rear panel) to reduce the 50MHz RF output of the system to approximately 9·5W – this ensures that the HF transceiver's ALC is the controlling signal.

50 & 70MHz

COMPONENT LIST FOR THE 28 – 50MHz RECEIVE CONVERTER AND LOCAL OSCILLATOR BOARD

RESISTORS
All 0·25W 2% or 5% (except as marked)
4R7	1	R19
6R8	1	R17
11R 2%	2	R5, 7
15R	3	R20, 21, 22
47R	1	R18
51R 2%	1	R9
100R	2	R3, 6
180R	2	R4, 8
360R 2%	1	R15
390R	1	R14
470R	1	R13
560R	2	R10, 16
4K7	1	R11
10K	1	R12
18K	1	R1
39K	1	R2

CAPACITORS
Sub-miniature ceramic plate, Philips 682 and 629 series
33p	1	C19
39p	2	C6, 10
68p	2	C8, 18
100p	2	C16, 17
120p	2	C14, 15
180p	2	C25, 28
220p	2	C7, 9
470p	2	C26, 27
10n	11	C2-5, 11-13, 20, 22-24
100n	1	C21
220n	1	C30

Solid aluminium electrolytic
10μ 16V	1	C29: STC code 57102B

Trimmer
10p	1	C1: Oxley solid-electrode type, or similar

SEMICONDUCTORS
BF981	1	TR1
2N918	1	TR2
BFR96	1	TR3
MSA0404	1	IC1
78L09	1	IC2
SRA-1/SBL-1	1	M1

INDUCTORS
Toko S18 series, ferrite core
L11, 12	2	Blue: code 301SS0600
L15, 16	2	White: code 301SS0800

Toko misc
L4, 5, 7, 8	4	code MC120/078
L13	1	code 113KN2K1026

Hand-wound
L1		4 turns on T50/10 toroid
L2		17 turns on L1
L3		10+10 turns bifilar on T37/6 toroid
L6, 9		20mm loop, 1mm enamelled wire
L14		10+10+10 turns trifilar on T37/6 toroid

Miniature moulded chokes with axial leads
10μH	1	L10
1mH	1	L17

MISCELLANEOUS
X1   22MHz HC33U Crystal
SMB sockets,
PC mounting   3   SK1, 2, 3

ALC AMPLIFIER

RESISTORS
All 0·25W 2% or 5%
10K	3	R4, 5, 6
22K	1	R2
220K	1	R1
330K	1	R3

CAPACITORS
Sub-miniature ceramic plate, Philips 682 and 629 series
10n	2	C5, 6

Tantalum
2μ2/16V	1	C1
10μ/16V	3	C2, 3, 4

SEMICONDUCTORS
CA3140	1	IC1
555	1	IC2
1N4148	2	D1, 2

POWER AMPLIFIER

RESISTORS
All 0·25W 2%
18R	1	R2
300R	2	R1, 3

CAPACITORS
Sub-miniature ceramic plate, Philips 682 and 629 series
10n	4	C2, 4, 6, 7

Solid aluminium electrolytic
33μ/16V	3	C1, 3, 5

SEMICONDUCTORS
78S09	1	IC1
M57735	1	IC2

INDUCTOR
L1	10 turns 4mm diameter, 0·4mm enamelled wire, close wound

MISCELLANEOUS
Heatsink   Siefert KL-139 75SW, black anodized finish· (Schaffner EMC, Micromark or GCA Electronics)
Solder tags and braid
Bolts and washers

50MHz TRANSVERTER

TRANSMIT CONVERTER AND T-R SWITCHING

RESISTORS
All 0·25W 2% or 5% (except as marked)

Value	Qty	Ref
1R0	1	R20
6R8	1	R10
10R	3	R12, 16, 19
47R	2	R17, 18
51R 2%	1	R6
68R	2	R22, 30
100R	3	R3, 11, 25
150R	1	R15
180R	1	R4
220R	1	R5
240R 2%	1	R13
330R	2	R21, 29
390R	2	R7, 8
470R	1	R14
560R	1	R9
1K2	3	R24, 26, 27
2K2	1	R1
2K7	1	R28
3K3	1	R23
180K	1	R2
1K	1	RV1 horizontal preset
2K2	1	RV2 horizontal preset

CAPACITORS
Sub-miniature ceramic plate, Philips 682 and 629 series

Value	Qty	Ref
6p8	2	C28, 30
15p	1	C29
39p	2	C12, 14
56p	2	C27, 31
68p	1	C13
100p	1	C10
220p	2	C11, 15
10n	22	C1-9, 16-26, 32, 33

Solid aluminium electrolytic
10µ/16V 1 C34: STC code 57102B

SEMICONDUCTORS

Part	Qty	Ref
2N3819	1	TR1
BFR96	1	TR2
BFG34	1	TR3
2N4400	2	TR6, 7
BD236	1	TR4
BD136	1	TR5
MSA0404	1	IC1
MSA0304	1	IC2
BA479	1	D1: PIN diode
1N4148	1	D2
SRA-1/SBL-1	1	M1

INDUCTORS
Toko S18 series

Ref	Qty	Description
L3	1	Blue, aluminium core: code 301AS0600
L12	1	Yellow, ferrite core: code 301AS0400
L4, 5, 7, 8	4	Toko MC120/078

Hand-wound

Ref		Description
L6, 9		20mm loop, 1mm enam wire
L10		10+10+10 turns trifilar on T50/10 toroid
L11		10 turns on T50/10 toroid

Miniature moulded chokes with axial leads
10µH 2 L1, 2

MISCELLANEOUS
SMB sockets,
PC mounting 3 SK1, 2, 3

VSWR & LOWPASS FILTER

RESISTORS
All 0·25W 2%

Value	Qty	Ref
220R	2	R1, 2
1K	3	R3, 4, 6
10K	1	R5

CAPACITORS
Sub-miniature ceramic plate, Philips 682 and 629 series

Value	Qty	Ref
15p	2	C7, 14
33p	2	C8, 13
39p	2	C9, 12
68p	2	C10, 11
10n	2	C1, 2

Solder-in feedthrough type
1n 4 C3-6

SEMICONDUCTORS

Part	Qty	Ref
2N4402	1	TR1
ND4891E	2	D1, 2: Schottky diodes
1N4148	2	D3, 4

INDUCTORS

Ref	Description
L1	5 turns 8mm diameter, close wound, 1mm enamelled wire
L2	5 turns 9mm diameter, close wound, 1mm enamelled wire

MISCELLANEOUS

Item	Qty	Ref
CX-120P coax relay		RL1: Cirkit code 46-90120
SMB sockets, chassis mounting	3	SK1, 2, 3
Teko box, 373		West Hyde Developments

ITEMS MOUNTED ON CASE

RESISTORS
0·25W 2%

Value	Qty	Ref
330R	1	R1, 2
1M	1	RV1: Front panel mounted, linear (POWER OUTPUT)
220k	1	RV2: preset, on rear of RV1 (can be 50k preset plus 100k fixed)
20k	1	RV3: Panel mounted on rear of case (EXTERNAL ALC O/P LEVEL)

LEDS

Colour	Qty	Ref
Green	1	D1: POWER ON
Red	1	D2: TRANSMIT

THE CRAY HIGH-PERFORMANCE 70MHz TRANSVERTER

by Dave Robinson G4FRE/WG3I

The *Cray* transverter described here is designed to provide a reliable 28–70MHz transverter with a good receive and transmit performance.

During past years my 70MHz experience has ranged from the quiet wilds of northern Scotland, through England and Wales, to the polluted band found in Gibraltar. In all these locations the limiting factor on the receiver's sensitivity at 70MHz is not its own noise floor but the noise level of the environment seen by the antenna. This was explained in Chapter 4, and can easily be demonstrated by plugging a resonant 70MHz antenna into the receiver and noting the considerable increase in background noise. Experience has shown that an overall noise figure of 3dB is more than adequate for a 70MHz receiver. My own operations have also confirmed that it is detrimental for the receive converter to have huge amounts of gain. One commercial transverter with 32dB of gain has proved itself a liability in contests! Given a reasonable 28MHz receiver with a noise figure of typically 10dB, a transverter gain of 16dB is a good compromise between low-noise performance

Fig. 9.25. Block diagram of the *Cray* 70MHz transverter

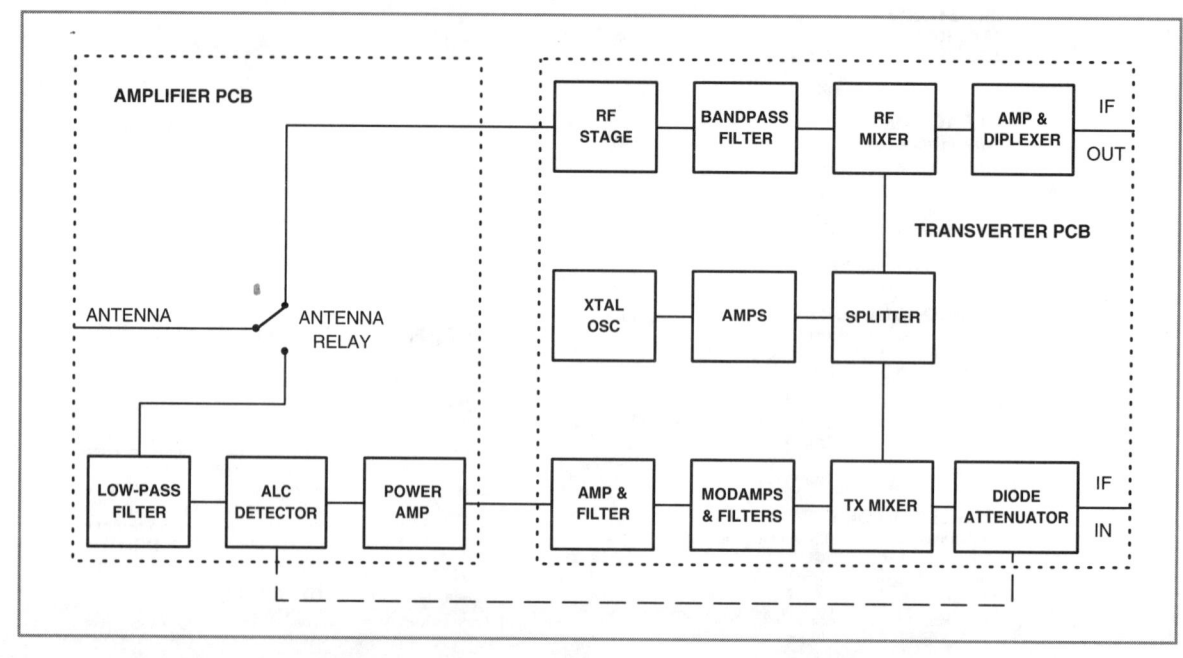

CRAY 70MHz TRANSVERTER

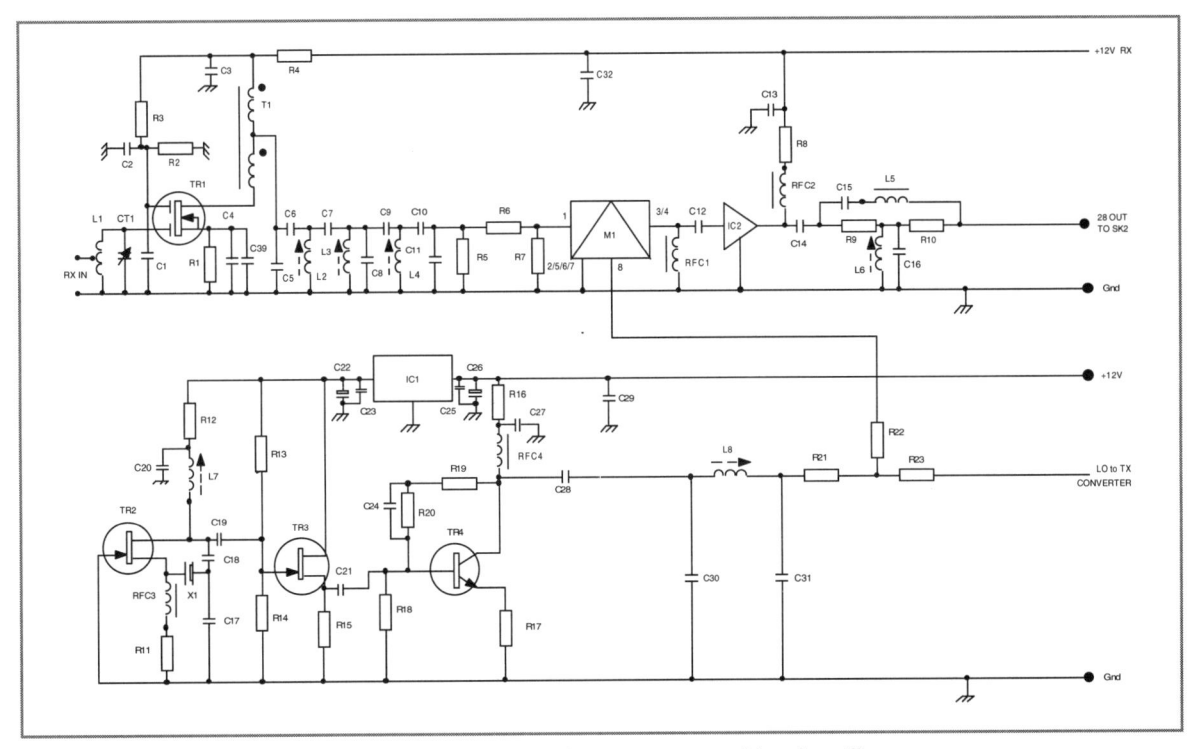

Fig. 9.26. Circuit of the 70MHz receive converter and local oscillator

and overload capability.

On the transmit side, the aim was to produce around 10W PEP output with good linearity. This would be adequate to drive a solid-state amplifier to full legal output, and more than adequate for a valve PA. Since 70MHz has limited international availability there are no PA modules available for this band, so the PA in the *Cray* has to use a discrete RF power transistor with variable tuning and matching capacitors. With care in alignment, the PA of the *Cray* is capable of better linearity than a fixed-tuned module – although it is also possible to align the PA incorrectly, and hence to transmit poor-quality signals. The transmit converter is intended to be driven by a transceiver capable of producing around 5mW at 28MHz. The transverter is activated on transmit by a ground-to-transmit PTT signal provided by the HF transceiver. Antenna changeover facilities are included in the transverter, although connection points are also provided to interface an external amplifier and higher-power antenna changeover relay, as described later.

CIRCUIT DESCRIPTION

A block diagram of the transverter is shown in Fig. 9.25. As far as possible, each section is designed to have 50Ω input and output impedances to allow individual stage checking if necessary. The circuit diagram of the local oscillator and receive converter is shown in Fig. 9.26. The crystal oscillator uses a 42MHz third-overtone crystal in a low-noise FET oscillator. The oscillator drive is buffered by a further FET amplifier and fed to a broadband transistor amplifier using a BFR96. This amplifier is followed by a lowpass filter to ensure that a clean spectrum is fed to the mixers. Two outputs each of +10dBm are provided from the two-way splitter (R21, R22 and R23), enabling the receive and transmit mixers to be driven simultaneously. Drive level is 3dB above that normally specified for

9 - 21

50 & 70MHz

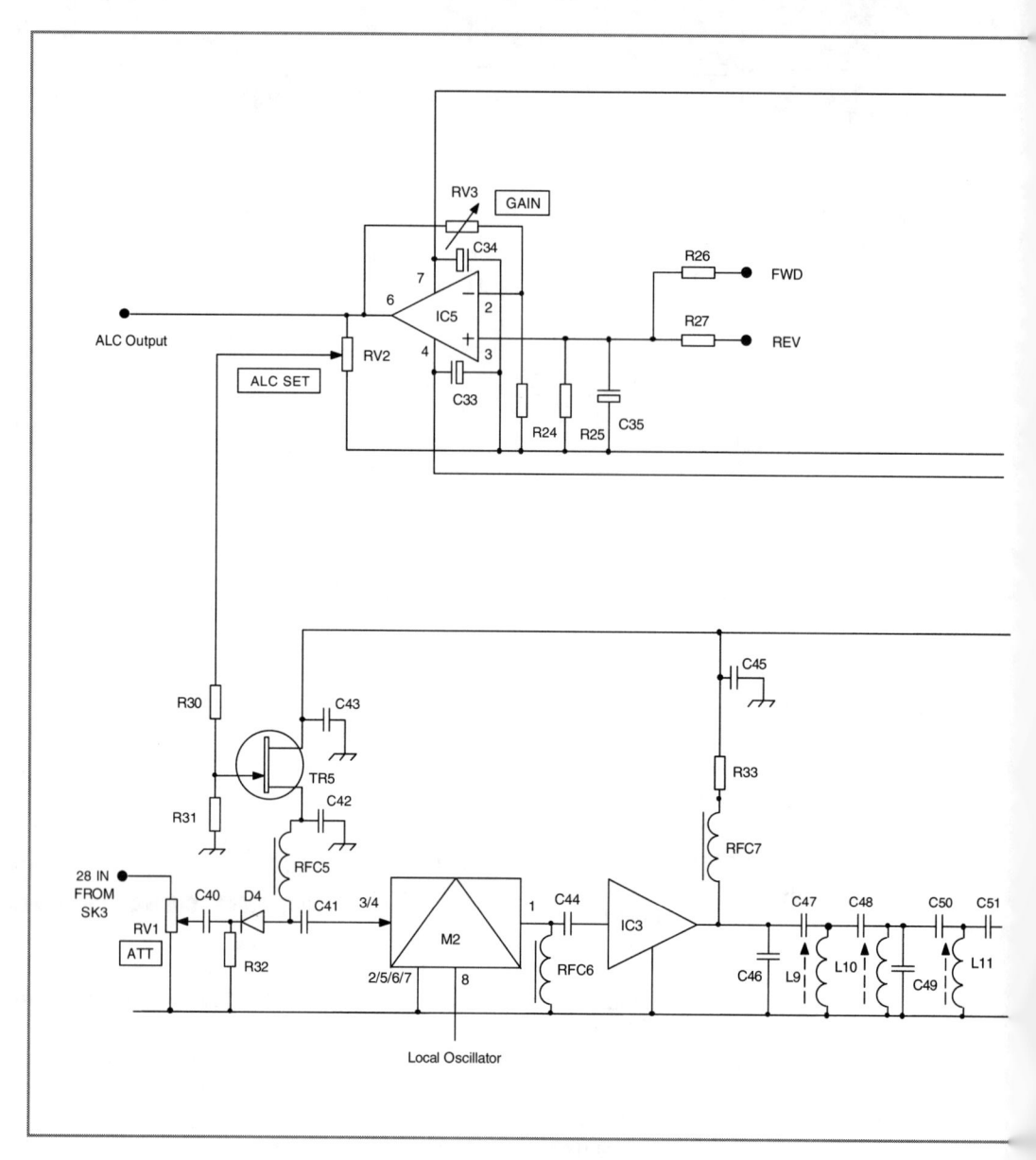

the mixers to improve the intercept point.

A dual-gate MOSFET is used in the RF amplifier stage to obtain high dynamic range and low noise figure. Although it is capable of sub-1dB noise figures on 70MHz, in this design a compromise is made between noise figure and gain to obtain the best strong-signal handling performance. The input circuit of the RF stage is an air-spaced coil in order to obtain a reasonable Q. The output load is a toroidal transformer, whose contained magnetic field isolates the input and output of the FET and helps avoid instability. The output from the RF stage passes through a 3-pole bandpass filter and a 13dB attenuator, to the mixer. The resistor values used in the attenuator may be adjusted to provide optimum performance in your particular situation. For further details see Chapters 4 and 5.

The mixer M1 is an SBL-1, which was

CRAY 70MHz TRANSVERTER

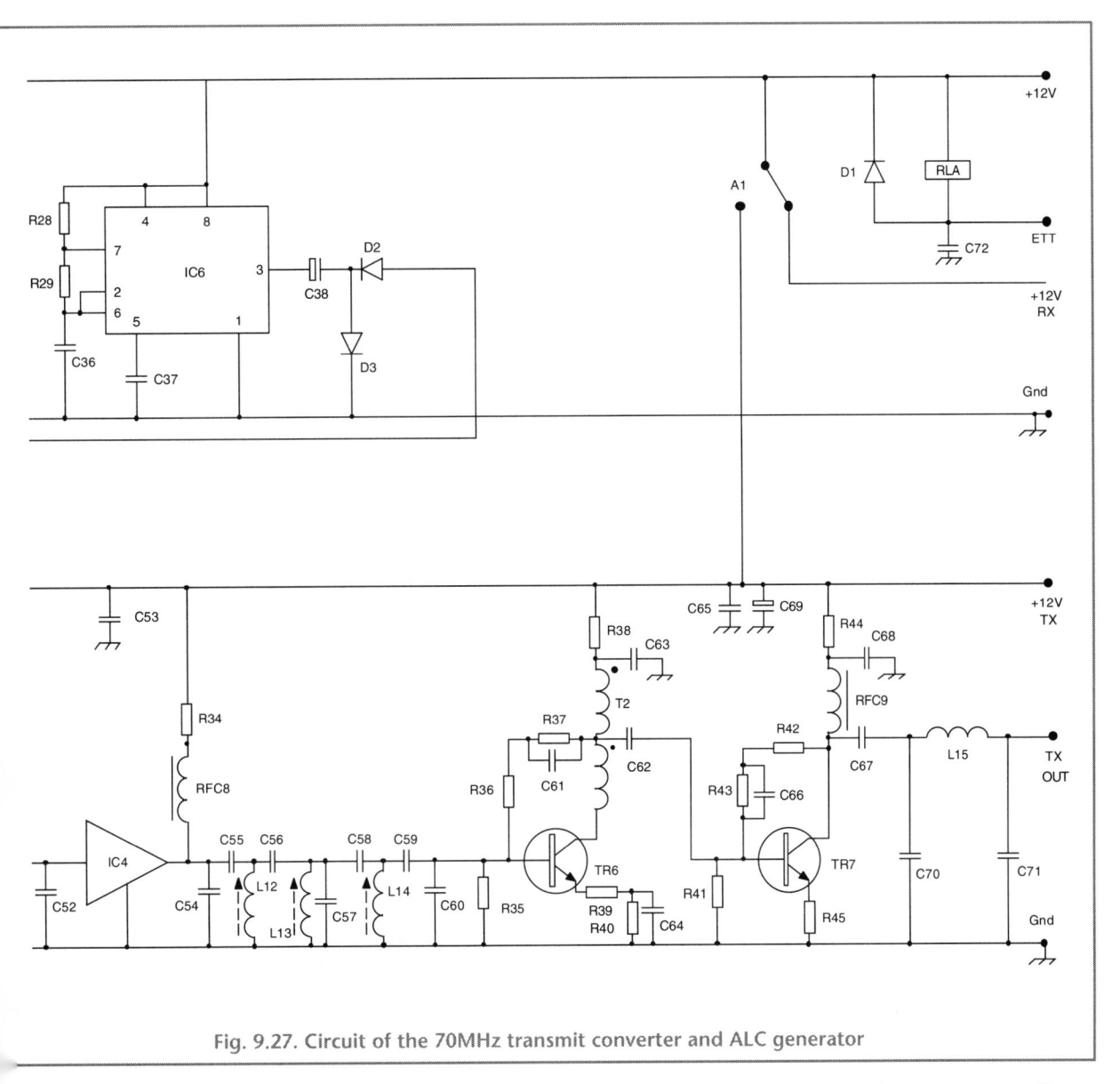

Fig. 9.27. Circuit of the 70MHz transmit converter and ALC generator

chosen for its high performance-to-cost ratio. If the more expensive SRA-1 is available, this may be used as a direct replacement. An MSA0404 MMIC is used to ensure a good wideband 50Ω termination for the mixer, and is provided with a bias current of 40mA. The output of this MMIC feeds the receiver output socket (SK2) via a diplexer. The diplexer has a broad frequency response, which gives little selectivity but ensures a good termination to the MMIC over a wide frequency range. This arrangement ensures that the receive converter's overall performance is not affected if the input impedance of the HF receiver is not 50Ω.

Fig. 9.27 shows the circuit diagram of the transmit converter, switching and ALC generator. The transmit converter requires a 28MHz drive signal at around +7dBm, which is fed in through the PIN diode D4. The function of this component is to act as a

50 & 70MHz

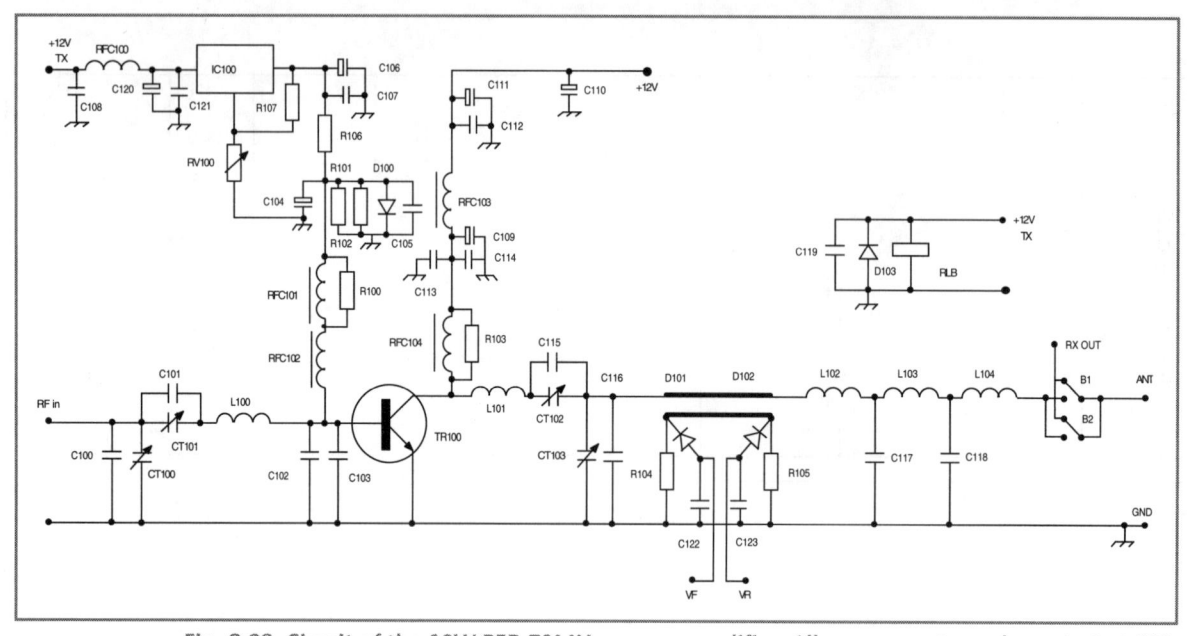

Fig. 9.28. Circuit of the 10W PEP 70MHz power amplifier. All component numbers start at 100

voltage-controlled attenuator. This is followed by M2, another SBL-1 (or SRA-1) mixer, which is also provided with +10dBm of local oscillator drive at 42MHz. This mixer too is terminated in a wideband manner by an MMIC, an MSA0404 biased at 40mA. The output is filtered by a three-pole bandpass filter and is then amplified by a further MMIC, an MSA0304 biased to 20mA. Following a further three-pole filter is a two-stage wideband feedback amplifier using a BFR91A and a BFR96S. Discrete amplifiers are used from this point since MMICs do not offer the necessary linearity and output capability. The BFR96S was chosen for its high collector-current rating – and hence higher linear power output – and is biased to a quiescent current of 70mA. It should not be replaced with a BFR96, which has inferior characteristics. The output stage is followed by a low-pass filter with a cut-off frequency of 80MHz to clean up the output spectrum.

A CA3140 operational amplifier (IC5) generates an ALC voltage from the forward and reverse voltages provided from the PA board. This ALC voltage is used to control the amount of attenuation provided by the PIN diode D5 in the 28MHz feed to the transmit mixer. This ALC provision reduces the risk of overdriving the transverter. IC6 (an NE555) is used to generate a negative voltage rail for the operational amplifier.

The circuit diagram of the power amplifier, filter and ALC generator are shown in Fig. 9.28. To improve its linearity and reliability the power amplifier uses an under-run TP2320 or 2N6083, either of which is rated at well over 10W output power. IC100, D100 and associated components provide a regulated low-impedance bias supply to maintain the PA transistor in the linear portion of its characteristic. Connected to the output of the PA stage is a directional coupler providing rectified samples of the forward and reverse powers. These are fed back to the transverter board to generate the ALC voltage. A further lowpass filter with a cut-off frequency just above the 70MHz band is included before the antenna changeover relay, to reduce any unwanted harmonics produced in the PA or directional coupler.

The complete components list for the transverter board is given on page 9–32, followed by the components list for the PA

CRAY 70MHz TRANSVERTER

board. To ease identification, all PA components are numbered from 100.

CONSTRUCTION

The transverter consists of two printed-circuit boards. The first contains the local oscillator, receive mixer, transmit mixer, ALC amplifier and negative-supply generator; this is designated the transverter board. The second board contains the transmit amplifier, ALC detector, low-pass filter and antenna changeover relay. This is designated the PA board. Both boards are etched on double-sided PC board.

The track layout of the transverter board is shown in Fig. 9.30, and the component overlay in Fig. 9.31. The top side of the transverter PCB is left copper-clad, the copper being counterbored from around the holes where non-grounded leads pass through. The track layout of the top side of the PA board is shown in Fig. 9.32; the underside is left completely copper-clad. All the PA components are mounted on the track side of the board as shown in Fig. 9.33, allowing it to be attached directly to a heatsink and case as shown in Fig. 9.34.

I chose to mount the whole transverter in a large die-cast box (Fig. 9.29) although a more conventional type of case would be equally suitable. Before inserting any components on the transverter and PA boards, their mounting holes (indicated by the letters 'H' in Figs 9.31 and 9.33) should be drilled in the case using the bare boards as templates. Further details of mounting the PA board and components are given later.

Begin assembly of the transverter board by fitting R23. Then cut the four tin-plate screens and solder them to the board, leaving a notch for R23. Copper-clad PCB does not provide adequate screening in this application and should not be used. Solder in all the resistors, followed by all the capacitors. Mount all the moulded coils next, oriented so that the long connection leg is grounded. The transformers are then wound as described in the components list (see page 9–10 for advice on winding toroids), and are secured to the board with a

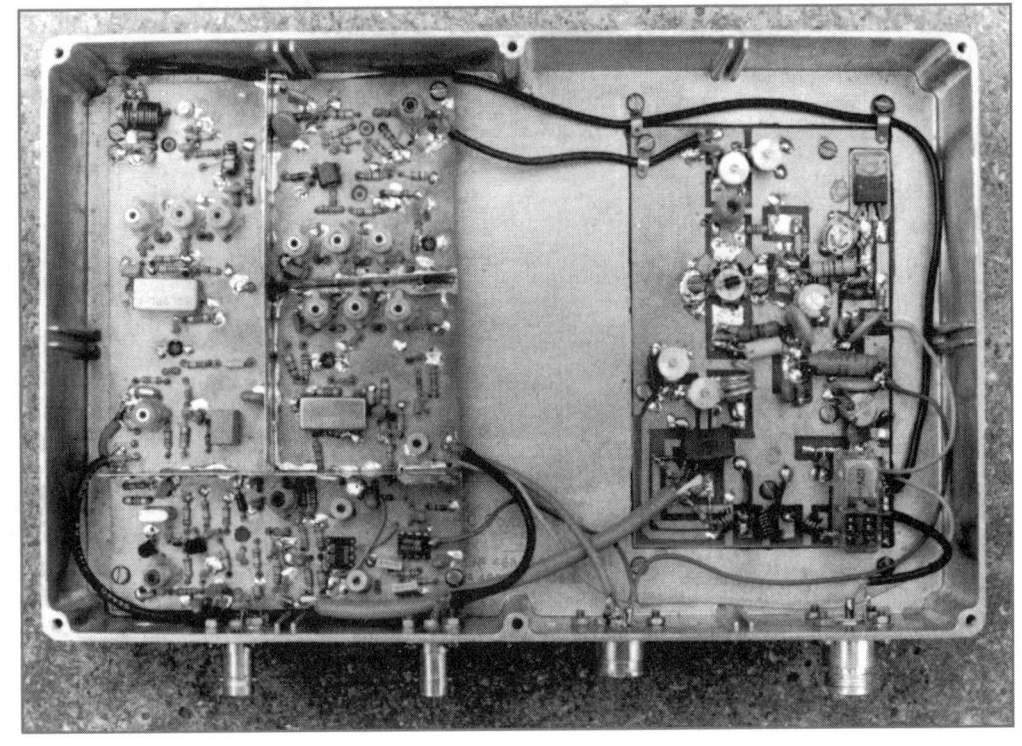

Fig. 9.29. Photograph of a prototype *Cray* transverter assembled in a die-cast box

50 & 70MHz

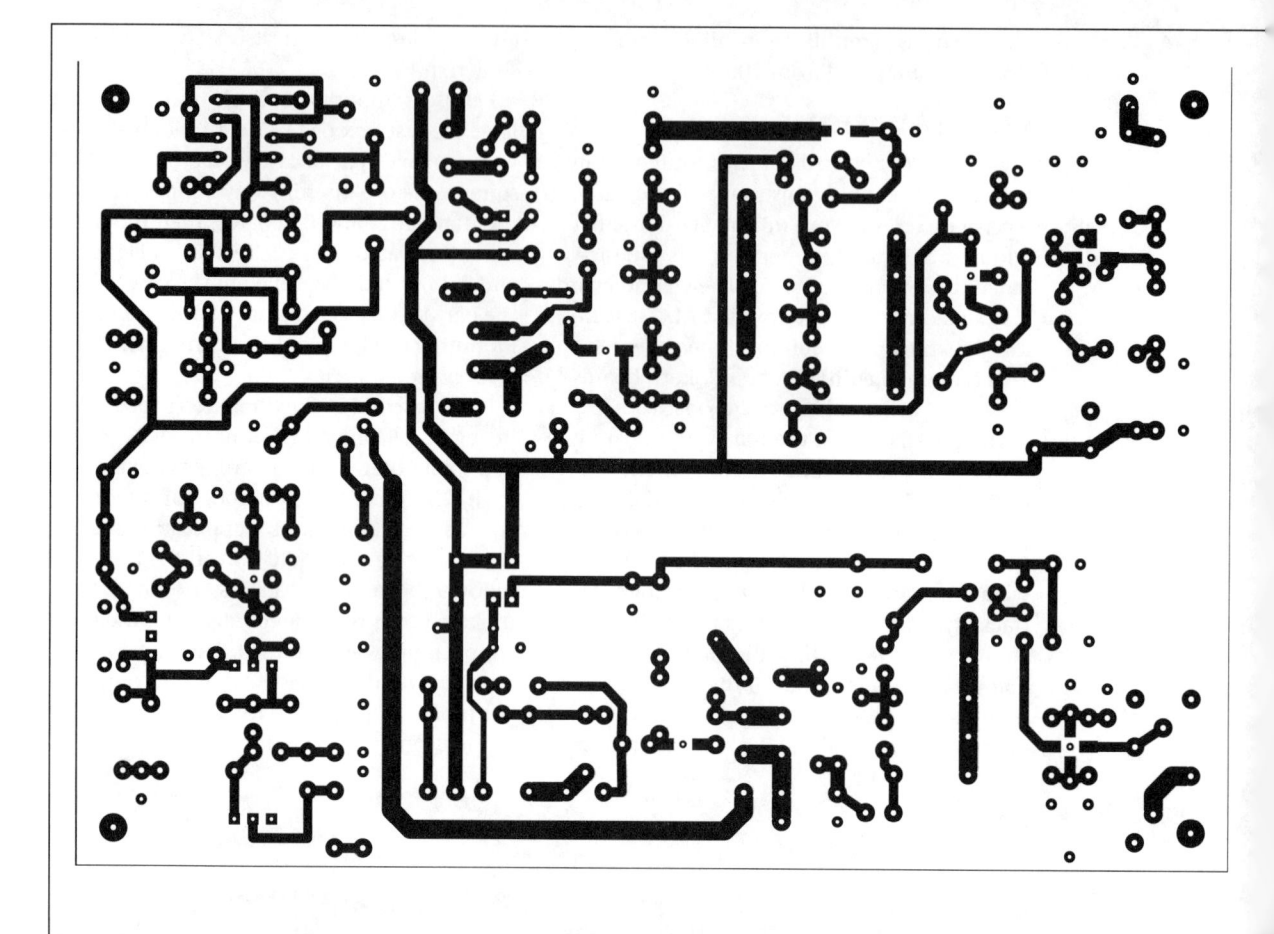

Fig. 9.30. Full-size PCB track layout of the *Cray* 70MHz transverter board (underside view; top side is un-etched copper groundplane)

small amount of adhesive. The semiconductors are then soldered in place; the pinouts are the same as shown in Fig. 8.13 for the *Suffolk* 144MHz transverter. Great care should be taken not to damage the fragile leads of the MOSFET and MMICs, all of which are soldered on the track side of the board. A clearance hole is drilled to locate the body of each device, using the centres marked on Fig. 9.30. The MMICs have their ground leads bent upwards and soldered to the top surface of the board. Both of the SBL-1 mixers are then soldered in place, the 'M' in the 'MCL' label on the can being over pin 2. The cans of the mixers should also be soldered to the top of the board; do this quickly using a large, hot iron.

Wire links are required to join the top and bottom surface on the transverter board as indicated by the letters 'X' in Fig. 9.31. To improve stability, RFC8 is soldered between the two relevant copper pads on the track side of the board. The board is mounted in the box on four 8mm M3 metal spacers.

The PA board is mounted on the heatsink using M3 x 6mm cheese-head screws, the heatsink having been tapped M3 in the corresponding positions using the PA board as a drilling template; the positions of the screws are denoted by letters 'H' in Fig. 9.33. Cut out the box to clear the whole of the PCB as shown in Fig. 9.34 and bolt the PCB to the box and the heatsink before mounting any

CRAY 70MHz TRANSVERTER

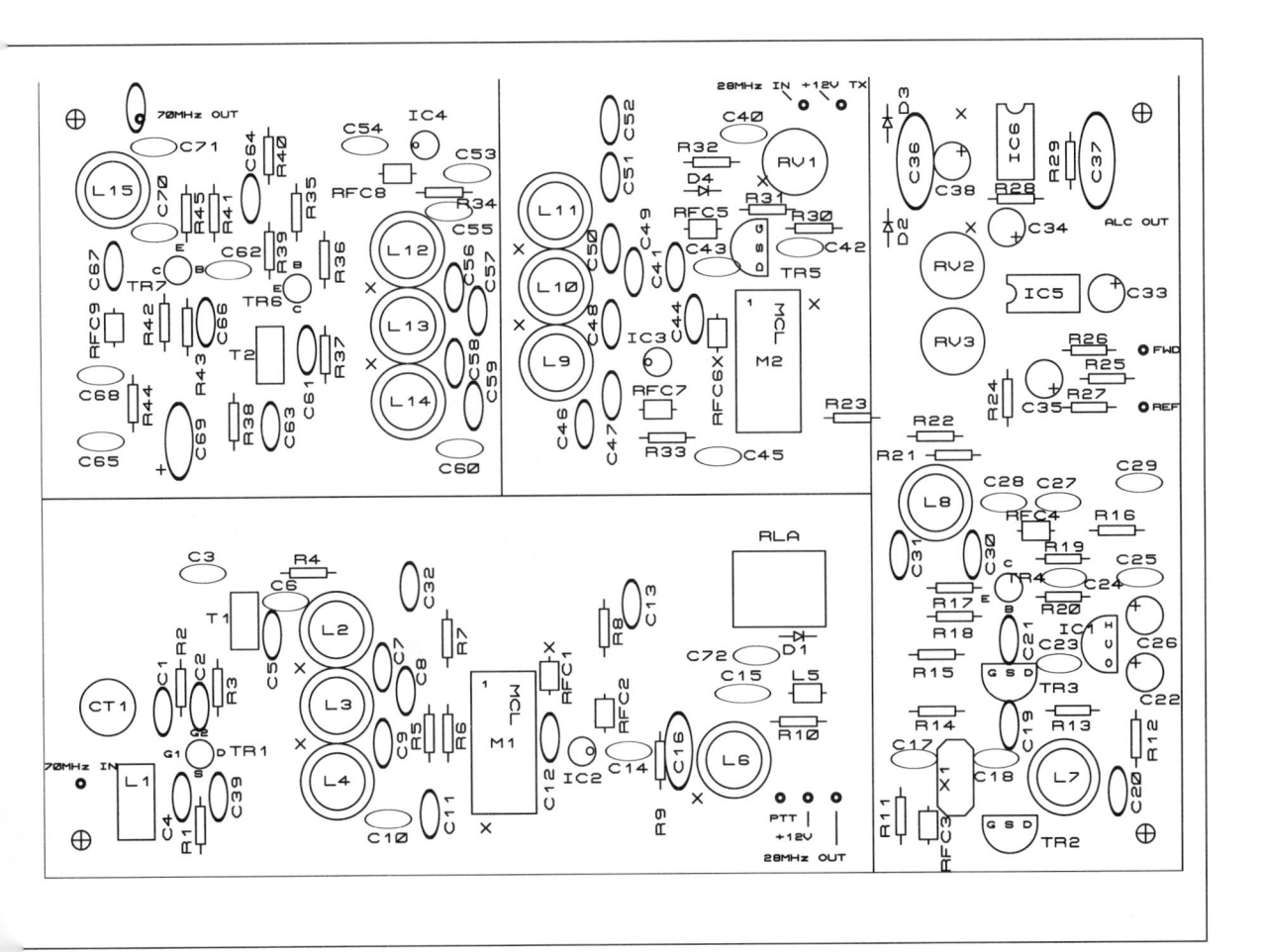

Fig. 9.31. Component layout of the *Cray* 70MHz transverter board

components. The hole for the PA transistor stud in the heatsink is 4mm diameter, while the hole in the PCB for the transistor is 10mm in diameter, allowing the metal shoulders of the stud to be in good thermal contact with the heatsink. The two screws adjacent to TR100 pass through holes drilled carefully through the emitter leads of the transistor, ensuring that the emitter leads are well grounded.

The bias regulator IC100 is mounted in a cut-out in the corner of the board, and is insulated from the heatsink by a mica washer. It is held in place by an M3 x 8mm nylon screw into a tapped hole. All components on the PA board must be soldered in place using the shortest possible lead lengths. D100 is placed on top of TR100, adequate thermal contact between the two being ensured by a small amount of heatsink compound. The orientation of coils L102, L103 and L104 on Fig. 9.33 should be carefully noted; these orientations are used to ensure minimum interaction between coils.

The interconnections between the two printed circuit boards and the coaxial sockets are shown in Fig. 9.35. All RF interconnections are made using miniature coax (RG174 or similar). To keep RF away from susceptible areas, the connection between the 5-pin DIN socket and the PTT terminal on the transverter board is made with single-core screened cable.

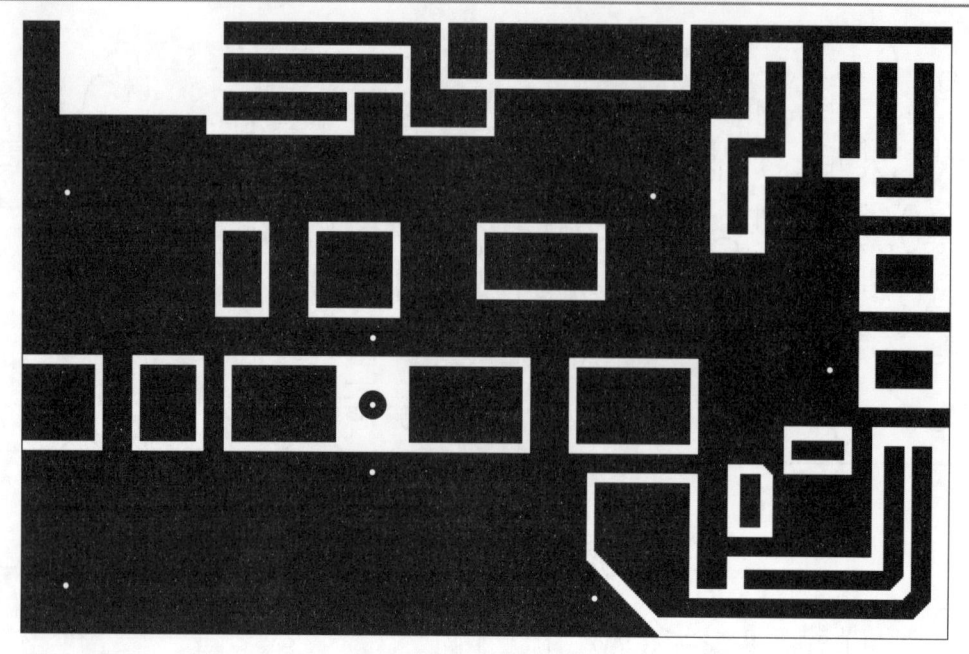

50 & 70MHz

Fig. 9.32. Full-size PCB track layout of the *Cray* 70MHz power amplifier board (top view; underside is un-etched copper groundplane)

Fig. 9.33. Component layout of the *Cray* 70MHz power amplifier board

● Denotes connection to groundplane   H Denotes mounting point

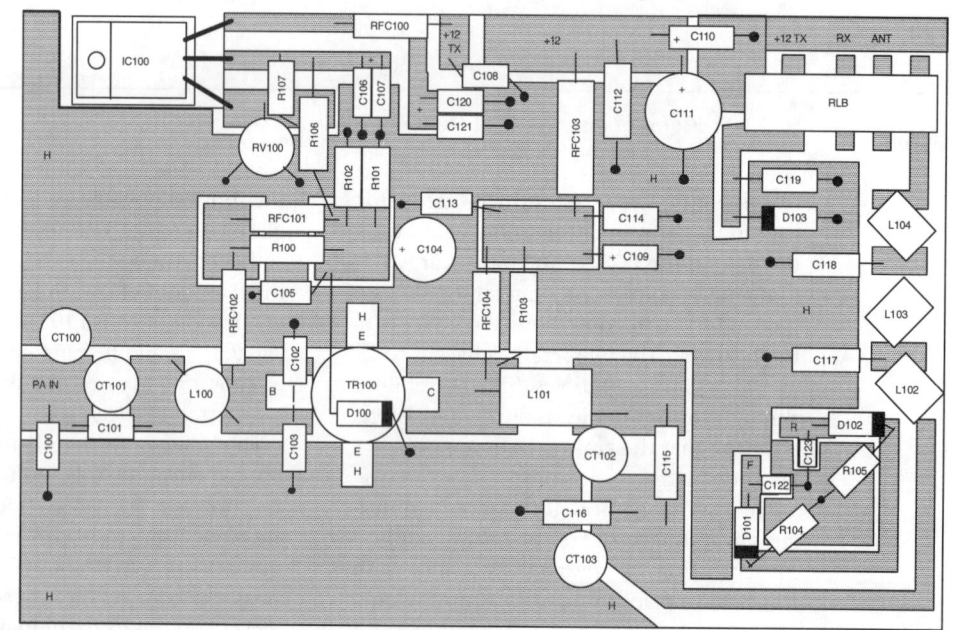

CRAY 70MHz TRANSVERTER

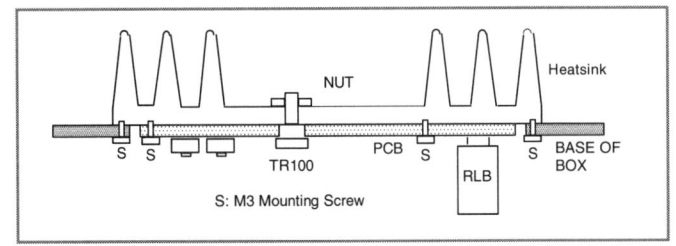

Fig. 9.34. Mounting of the 70MHz PA board and power transistor to the case and heatsink. Note that TR100 is mounted directly to the heatsink, through a large cut-out in the box

Similarly the ALC signals are interconnected with twin-core screened cable. All the +12V power connections are made using thick wire, particularly between the supply terminal and the PA board. In the normal configuration, two terminals on the PA board (12V TX and 12V RL) are connected to the +12V TX terminal on the transverter board.

ALIGNMENT

To align the *Cray* transverter, the following test gear will be required as a minimum:

- Absorption wavemeter to cover 25 to 150MHz
- Absorption power meter (or through-line power meter and 50Ω load) capable of reading 100mW to 10W at 70MHz
- Multimeter.

Connect a +12–13·8V regulated power supply to the transverter board only, *via* the multimeter set to its highest current range. If the power supply has current limiting, set it to 250mA. Switch on the power supply; the meter should indicate no more than 200mA. If the current is higher than that, switch off and look for signs of distressed components and short-circuits. When the current is within limits, check that the negative-voltage generator IC6 is supplying –8 to –10V at pin 4 of IC5. Ground the PTT line and check that RLA operates and switches the correct voltage to the 12V RX and 12V TX lines. Check that the DC voltages at the outputs of IC1-IC4 are all +4·5 to 5·0V when energized, and that the MMICs are drawing the correct bias currents as indicated by the voltage drops across their feed resistors. If all is well, alignment can commence.

Fig. 9.35. Interconnection diagram for the *Cray* 70MHz transverter

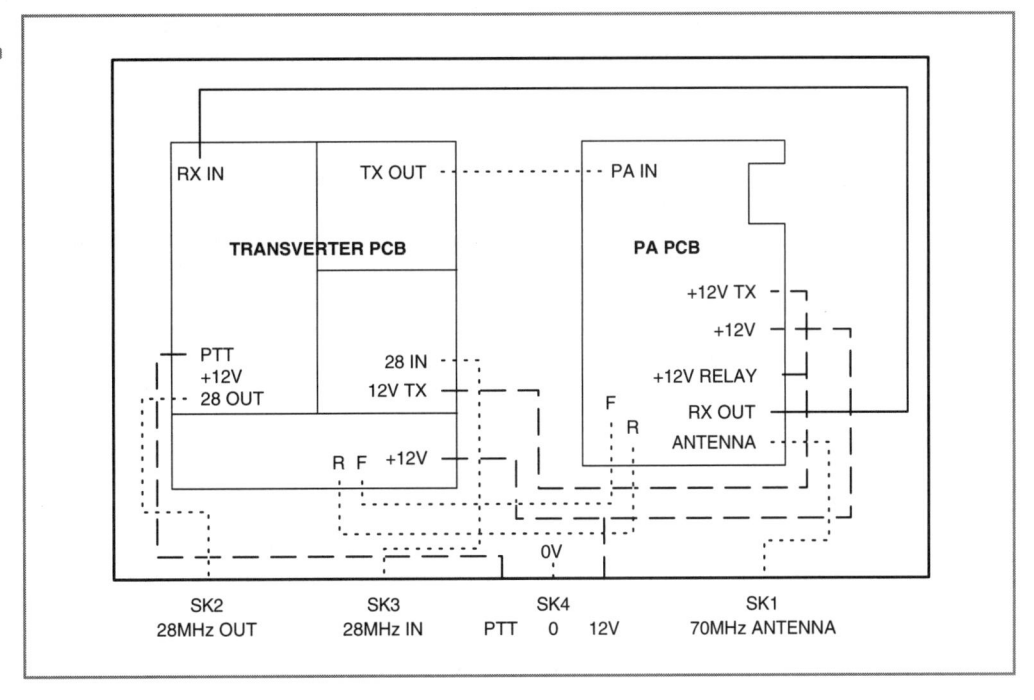

Set the wavemeter to 42MHz and loosely couple it to L7. Adjust the core of the coil for the maximum reading on the wavemeter. This will also correspond to a slight increase in the current consumption, indicated on the meter. If a frequency counter is available, L7 can be used to set the oscillator to exactly 42·0000MHz. The only other adjustment in the local oscillator is to tune L8 for maximum 42MHz indication.

Alignment of the receive converter is best accomplished using a signal generator. However, the large amount of available antenna noise mentioned earlier may be used instead. Connect a receiver tuned to 28·25MHz to SK2 and a resonant antenna to the antenna socket. Initially set the cores of L2, L3 and L4 to 1mm below the top of their respective coil formers. Adjust CT1 and then L2, L3 and L4 (in that order) for maximum noise output from the receiver. If a local signal or beacon can be found, CT1 can be adjusted for best audible signal-to-noise ratio.

Next the transmit converter can be aligned. Adjust RV1 so that its wiper is at the SK3 end of its travel. Connect a 28MHz transmitter capable of producing 5mW to SK3 and temporarily connect the output of the transverter board directly to the absorption power meter. The +12V and 12V TX connections between the transverter and PA boards should be disconnected for the time being. Set the positions of the cores in L9–L14 at 1mm below the tops of the formers. Adjust the current limit on the power supply to 500mA and switch on. Ground the PTT terminal and note the current indicated on the multimeter. If it is greater than 400mA, turn off the supply and fix any short-circuits.

Apply the 28MHz drive and tune L9–L14 for maximum output, as indicated when the wavemeter is loosely coupled to each coil in turn. Final tuning is accomplished by using the absorption power meter connected to the output. To ensure that the output is at 70MHz, turn off the 28MHz drive (with the PTT still grounded) and ensure that the output signal *totally* disappears. To avoid overdriving the transverter, RV1 should be adjusted so that the power indicated on the output power meter is just below maximum output. The output from the transverter board should be in the region of 100mW.

Connect the output of the transverter board to the input of the PA board, and the antenna socket (SK1) to the absorption power meter. Reconnect the +12V and +12V TX supplies to the appropriate points on the PA board, but do not connect the 'forward' and 'reverse' voltage outputs to the transverter board at this stage. Insert an ammeter in the collector lead of the PA transistor; this may be achieved by de-soldering RFC103 at its C112 end and inserting an ammeter with short leads. Turn the wiper of RV100 to the ground end of its track.

Switch on and note that the PA transistor draws at most a few milliamps of current. Ground the PTT terminal and check that the antenna changeover relay operates; the PA current should not change significantly. Now slowly rotate RV100 until the ammeter indicates 30mA. The current should increase gradually; any sudden jumps are a sign of instability, which should be remedied before proceeding. Apply 28MHz drive to the transverter, which should result in a slight increase in collector current and possibly indications of output power. Adjust CT100 and CT101 for maximum collector current, and then CT102 and CT103 for maximum RF output power. None of the capacitors should tune sharply, but all should show a peak. The available output should be something over 10W; however, to minimize the risk that the transverter is being overdriven, adjust RV1 until the output power shown on the power meter has just started to drop, and leave it at that setting (see panel on page 6-7).

The last task is to set up the ALC circuit. Firstly ensure that the detector circuits are working by measuring VF and VR, the 'forward' and 'reverse' voltages from the VSWR detector, when the transverter is producing full power output into 50Ω. VF should be over –0·5V and VR should be much less than VF. Now connect VF and VR to the appropriate points on the transverter board. Adjust RV3 (GAIN) for maximum voltage of –8V at the output of IC5, and then back it off slightly to a reading of say −7·8V. Now adjust RV2 (ALC SET) and note that the output power starts to fall as TR5 reduces the bias current through D4. Adjust RV2 for the desired RF output, **which should not be more than 10W**.

INTERFACING

Interfacing of the *Cray* transverter to HF transceivers is exactly the same as for the other transverters in this book. The description of the *Suffolk* 144MHz transverter on page 8-26 gives full details. If your HF transceiver can only provide about −10dBm (100µW) from its transverter socket, additional amplifier stages will be required in the TX IN feed. This may take the form of further MMICs employing the same circuit as IC3. As the circuit is operating at low signal level, it may be built on a small piece of PCB and mounted in the transverter box.

To interface an external amplifier to the *Cray* transverter, a few configuration changes will be needed. The 70MHz drive to the external PA's input may be obtained from the antenna socket as normal. The coax lead connected to the RX OUT terminal of the PA board should be transferred to a coaxial socket mounted on the rear of the transverter box, and the receive signal from the PA's own external antenna changeover relay is connected to this socket. If the external changeover relay has a 12V coil, a suitable switched voltage may be obtained from the 12V TX terminal on the transverter board via a spare pin on SK4.

PERFORMANCE

The measured noise figure of a series of receive converters has been found to be around 2·2dB, suggesting that R5, R6 and R7 could be changed to provide even more attenuation in the receive path. Rejection of a 14MHz image signal (42MHz minus 28MHz) is around 80dB. The receiver has been found to be stable under a wide variety of conditions including an open-circuit input. All harmonics and spurious transmitter products are greater than 60dB down on the wanted signal, which is sufficient when the transverter is driving an antenna directly. It is advisable to use an additional lowpass filter if a further stage of amplification is used, or if the transverter board is used without the PA board. A copy of the lowpass filter on the PA board would be suitable for 100W output or more, if UNELCO or SEMCO cased-mica capacitors are substituted for the ceramic-plate types.

Two-tone measurements of the linearity of the transverter demonstrate that, with care in alignment, the third-order intermodulation products are 38dB down on 10W PEP. This value is quite fair, and is better than that of modular PAs used on the other bands. To improve on this figure the transistor would have to be biased in class A, or else a bipolar power transistor or a VHF power FET would have to be used with a higher-voltage supply.

50 & 70MHz

CRAY 70MHz TRANSVERTER BOARD COMPONENTS

RESISTORS
All 0·25W miniature carbon film or metal film

2R2	1	R44
10R	3	R12, 16, 39
15R	1	R45
18R	3	R21, 22, 23
22R	1	R17
33R	1	R38
39R	2	R6, 40
47R	1	R1
51R	2	R9, 10
100R	1	R4
120R	1	R32
150R	2	R5, 7
180R	2	R33, 34
220R	5	R8, 11, 15, 18, 41
510R	2	R19, 43
560R	1	R36
680R	2	R20, 42
1k0	1	R35
2k7	1	R31
3k3	1	R37
10k	4	R2, 24, 28, 29
15k	1	R3
33k	1	R27
100k	1	R14
120k	1	R26
180k	1	R30
330k	1	R25
470k	1	R13

Horizontal 6mm round cermet trimmers

470R	1	RV1
1k0	1	RV2
1M0	1	RV3

CAPACITORS
Low-K ceramic plate, eg Philips 632 series

3p9	6	C7, 9, 48, 50, 56, 58
6p8	1	C19
33p	3	C8, 49, 57
39p	2	C70, 71
47p	7	C6, 10, 18, 47, 51, 55, 59
68p	2	C30, 31
120p	6	C5, 11, 46, 52, 54, 60
330p	1	C17

High-K ceramic plate, eg Philips 629 series

4n7	24	C1, 3, 12, 14, 21, 23, 24, 27, 28, 29, 32, 39, 44, 45, 53, 61, 62, 63, 64, 65, 66, 67, 68, 72
10n	4	C40, 41, 42, 43
100n	7	C2, 4, 13, 17, 20, 23, 25

Polystyrene 2%

22p	1	C15
560p	1	C16

Miniature polyester

10n	2	C36, 37

Tantalum bead electrolytic, 35V

1µ0	3	C22, 26, 35
10µ	3	C33, 34, 38

Solid aluminium electrolytic, min 16V

4µ7	1	C69

Miniature ceramic or air-spaced trimmer

25p	1	CT1

INDUCTORS
Toko S18 red, aluminium core, 301AS0200

1	L6

Toko S18 green, no core, 301ASS0500

1	L15

Toko S18 blue, aluminium core, 301AS0600

9	L2, 3, 4, 9, 10, 11, 12, 13, 14

Toko S18 violet, ferrite core, 301SS0700

2	L7, 8

L1	10 turns 22SWG, 6·5mm ID, tapped 2 turns from ground
T1	7+7 turns 26SWG, bifilar wound on Amidon T37-12 toroid
T2	9+9 turns 40SWG, bifilar wound on Fair-Rite 26-43002402 toroid

Moulded axial chokes

15µH	5	L5, RFC5, RFC7, RFC8, RFC9
22µH	2	RFC3, RFC4
33µH	1	RFC2
220µH	2	RFC1, RFC6

TRANSISTORS

BF981	1	TR1
J310	2	TR2, TR3
BFR96	1	TR4
J309	1	TR5
BFR91A	1	TR6
BFR96S	1	TR7

INTEGRATED CIRCUITS

78L08	1	IC1
MSA0304	2	IC2, IC3
MSA0404	1	IC4
CA3140E	1	IC5
NE555N	1	IC6

DIODES

1N914, 1N4148	3	D1, D2, D3
BA479	1	D4: PIN diode

MISCELLANEOUS

SBL-1, SRA-1	2	M1, M2
42·0000MHz HC33/U 3rd overtone crystal	1	X1
Relay 12V SPCO, OVC style	1	RLA

CRAY 70MHz TRANSVERTER

CRAY 70MHz POWER AMPLIFIER COMPONENTS

(NUMBERS START FROM 100)

RESISTORS

0·25W miniature carbon film or metal film
10R	1	R100
220R	3	R104, 105, 107

0·5W carbon film or metal film
10R	2	R101, 102

1W carbon film or metal film
27R	1	R106
47R	1	R103

Miniature horizontal cermet preset with carbon wiper
1k0	1	RV100

CAPACITORS

Low-K ceramic plate, eg Philips 632 series
22p	1	C101
120p	3	C100, 102, 103

High-K ceramic plate, eg Philips 629 series
4n7	5	C105, 114, 119, 122, 123
22n	2	C107, 121
47n	3	C108, 112, 113

Silver-mica
30p	1	C116
56p	2	C117, 118
120p	1	C115

Single-ended electrolytic, min 16V
220µ	2	C104, 111

Solid aluminium electrolytic, min 16V
4µ7	2	C109, 110

Tantalum bead electrolytic, min 16V
1µ0	2	C106, 120

Variable, film dielectric
65p	4	CT100, 101, 102, 103

INDUCTORS

Toko S18 green, no core
L100	1	code 301SS0500
L101		3 turns 16SWG, 10mm ID close wound
L102, L104		4 turns 22SWG, 5mm ID close wound
L103		7 turns 22SWG, 5mm ID close wound

Moulded axial chokes
0·33µH	1	RFC102
2·2µH	1	RFC100
33µH	1	RFC101
1µH 2·5A	1	RFC104: (RS 228-113)
4µH	1	RFC103: (RS 238-277)

INTEGRATED CIRCUITS
LM317T	1	IC100
TP2320,		

TRANSISTORS
2N6083	1	TR100

DIODES
1N4004	1	D100
1N914,		
1N4148	3	D101, D102, D103

MISCELLANEOUS

Relay type
OM1	1	RLB
N-type socket	1	SK1
BNC sockets	2	SK2, SK3

5-pin 240°
DIN socket	1	SK4

Die-cast box 273 x 171 x 50mm
Heatsink   Redpoint 6M1 size 152 x 94 x 8mm
TO220 insulating washer, nylon M3 screw
Heatsink compound
Veropins
Tin-plate strips 25mm wide
M3 brass cheese-head screws
M3 x 8mm tapped metal spacers

SOLID-STATE POWER AMPLIFIERS

Chapter 6 is pretty scathing about the vast majority of solid-state power amplifiers, and quite rightly so. And yet it *is* possible to build transistor PAs of quite acceptable linearity, with a little care and attention. The Big Mistake is to try and do it on the cheap. Manufacturers for the amateur market say that we, their customers, expect solid-state PAs to be cheap, so we only get what we're prepared to pay for – and there's some justice in that remark. As a breed, we often seem to expect professional products at amateur prices.

Happily, that vicious circle is now beginning to break. There is a concerted drive to raise the standards expected from power amplifiers of all kinds, and information on how to build good solid-state PAs is beginning to appear. John Matthews G3WZT has written two excellent articles on how to design and build solid-state power amplifiers for 50MHz and 144MHz [1,2]. Even a cursory glance will reveal the vast difference between his designs and the average "black brick". The devices are run well within their linear range, using 28V or 50V supplies, and higher powers are achieved by combining identical amplifiers. Although transistors still cannot reach the ultra-smooth intermodulation standards of a really good pair of tetrodes (Chapter 6), a well-designed solid-state power amplifier can be as clean as the normal run of valve PAs. But we repeat: don't expect it to be cheap.

Another exponent of high-quality solid-state PAs is Dale Harvey G3XBY, who provided the following design information.

COMPONENTS

Use only new transistors – they are worth the money. Some transistors obtained from surplus sources are manufacturers' drop-outs or early versions. Select a device which is designed for your intended purpose. For example, transistors for Class-C (FM) operation may produce lots of power in Class AB for SSB, but they seldom produce more than around a quarter of their rated Class-C power with acceptable linearity. For the lower VHF bands at least, you can use transistors specifically designed for SSB operation. If you really must use surplus devices, at least make sure that they're the right ones for the job!

Try to get away from the amateur's insistence on 13·8V supplies. Devices intended for 28V or even 50V supplies are more efficient, have more gain, are more linear and are easier to match. Also, the lower-current power supplies are easier to make. How much high-power SSB mobile operation do you intend to do anyway?

Produce a gain budget before you start and always try to have a few dB in hand. It's good to be able to put a 1–3dB attenuator pad at the input of the amplifier, since this will reduce the effects of varying input impedance upon the previous stage.

Use passive components which have adequate Q at the frequency of interest. For example, Philips film-dielectric trimmers are often satisfactory in 200W amplifiers! Mica compression trimmers are lossy; although they survive because they can also dissipate more heat, they are not a good choice unless you really need a big capacitance variation. Padding capacitors should have low loss and low inductance. ATC transmitting chip capacitors are certainly good up to 300W.

Try to keep coils "in the clear". The field from the collector series coil in a 70MHz

SOLID-STATE POWER AMPLIFIERS

BLW96 PA was enough to melt a film trimmer placed near the open end of the coil. Use large-diameter wire for the collector RF coil – even 0·141" semi-rigid coax gets hot owing to the large circulating currents and the direct thermal connection to the transistor. On the other hand, the collector DC feed coil carries much lower circulating currents and is much less critical.

RF BYPASSING

RF power transistors require thorough bypassing of their collector and base bias supplies, at all frequencies from AF to UHF regardless of the actual frequency of operation. Otherwise they will oscillate – and that's a cast-iron guaranteed promise. To achieve a low impedance over such a wide frequency range, several different types of capacitors need to be connected in parallel. Starting at the VHF/UHF end of the spectrum, use one or more physically large ceramic capacitors of 1nF or less, preferably chip types but certainly with the shortest possible leads. Take care of the middle frequencies by connecting a 100nF polycarbonate capacitor in parallel, plus possibly a 10nF ceramic for good luck.

Finally, add a 100µF tantalum or solid-aluminium capacitor for LF and AF – not an ordinary aluminium-foil electrolytic because its inductance is too large for effective operation beyond AF. However, since even tantalum electrolytics have appreciable inductance, parallel resonances with the smaller capacitors could produce 'holes' in the broadband bypassing capability you're trying to achieve. So the tantalum capacitor should always be separated from the rest of the RF components by a ferrite bead to kill any HF/VHF resonances.

BIAS CIRCUITS

Many transistor PAs are let down by their base-bias circuits, which fail to provide enough current on peaks of drive. For example, if the peak collector current is 10A, the bias circuit must provide its stabilized voltage at something in the region of 0·2–0·5A. In the traditional circuit, a simple resistor-and-diode voltage divider, well over an amp of standing current has to flow down the bias chain in order to provide anything near adequate regulation. Most amplifier circuits just don't do that, so the bias voltage starts to fall on peak current demand. The operating point of the transistor then moves towards cut off and the collector current no longer rises in response to the RF drive. The result is gain compression and splatter on every speech peak.

As the power transistor heats up, the voltage drop across its base-emitter diode decreases and the collector current will start to rise. This in turn heats up the transistor, the collector current rises still more, and before you know it the transistor has gone into thermal runaway and fried itself. So part of the function of the base-bias circuit is to prevent any tendency to thermal runaway by reducing its output voltage as the power-transistor junction temperature increases. This so-called 'thermal tracking' is usually achieved by using a semiconductor junction (often simply a diode) in the bias circuit as a temperature sensor.

A lot of diodes have been trailed over the tops of power transistors in efforts to improve thermal tracking. But it's the temperature of the RF power-transistor *junction* that really matters. Outside the semiconductor wafer itself, the best thermal contact with the transistor junction is at the mounting flange or on the heatsink. The temperatures in those locations can be sensed by screwing a tab-mounting power transistor directly to the metalwork. If the bias circuit calls for a diode, use a PNP transistor with its base and collector tab grounded. A bias supply with good thermal tracking can help prevent the slow onset of thermal runaway caused by long periods of full-power operation. However, if the power transistor decides to run away suddenly – for whatever reason – the bias supply can't do a thing to stop it. In such a case, the underlying problem is thermal, not electronic. Either the heatsink isn't big enough, or the late lamented RF power transistor wasn't in proper thermal contact with it.

50 & 70MHz

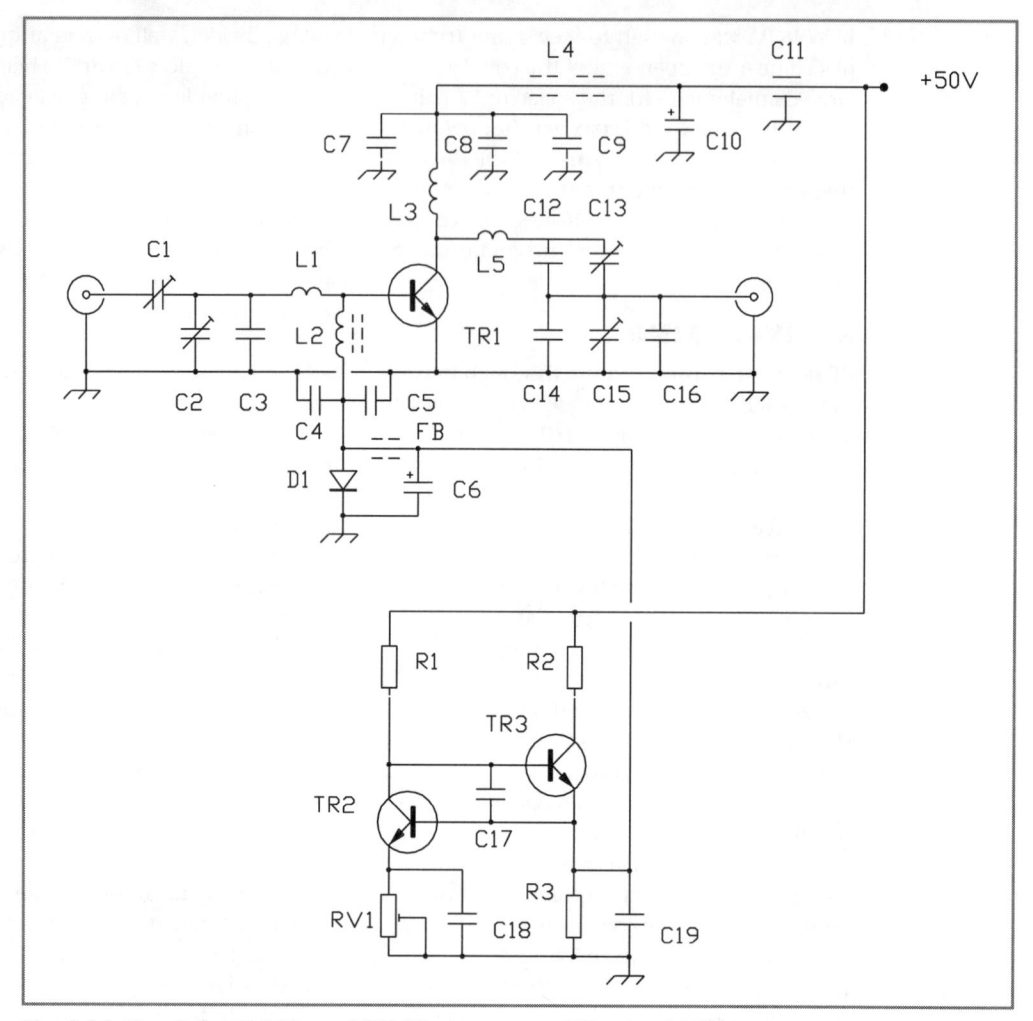

Fig. 9.36. Circuit for 50MHz and 70MHz power amplifiers by G3XBY

The simple and reliable base-bias circuit used by both G3WZT and G3XBY (Fig. 9.36) originated with Mullard Ltd. The circuit uses two LF power transistors in TO-126 or TO-220 packages, which can be mounted in good thermal contact with the heatsink. TR2 is the temperature-sensing transistor, and should be mounted as close as possible to the RF device TR1.

A large tantalum or solid-aluminium (not foil) electrolytic capacitor C6 provides low-frequency bypassing and also acts as a reservoir capacitor to improve the dynamic voltage regulation.

PC BOARDS

When building transistor PAs, the most successful way G3XBY has found for producing one-off PC boards is as shown in Fig. 9.37.

Use standard double-sided glassfibre board, and remove no more copper than the minimum necessary. Lay out the board with all components mounted on the top side, apart from the power transistor itself, and mark out the areas of copper to be removed. Mill these out using a small hobby cutter, or better still a dentist's drill bit. G3XBY uses a drill press with a cross-head vice to move the board beneath

SOLID-STATE POWER AMPLIFIERS

the cutter. You can of course etch the copper instead. Leave plenty of clear insulation around the output end of the collector coil, where high RF voltages are developed.

Mount the board flat on to the bottom of a die-cast box, with the lid acting as a screening cover. Don't even think of using the box itself as a heatsink; cut away both board and box so that the flange of the transistor bolts directly to the heatsink. If the heatsink has one flat side which is bolted to the base of the box, the thickness of the box plus that of the board is just right for a standard flange-mounting RF power transistor (see below).

Pay particular attention to the grounding arrangements close to the power transistor. To get the expected power out of the device, good emitter grounding is crucial. The scheme in Fig. 9.37 works well; G3XBY has used it in seven amplifiers in the 50-150W range, and every one has produced the expected gain within 1dB.

MOUNTING POWER TRANSISTORS

Mount the power transistor(s) directly to a heatsink of the correct rating [3], with the recommended mounting hardware. For the most efficient heat transfer, make sure that both mating surfaces are flat. Remove any high spots from the heatsink and the power transistor flange using wet-and-dry paper on a flat supporting block. Apply only a very thin film of heatsink compound, just enough to fill any irregularities – and use the proper stuff, not ordinary grease! Finally, apply the correct torque to the mounting screws. If you don't have a torque screwdriver, tighten the screws 'moderately', but not too hard. The screws should be merely holding the transistor in place; if the mating surfaces are properly flat, it shouldn't be necessary to pull the transistor down on to the heatsink.

Make sure that the depth between the PC board and the heatsink is not too great (see above). Otherwise you'll split the top off the ceramic package as you tighten down the device. This could also happen if you solder the device into place before tightening the mounting screws. A broken power transistor is not only expensive but also dangerous – all RF power transistors contain beryllium oxide, which is toxic if ingested, inhaled or rubbed into cuts. In case of problems, DO NOT blow out the dust! Mop it up with a damp cloth, and dispose of the cloth and the device in an industrial-waste skip. A useful tip to aid the removal of transistors without breaking them open is to bend up one corner of each tab before soldering, so that you can use pliers to peel it back if necessary.

POWER SUPPLIES AND CONTROL CIRCUITS

Always use regulated power supplies. If the voltage sags on peak current demand, you will

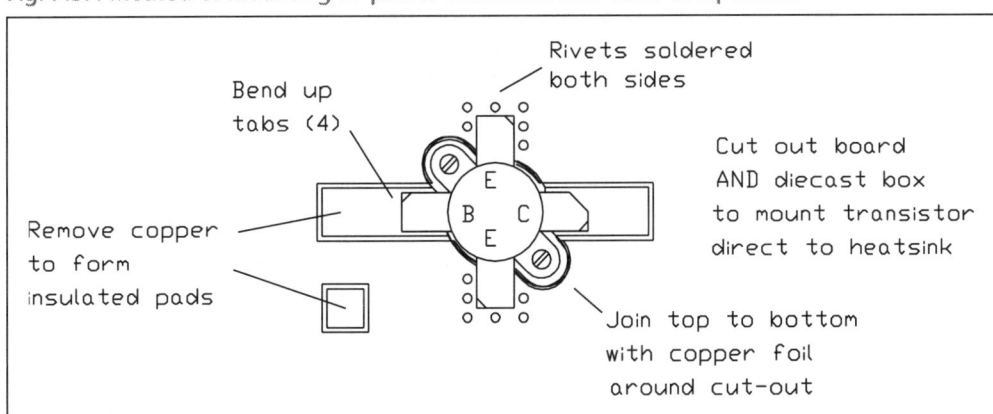

Fig. 9.37. Method of mounting RF power transistors and other components

lose power and the linearity of the amplifier will be limited. An overvoltage 'crowbar' circuit is a very worthwhile feature in the power supply. The BLW96, for example, has a V_{CE} rating of 110V (DC+RF), which means that it will run quite safely from a 50V supply, but anything above 55V is likely to blow the device when RF drive is applied.

Control circuits should take care of the transmit/receive sequencing and also protect the transistor against excessive VSWR. The articles by G3WZT [1,2] contain some very good ideas about power supplies and control circuits for solid-state power amplifiers.

IMPEDANCE MATCHING

Tune everything up in a 50Ω environment. For intermediate amplifiers, pay careful attention to the input matching of the succeeding stage; and for the power amplifier, make sure that the antenna is well-matched. Otherwise, all the tuning and matching adjustments will interact, your power meters will not read correctly and you can never understand what is going on. If you are using combined amplifiers, matching errors will cause unequal power sharing between the two amplifiers. So check the VSWR into and out of every stage; if it's worse than about 1·4, do something about it.

POWER AMPLIFIERS FOR 50MHz AND 70MHz

These two amplifiers are each capable of over 100W PEP output. They use a 50V transistor specifically designed for linear operation, the Philips BLW96. Fig. 9.36 showed the circuit, which is essentially the same for both bands. Note that this is only a bare circuit, presented to give you a few ideas. Having invested a large sum in the RF power transistor, you are expected to look after it by providing switching and VSWR protection as described above. The coil dimensions may require a little experimentation, because the values of these small inductances depend very much on the arrangement and lengths of their leads. Similarly it may be necessary to add or remove 47pF chip capacitors to achieve the correct

COMPONENTS FOR 50MHz AND 70MHz POWER AMPLIFIERS

RESISTORS
15R	1	RV1: Multi-turn trimpot
68R	1	R3: 0·5W
100R	1	R2: 20W
5k6	1	R1: 1W

CAPACITORS
47p	4	C3, 12, 14, 16: 300V transmitting chip capacitors, ATC or Waycom
57p	4	C1, 2, 13, 15: 300V foil trimmers, Philips type 200-8090-8003
1n	3	C4, 7, 8: ceramic chip
1n	1	C11: ceramic feedthrough
100n	5	C5, 9, 17, 18, 19: polycarbonate
100μ	1	C10: 64V tantalum electrolytic
220μ	1	C6: 6·3V tantalum electrolytic

INDUCTORS
L1	50MHz	2 turns
	70MHz	1·5 turns
		6mm inside diameter
		18SWG tinned copper wire
L2		6 turns through Philips FX1898 six-hole ferrite bead
L3	50MHz	8 turns, 12mm inside diameter, 18SWG enamelled
	70MHz	5 turns, copper wire, close wound
L4		18SWG wire through 2 ferrite beads
L5	50MHz	4 turns
	70MHz	3 turns
		12mm inside diameter, 12SWG copper or 0·141" solid coax, preferably silver-plated, turns spaced 1 wire diameter

SEMICONDUCTORS
BLW96	1	TR1: Philips (Gothic Crellon Ltd)
BD135	1	TR2
BD228	1	TR3
1N4007	1	D1

impedance transformations. In other words, high-power solid-state amplifiers at these frequencies are still an area for the more advanced experimenter.

ANTENNAS FOR 50MHz AND 70MHz

Since even a minimum-sized DL6WU long Yagi is very long indeed at a wavelength of 6m or 4m, this chapter gives some designs for shorter Yagis.

The mechanical construction adopted for 50MHz and 70MHz Yagis depends largely on the weather conditions they are likely to have to experience. For portable use, mostly in summer, a very lightweight form of construction can be used, but building a Yagi to survive several years of Scottish winters is quite another matter! The Yagis described here use 12·7mm (0·5") diameter elements and 25mm (1") square booms. They are light enough for portable use, but should also stand up to the rigours of winter in most of the UK. In parts of the world which are subject to ice storms the construction might need to be strengthened to withstand thick layers of heavy ice.

TUNING THE ANTENNA

For the best DX performance, you will always need to check the performance of your antenna in your particular installation. Especially at these longer wavelengths, the antenna is likely to be quite close to ground. It may also have antennas for higher frequencies inside its capture area (Chapter 7), and these too can cause detuning.

For any antenna, the effective electrical length of the elements depends on the element diameter and the method of mounting on the boom. As explained in Chapter 7, these effects can cause the antenna to be completely off-frequency. This leads to incorrect polar patterns, poor sidelobe performance and probably reduced gain, the gain falling off more rapidly on the HF side of optimum. Electrically short Yagis can be very frequency-sensitive, unlike their longer counterparts used on the higher bands. Since 50MHz and 70MHz DX operation takes place in narrow frequency segments, an error of less than 1% in element lengths can shift the optimum performance right outside of the DX sub-band. And a 1% error in the length of a Yagi element represents only 30mm at 50MHz, or 20mm at 70MHz.

It is a common misconception that an antenna must be 'resonant' or 'on frequency' if the VSWR is good. This is *untrue*. Even if the antenna has been adjusted to a perfect 50Ω match it can still be hopelessly mistuned for pattern and gain, as is regrettably demonstrated by many commercial Yagis. The *only* way to get the antenna working properly on the frequency you want is to alter *all* the element lengths. When you do this, the VSWR is bound to become worse until you have also re-adjusted the matching, but you shouldn't regard that as a problem.

The easiest way to check the performance is to compare the rearward polar pattern with its predicted shape. The rearward lobes of all kinds of electrically short Yagis change with frequency in a very similar way, as shown in the seven polar plots comprising Fig. 9.38a-g. The first thing to note is that although the plots cover a span of only ±1·5% in frequency, the patterns change dramatically within that range, so they give a very good indication of how the antenna is tuned.

When the antenna is operated far below its optimum frequency, the front/back ratio is poor (Fig. 9.38a). As the frequency increases, the backlobe begins to split (Fig. 9.38b-c) and there is generally one frequency at which the

50 & 70MHz

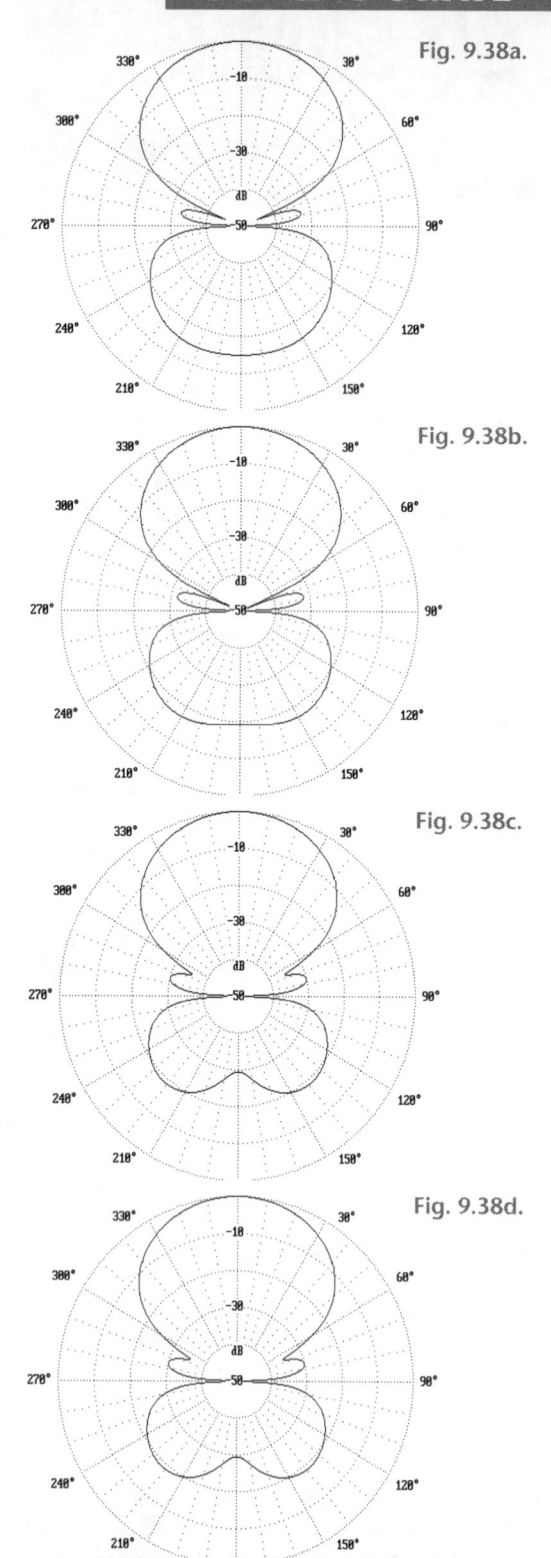

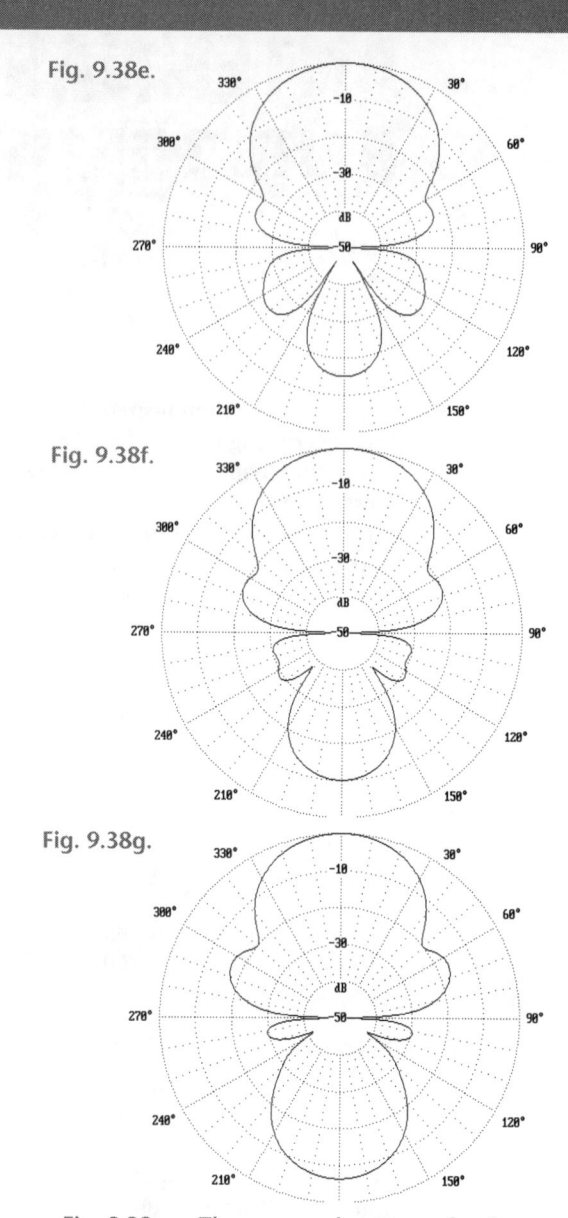

Fig. 9.38a.

Fig. 9.38b.

Fig. 9.38c.

Fig. 9.38d.

Fig. 9.38e.

Fig. 9.38f.

Fig. 9.38g.

Fig. 9.38a-g. The rearward pattern of a short Yagi changes rapidly with less than ±1% change in frequency. The Yagi can be tuned to place the optimum patterns for interference rejection (c-d-e) in the DX segment of the band.

Fig.	MHz	Ratio	dBd
9.38a	49.35	−1.5%	7.79
9.38b	49.60	−1.0%	8.02
9.38c	49.85	−0.5%	8.38
9.38d	50.10	optimum	8.61
9.38e	50.35	+0.5%	8.87
9.38f	50.60	+1.0%	8.85
9.38g	50.85	+1.5%	8.21

Results calculated by NEC program

50MHz & 70MHz ANTENNAS

Fig. 9.39. Computed pattern of the 8-element 70MHz Yagi

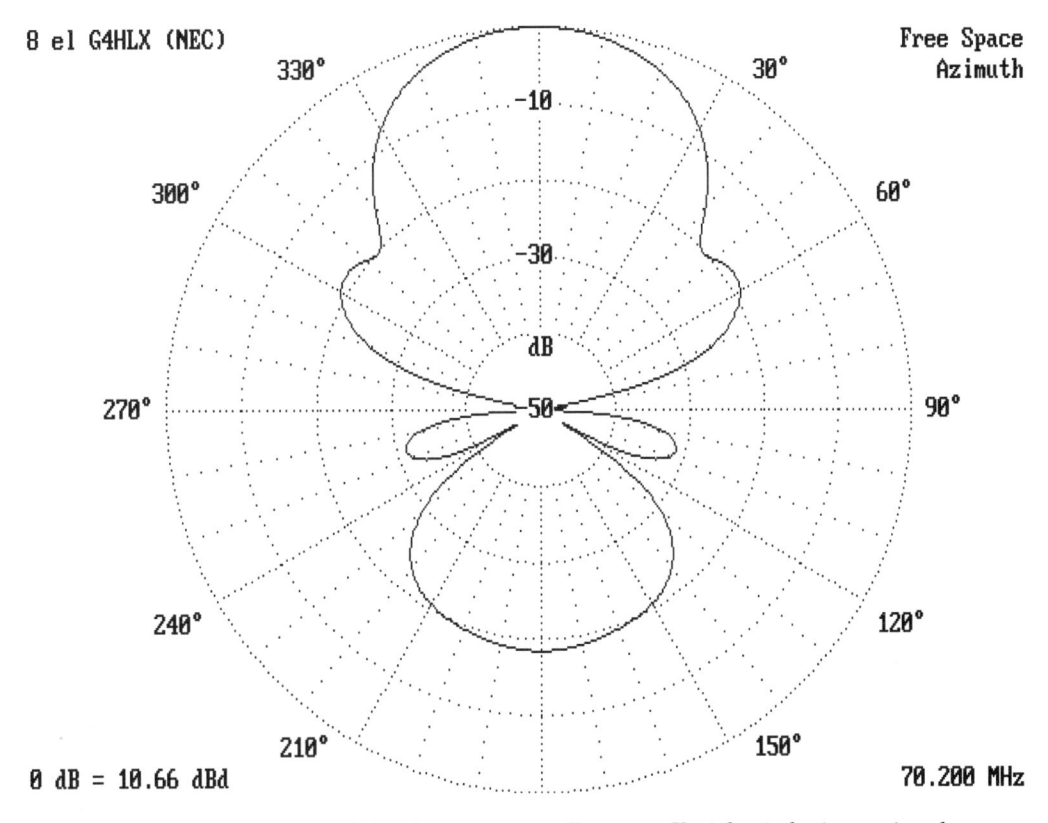

8 el G4HLX (NEC)   Free Space Azimuth
0 dB = 10.66 dBd   70.200 MHz

split becomes deep and the front/back ratio is very high. A small further increase in frequency produces a new 'HF' rear lobe (Fig. 9.38d) which rapidly grows outwards at 180° (Figs 9.38e-f-g), while the remains of the old 'LF' backlobe wither away. Meanwhile the forward gain has been gradually increasing with frequency, although it will eventually begin to drop very sharply. If you go even further HF, the first director will begin to act as a reflector, and the antenna beams backwards!

The true 'optimum frequency' is debatable. Optimizing for the best pattern will generally involve the sacrifice of a few tenths of a dB of forward gain, but will greatly improve the signal-to-interference ratio. The pattern of Fig. 9.38d gives quite good suppression of unwanted signals from the rearward direction in general, and also provides a number of deep nulls which can be very useful in eliminating QRM. In fact, any of the three patterns in Fig. 9.38c-d-e would be pretty good in practice, and between them they will nicely cover the DX portion of any VHF amateur band.

To tune a Yagi, begin by inspecting the rearward pattern using test signals from another amateur at various points across the band. Having located the optimum frequency range, adjust all element lengths in the same proportion to bring the optimum pattern into your chosen part of the band. Ignore the VSWR until you've got the pattern right, and then finally adjust the matching if necessary.

FOUR- AND EIGHT-ELEMENT YAGIS

The table below shows the dimensions for a 4-element Yagi for 50MHz and 4- and 8-element designs for 70MHz. The 4-element design (Fig. 9.40) has about the highest gain obtainable from that number of elements, and is quite unusually long. It is taken from the excellent book on Yagis by W2PV [4], and the patterns in Fig. 9.38 were computed for this particular design. At the frequency of optimum pattern (50·1MHz, Fig. 9.38d), the predicted forward gain is a very respectable 8·6dBd, which is significantly better than for shorter 4- or 5-

9 - 41

50 & 70MHz

Fig. 9.40. The 4-element Yagi for 50MHz. Sitting above the 432MHz beam is a plastic Little Owl (*Polythene Noctua*) to keep the starlings away!

element designs.

The 8-element Yagi by G4HLX is based on designs by DL6WU, and has a forward gain of about 10·6dBd. The computed pattern at its optimum frequency is shown in Fig. 9.39. Dimensions are given for 70MHz only, though the design can of course be scaled up to 50MHz if you're really ambitious!

Figs 9.41-9.43 show constructional details, which are essentially the same for both bands. The clamps for a 25·4mm (1·00") square boom are commercially available (*eg* from Sandpiper Communications). Each element is made from two lengths of 12·7mm (0·50") OD aluminium

Fig. 9.41. Element construction, mounting method and gamma match for the 50MHz and 70MHz Yagis

50MHz & 70MHz ANTENNAS

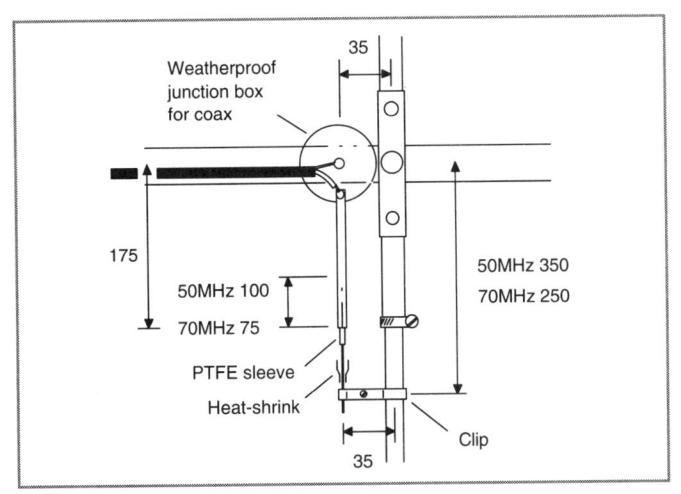

Fig. 9.42. Approximate dimensions (mm) of gamma matches for 50MHz and 70MHz, using the constructional methods described in the text. Adjust for minimum VSWR as shown in Fig. 9.43, and optionally cut off excess rod beyond the clip

tubing, telescoped into a 380mm (15") centre section of 15·9mm (0·625") tubing for strength. Four slits are cut in the ends of the centre section with a hacksaw, and the joints are secured with hose clamps. This permits easy adjustment to frequency. The 8-element design is not suitable for heavy weather without extra reinforcement to stiffen the centre section against side-winds, plus supporting lines from each end of the boom up to an extension of the mast.

Gamma matches have been used for all these designs, and if constructed with care

Fig. 9.43. Adjusting the VSWR of the 4-element Yagi for 50MHz, with the antenna pointing straight up. No further adjustment was needed after the antenna was installed.

YAGIS FOR 50MHz AND 70MHz

	50MHz 4-element		70MHz 4-element		70MHz 8-element	
	Spacing	Length	Spacing	Length	Spacing	Length
Reflector	0	2932	0	2090	0	2095
Driven	1493	2860	1066	2055	920	2000
D1	2986	2736	2130	1940	1337	1930
D2	4479	2736	3196	1940	2113	1910
D3					3033	1895
D4					4098	1880
D5					5293	1870
D6					6573	1850

All dimensions are given in mm, and the element lengths apply only to the method of construction described.

there should be no noticeable distortion of the polar pattern. Fig. 9.41 shows the general arrangement, and Fig. 9.42 gives approximate dimensions for the two bands. The spacing between the rod and the driven element is important, as also is the ratio of their diameters, so the details in Fig. 9.42 apply only to a $1/_8$" diameter gamma rod and the particular method of element construction described above. To provide the necessary series capacitance, the gamma rod is telescoped inside a ¼" diameter tube about 175mm (7") long, the tube being lined with PTFE sleeving which acts as a dielectric and provides a sliding fit. Coaxial cable is connected to the gamma rod in a weatherproof junction box, and the outer of the cable is securely connected to the centre-line of the boom, as close as possible to the driven-element mounting.

The position of the shorting clip and the length of rod inside the tube are adjusted for the lowest VSWR, preferably with the antenna pointing straight up (Fig. 9.43). After adjustment the joint between the tube and the gamma rod is waterproofed with heat-shrink sleeving. The 8-element Yagi has a feed impedance close to 50Ω, so as an alternative to the gamma match the driven element could be centre-fed *via* a 1:1 balun, or made into a folded dipole and fed *via* a 4:1 balun.

The 4-element design has been tested at 50MHz, and the 8-element at 70MHz. All the Yagis should be on-frequency if built exactly as described, though in some situations the element lengths may require further correction as described above.

REFERENCES FOR CHAPTER 9

[1] John Matthews G3WZT, A single-stage linear amplifier for 50MHz. *Radio Communication*, June 1986

[2] John Matthews G3WZT, Guidelines for the design of semiconductor VHF power amplifiers. *Radio Communication*, September and December 1988

[3] In Practice: Heatsinks. *Radio Communication*, October 1989

[4] James L Lawson W2PV, *Yagi Antenna Design*. ARRL (1986) ISBN 0-87259-041-0

CHAPTER 10

432MHz

A DXER'S TRANSVERTER FOR 432MHz

by John Wilkinson G4HGT

I began thinking about designing a new 432MHz transverter during the construction of a K2RIW legal-limit amplifier. Until then my 10W-output, 2dB-noise-figure system had worked well enough, but I had noticed a few problems with local strong signals, and a preamplifier to match my increased transmitter power would only have made things worse. So a new transverter was a serious requirement. The design started life as a true home-brew project, unconstrained by the rigours of publication. However, at some stage I became a contributor to this book...

My main concern in passing on this design has been about its reproducibility. The majority of amateurs involved in UHF construction do not have access to professional (or probably even adequate) testgear, so subjective opinions frequently take over. The best way to avoid problems of reproducibility and lack of testgear is to put some quality into the design itself. The more areas which can be closely and reliably specified, the better the chances of the whole thing working well for other builders. This approach costs a little more because you're paying the component manufacturers to worry about their tolerances, but it greatly reduces the number of things

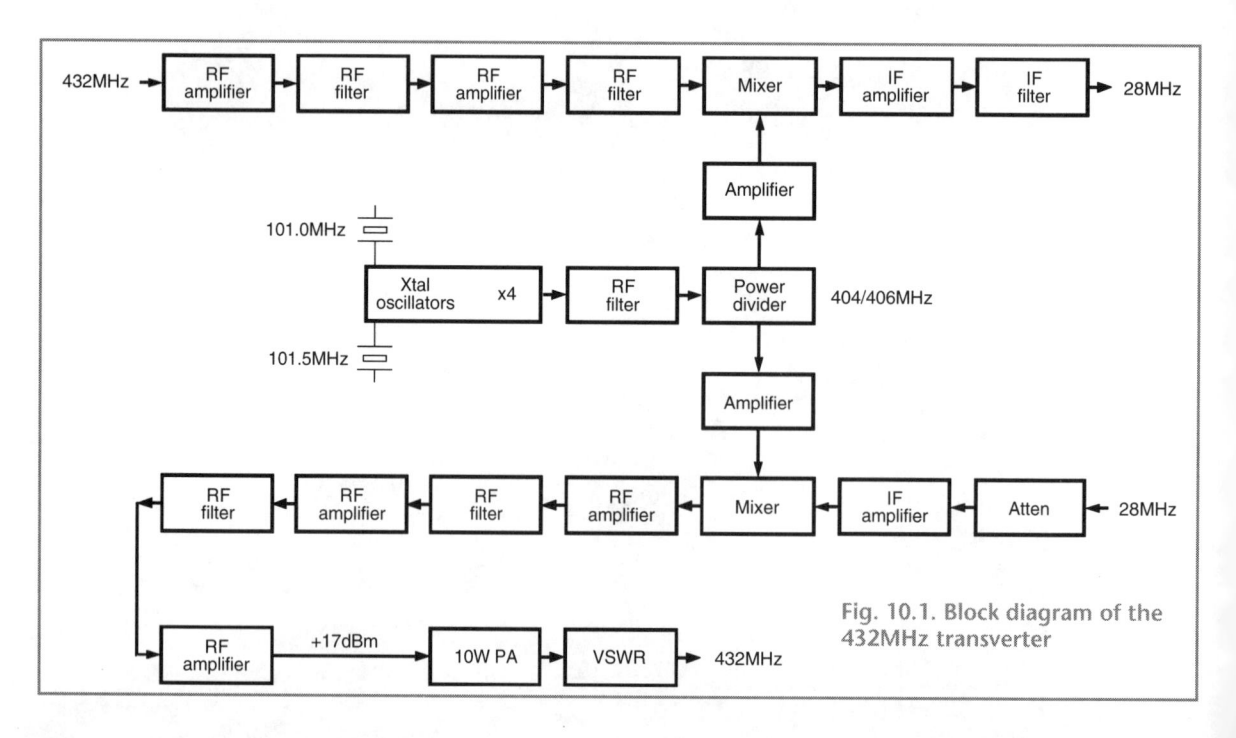

Fig. 10.1. Block diagram of the 432MHz transverter

DXER'S 432MHz TRANSVERTER

that you yourself need to optimize or adjust.

Even so, a high-performance 28–432MHz transverter is no project for a novice. In spite of all my attempts to make the design tolerant and reliable, there are still a few tricky areas, so my description concentrates on these. I have included only the basic outlines of interfacing, control circuitry and mechanical construction. If you feel confident enough to build this transverter from my description, perhaps taking a few hints from G4DDK's *Suffolk* 144MHz transverter design in Chapter 8, wiring up the LEDs and drilling a few holes should be no problem to you.

For alignment, a multimeter and milliwattmeter (Chapter 12) are required, together with a wavemeter or frequency counter. A stepped attenuator is also useful, though not essential. In case everything does not go smoothly, you also need to be reasonably confident about fault-finding using the aforesaid test equipment. The modular design approach will help in determining any problems quickly.

DESIGN CONSIDERATIONS

I began the design exercise with a survey of existing transverter and preamplifier circuits. Surprisingly, there aren't that many [1-7] and only a couple of them are actually worth serious consideration. Also, it is very suspicious to see claims of system noise figures comparable with the NF of the front-end device itself, or of power-amplifier gains which are higher than those in the device manufacturer's published data! I have therefore taken a pessimistic approach to noise figures and stage gains, in the hope that anyone else building these circuits will obtain results that will definitely meet my design specification and hopefully exceed it.

What do you want from this design? For DX-chasing, low noise figure is important, although with modern devices this is not difficult to achieve. At 432MHz a system noise figure of 1·8dB is a reasonable design target (see Chapter 4). For contest operation a high dynamic range is essential, even at the expense of some sensitivity. At UHF, diode-ring mixers are the simplest and best way of achieving the target of 90dB for the spurious-free dynamic range (Chapter 5). Finally, excellent filtering of all out-of-band receiver responses is needed to keep out the image response at 376MHz, direct breakthrough of 28MHz, and every other frequency except for the desired 432–436MHz. Commercial helical filters have been used for both the signal frequency and the 404/406MHz local oscillator. The problems in developing home-made UHF bandpass filters, not to mention describing how you could duplicate them, were just too numerous to contemplate.

On the transmit side, output levels of 3W, 10W and 25W PEP are the common standards for commercial gear. The problems of the power-amplifier stage have largely been overcome by the availability of hybrid modules by Mitsubishi. Two variants are available, with linear output powers of 10W and 25W PEP, and either could be used here. Their only limitation is their intermodulation distortion, which is adequate but not outstanding (Chapter 8). Most serious DXers will actually choose the lower-power 10W module because it is more than adequate to drive a K2RIW amplifier in Class AB1.

This transverter could form part of at least four different types of 432MHz system, intended for:

1 Terrestrial DX from home station (good all-round performance).
2 Contests and DXpeditions (high quality throughout).
3 Satellite working (435–436MHz coverage required, and also good receiver sensitivity).
4 EME (ultimate high performance, especially in receiver sensitivity).

Receiver sensitivity is important in all four cases, and the best way to achieve it is to use a preamplifier mounted close to the antenna; this will also relax the sensitivity requirements for the transverter. But there are many amateurs who either do not have easy access to the antenna or prefer to have all the electronics indoors. For them, the transverter itself must have a low noise figure.

If you subsequently add a masthead preamp

432MHz

to a sensitive transverter which already has quite a lot of front-end gain, you are likely to find problems with strong signals. The logical solution is to have the ability to bypass the first RF stage in the transverter if desired, so that if the preamplifier fails you can get back on the air with a sensitive system before the end of the opening or the contest. Along the same lines, I have kept coaxial relays out of the transverter design and simply provided the necessary switched DC supplies. Some operators will want the coax relays at the masthead, while others prefer to mount them on the power amplifier or even on the shack wall – it's your choice.

The overall block diagram is shown in Fig.

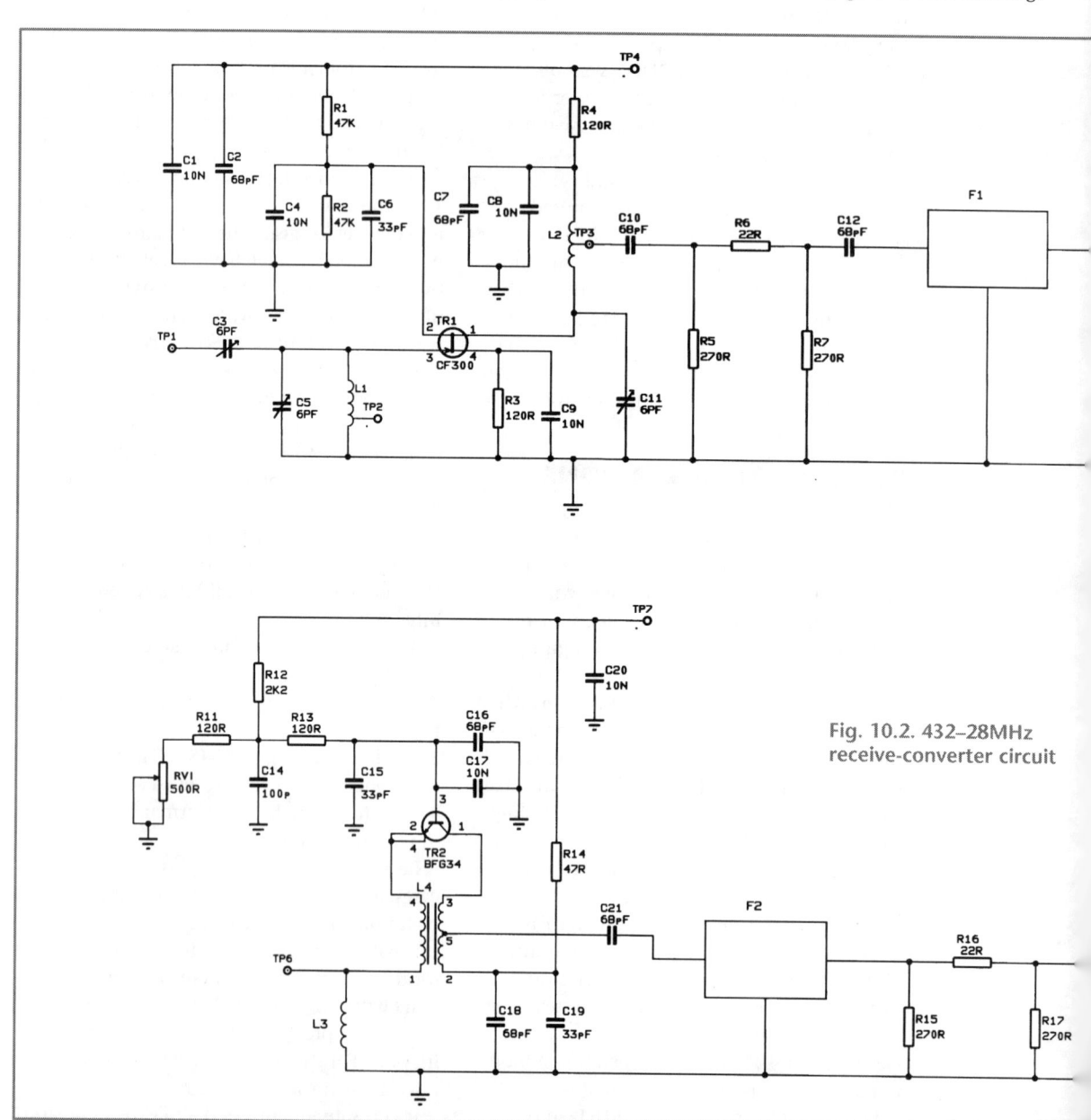

Fig. 10.2. 432–28MHz receive-converter circuit

DXER'S 432MHz TRANSVERTER

10.1, and the modular approach is taken one step further by dividing the entire transverter into functional blocks with 50Ω interfaces. A lot of constructional articles recommend building the whole thing and then 'smoke testing', but I prefer to build and test each stage before proceeding to the next. This makes it easier to deal with any problems as they arise. It's also very encouraging to see your project showing signs of life within the first few hours after starting construction.

The transverter is constructed on four double-sided fibreglass printed-circuit boards –
1. Receive converter
2. Local oscillator
3. Transmit converter, with TX/RX switching

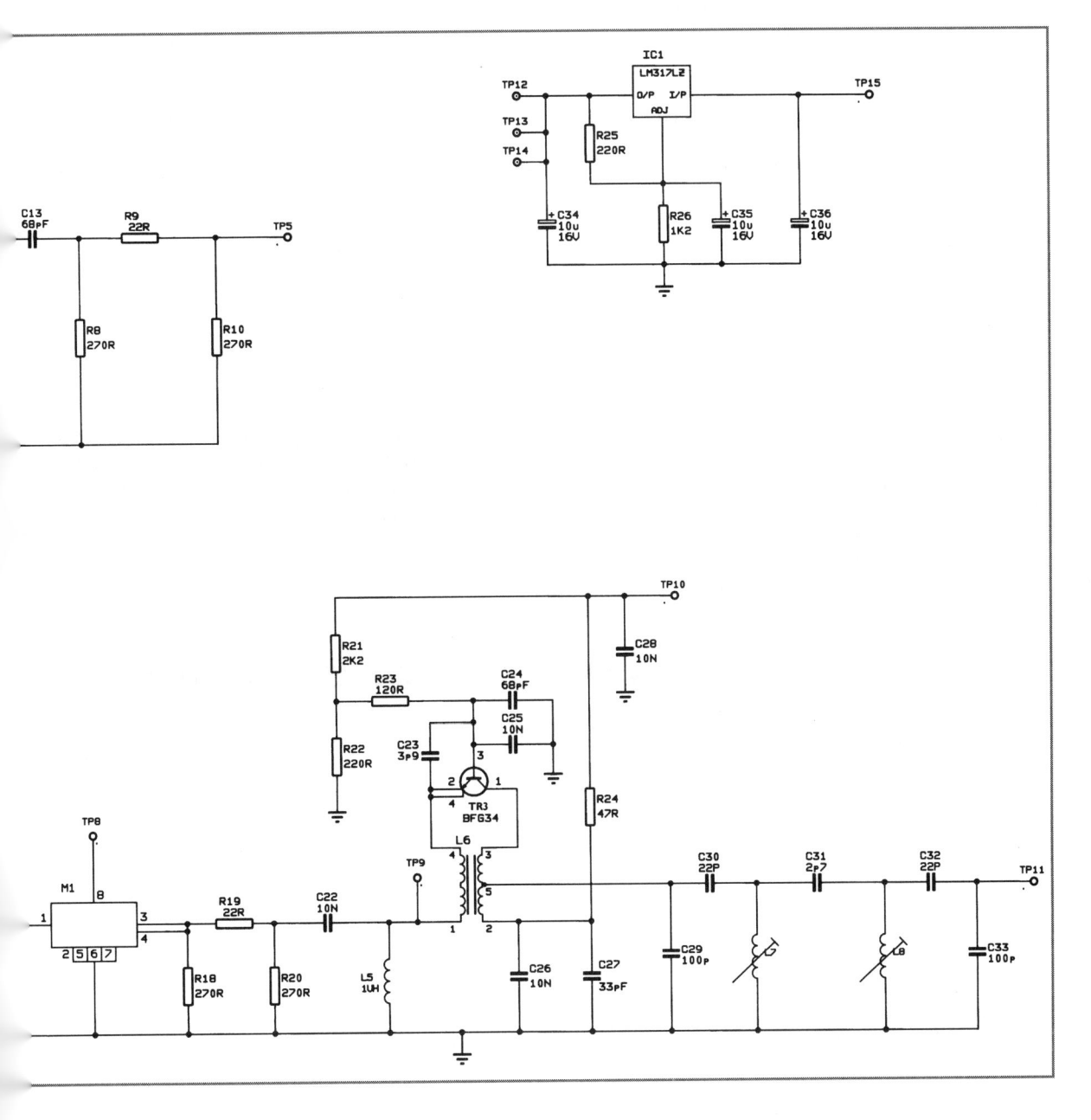

432MHz

and ALC provision

4 Power amplifier and power/VSWR meter.

All the boards have been designed to fit into commercial tin-plate screening boxes if required. A BICC-Vero 'G-line' case houses the whole unit; its size was mainly determined by the dimensions of the rear-mounted heatsink for the power amplifier, so the case looks rather empty inside. Even so, I feel that the extra space and ease of access far outweigh the compactness of a smaller case. All connections have been made to the rear panel of the unit, with the front panel reserved for switches and LED indicators.

Throughout this design, the aim has been for repeatable performance, keeping well within the specifications of the component manufacturers. If you want it to work as specified, you **must** buy full-spec components from a reputable dealer – surplus components are generally cheap for good reasons. I have avoided rare components; the passive components are generally available from a variety of dealers, and most of the semiconductors from at least two sources. You can sometimes make substitutions if you know exactly what you're doing – but if your substitutions don't work, you're on your own!

RECEIVE CONVERTER

The first band for which the majority of radio amateurs acquire equipment is 144MHz. Some move on to the HF and lower VHF bands, while others aspire upwards to 432MHz. The obvious temptation is therefore to buy a 144 to 432MHz transverter. But the quality of commercial 144MHz transceivers is still not really outstanding, especially in the strong-signal department (Chapter 5), and adding a UHF transverter would result in a system completely limited by the performance of the 144MHz rig. Transverting directly between 144MHz and 432MHz also leads to severe problems with mixer intermodulation products and spurious responses appearing in-band, not to mention direct leakage of 144MHz signals into and out of the main transceiver, so the 144-432MHz idea was swiftly abandoned.

The performance of modern HF rigs is much more conducive to using transverters in front of them. Chapters 4, 5 and 8 outline the arguments for this approach, and I assumed the same high-quality HF receiver performance in my noise/intermodulation analysis for this transverter design. After examining several system block diagrams with the aid of the TCALC program (Chapter 5), I settled on the design whose circuit is shown in Fig. 10.2. The following table shows the parameters used in the TCALC analysis, and the predicted results. Note the frequent use of 1dB attenuators to stabilize the circuit impedance close to 50Ω, ensuring correct operation of the filters,

TCALC ANALYSIS OF 432MHz TRANSVERTER

STAGE PARAMETERS		SYSTEM RESULTS			
	T (K)	NF (dB)	Gain (dB)	Level (dB)	IPi_3 (dB)
Rest of system IP_3 = +20dBm	2610.0	10.00	0.0	18.5	+1.5
28MHz filter	3806.4	11.50	-1.5	20.0	
BFG34 Norton amplifier	612.8	4.93	7.0	11.5	+12.0
1dB attenuator	846.5	5.93	6.0	12.5	
SBL-1 ring mixer	4645.7	12.31	-0.5	19.0	-3.0
1dB attenuator	5923.7	13.31	-1.5	20.0	
432MHz filter	12107.9	16.31	-4.5	23.0	
BFG34 Norton amplifier	1879.9	8.74	4.0	14.5	+9.0
1dB attenuator	2441.7	9.74	3.0	15.5	
432MHz filter	5160.6	12.74	0.0	18.5	
1dB attenuator	6571.8	13.74	-1.0	19.5	
CF300 RF amplifier	148.8	1.80	18.5	0.0	+1.0
WHOLE RECEIVER					-5.7
Noise floor (2.5kHz) = -142.9 dBm. Dynamic range = 91.5 dB					

DXER'S 432MHz TRANSVERTER

NORTON FEEDBACK AMPLIFIERS

Application of negative feedback around an amplifier is a very powerful technique for ensuring repeatable results in spite of variations between active devices. The purpose of feedback is to set the gain, input and output impedance of the amplifier, regardless of any gain variations in the active device. Resistive feedback is widely used for this purpose, for example in MMICs where the necessary components are integrated on the chip. But resistive feedback has the disadvantage of degrading the noise figure; hence the mediocre noise figures of MMICs. A technique which does not suffer from this problem is 'lossless' negative feedback using a transformer, first described and patented by Norton [10] and further developed for amateur VHF/UHF applications by DJ7VY [6]. This technique produces controlled gain, low noise figure and high dynamic range, all at the same time.

As explained earlier (appendix to Chapter 5), the maximum output power P_O from a Class-A amplifier is a function of both collector load impedance R_L and collector current I_C, where –

$$P_O max = I_C^2 / 2R_L$$

A tapped transformer at the collector provides a suitably high load impedance, and also acts as the primary winding of a feedback transformer (Fig. A).

The transistor is operated in common base. To analyse the circuit it is necessary to assume a very low input impedance, very high output impedance and unity current gain. In order to match both input and output to the same circuit impedance Z_O, the transformer turns ratio (Fig. A) must then be:

$$n = m^2 - m - 1$$

for a single-turn feedback winding (f = 1). The resulting power gain will then be m^2, and the load impedance at the collector is (m x n) times Z_O.

The only practical solutions involve small whole numbers of turns; for m = 2, 3 and 4, theoretical gains of 6, 9.5 and 12dB result. In practical terms, some departure from theoretical performance must be expected and incorporated in the system calculations. Some losses are also unavoidable, especially at UHF, so the 432MHz transverter design assumes that the nominal 9.5dB amplifiers used in this design actually deliver 8.5dB of gain.

When constructing a Norton amplifier, glue the tiny core of the feedback transformer to the PC board before you attempt to put on the windings. The board is *far* easier to handle than the bare core – especially when you drop it amongst the mess on the shack floor! Fig. B shows details of the windings. For clarity, only one turn through the core is shown for each winding in Fig. B; for the amplifiers in this chapter, the actual numbers of turns are m = 3, n = 5 and f = 1. The 432MHz and 28MHz amplifiers use identical transformers.

To wind the transformers, use thin 'self-fluxing' enamelled copper wire, and begin by twisting the middle of the wire to make the tapping point. Wind the 'm' and 'n' windings in opposite directions away from the tap, making sure that the twisted wires are soldered together right up to their junction at the core. Then insert a hairpin loop to make the single-turn 'f' winding. Finally, make absolutely certain that you have identified the connections exactly as shown in Fig. B.

At 432MHz it is vital to use only the recommended components and layout. In particular the two-hole ferrite cores suggested by DJ7VY are essential. My main area of experimentation while building these amplifiers for 432MHz and 28MHz was in the type and value of the decoupling capacitor at the base of the transistor. This seems to be most critical for obtaining the best gain. DJ7VY's use of leadless discs is good practice, though after trying a variety of capacitors I settled for two disc ceramics in parallel. Both capacitors must be of the specified values and type, in order to provide a solid broadband RF ground at the base of the transistor. If the amplifier oscillates, base grounding is almost certainly the problem – unless you've connected the transformer wrongly!

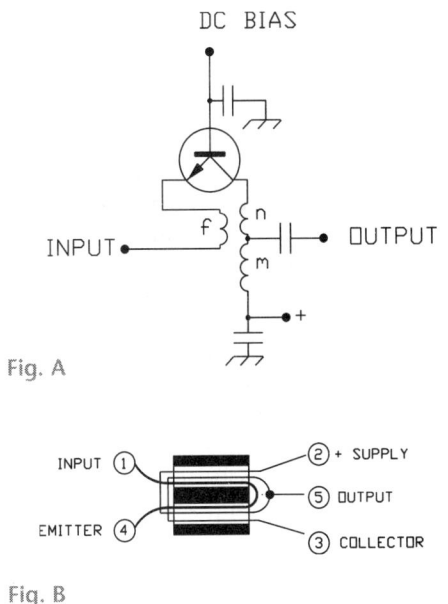

Fig. A

Fig. B

432MHz

mixer and feedback amplifiers. The target noise temperature of 150K is achieved almost exactly, and with a good HF receiver the dynamic range is very respectable.

The first RF stage (Fig. 10.2) is a dual-gate GaAsFET, providing a good balance between performance and cost. Whilst microwave single-gate GaAsFETs have excellent noise performance at UHF, the improved sensitivity is not really necessary unless EME is contemplated. The high associated gains of microwave GaAsFETs at lower frequencies also increase the likelihood of instability – possibly in the gigahertz region – and the high input impedance can also cause problems in achieving a low-loss match [8, 9]. Therefore an entirely sufficient solution for 432MHz terrestrial DXing is the modern dual-gate GaAsFET. As more of these robust devices are designed into consumer products, so the price will fall, making them an even more sensible choice. Virtually all of the common devices available (CF300, MGF1200, S3030, 3SK97 etc) should work well in the design adopted, all giving gains around 16–20dB and device noise figures below 1dB. This low device noise figure can then be traded-off in the interests of improved dynamic range for the system as a whole, as explained in Chapters 4 and 5.

For both input and output circuits, a length of silver-plated wire is used as the inductor. Both ends are bent down to fit through holes in the PC board and the inductor is spaced 5mm above the board. Taps on the inductors are used for matching, allowing optimization of the gain and noise figure. As an alternative, provision has also been made on the PCB for capacitive coupling to the top of the input tuned circuit. A simple tin-plate screen is needed to isolate the two tuned circuits and prevent oscillation. The output of the first RF stage is fed via a resistive pad to a Toko pre-tuned helical filter which provides significant selectivity. In the TCALC runs, the loss value used for these filters has been taken as 3dB – the worst case according to the manufacturer. I have measured the loss as typically 1·7dB, with a 376MHz image rejection of 41dB for a single two-cavity filter.

Fig. 10.3. Full-size PCB layout on the underside of the 432MHz receive-converter board. The top of the board is a continuous copper groundplane, etched or counterbored as shown

Fig. 10.4. Component layout of the 432MHz receive-converter board

DXER'S 432MHz TRANSVERTER

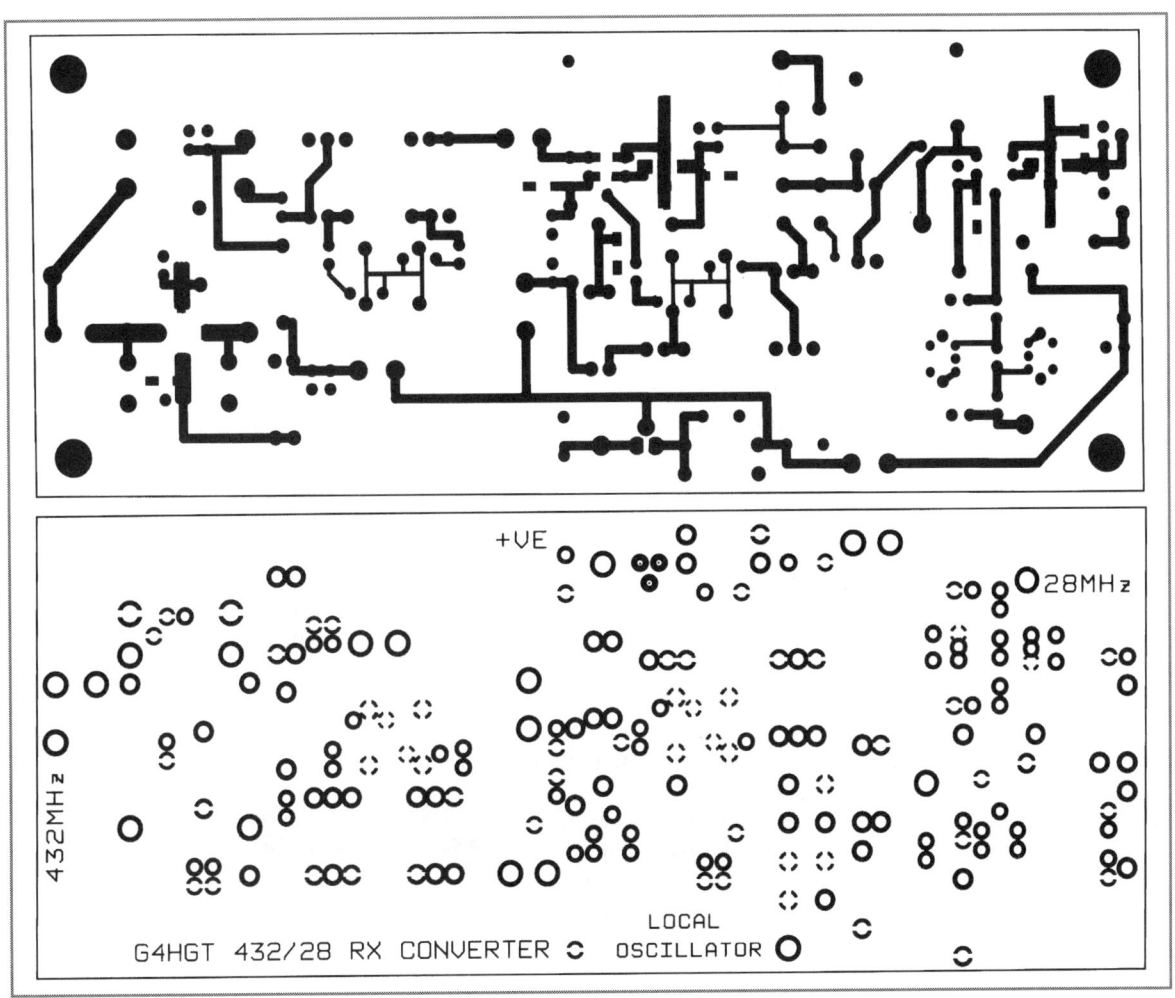

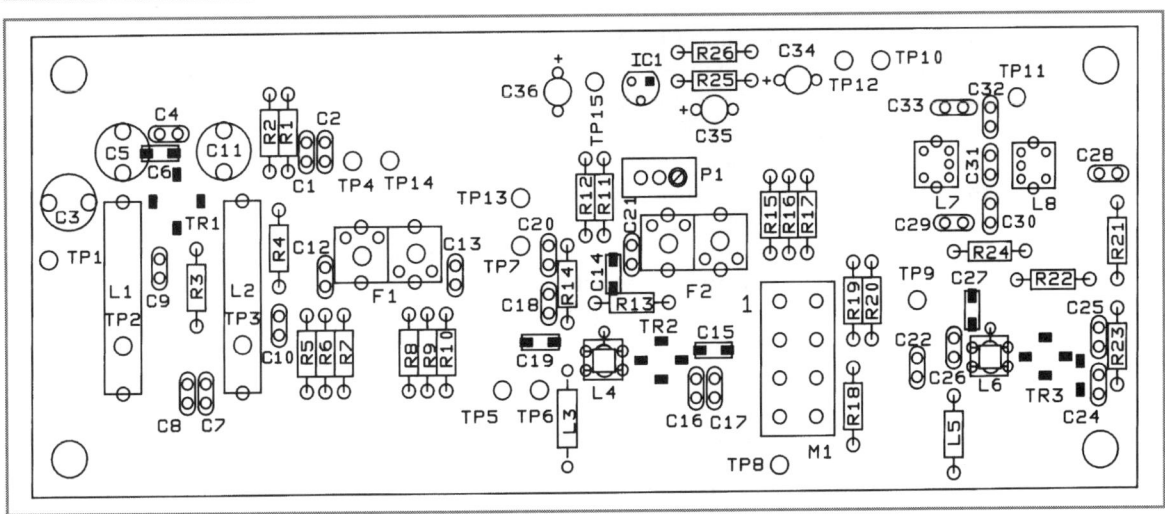

With flexibility in mind, the input to the second RF stage can be fed either from the front-end on the board or directly from the input BNC connector. After the success of MMICs in the transmit converter, I made many attempts using TCALC to design a receiver of equal simplicity. The problem with the available range of MMICs is that they offer either low noise figures with high gain or medium gain with high noise figures, whereas this system needs low noise figure with quite low gain. Therefore the second RF stage uses a BFG34 bipolar transistor in the 'lossless' negative-feedback configuration developed by Norton [10] and made popular in amateur circles by DJ7VY [6] – see the panel on page 10-7. After considerable research I am convinced that this form of amplifier is presently the only one capable of both low noise and good strong-signal handling at 432MHz. The Norton configuration offers only a limited choice of stage gains; the option giving a practical gain of 8·5dB was adopted for the second RF stage because TCALC showed this to give the best balance between system noise figure and overall gain.

The second RF stage is followed by another helical filter which tightens up the out-of-band response yet further. Provision is then made for a resistive attenuator, to improve the matching between the filter and the mixer if desired and to shed any excess RF gain.

Virtually all the commonly available diode-ring mixers will fit the same PCB layout. Interestingly, the expected increase in dynamic range obtained by using higher-level mixers requiring more local oscillator (LO) power is not as significant as might be first thought, since other circuit areas also limit the overall performance. I believe that the standard-level mixer (+7dBm LO injection) will suffice for most applications; even if you can afford a better mixer, you might do better to spend the money on more antennas or a masthead preamp. The dynamic range of the standard mixer is improved by increasing the local oscillator drive to around +10dBm, and the LO board provides sufficient output for that purpose.

Accurate termination of double-balanced mixers has been stressed many times. The simplest method is to feed the output into an attenuator (say 10dB) but this method is wasteful of noise figure, and more elegant methods have to be adopted at 432MHz. By far the *worst* method is to feed the mixer output directly into a high-Q bandpass filter. It might look like 50Ω at the selected frequency (if you're lucky) but it's anyone's guess what sort of impedance will be presented outside the passband. The mixer has to see 50Ω at **all** expected output frequencies, so a wideband termination is essential. In this case, although the desired output is at 28MHz, the input impedance of the termination has to be good to 500MHz and beyond. Again I used a Norton negative-feedback amplifier to meet the system requirements. Whilst providing a good wideband match, an active terminator also has the less desirable property of amplifying everything that comes out of the mixer, so the final element on the receive board is a 28MHz bandpass filter which reduces the level of any residual signals.

CONSTRUCTION AND ALIGNMENT

Fig. 10.3 shows the track layout for the underside of the receive-converter board, together with the etching diagram for the copper groundplane on the top of the board. If you choose to etch the top of the board where insulated lead-throughs are required rather than counterboring the holes by hand, take care to align the two masks correctly.

The receive-converter board is best set up by first installing all the passive components and the filters, following the component layout in Fig. 10.4, and then working backwards from output to input. Remember to solder the grounded leads of components and the screening cans to the groundplane. See the panel on page 10-7 for details of the Norton amplifiers and construction of the feedback transformers. As a matter of fact, winding these two transformers is the trickiest part of the whole project.

To begin the alignment, check that the

voltage regulator is working. Then install TR3, reconnect the supply and check the DC conditions. This stage can be checked at RF by using it as a simple preamp into the 28MHz receiver, the antenna replacing the mixer. If TR3 oscillates, install the small capacitor C23 beneath the board, directly between base and emitter with zero lead length; if obtainable, a chip capacitor is best. If oscillation persists, double-check the windings on the transformer L6 with a powerful magnifying glass, and make sure you have used the correct bypass capacitors with zero lead lengths. The output filter L7-L8 is reasonably broadband and should need little adjustment.

Before you can go any further with aligning the receive converter, you need to have built and aligned the local-oscillator board as described in the next section. Having done that, solder in the receive mixer. As well as soldering the pins beneath the board, carefully solder the corners of the mixer case to the PCB groundplane, working quickly with a large, hot iron. Connect up the local oscillator, and with just a simple antenna connected to the input you may be able to hear local 432MHz signals or the third harmonic of your 144MHz transmitter.

Install TR2 and set potentiometer RV1 to minimum resistance. Connect power to TP7 and set up the correct bias point (I_C = 20mA) using RV1. Signals should be audible with an antenna attached to TP6. Finally install TR1. Modern GaAs devices are quite rugged, and fanatical care to protect against static damage is unnecessary. Power up and check the drain current by measuring the voltage across R3. The positions of the taps on the inductors should be acceptably close to the optimum for gain and noise figure. If the tap is used on the input, C3 must be omitted; alternatively, use C3 for capacitive input matching and omit the tap. Final adjustment for optimum noise figure is best carried out using a proper noise generator and NF indicator (Chapter 12), but if these are not available you can optimize the signal/noise ratio of a weak but consistent off-air signal (see Chapter 5 for a method using an FM signal).

I noticed a slight shift in tuning when I moved the board away from the workbench, so it is sensible to do a final tweak when the board is installed in its box. The lead to the input BNC connector should be as short as possible, and coax's braid well soldered to the PCB groundplane without any 'tail'.

LOCAL OSCILLATOR

While much time is spent discussing the merits of receiver front-ends and transmitter linearity, the local oscillator is frequently overlooked in transverter designs. In fact its specification needs to be determined quite carefully since the LO is responsible for frequency accuracy and stability. A local-oscillator signal for 432MHz will normally be derived from an oscillator running at a lower frequency which is then multiplied to the desired harmonic, so any frequency errors and instability will be multiplied too.

The right design of crystal oscillator will ensure excellent stability with temperature and time, as well as the lowest phase-noise characteristics. One or two designs I tried have only worked with the output frequency slightly offset from the value marked on the crystal, and this undesirable offset is multiplied four-fold at 432MHz. G4DDK recommended the Butler oscillator configuration in Fig. 10.5, and I have been very satisfied with the results. At 404MHz the frequency can be shifted ±10kHz without loss of oscillation, and with only a minimal change in output level.

Stability of frequency with supply voltage and load variations is also important. An integrated-circuit voltage regulator provides better voltage stability than the commonly used Zener-diode arrangement, and each oscillator has its own regulator. By using separate buffer amplifiers to feed the transmit and receive mixers, the load presented to the oscillator output is more constant, reducing the occurrence of frequency pulling due to load variations.

Two entirely separate crystal oscillators in the 101MHz region are provided for terrestrial and satellite operation. Selection is made using DC control to permit split-frequency opera-

432MHz

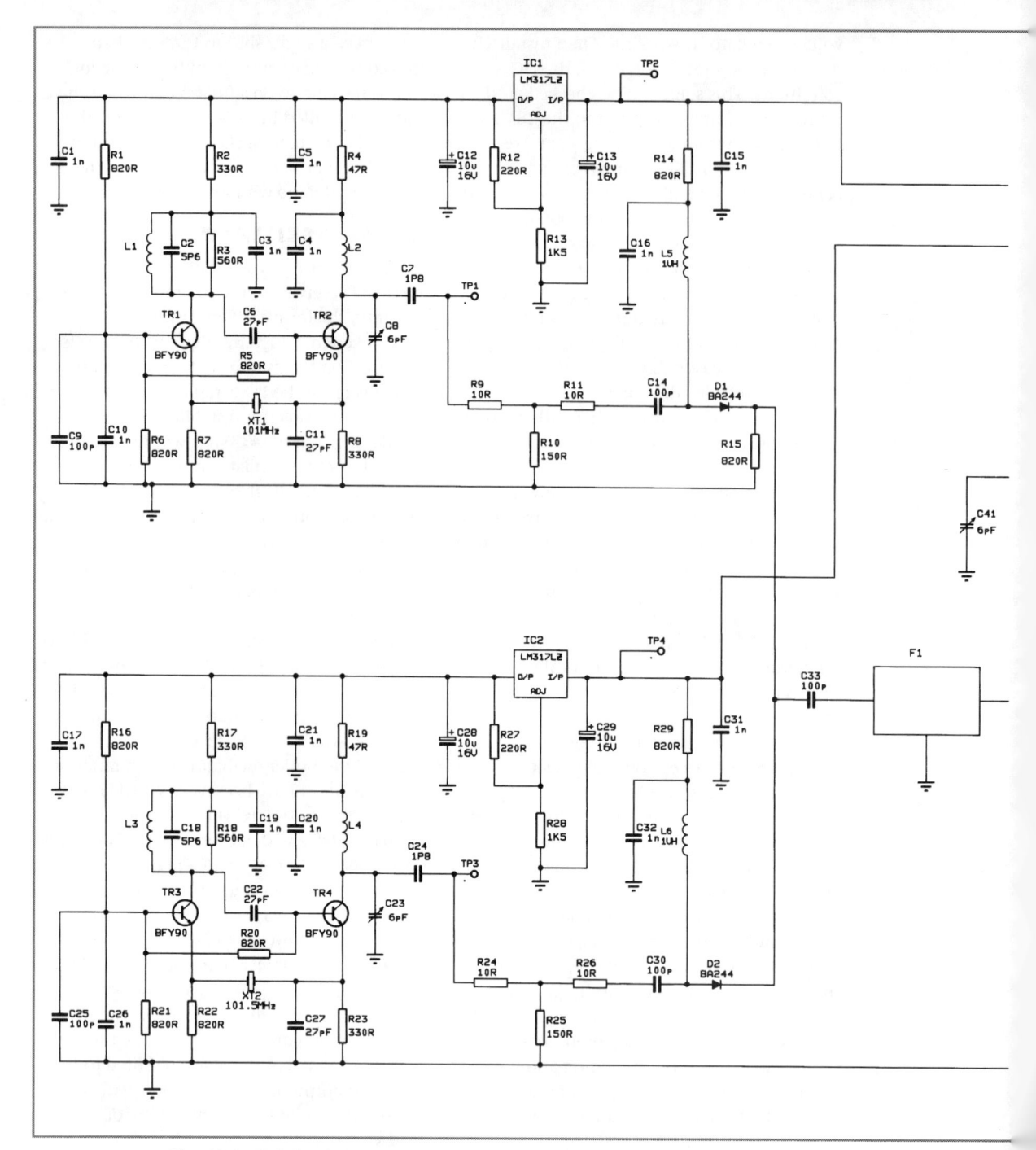

Fig. 10.5. Circuit of the 404/406MHz local oscillator for the 432MHz transverter

DXER'S 432MHZ TRANSVERTER

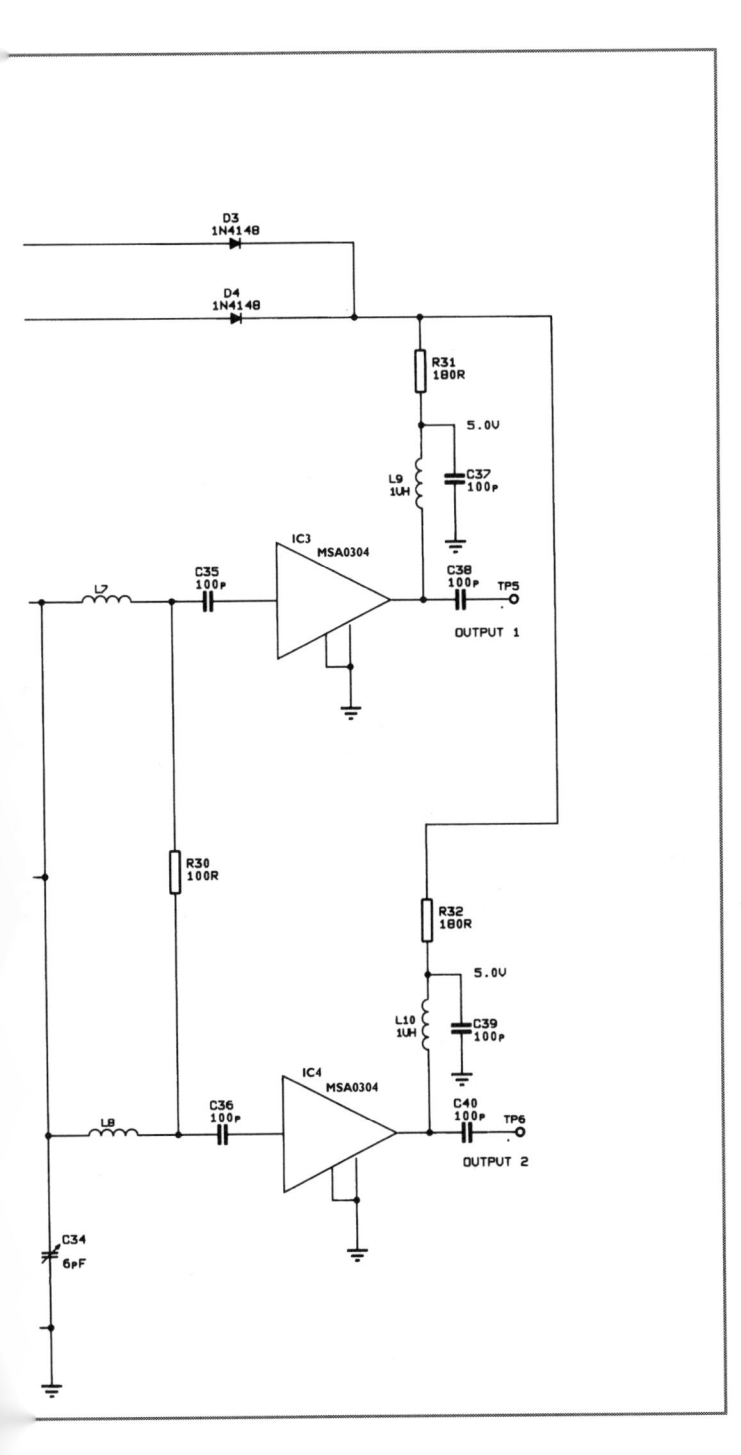

tion, and circuits can easily be devised to connect the two oscillators to the TX/RX switching if desired. A helical filter at 404MHz selects the desired fourth harmonic and attenuates all other signals to –50dBc typical. As with the other parts of the transverter, using a pre-tuned filter makes complicated testgear unnecessary. The output of the filter feeds a Wilkinson power divider, each output of which is followed by an MSA0304 MMIC. These provide the gain required to produce two output signals of +10dBm at 404MHz.

Fig. 10.6 shows the etching masks for the underside and topside of the 404/406MHz LO board. Assemble all the resistors, capacitors and inductors on the PC board, following Fig. 10.7, and then insert and test both IC voltage regulators. Install TR1 and TR2, and check their DC conditions by measuring the voltages across R2 and R4. Adjust the core of inductor L1 until oscillation starts. The presence of oscillation can quickly be checked at 101MHz using an FM broadcast receiver, although the frequency must be set accurately with a counter when the rest of the board is complete. The signal at the test point TP1 is able to feed a fairly low impedance, so coax cable can be used to connect a broadband milliwattmeter (Chapter 12); a reading of the order of 0dBm can be expected. Install TR3 and TR4, change over the power connections and repeat the setting-up process on the second crystal oscillator.

Next install the PIN diode switching which selects between the two oscillators, and check that the DC conditions are correct before soldering in the helical filter. With the power meter connected to the output of the filter, some signal should be obtainable from both oscillators; adjust trimmers C8 and C23 for maximum level.

Finally, install the output MMICs. After checking that their bias currents are correct (7·8V across R31 and R32), connect the power meter to each output port in turn; you should obtain around +10dBm from each. Adjust the capacitors C34 and C41 on the Wilkinson power divider to obtain the same level from both ports. If this is not possible, note which

432MHz

Fig. 10.6. Full-size PCB layout of the underside of the 404/406MHz local-oscillator board. The top of the board is a continuous copper groundplane, etched or counterbored as shown

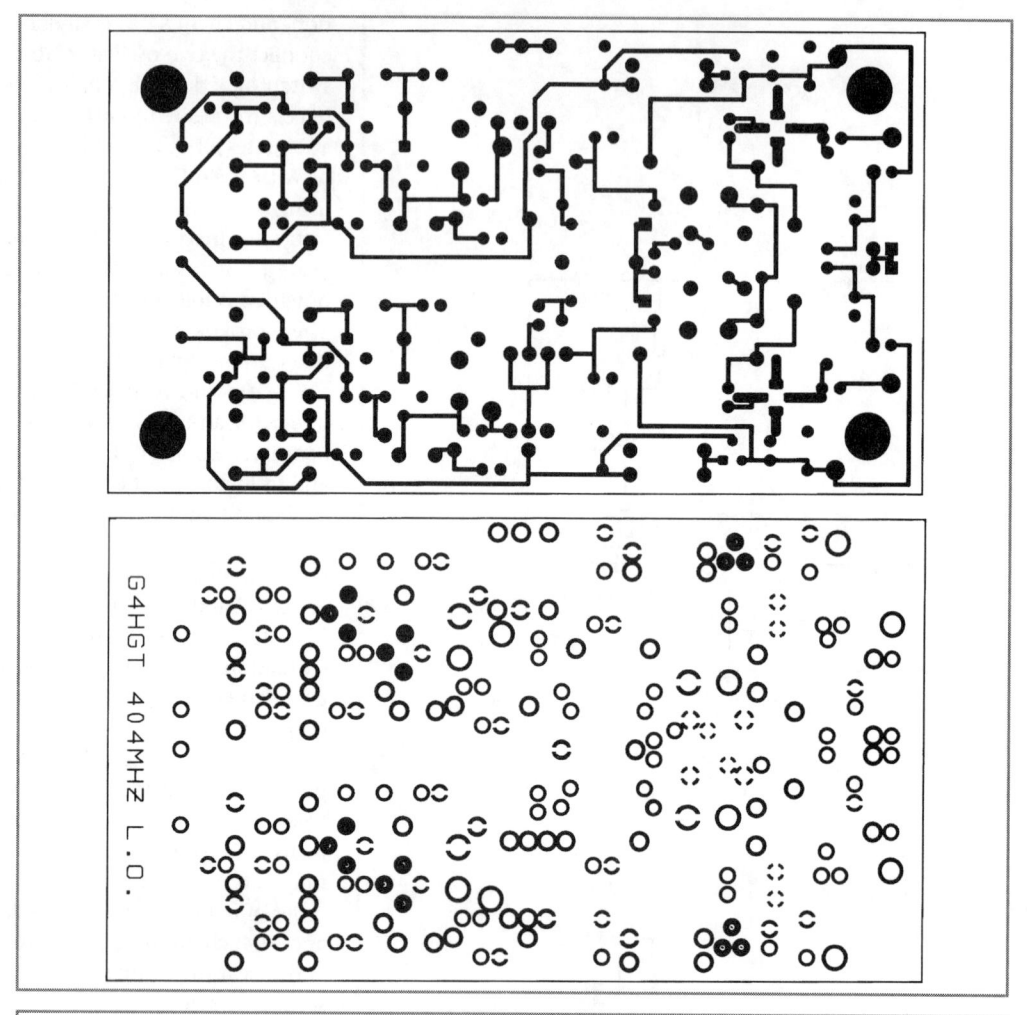

Fig. 10.7. Component layout of the 404/406MHz local-oscillator board

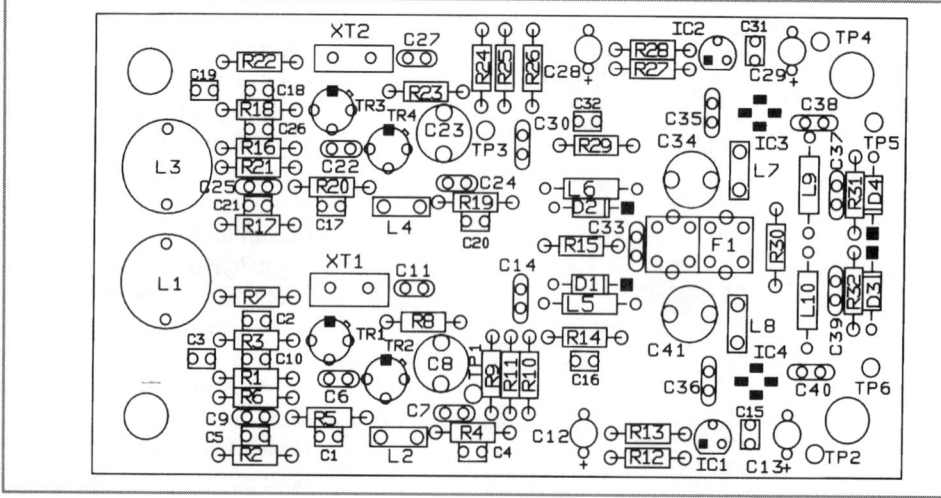

DXER'S 432MHz TRANSVERTER

is the higher and use it for the receiver local oscillator.

TRANSMIT CONVERTER

This section of the design takes an input signal in the range 28–30MHz and gives an output at 432–434MHz, or 434–436MHz with the alternative LO frequency. There is no reason why almost any IF in the range 20–70MHz could not be used, although consideration of harmonics and other intermodulation products should exclude certain frequency plans. The transmit strip is essentially broadband in both its LO and IF sections, while the 432MHz amplifier chain is very narrowband with a great deal of filtering (Fig. 10.8). So long as the frequencies and levels at the LO and IF inputs are correct, the output will be on 432–436MHz and nowhere else.

The design of the transmit strip can be approached from either the input or the output, but if linearity is of importance you should start by thinking about the output level. In this way you will avoid having to squeeze every last dB out of the final stage. For compatibility with the Mitsubishi power-amplifier modules, an output power of +17dBm (50mW) is required. It is a reasonable assumption that third-order intermodulation products will be about –30dBc when an amplifier is operating at its 1dB compression point. Following the well-known rule of thumb (think of a number and double it, *ie* add 3dB), I therefore aimed for an output compression point of +20dBm, so that all intermod products from this stage would be well down.

The BFG34 is a good device for the output stage. It has a maximum collector current of 150mA, so it can easily handle the RF power level required, and being a modern device it has a useful power gain of 17dB at 432MHz. Any RF transistor operated at the +20dBm power level will have input and output impedances which are lower than 50Ω. In the case of the BFG34, the *s*-parameters given in the data sheet were converted into the equivalent real and imaginary components. Networks designed for matching these values of Z_{IN} and Z_{OUT} to 50Ω were then extracted from the tables in [11], although some experimentation was still needed to obtain the maximum output from the stage. DC is fed to the collector via an small RF choke L7; it may be possible to replace this component by a few turns of enamelled copper wire wound around a low-value resistor. The base bias voltage is derived using a circuit used many times elsewhere, and temperature compensation can be arranged (or at least attempted) by thermally coupling the diodes to the transistor. The voltage is fed to the base via RF choke L6 and resistor R21, both of which are needed to ensure stability.

The BFG34 output stage needs to be driven at around +3dBm. It is preceded by a helical filter (assume 3dB loss) so the driver needs to be capable of +6dBm output. Once again adding a 3dB margin for linearity leads to a desired 1dB compression point of +9dBm. The Avantek MSA0304 MMIC fits the bill nicely with a 1dB compression point of +9dBm and 12dB of gain.

Continuing backwards from IC4 (Fig. 10.9) another helical filter is preceded by another MSA0304 (IC3) which is again operated well inside its linear region, and this is fed by the mixer. Using an MMIC directly after the mixer is a good way of providing an accurate wideband termination, and its noise figure is low enough to ensure a very low level of wideband noise in the transverter. In the final design I included a small resistive attenuator between the mixer and IC3 to adjust the overall gain to the desired value.

With so much spare gain in the amplifier stages, it is possible to run the mixer at a very low signal level to reduce its contribution to intermodulation. With the sequence of stages described, the RF drive to the mixer input can be kept below –8dBm. On the input side of the mixer is a simple DC-controlled RF attenuator network. Input drive at 28MHz is first attenuated by the preset potentiometer RV1 and then passes through the PIN diode D1. The forward DC current through this diode controls its RF attenuation. This current is provided by TR1 which is wired as an emitter

432MHz

follower to provide a relatively high impedance for the control-voltage line. After the attenuator, some amplification at 28MHz is required to restore the signal level; IC2 (MSA0404) gives around 8dB of gain. RV1 should adjusted to set the transmit strip output level to just over +17dBm. If you require an ALC loop to control the output from the transmit strip, refer to the circuit of the *Suffolk* transverter described in Chapter 8.

Also on the transmit-converter board is the TX/RX switching (Fig. 10.8). I have deliberately kept this as simple as possible, and you may wish to expand this part of the circuit to interface with your existing equipment. Chapters 8 and 11 between them give comprehensive advice. For both transmit and receive, pairs of switching transistors give either low

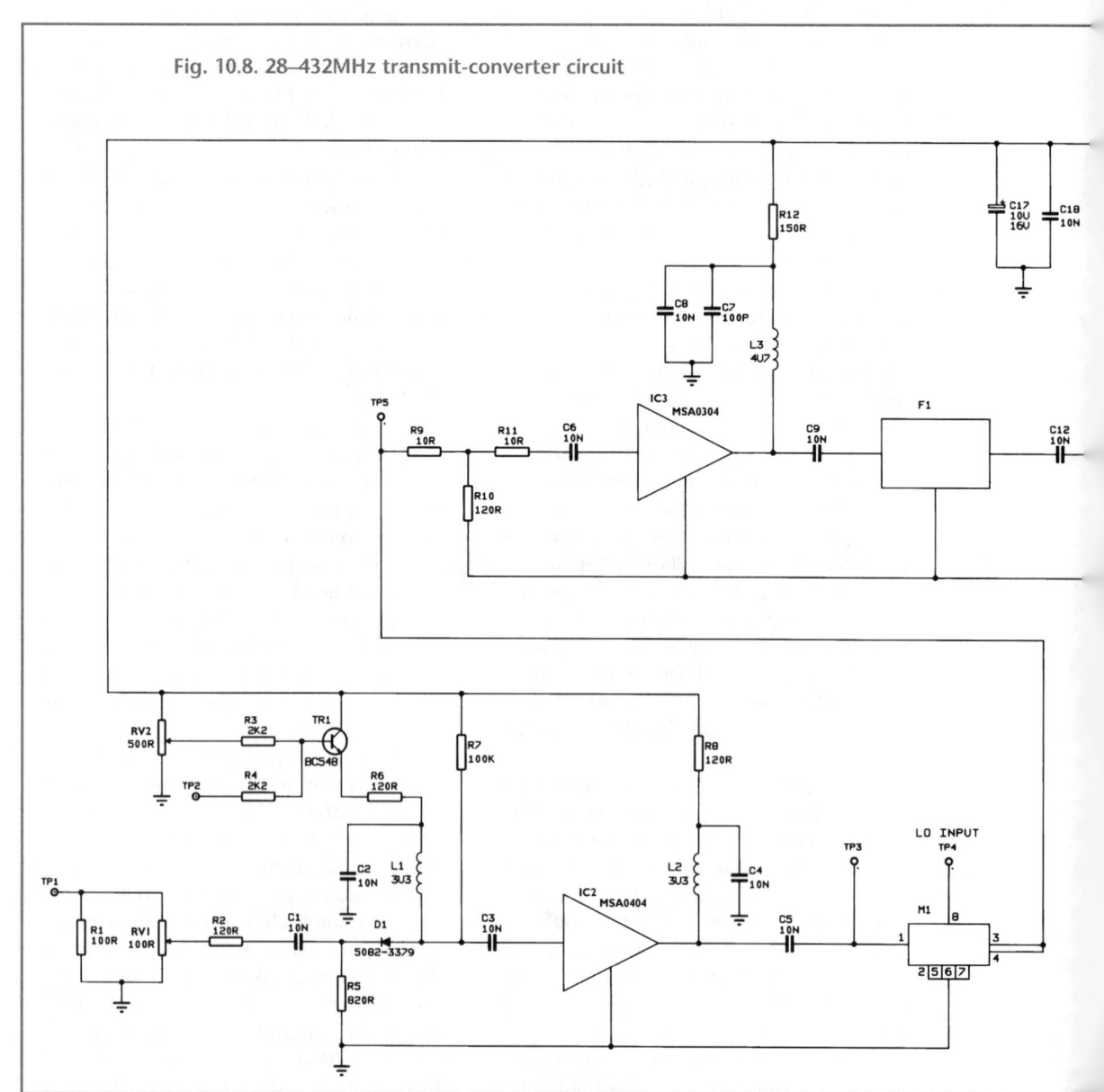

Fig. 10.8. 28–432MHz transmit-converter circuit

DXER'S 432MHz TRANSVERTER

(50mA) or high (200mA) current capability. I have used the former for control of the output relays and indicator LEDs, with the high-current devices powering the transmit and receive boards.

Fig. 10.9 shows the etching masks for the underside and the top of the transmit converter board. Assemble all the resistors, capacitors and inductors on the PC board, following Fig. 10.10. Use surplus wire to provide through connections from the groundplane to the terminals of IC2–4, TR2 *etc*, and remember to solder all ground connections to the groundplane.

Install the regulator IC1 and check that its output voltage is satisfactory. Next install IC2 and ensure it draws the correct current by checking the voltage across R8; for 50mA

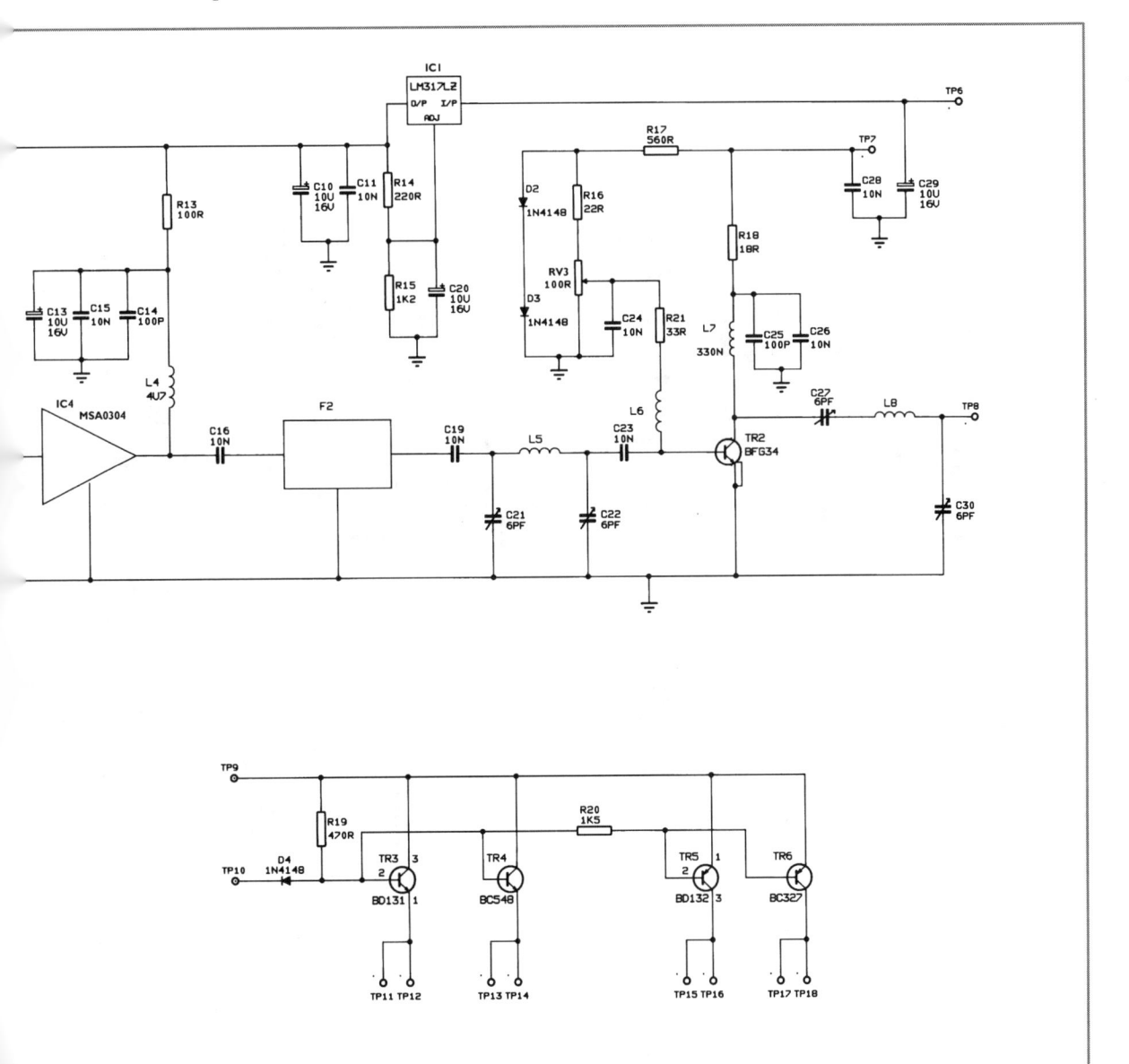

432MHz

standing current this should be 6·0V. Solder in TR1 and the PIN diode D1. Set RV1 to midway. Apply DC power and measure the voltage across R5 to check that RV2 controls the current through D1. Leave RV2 at its maximum setting, corresponding to minimum attenuation through the diode.

It is important to set the HF drive level before installing the mixer. Begin by using a milliwattmeter to set the drive from the HF rig to 0dBm (yes, only 1mW) maximum, and install a permanent attenuator if the level is too high. If the available HF drive level is less than 0dBm, simply set the drive to the maximum available. Connect the milliwattmeter to TP3 (Fig. 10.10) using a coaxial lead. Apply 28MHz drive and some output at TP3 should be observable. Adjust RV1 until the level is around −5dBm. Now slowly decrease RV2 and the signal will start to reduce; stop when the signal level has fallen by around 1dB. From this point onwards, any negative-going ALC voltage applied to TP2 will reduce the drive to the mixer.

Finally, adjust RV1 to limit the mixer drive level at TP3 to around −8dBm. The input

Fig. 10.9. Full-size PCB layout on the underside of the 432MHz transmit converter board. The top of the board is a continuous copper groundplane, etched or counterbored as shown

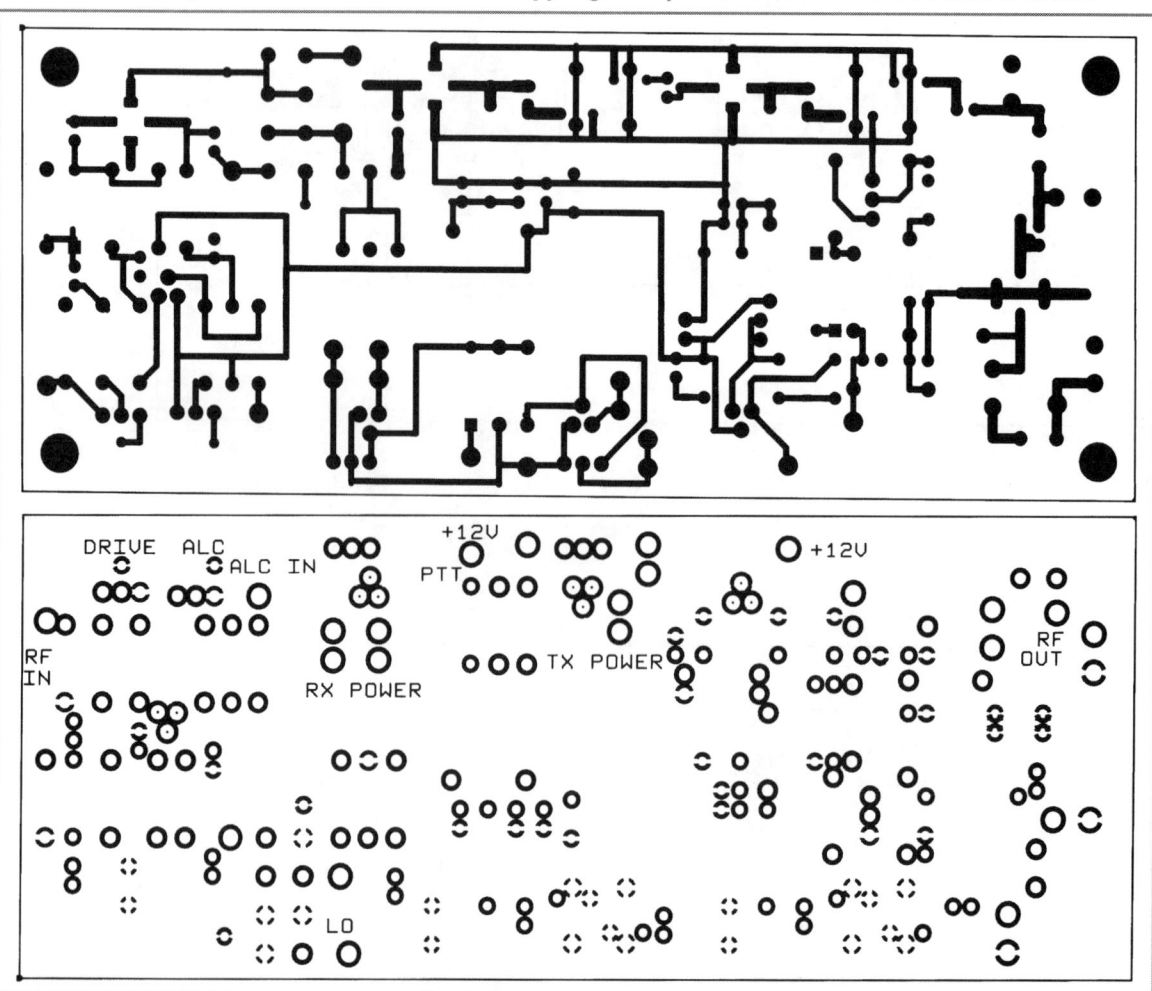

DX'ERS 432MHZ TRANSVERTER

circuitry is now set up and no further adjustments to RV2 should be needed. If the 28MHz drive level ever needs to be changed, readjust only RV1 to control the drive to the mixer.

Remove the milliwattmeter from TP3 and install the mixer, soldering its case carefully to the PCB groundplane. With the local oscillator running and connected to the mixer port, the rest of the transmit strip can then be built up stage-by-stage. Having first checked the DC conditions on the MMIC IC3 (40mA, *ie* 6·0V across R12), connect the power meter to its output point and observe the presence of a signal when drive is applied. Move on stage by stage, until a signal is obtained from the output of F2.

Next solder in D2, D3 and TR2 and set the wiper of RV3 to ground. Connect the power meter to TP8 and the DC supply voltage to TP7. With no RF drive, increase RV3 to give 1·5V across R18, corresponding to a standing current of 80mA. Any fluctuation in the voltage across R18 as the potentiometer is varied could be due to instability and must be investigated. Leave the circuit powered-up for a few minutes and then reset the standing current if it has changed significantly.

Apply 28MHz drive cautiously and check for output on the power meter. The four trimmer capacitors are simply peaked for maximum signal output. If the above procedure has been followed, it should be possible to adjust RV1 to give an output level of +20dBm before there are any signs of limiting. Finally adjust RV1 to reset the output to +17dBm (50mW), the correct level for the modular power amplifier which follows.

MODULAR POWER AMPLIFIER

One of the more time-saving new products to emerge recently has been the power-amplifier module, with performance aimed specifically at the mobile-radio and amateur markets. At the time of writing, Mitsubishi Semiconductors supply a range of modules covering 50–1300MHz, with output powers as high as 30W. Versions are usually available for either FM or SSB, the latter simply containing extra circuitry to bias all the transistors into Class AB1.

As with a lot of convenience products, one or two parameters have to be compromised. In the case of these PA modules, the compromise is that their linearity is not as good as one

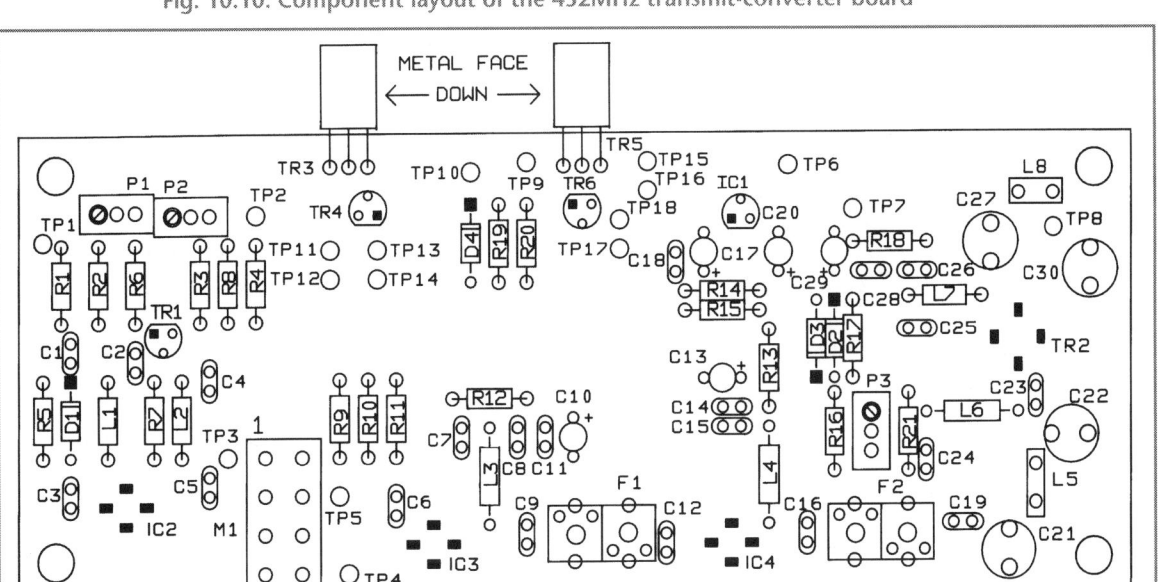

Fig. 10.10. Component layout of the 432MHz transmit-converter board

432MHz

might like. At their rated PEP outputs, these modules have third-order intermod products of typically –32dB, with equally reasonable levels of higher-order products. Although this is quite good going for a transistor PA, at lower drive and output levels these intermod products do not fall away as rapidly as you would expect. It follows that you cannot improve the linearity of a modular PA merely by under-driving it. In contrast, a discrete-component PA with lots of variable capacitors can generally be optimized at any chosen power level, producing a much cleaner signal – but it is equally possible to get the adjustments totally wrong! At least with the module you can't go too far astray.

The manufacturer's data show that these modules can also be driven much harder, to give output powers up to 50% higher than their specified maximum. But the intermod products will then be rapidly degrading and the extra output power will simply contribute to the width and splatter of your signal. Don't be greedy – settle for the manufacturer's rated linear power.

This design uses the Mitsubishi M57716 430MHz amplifier which is rated at 10W. This is more than enough to drive a K2RIW amplifier to full output power on SSB. When you purchase the module, also obtain the full five-page data sheet. The PA module will be the most expensive single item in your transverter, so take your time with this section of construction to avoid introducing costly problems.

The modules are hybrid circuits containing discrete components for matching and biasing, and, thanks to some clever production tolerancing, no adjustments are necessary when the module is installed. The components are encased by a black plastic housing attached to an aluminium heat-transfer plate. In the 430MHz devices there are three transistors which have to be cooled via this plate, so the heatsink needs to be somewhat larger than those for the two-transistor modules used for lower frequencies. A thermal resistance less than 1°C/W is necessary for a 100% duty cycle and for high reliability under contest, expedition and EME conditions. It is always worthwhile to use a large heatsink – select the transverter case to suit the heatsink, not the other way round!

DC supplies could hardly be simpler, as Fig. 10.11 demonstrates. Supply voltage is applied directly to the final transistor stage, pin 4

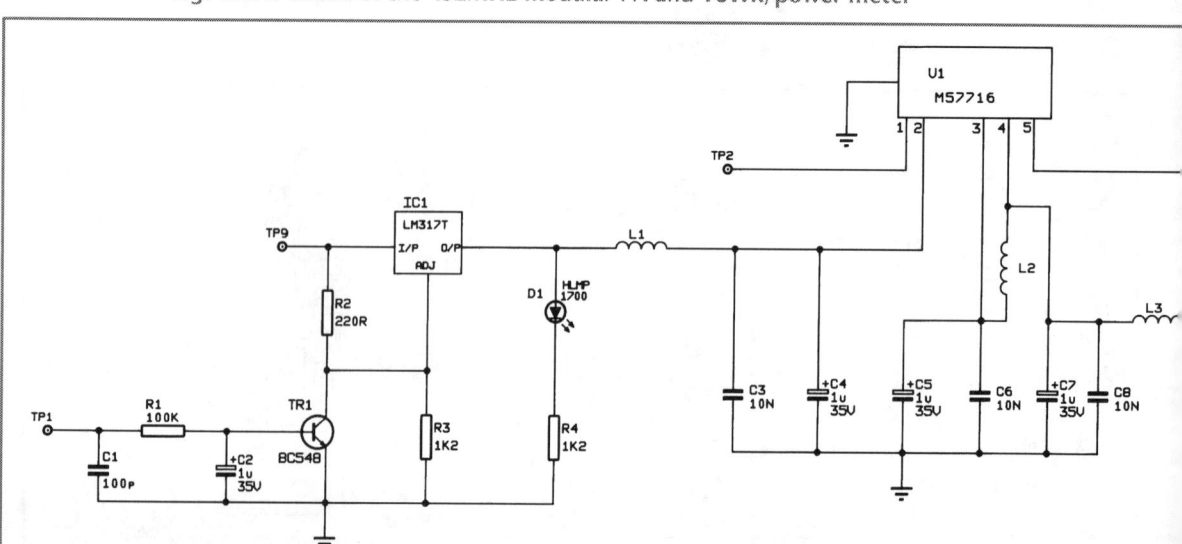

Fig. 10.11. Circuit of the 432MHz modular PA and VSWR/power meter

DXER'S 432MHz TRANSVERTER

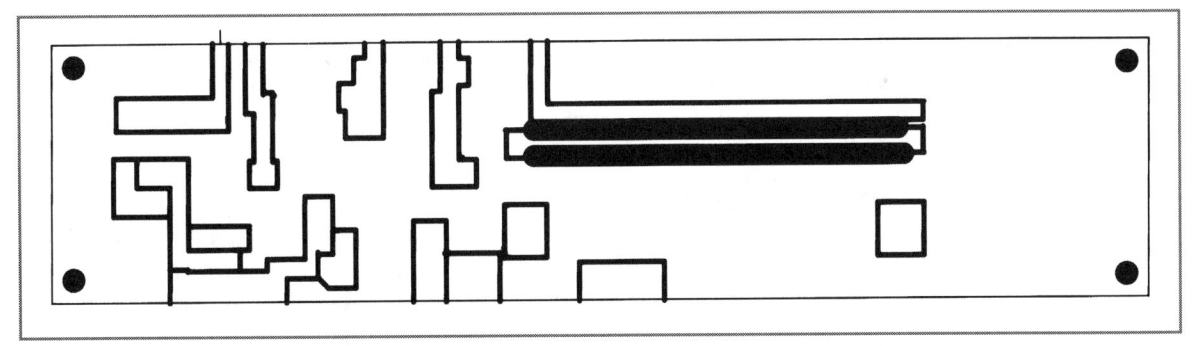

Fig. 10.12. Full-size PCB layout on the top-side of the 432MHz PA/VSWR board (remove copper from black areas). The underside of the board is a continuous copper groundplane

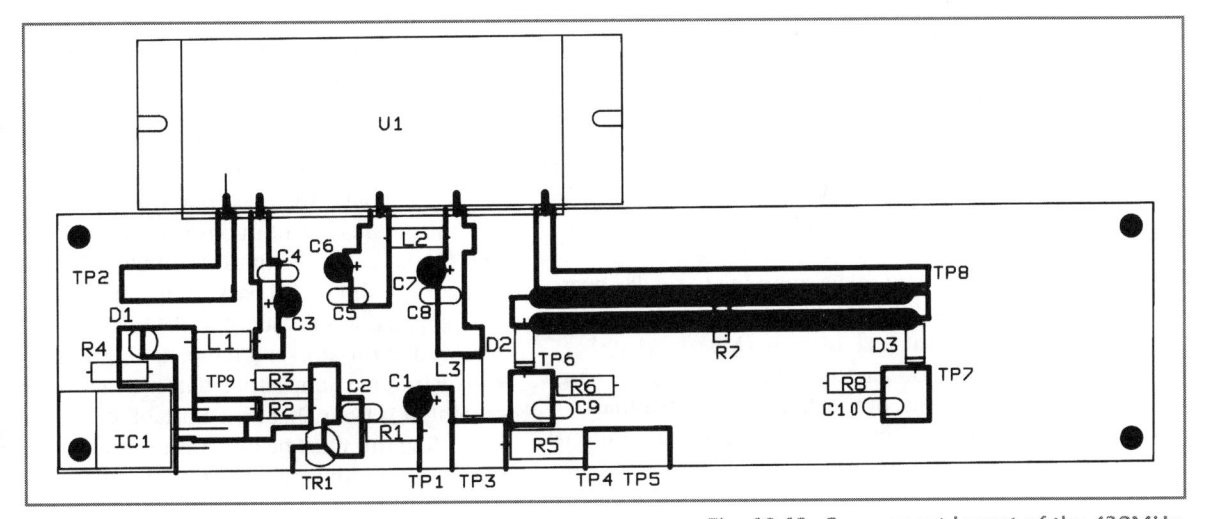

Fig. 10.13. Component layout of the 432MHz PA/VSWR board

(V_{CC2}); the other two device collectors are connected together internally and are brought out to pin 3 (V_{CC1}). This is connected to supply *via* an RF choke. The voltage is relatively uncritical, though the manufacturer's recommendation of 12·5V should be followed if possible. The three devices have their base bias circuits commoned together at pin 2 and this requires 9·0V at typically 600mA.

A properly regulated supply with low output impedance is required and Fig. 10.11 shows the arrangement used, together with rest of the circuit and interfacing details. I have included an LED (D1) to give an indication of bias being available; a transistor (TR1) is provided to allow DC switching of the amplifier.

The supply current can be monitored using a voltmeter connected across the low-value resistor (R5) in the supply lead.

VSWR/POWER METER

Connected directly to the output of the PA module is a printed-circuit VSWR/power meter. Diodes D2 and D3 (Fig. 10.11) rectify the sampled signal, and must be Schottky-barrier devices for correct operation. The 'forward' and 'reflected' DC voltages can be fed either directly to a meter, or to an ALC circuit as described in Chapter 8 if preferred.

The power-amplifier PC board is single-sided (Fig. 10.12), the reverse being un-etched copper. Multiple ground wires connect the grounded tracks on the component side to the groundplane. The board is attached to the rear panel by four M3 screws and held clear by plastic washers.

Before you solder anything to the PC board, try assembling the board and the PA module to the heatsink, using two M3 screws and tapped holes in the heatsink. Avoid stressing the package of the module since external forces or mechanical shock can cause internal damage. A torque of 5kg/cm on the screws is all that is required to ensure sufficient thermal contact to the heatsink. Since this may not be enough to retain the screws against vibration, use lock washers or 'Loctite' on each screwhead. A couple of short ground-straps made from wide copper strip (or solder tags and the braid from UR43 coaxial cable) are used to provide low-impedance connections between the PCB groundplane and the aluminium cooling plate, which also forms the module's RF ground.

Once you are happy that the mechanics of the module and PC board are correct, remove the module and assemble the rest of the components on the board (Fig. 10.13). One of the mounting screws to the rear panel is used to mount the regulator IC, and insulating washers are necessary to keep the heatsink tab isolated. Connect +12·5V DC to TP9 and check that the bias supply gives 9·0V, adjusting the value of R3 if necessary. Also check the operation of the DC switching by connecting TP1 to +12·5V; the bias voltage should fall to less than 2V and the LED should go out.

The PA module can then be installed. Remember to place a thin smear of heatsink compound on the flange, and tighten the screws *before* soldering the leads. Connect a dummy load and power meter to the output connector, and monitor the supply current as you apply DC power. Little or no current should be drawn in this state. Connect TP2 to a +12·5V DC supply; the module should draw approximately 300mA. Now connect the output of the transmit converter to the PA input using thin 50Ω coaxial cable (RG174). If you have set the transmit converter output to +17dBm maximum, it will not be possible to overdrive the PA. Apply enough 28MHz drive to obtain about 10W RF output; the supply current should be continuously monitored and should not go significantly above 2·5A. The current should fall back to the 300mA standing-bias level as soon as the drive is removed. If this does not happen, or if there are any sudden jumps in RF output or DC current, it is likely that some instability is occurring; this must be investigated and cured before putting the transverter on the air.

DXER'S 432MHz TRANSVERTER

COMPONENTS LIST – 432MHz RECEIVE CONVERTER

RESISTORS
All 0·25W 2% or 5%

22R	4	R6, 9, 16, 19
47R	2	R14, 24
120R	5	R3, 4, 11, 13, 23
220R	2	R22, 25
270R	8	R5, 7, 8, 10, 15, 17, 18, 20
1k2	1	R26
2k2	2	R12, 21
47k	2	R1, 2
500R	1	RV1 multi-turn preset

CAPACITORS
Disc ceramic, Philips 632 series

2p7	1	C31
3p3	1	C23: see text
22p	2	C30, 32
68p	11	C2, 7, 10, 12, 13, 16, 18, 21, 22, 24, 26
100p	3	C14, 29, 33
10n	8	C1, 4, 8, 9, 17, 20, 25, 28

Leadless chip ceramic

3p9	1	C23
33p	4	C6, 15, 19, 27

Tantalum

10μ 16V	3	C34, 35, 36

Trimmers

2-6p	3	C3, 5, 11 (Farnell 148-157)

INDUCTORS, RF CHOKES AND FILTERS

L1, 2	16SWG silvered copper, spaced 5mm above PCB groundplane
L3	5 turns 4mm diameter, 22SWG enamelled copper
L4, 6	Transformers wound on Siemens B62152A8-X17 core (Electrovalue) – see panel, page 10-7
L5	10μH moulded choke with axial leads
L7, 8	1·4μH 5mm-format inductor, Toko 215PN0874N
F1, 2	435MHz, Toko 252MX-1549A

SEMICONDUCTORS

CF300	1	TR1
BFG34	2	TR2, 3
SBL-1	1	M1
LM317LZ	1	IC1

MISCELLANEOUS
Box – Teko TEK S394 (West Hyde)
Tin-plate for screen

COMPONENTS LIST
LOCAL OSCILLATOR FOR 432MHz TRANSVERTER

RESISTORS
All 0·25W 2% or 5%

10R	4	R9, 11, 24, 26
47R	2	R4, 19
100R	1	R30
150R	2	R10, 25
180R	2	R31, 32
220R	2	R12, 27
330R	4	R2, 8, 17, 23
560R	2	R3, 18
820R	11	R1, 5, 6, 7, 14, 15, 16, 20, 21, 22, 29
1k5	2	R13, 28

CAPACITORS
Disc ceramic, Philips 632 series

1p8	2	C7, 24
5p6	2	C2, 18
27p	4	C6, 11, 22, 27
100p	11	C9, 14, 25, 30, 33, 35-40
1n	14	C1, 3, 4, 5, 10, 15, 16, 17, 19, 20, 21, 26, 31, 32

Tantalum

10μ 16V	4	C12, 13, 28, 29

Trimmers

2-6p	4	C8, 23, 34, 41 (Farnell 148-157)

INDUCTORS, RF CHOKES AND FILTERS

L1, 3	Toko S18 301AS0500, green (5·5 turn) coil with aluminium core
L5, 6	1μH moulded chokes with axial leads
L2, 4, 7, 8	0·5 turn 4mm diameter, 22SWG enamelled copper
L9, 10	10 turns, 3mm diameter, 22SWG enamelled copper
F1	405MHz, Toko 252MX-1547A

SEMICONDUCTORS

BFY90	1	TR1-4
LM317LZ	2	IC1, 2
MSA0304	2	IC3, 4 Avantek
BA244	2	D1, 2
1N4148	2	D3, 4

MISCELLANEOUS

Crystals	101MHz, 101·5MHz (HC33-U)
Box	Teko TEK S393 (West Hyde)

432MHz

COMPONENTS LIST – 432MHz TRANSMIT CONVERTER

RESISTORS
All 0·25W 2% or 5%

10R	2	R9, 11
18R	1	R18
22R	1	R16
33R	1	R22
100R	2	R1, 13
120R	4	R2, 6, 8, 10
150R	1	R12
220R	1	R14
470R	1	R19
560R	1	R17
820R	1	R5
1k2	1	R15
1k5	1	R20
2k2	2	R3, 4
100k	1	R7

CAPACITORS
Disc ceramic, Philips 632 series

100p	3	C7, 14, 25
10n	18	C1-6, 8, 9, 11, 12, 15, 16, 18, 19, 23, 24, 26, 28

Tantalum
10µ 16V	5	C10, 13, 17, 20, 29

Trimmers
2-6p	4	C21, 22, 27, 30 (Farnell 148-157)

INDUCTORS, RF CHOKES AND FILTERS

L1, 2	3·3µH moulded choke with axial leads (Cirkit or Bonex)
L3, 4	10 turns, 3mm dia 22SWG enamelled copper
L5, 8	1·5 turns, 4mm dia 22SWG enamelled copper
L6	3 turns on ferrite bead (ie 3 times through hole)
L7	330nH moulded choke with axial leads (Cirkit or Bonex)
F1, 2	435MHz, Toko 252MX-1549A (Cirkit or Bonex)

SEMICONDUCTORS

BC548	2	TR1, 4
BFG34	1	TR2
BD131	1	TR3
BD132	1	TR5
BC327	1	TR6
HP5082-3379	1	D1: PIN diode (Farnell)
1N4148	3	D2, 3, 4
SBL-1	1	M1 (Cirkit or Bonex)
LM317LZ	1	IC1
MSA0404	1	IC2 Avantek
MSA0304	2	IC3, 4 Avantek

MISCELLANEOUS
Teko TEK S394 (West Hyde)

COMPONENTS LIST – 432MHz MODULAR PA, INTERFACING CIRCUITS AND HARDWARE

RESISTORS
All 0·25W 2% or 5% unless noted

0R33	1	R5: 1W wirewound
51R	1	R7: 0·125W chip if possible
220R	1	R2
1k2	1	R3, 4: adjust R3 if necessary
10k	2	R6, 8
100k	1	R1

CAPACITORS
Disc ceramic, Philips 632 series

100p	1	C1
10n	5	C3, 6, 8, 9, 10

Tantalum
10µF 25V	4	C2, 4, 5, 7

INDUCTORS

L1, 2, 3	5 turns 6mm diameter, 22SWG enamelled copper

SEMICONDUCTORS

M57716/M57745	1	U1 Mitsubishi PA module
HP5082-2835	2	D2, 3: Schottky diodes (Farnell)
LM317	1	IC1 1A regulator
BC548	1	TR1
LED	1	D1

MISCELLANEOUS
Screws, nuts, plastic and metal washers
Heatsink grease
Switches
UR43 and RG174 coax cable
Wire

Heatsink	Siefert KL-144 75SW (Schaffner EMC, Micromark or GCA Electronics)
Case	BICC-Vero G-line (Verospeed)

A LOW-NOISE 432MHz GaAsFET PREAMPLIFIER

by Ian White G3SEK

This preamplifier (Fig. 10.14) is suitable for a wide range of GaAsFET devices, and is based on a highly successful design by G3WDG. Prototypes have been used for both EME and tropo working, and with a suitable GaAsFET the noise temperature can be well below 50K – which is low enough to challenge the accuracy of expensive professional test equipment! The gain is about 15dB, depending on the GaAsFET and the way the amplifier is aligned, and this should be sufficient for most receiving systems. Apart from the low-loss input circuit, which is essential to avoid degradation of the noise temperature, the main feature of this amplifier is its simplicity. However, there are a few electronic and mechanical subtleties which you need to bear in mind when copying or adapting the design.

The two big problems with GaAsFET amplifiers are stability and protection against supply-voltage transients. Although GaAsFETs are capable of oscillation at any frequency from AF to SHF, they can be tamed with a little care. In this design the input and output circuits are isolated by a screen through which only the gate lead of the GaAsFET passes. Another important contributor to stability is the source bypassing, and leadless disc or chip capacitors provide perfectly adequate source bypassing at frequencies up to several GHz. This design uses two leadless disc-ceramic capacitors soldered directly to the dividing screen, with the GaAsFET mounted between them (Fig. 10.14).

The untuned output circuit is also a considerable aid to stability; T1 is 2+2 turns bifilar wound on a ferrite bead or miniature dust-iron toroid [12] and amplifiers using this form of output coupling will be stable unless seriously mistuned or misterminated. If a previously stable GaAsFET preamplifier starts to oscillate, don't tear it apart before you've checked the rest of the RF system – maybe the antenna impedance has changed or a connector has come loose.

Fig. 10.14. Circuit and mechanical sketch of the 432MHz GaAsFET preamplifier (see text and Fig. 11.12 for further details)

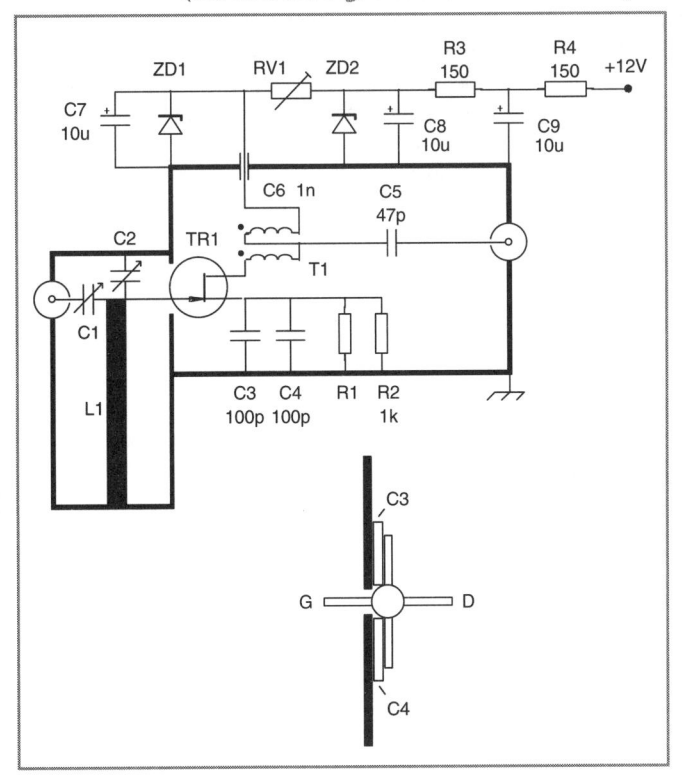

432MHz

GaAsFETs also have an undeserved bad reputation for vulnerability to RF damage and voltage transients. In fact we've been carelessly subjecting our old valve and bipolar RF stages to such insults for many years, and the problem is that GaAsFETs won't tolerate that kind of bad engineering. If any of the applied DC or RF voltages exceed the breakdown limit – even for an instant – it's goodbye GaAsFET. Adequate coaxial relays and proper sequencing of transmit-receive change-over can totally prevent problems of RF burnout, and the problems of supply-voltage transients can equally be avoided by proper design. Chapter 11 gives full details of the power supply and station-control features required to protect GaAsFET preamps so this really shouldn't be a problem.

CONSTRUCTION

The low-loss input inductor L1 consists of an 86mm length of 0.25" diameter brass rod, inside a cavity made from standard 22mm outside-diameter copper heating pipe. The cavity connects with a small screening box for the rest of the RF components, *via* a 2mm hole through which the gate lead of the GaAsFET protrudes. The leadless disk capacitors C3 and C4 are soldered as close as possible to the hole, leaving just enough space for the body of the GaAsFET (Fig. 10.14).

Particular care is required to minimize contact losses in the input cavity, especially at the joint between L1 and the disc which forms the base. Ordinary solder is less effective than a knife-edge joint formed by counterboring the end of L1 and drawing the joint up really tight with a steel screw. It is more difficult to avoid solder in the joint between the base disc and the copper tube, though this is a somewhat less critical point. For long-term performance, the whole assembly should be silver-plated [13]. For less demanding tropo use, a quite satisfactory input circuit can be made using a 22mm capillary end-cap instead of the turned base disc, with soldered joints throughout.

Having taken all that trouble with the cavity itself, it would be a shame to spoil the unloaded Q by using poor-quality trimmers for C1 and C2, or a lossy and unreliable input connector. The circuit has been designed to require only small tuning and coupling capacitances, again in the interests of low losses, so C1 and C2 must each have a low ***minimum*** capacitance and a high Q. C1 can typically be a miniature 0.5–1.5pF PTFE trimmer, and C2 a 0.5–3pF glass or quartz trimmer – although you can really go to town with microwave sapphire trimmers if you want. To avoid degrading the noise figure, both of these capacitors must have solid low-loss connections.

The input connector should be a type N, SMA or TNC; a BNC connector would soon start to become unreliable in this critical application. Another alternative is to eliminate unnecessary coaxial connectors by building the protection relay (Chapter 11) directly into the preamplifier.

Electrically and mechanically, this preamplifier design can do justice to the best available low-noise GaAsFETs. It should be adjusted for lowest noise temperature using a G4COM-type noise figure indicator with a well-matched 50Ω noise source (Chapter 12). For the best possible results, the preamplifier should be given its final adjustments when installed at the antenna with all its normal relays and cables; this can be done by injecting the noise through another antenna close by.

THE K2RIW POWER AMPLIFIER

by John Nelson GW4FRX and Ian White G3SEK

At the dizzy heights of 432MHz you have a regrettably limited choice of high-power amplifier designs. If you intend to use a pair of 4CX250 tetrodes, only one published design is routinely duplicated; it is the celebrated 'K2RIW' which was first featured in the April and May 1972 editions of *QST* under the heading "A Strip-Line Kilowatt Amplifier for 432MHz". Dick Knadle K2RIW was a pioneer of "real DX" on 432MHz; in addition to this power amplifier, his 19-element Yagi and stressed-dish antenna designs are still frequently used today.

Unlike the push-pull W1SL amplifier for 144MHz (Chapter 8), the K2RIW is a parallel design. The two valves are driven in parallel and excite the anode tuned circuit in the same way. Knadle claimed several advantages for this configuration. He stated that:

"Most designers of transmitters for the higher frequencies have felt that push-pull was necessary in order to minimize the effects of tube and circuit capacitance. Except where a true resonant cavity or properly designed strip-lines are used, this is still true. In conventional coil-and-capacitor circuits especially, push-pull

Fig. 10.15. Beneath the lid of an ARCOS K2RIW

432MHz

has marked advantages over parallel on frequencies near the maximum at which these techniques can be used. But with the stripline, the inductance can be made as low as desired, simply by making the line wider. Any number of tubes can be put in parallel and resonated, so long as the intertube resonances (push-pull modes) can be controlled...

"In this amplifier the parallel grid and plate networks force the RF voltages on the two tubes to be identical. If one tube has higher emission than the other, it may draw slightly higher current, but this is of little consequence because it happens to be the condition under which the amplifier will produce the greatest output with this pair of tubes".

The first time you look inside a K2RIW amplifier (Fig. 10.15), you might suspect that some portions of it are missing. From above, there's no sign of variable capacitors for tuning or loading; you only see a pair of valve anodes (with home-made chimneys, even) and a flat sheet which you presume is the anode inductor. The capacitors for anode tuning and loading are in there, however; they are 'flappers' made from flexible phosphor-bronze and operated by plastic fishing line, and they live underneath the plate line where you can't see them. There are no sliding contacts to cause losses and erratic operation. The grid compartment is almost equally bare, although the tuning and loading capacitors are more conventional.

Knadle's original write-up gives the impression that the K2RIW is very easy to build and use, and also that it was a fully-developed design. However – and putting this in the nicest possible way – it isn't and wasn't, and the 'K2' is seldom "...routinely duplicated" in its original form despite Knadle's use of precisely those words in the article. Although the original was basically sound, there was room for improvement. But the 'K2' is a subtle and individual device, and some of the attempted variations on the original have produced mixed results. The basic problems are in the tuned circuits and the cooling, although by 1992 Varian-Eimac had decided that the latter wasn't such a problem after all.

The K2RIW variant described here was developed by the late Fred Merry, W2GN, and large numbers were sold in kit form by his company (ARCOS) until Fred retired for the second time at the age of 75! The ARCOS design benefited from considerable user feedback, two extremely satisfied users being W1SL and the editor of this book. Another variant which works well is produced by

Fig. 10.16. Circuit of the K2RIW 432MHz power amplifier. See Fig. 11.5 for additional screen-grid protection components

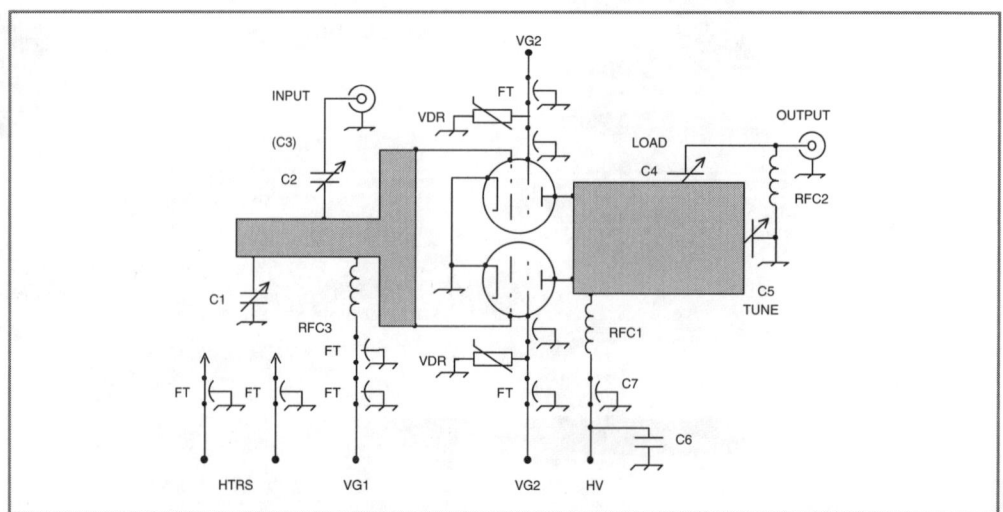

K2RIW POWER AMPLIFIER

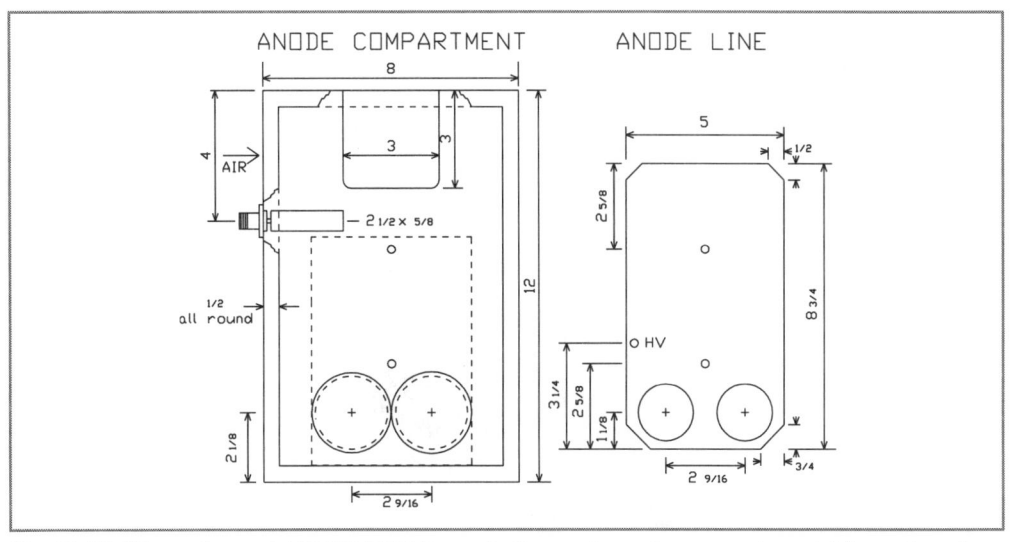

Fig. 10.17. Dimensions of ARCOS K2RIW anode line and anode compartment (dimensions in inches). The size of the large anode-line holes depends on the type of finger stock used

Gordon High Precision in the UK.

The original circuit is shown in Fig. 10.16, and is about as simple as you can get. Note, however, that extra screen-grid protection components are required, as recommended in Fig. 11.15.

Many builders have found that anode lines built to the dimensions originally given by

Fig. 10.18. K2RIW anode compartment, with anode line removed to show the tuning and loading flappers

K2RIW will not resonate. Part of the problem was that the original article did not specify exactly how the corners were to be rounded off, which makes a big difference. Considerable difficulties have also been experienced with anode lines made from double-sided PCB (as recommended by K2RIW) to the extent that two apparently identical lines made from the same material behaved quite differently – one worked and the other wouldn't. However, anode lines made from silver-plated solid brass [13] to the dimensions of Fig. 10.17 will just tune up and go.

With the correct anode-line dimensions, the dimensions of the tuning and loading flapper capacitors shown in Fig. 10.17 should also be correct. Fig. 10.17 also shows where these two flappers are located, and Fig. 10.19 shows that end of the anode compartment in close-up. The flappers are made of springy beryllium copper, and their natural rest positions are parallel with the anode line; this represents maximum capacitance. To tune and load the amplifier, each flapper is pulled away by a piece of plastic fishing-line wound around the shafts of the tuning and loading controls. There have been some problems in this area, mainly with melting of the nylon line in the intense RF field. One constructor's solution

432MHz

Fig. 10.19. Close-up of the end of the anode line. The output flapper capacitor and its safety RF choke are mounted on a PTFE block. Also visible are the HV RF choke (top left), the screened blower inlet and the tuning flapper (bottom centre)

stand-off insulator with the metal parts removed is ideal, and the mushroom-shaped PTFE buttons used in the ARCOS variant are just visible in Fig. 10.19.

The main problems with the grid circuit have been down to the use of an unsuitable size of grid box and also poor substitutions for the miniature tuning and coupling capacitors used by K2RIW. The ARCOS version uses a 2¾"-deep grid box with its own cover inside a 3"-deep base, and a grid line of the original K2RIW dimensions (Fig. 10.20) resonates very nicely with the original miniature variable capacitors.

Difficulties with C1 and C2 have always been traced to excessive minimum capacitance, owing to inappropriate substitutions for the original miniature air-spaced variable capacitors (Figs 10.21 and 10.22). If the capacitance of C1 is too high, the line will not tune. If C1 will tune but C2 is too high, the driver stage will see an excessive VSWR, resulting in excessive drive requirements and enthusiastic intermodulation in the driver stage. One solution is to convert C2 to a flapper like C4; another (originally adopted by K2RIW) is to place another capacitor, C3, in series with C2. But however you do it, make sure that the VSWR on the coax from the driver is no more than 1·5 at all drive levels. It can be done – and if you already own a K2RIW, make sure that yours will do it too.

Above all, don't forget what we said in

has been to use brass rods, which seems like a sure-fire way of conducting RF into the sub-chassis space and causing all manner of problems! The best solution is actually quite simple: space the plastic line away from the metal surface, where the voltage gradient is most intense, by attaching a PTFE spacer to the underside of the flapper. A long PTFE

Fig. 10.20. Dimensions of the K2RIW grid line and grid compartment (dimensions in inches)

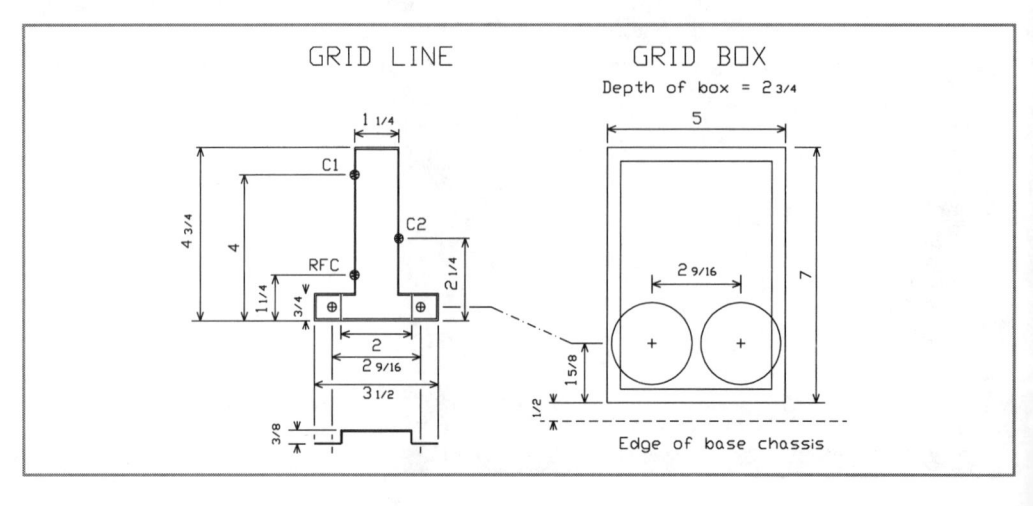

K2RIW POWER AMPLIFIER

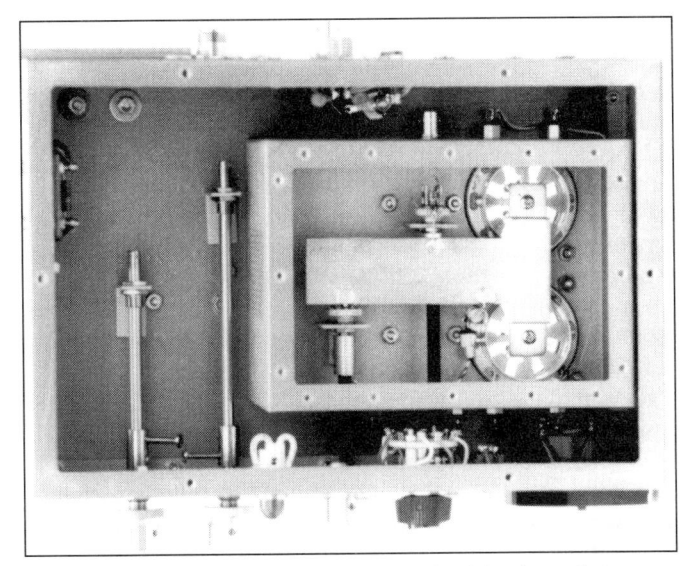

Fig. 10.21. General underside view of the K2RIW

Chapter 6; unless you believe you know exactly how to use a network analyser to stabilize a power amplifier (and are prepared to be proved wrong) do *not* build a K2RIW with any bases other than the Eimac SK620/630 series.

Another potential problem with K2RIWs is the cooling scheme, with air blown into the anode compartment and some of it allowed to

Fig. 10.22. Close-up view inside the K2RIW grid compartment. Following the guidelines in Chapters 6 and 11, a VDR is connected directly from each screen-grid to ground

flow the 'wrong way' downwards through the grid box. As discussed in Chapter 6, this has both advantages and drawbacks. Measurements made with temperature-sensitive paint suggest that the grid spigots of '250s 'cooled' with what should be an adequate blower can reach something in excess of 400°C (the datasheet limit being 250°C). Unfortunately, the physical design of the K2RIW makes it very difficult to blow into the grid compartment, because the box is not deep enough to accept a large air duct into the grid compartment from the rear. Either you choke the air flow and lose the benefits of proper cooling, or you change the dimensions of the grid compartment and risk big problems with RF performance.

The solution seems to be to use the K2RIW cooling scheme, but to increase the airflow downwards through the base. This depends critically on the size of the exit aperture in the grid compartment. The ARCOS arrangement detailed in the construction notes allows considerably more downward airflow than did the original K2RIW design, and has kept more than one pair or 4CX250Bs alive through several years of EME operation at the kilowatt level.

Like sex and neutralizing, getting a new K2RIW to work is easy provided you've done it before. You won't go far wrong if you follow the dimensions given here, use a decent power supply (see Chapter 11 – K2RIW's 1972 recommendations on power supplies are exceedingly obsolete for modern 432MHz operation, especially as far as Class-AB1 working is concerned) and follow the general commissioning guidelines given in Chapter 6. Capacitive output coupling can lead to inadequate loading if the flapper is too small or too far from the anode line, so you may wish to start off with an oversize output-coupling probe with dimensions of 2½" by 1". Many K2s seem to need this size of probe anyway, and it'll save you a flashover or seven if yours is one of them. Watch the screen current and NEVER let the coupling decrease to the point where screen current goes positive, even at low drive levels. If you do, the

432MHz

screen current can run away very quickly and the result will be another flashover.

Many K2RIWs seem to require excessive drive. At the supply voltages recommended in Chapter 6 for 4CX250s, an output power of 400W should require just 7W of drive. If yours requires more, it's got a problem somewhere.

All in all, Richard T. Knadle Jr's brainchild isn't perfect, but if you want to use a pair of 4CX250 tetrodes at 432MHz it's the only game in town. If you make unwise substitutions for critical dimensions and components, be prepared for a mighty struggle to get it to behave. Follow the ARCOS dimensions given here and it'll run like a charm. 4CX250Rs, BMs and Bs all work well in the K2, as does the later 8930 (electrically identical to the 4CX250R but with a larger anode-radiator assembly, so the anode dissipation goes up to 350W). However, forget the 4CX350 family – they really don't want to know at this frequency.

CONSTRUCTION NOTES

The anode compartment is an inverted chassis, 12" x 8" x 3", made of 16SWG (14AWG) aluminium. Corners should be soldered, welded or fully overlapped with aluminium angle inside. The top plate is secured to the flanged edge by at least 24 screws; use bushes and machine screws in preference to self-tappers if possible. For the K2RIW cooling arrangement, the anode box must be airtight.

The base for the amplifier is a similar chassis the right way up, with a base plate, and the top cut out to leave a ½" flange all round. The grid compartment is a 5" x 7" x 2¾" box with its own base-plate.

The anode line is made from solid 16SWG brass, preferably silver-plated, and supported on 1" ceramic or PTFE standoffs in the positions shown. Drill pilot holes for the valves in the anode line and use it as a template to drill accurately aligned holes through the top cover, the base of the anode compartment and the grid box, having first bolted the two boxes together. RFC1 mounts vertically beside the anode line and connects to the HV feedthrough C7.

Flapper capacitors are made from thin, springy phosphor-bronze. C5 (anode tuning) is secured to the end wall by clamps or blocks, with several screws. C4 (loading) may require support besides the soldered joint to the output socket (type N). The dimensions of the tuning and loading flappers should be as shown, though some experimentation with the width of the loading flapper may be necessary; see main text. The rest positions of the flappers should be ½" below the anode line and parallel to it.

The input socket (type N or BNC) is mounted on the grid box, with a hole through the rear drop of the base. Use miniature components with low minimum capacitance for C1 and C2. C1 is mounted on an L-shaped metal bracket, and C2 on a piece of insulating fibreglass board (Fig. 10.22). To achieve the correct reactance for C2, you may need to use a series capacitor (C3) and/or thin wire leads connecting C2 to the input socket and the grid line.

Heater, screen and control-grid supplies are brought into the grid compartment *via* feedthrough capacitors. By rotating the valve bases and positioning the feedthroughs appropriately, all wiring in the grid compartment can be eliminated (Fig. 10.22). A V275LA40B VDR is soldered directly to the screen-grid pin of each valve base. RFC3 is mounted vertically from the edge of the grid line and connects to a feedthrough capacitor soldered to the flange of one valve base; this in turn connects directly to the feedthrough capacitor in the box wall.

When using the K2RIW cooling arrangement, the size of the exit vents for the grid airflow is critical. One vent should be in the end of the grid box furthest from the valve bases, with another identical opening in the end wall of the base directly opposite. Fig. 10.22 shows a pattern of 90 holes of $\frac{1}{8}$" diameter – a tedious job but the result will be well-cooled valves. The blower inlet should be the largest possible screened hole in the position shown. Chimneys are made from PTFE sheet, rolled around the anode coolers and secured with Mylar tape. Exhaust holes in

K2RIW POWER AMPLIFIER

the top plate are covered with metal screening, held in place by a thick metal sheet with two holes the same diameter as the valve anodes. The chimneys are lightly held between the top of the anode line and the holes in the metal sheet. Use very light screening material which provides minimal obstruction to the airflow – not 'expanded metal' which has only about 50% open area at best. Shafts for the flapper tuning lines can be metal or plastic, and will require slow-motion drives and/or friction locks.

All in all, getting a K2RIW going is well worth the trouble even though it can occasionally turn into something of a character-building exercise. Construct it with care and attention to detail and provide it with a decent power supply, however, and we guarantee you'll be delighted by the results.

ANTENNAS FOR 432MHz DX

DL6WU long Yagis are hard to beat for 432MHz DX. There are a number of computer-optimized designs which you could consider [14-17], but most are only optimized for specific boom lengths and numbers of elements. The DL6WU family of long Yagis, on the other hand, can be made for 432MHz in boom lengths starting at 2·8m and increasing in steps of about 278mm up to any length you like, and their performance will be very close to the optimum achievable for your chosen boom length. An example is shown in Fig. 8.28, which also demonstrates that the same basic designs, construction and method of feeding can be used from 144MHz up to at least 1·3GHz.

The table on page 7-24 shows the possible boom lengths in wavelengths, and the designs with 14/15, 19/20, 24/25, 30/31 elements and so on are particularly favourable. These Yagis have 'naturally' good front/back ratios and potentially higher gain in the forward lobe. Such generalizations apply to DL6WU-type Yagis only; other types of long Yagis will show similar behaviour, but possibly at different boom lengths and numbers of elements. If one of these designs suits your requirements for gain and overall boom length, use it; otherwise, go for the longest Yagi you can possibly manage.

The table opposite gives more detailed dimensions for 432MHz DL6WU Yagis up to 30 elements. The data were generated using a computer program [18] written by KY4Z/6, W6NBI and G3SEK, which has been validated by DL6WU. The table also shows the estimated gains and recommended stacking distances at intervals of five elements; you can judge the intermediate values for yourself. As noted above, the particularly favourable lengths may produce a little more gain.

Element spacings are quoted to the nearest 0·1mm, but are nowhere near as critical in practice. By making all measurements from one datum point – the position of the reflector – you should easily be able to achieve an accuracy of a millimetre or two along the entire length of the boom. Square-section booms are recommended for mechanical convenience, although round booms have a better ratio of strength to windload. The longer booms will require support and strengthening against sag and side-winds, but this must not affect the electrical behaviour of the elements.

Element lengths are given for a number of element diameters, boom thicknesses and mounting methods. If you need a different combination, you can easily run the computer program for yourself, and this facility is often provided on the RSGB VHF Committee stand at VHF conventions. Element mounting methods can be the same as those shown in Fig. 8.30, and the appropriate length corrections *must* be applied as shown in the table. Having made the corrections, it is quite safe to round the element lengths to the nearest whole millimetre. K1FO has reported noticeable effects on the resonant frequencies of elements due to chamfering the ends [15], and a 0·5–1mm 45° chamfer might be beneficial in reducing the LF detuning which affects all Yagis in wet weather. The fingers of the recommended Heyco nylon bushes are too long to fit inside a 15mm square boom, so it is necessary to cut two opposite fingers off each bush.

The driven element should be a folded

ANTENNAS FOR 432MHZ DX

dipole about 335mm long overall; within reason, conductor diameter and separation are not critical. The dipole should be fed with a half-wave 4:1 balun (Fig. 7-17), which will generally give a good match to 50Ω. As explained in Chapter 7, the match can be improved if necessary by making *small* alterations to the length of the driven element and/or the first director. The boom passes between the limbs of the folded dipole but is not connected to it since this may cause asymmetric excitation; hence the size of the boom has no effect on the dimensions of the folded dipole.

The losses in a half-wave balun can be quite appreciable at 432MHz unless the balun loop is made from low-loss coax. Since URM67 and RG213 are difficult to form into the necessary small loop, a thinner PTFE hard-line such as UT141 (RG402) is recommended. The correct length for UT141 is 233mm, measured over the shielding, plus a 5mm length of dielectric exposed at each end [16]. The shield of the main feedline can be grounded symmetrically to the boom at the balun if desired.

Weatherproofing and mechanical stability can be problems with the construction of Yagi feedpoints. Although some constructors seem to manage with a balun and coaxial connector out in the open [14–16] it is better to use some form of plastic enclosure for the connections between the driven element, balun and feedline. These are often available from the smaller antenna manufacturers who sell spare parts, and for waterproofing they can be filled with closed-cell 'aerosol' foam which is available from boating stores. If the antenna is fitted with a coaxial connector, this may be another point of moisture ingress; in many installations it is better to dispense with the connector and bring the cable straight into the waterproofed junction box.

When mounting a Yagi for 432MHz, try to keep the mast and any supporting struts as close as possible to the centre-line of the boom to avoid distorting the pattern. Do not use a mast diameter greater than 40mm (1·5") and do not use a clamp which stands the boom a long way off the side of the pole. And having taken so much care with the metalwork, don't ruin it all by asymmetrical positioning of the coaxial cable.

DIMENSIONS FOR 432MHz DL6WU YAGIS

Element	Spacing (mm)	Length (mm)			Estimated gain (dBd)	Stacking* (mm)	
		\multicolumn{3}{c}{Element diameter}		Vert	Horiz
		3/16"	4·0mm	1/8"			
Refl	Zero	340·4	340·6	340·9			
Driven	138·8		– see text –				
D1	190·8	300·8	302·2	305·3			
D2	315·8	296·7	299·1	302·3			
D3	465·0	292·9	295·6	298·9			
D4	638·4	289·4	292·2	295·6			
D5	832·8	286·3	289·1	292·7			
D6	1040·9	283·6	286·5	290·2			
D7	1259·5	281·3	284·2	288·0			
D8	1488·6	279·2	282·2	286·1			
D9	1728·0	277·4	280·5	284·4			
D10	1977·8	275·7	278·9	282·9			
D11	2238·0	274·2	277·4	281·5			
D12	2508·7	272·9	276·1	280·3			
D13	2786·3	271·6	274·9	279·1	13·9	1250	1300
D14	3063·9	270·4	273·8	278·0			
D15	3341·4	269·4	272·8	277·0			
D16	3619·0	268·3	271·8	276·1			
D17	3896·6	267·4	270·9	275·2			
D18	4174·2	266·5	270·0	274·4	15·3	1450	1500
D19	4451·8	265·6	269·2	273·6			
D20	4729·4	264·8	268·4	272·9			
D21	5007·0	264·1	267·7	272·2			
D22	5284·5	263·3	267·0	271·5			
D23	5562·1	262·6	266·3	270·9	16·3	1650	1700
D24	5839·7	262·0	265·7	270·3			
D25	6117·3	261·3	265·0	269·7			
D26	6394·9	260·7	264·5	269·1			
D27	6672·5	260·1	263·9	268·6			
D28	6950·1	259·5	263·3	268·0	17·0	1850	1900
D29	7227·6	259·0	262·8	267·5			
D30	7505·2	258·5	262·3	267·0			

* for maximum gain

Element mounting methods
(1) Insulating boom, or elements mounted on insulators at least 1 boom radius clear of boom – use the above element lengths
(2) Elements mounted in full electrical contact through a square boom, through the middle of a round boom, or across the top of a square boom –
  for a 25·4mm boom, add 15·6mm to the above element lengths
  for a 15mm boom, add 6·6mm to the above element lengths
(3) Elements mounted on insulating bushes through the middle of a round or square boom –
  for a 25·4mm boom, add 7·8mm to the above element lengths
  for a 15mm boom, add 3·3mm to the above element lengths

432MHz

REFERENCES

[1] Greg McIntyre AA5C, The AA5C 432MHz transverter. *Proceedings of the 21st conference of the Central States VHF Society, 1987*. Published by ARRL, 1988.

[2] M Lass DJ3VY, Modern receive converter for 70cm. *VHF Communications* 3/1980 page 148.

[3] A simple 70cm converter using a ring mixer and S3030 GaAs MESFET. *UHF Compendium parts 3 & 4* page 470, Editor K. Weiner DJ9HO.

[4] H W Scheffler DC9CS, 432MHz transverter for 70cm (transmitter). *DUBUS-Technik* page 221. Published by DL7QY.

[5] Shichiro Mori JA6CZD and M. Martin DJ7VY, 70cm moonbounce preamp with 0·5dB noise. *DUBUS-Info* 3/1979 page 128. (NB – Reference 12 explains the shortcomings of this design.)

[6] M. Martin DJ7VY, New type of preamp for 145 and 435MHz receivers. *VHF Communications* 1/1978 page 30.

[7] Kent Britain WA5VJB, Dual-gate GaAsFET preamps for 432MHz. *ARRL Handbook*, various editions.

[8] Dragoslav Dobricic YU1AW, Preamplifier – pros and cons. *VHF Communications* 4/1987 page 219.

[9] Dragoslav Dobricic YU1AW, Low –noise 144MHz preamp using helical tuned circuits. *VHF Communications* 4/1987 page 243.

[10] D Norton, High dynamic-range transistor amplifiers using lossless feedback. *Microwave Journal* May 1976 page 53.

[11] *Matching Network Designs with Computer Solutions*. AN-267, RF Device Databook, Motorola Ltd.

[12] Rainer Bertelsmeier DJ9BV, How to achieve a low system temperature on 432MHz EME. *Third International EME Conference, Thorn (NL)*, 1988.

[13] J.S. Gurske K9EYY, Safe, sensible silverplating. *Ham Radio*, February 1985. (Electroplating using spent photographic fixer solution).

Roger Blackwell G4PMK and Ian White G3SEK, More gain from 1·3GHz power amplifiers. *Radio Communication*, June 1983. (Appendix B describes 'dry' silverplating using a rub-on powder)

[14] Joe Reisert W1JR, VHF/UHF World: A high-gain 70cm Yagi. *Ham Radio*, December 1986.

[15] Steve Powlishen K1FO, High-performance Yagis for 432MHz. *Ham Radio*, July 1987.

[16] Steve Powlishen K1FO, An optimum design for 432-MHz Yagis. *QST*, December 1987 and January 1988.

[17] Rainer Bertelsmeier DJ9BV, High gain Yagis for 432MHz. *DUBUS* 2/1991.

[18] The public-domain BASIC program is available via G4ASR on IBM PC disk or as a printed listing. SASE with all enquiries, please.

CHAPTER 11

POWER SUPPLIES & CONTROL UNITS

POWER SUPPLIES & CONTROL UNITS

by John Nelson GW4FRX

In this chapter I'll be presenting circuit ideas and detailed designs for all the power supplies for a big tetrode amplifier. I'll be referring mainly to a pair of 4CX250s or '350s, although the same ideas are adaptable to any tetrode and the HV supplies are equally suitable for beefy triodes. Most of this chapter is structured around a comprehensive power supply and control unit (PSCU) – this device will take care of all the routine chores of stop/start and transmit/receive switching, and will look after both the PA and itself in the event of a fault. I'll also throw in some ideas about simple T/R switching for use with masthead preamplifiers. Incidentally, what you *won't* find here are circuits for ordinary 13·8V DC PSUs and what-have-you – there are plenty of good designs for things like that in print already, so I'll concentrate on the specialized stuff.

First of all, let's recall from Chapter 6 the requirements for powering a pair of 4CX250s or '350s.

1 Anode HV: typically +2000V at 700–800mA with good regulation.
2 Screen grids: +350V for class AB1, +250V for Class C; able to source or sink 50mA per valve, with regulation better than ±50mV; preferably grounded on receive. The screens are never allowed to float when anode voltage is applied.
3 Control grids: variable around –55V for Class AB1, around –90V for Class C; able to sink a few mA in Class AB, and maybe 20mA for Class C with good regulation; switchable to about –120V on receive. The control grids are never allowed to float.
4 Heaters: 6V AC/DC at about 6A for a pair of valves.

I'll consider these requirements in turn – but first **AN IMPORTANT WARNING:**

Amateur radio, working DX and building big valve amplifiers are all enormous fun – but read Requirement 1 again and think about it. If you connect yourself across a 2000V supply, the very best you can hope for is something like a serious heart attack with severe burns thrown in for good measure.

It's much more likely that you'll be dead.

So when you build high-voltage supplies you **MUST** give full thought to the safety of yourself and others. Even if the rest of your station is built out of the proverbial chewed-string and sealing wax, don't skimp on proper metalwork grounded in the right places, good high-voltage wiring with proper insulation, good insulators and the very best components you can get.

Most of all, you **MUST** – not should or ought, but **MUST** – take very great care when testing and measuring.

ALWAYS follow the HV safety precautions in the box opposite.

ANODE HV SUPPLY

The anode HV supply is basically very simple. It doesn't need to be stabilized or homogenized or anything clever; all you want is a 2kV supply whose output doesn't sag too much when half an amp is abstracted from it, and which doesn't self-destruct if you have the occasional short-circuit or flashover in the amplifier. Switched-mode supplies delivering this order of voltage and current are still a bit

ANODE SUPPLIES

tricky for the average amateur, so we're essentially looking at a very simple circuit: a big HV transformer, a rectifier and a reservoir capacitor.

COMPONENTS

What kind of transformer do you need? More precisely, what kind of transformer can you *find?* If money was no object you'd design the supply and order a transformer to your exact requirements – but the odds are that you're not a prince of the realm or an Arabian oil sheikh, so you'll have to search the surplus market like the rest of us. It's a case of staying sharp and grabbing anything which looks likely to generate about 2kV when it's rectified and smoothed. Some of the alternatives are:

AC secondary voltage	Rectifier type	Smoothing
1400–0–1400 to 1800–0–1800	Full wave (biphase)	Capacitor–input
700–0–700 to 900–0–900	Bridge	Capacitor–input (ignore centre tap)
0–1400 to 0–1800	Bridge	Capacitor–input
1800–0–1800	Full wave (biphase)	Choke–input
900–0–900 to 1100–0–1100	Bridge	Choke–input (ignore centre tap)
0–1800 to 0–2200	Bridge	Choke–input

One other possibility for the HV transformer is a custom-wound toroidal type. These are at least affordable as single orders, and come even cheaper if you can get together with like-minded amplifier-bashers and order more than one at a time. Toroids are much smaller than the equivalent conventional transformers but they're mechanically more fragile; they have a number of other minor disadvantages as well but these needn't concern us for now.

The choice of smoothing circuit will depend entirely on whether you can find a suitable choke. Unless a companion choke is offered together with the transformer (and it will be about as big and heavy as the transformer itself) capacitor-input smoothing is going to be the only game in town – which is a bit unfortunate since it's a lot harder on the transformer than the good old choke-input variety. The transformer secondary voltage

---

HV SAFETY

● **KEEP ONE HAND BEHIND YOUR BACK.**

● **NEVER WEAR HEADPHONES** – or anything else connecting you to ground.

● **DON'T ADJUST TEST PROBES ON A LIVE HV SUPPLY** – switch off, clip on the probes, step back, pause for thought, and then switch on.

● **DON'T WORK WITH HV IF YOU'RE TIRED OR IN A HURRY.** When you've spent all evening finishing the HV supply, leave the testing till tomorrow. And *don't* decide to do it when you get back from the pub!

● **DON'T FORGET THAT LOWER VOLTAGES ARE DANGEROUS TOO** – the screen- and control-grid supplies will kill you if you let them. The 240V AC mains supply can also see you off prematurely if you're careless.

It's your life – and you only have one...

---

will dictate whether the rectifier configuration will be biphase or bridge. Avoid voltage doubling unless you know the transformer was specifically designed for that application, with an extremely low secondary resistance. Otherwise the rectified and smoothed HV will sag badly on load, which in turn will do your intermod performance a power of no good.

For rectifiers, do choose something better than boring 1A devices like the BY127 or 1N4007 – these have poor surge performance and are generally antique. The 3A 1N5408 is nicer all round. Twenty 1N5408s arranged as a 2 x 10 biphase or a 4 x 5 bridge with 680kΩ 2W equalizing resistors and 0·01μF 500V shunt capacitors (10% or better tolerance) across each diode will make a super rectifier stack [1].

For HV reservoir and/or smoothing capacitors, professionals would probably use oil-filled paper jobs – lucky them. Nowadays these are specialist items and priced accordingly, but you might be able to get hold of some surplus examples if you're exceedingly lucky. For the rest of us, it's down to a bank of electrolytics in series. The actual value of the resulting composite HV capacitor isn't particularly critical and anything above about 30µF will do. What working voltage do we need? Remember that in a capacitor-input filter the peak voltage on the reservoir capacitor is 1·414 times the transformer secondary voltage – so a 0–1800V AC secondary, for example, will apply no less than 2545 volts to the capacitor. A margin of at least 25% is required for safety, so you need the working voltages of the capacitors to add up to at least 3200V. This means that you're looking for something like eight identical capacitors of more than 240µF and at least 400V working.

Anything else we need to specify about the reservoir capacitor? Yes, ripple current – an important point, and one neglected by several commercial valve-amplifier manufacturers. The ripple-current rating of the capacitor should be about 1·3 times the DC current to be drawn from the supply, *ie* at least 650mA for a pair of 4CXs. Inadequate ripple-current rating is a major reason why amateur HV capacitor stacks blow up – there's a lot of internal heating which culminates in a loud explosion, fountains of white stuff everywhere and one of the most revolting smells on earth. So a safety factor of at least 100% in ripple-current rating is no bad idea.

Bearing in mind that you're after reliability and safety with no drama, the HV reservoir-capacitor stack is another area where I'd strongly recommend you to grit your teeth, smash the piggy-bank and buy new components. Try the STC Electronic Services catalogue for some excellent modern specimens; eight 330µF 400V capacitors cost about £60 in 1992, and ought to last for the next 50,000 hours or so. Their ripple current rating is 1·7A at 85°C, which is ample. To equalize the voltage which appears across each electrolytic, so that each one bears its proper burden and no more, each capacitor requires a resistor connected in parallel with it. The value should be about 100kΩ (not critical, so long as they're all the same) and the wattage rating should be as high as you can fit in. The reason for the conservative rating is that failure of an equalizing resistor will probably involve failure of one or more electrolytics in the stack, with the smelly and tiresome consequences already discussed. New good-quality wirewound or 2W carbon-film or metal-film resistors for this job don't cost much – why tempt Fate?

Even though the equalizing resistor chain will act as a bleeder resistor for the HV reservoir capacitors, and so will the resistors across the rectifier diodes, it's still best to play safe and fit a back-up of about 1MΩ across the entire capacitor stack. Since a 1MΩ resistor will dissipate about 5W at this high voltage, and normal resistors will flash over at more than a few hundred volts, you need to use a chain of resistors – in fact a near-duplicate of the chain of equalizing resistors. If you're using a single large HV capacitor, pay even more attention to the bleeder chain and connect it directly across the capacitor, independently from any other wiring. An isolated, fully-charged paper capacitor can remain lethal for **weeks**...

CIRCUITRY

Fig. 11.1 is a typical circuit for a HV power supply. The precise details will depend on what kind of transformer you can find, but all the rest of the circuit is pretty standard. I'll concentrate on features that help ensure a long and happy life for the two most important items – the HV transformer and yourself.

In HV supplies, fuses cannot give perfect protection – all they can do is to confer some degree of safety. At the end of the day the rectifiers, reservoir capacitors and all the other bits and pieces can be replaced if the HV power supply throws a major wobbler, but if you lose the HV transformer you're in for a prolonged period of depression and mourning. No-one will rewind it and you don't have a spare; to make matters worse, the new one you get after several months of hard searching won't fit in the existing box, so the entire supply will have to be rebuilt. Meanwhile, you've missed the Perseids, the autumn tropo and the only aurora this year. So do your best to protect everything else in the HV box – but be **totally neurotic** about protecting the HV transformer.

Protect the primary against overvoltages and

ANODE SUPPLIES

COMPONENTS

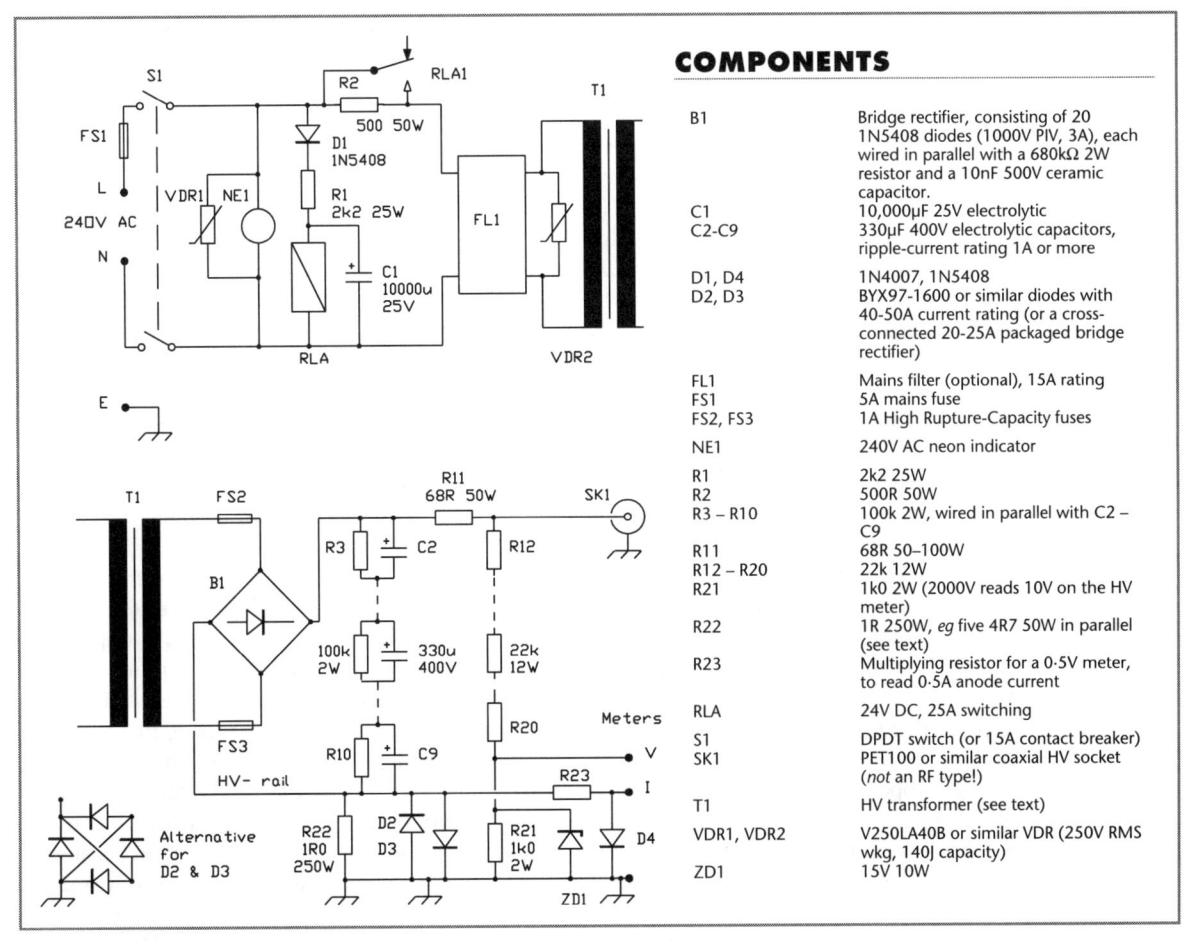

B1	Bridge rectifier, consisting of 20 1N5408 diodes (1000V PIV, 3A), each wired in parallel with a 680kΩ 2W resistor and a 10nF 500V ceramic capacitor.
C1	10,000µF 25V electrolytic
C2–C9	330µF 400V electrolytic capacitors, ripple-current rating 1A or more
D1, D4	1N4007, 1N5408
D2, D3	BYX97-1600 or similar diodes with 40-50A current rating (or a cross-connected 20-25A packaged bridge rectifier)
FL1	Mains filter (optional), 15A rating
FS1	5A mains fuse
FS2, FS3	1A High Rupture-Capacity fuses
NE1	240V AC neon indicator
R1	2k2 25W
R2	500R 50W
R3 – R10	100k 2W, wired in parallel with C2 – C9
R11	68R 50–100W
R12 – R20	22k 12W
R21	1k0 2W (2000V reads 10V on the HV meter)
R22	1R 250W, eg five 4R7 50W in parallel (see text)
R23	Multiplying resistor for a 0·5V meter, to read 0·5A anode current
RLA	24V DC, 25A switching
S1	DPDT switch (or 15A contact breaker)
SK1	PET100 or similar coaxial HV socket (*not* an RF type!)
T1	HV transformer (see text)
VDR1, VDR2	V250LA40B or similar VDR (250V RMS wkg, 140J capacity)
ZD1	15V 10W

Fig. 11.1. Circuit of a 2kV HV supply with soft-start and bridge rectifier. Current is measured using the 1Ω resistor R22 which connects the HV-minus rail to ground. The bleeder resistor chain R12–R21 acts as a 1/200 voltage divider for the HV meter, and can also be used to provide an 'HV present' logic signal

nasty spikes by mains filtering and a suitably-rated VDR, which will keep the rectifiers happy as well. Incidentally, does the shack light go dim and the transformer go "thunnnnnng" when you hit the switch? That awful thump is associated with a fair amount of mechanical and electrical stress on the secondary's insulation. It's much better to use a soft-start system like the one in Fig. 11.1, so you just hear a purposeful hum which increases in amplitude as the high voltage gently comes up.

Fuses, as I was saying... First of all, fuse the primary of the mains transformer with something suitable. A soft-start circuit will avoid the heavy initial current surge, and will allow you to fit a standard fuse (**not** anti-surge) of around 5A rating. That will give some protection to the components upstream of the transformer if, say, the primary goes short-circuit to the core. Regrettably, this failure mode isn't uncommon in elderly HV transformers which have had a hard life in the hands of their previous owners. Next, fit a fuse in each leg of the rectifier stack (Fig. 11.1). They will protect the transformer secondary if

POWER SUPPLIES & CONTROL UNITS

the rectifier stack decides it's tired of life, or if the reservoir capacitor takes it into its head to go dead short-circuit – which block paper capacitors sometimes do, unfortunately. These fuses must be of the 'high rupture-capacity' (HRC) type, with a ceramic body which is filled with sand or something similar. The older 1·25" fuse is best for this particular application, though a modern 20mm UK mains fuse will just about manage. DO NOT try to use a glass-bodied fuse in this position. If you do, and there's a short-circuit fault, two things will probably happen. The first will be a beautiful whitish-blue arc inside the fuse; this arc may drop a few tens of volts, but it certainly won't be the open circuit that all the hard-pressed components are begging for. The arc will last for a second or so before the fuse explodes, spraying fine glass particles everywhere. By that time, all the rectifiers in the stack will probably have resigned in protest, possibly taking with them a few equalizing resistors and even some of the reservoir capacitors. At this point the phone will ring and someone will tell you that it looks as though tropo is opening up to HB9 and Italy. Such a pity that it's just turned 6pm on Saturday evening and you've got no spares...

There's not a lot of advantage in fitting a fuse in the DC output of the supply itself – well, that's not strictly true but, amongst other things, you'd need a rather specialized high-voltage HRC item which is difficult to get. Standard 250V AC cartridge-type HRC fuses in this position may work or they may just arc and explode under short-circuit conditions; they work well enough in the HV rectifier stack because of the quasi-AC waveform, but straight DC is usually too much for them. The best practical solution is to fit a large wire-wound resistor in the DC output to limit the worst-case short-circuit current until the other fuses blow. For a number of years my standby 2·5kV supply has used a 68Ω 100W resistor in the output, with a 1A anti-surge ceramic HRC fuse at each end of the transformer secondary as I mentioned earlier. This much-loved box has survived countless deliberate and accidental flashovers, short-circuits and all manner of assorted abuse – and all that's ever needed replacement have been the two fuses.

While we're pondering the HV supply, how are you going to measure the anode current of the valves? Bit tricky, this. For a start, you'll have to forget about trying to measure the separate anode currents of a two-valve amplifier directly because both anodes are hooked straight to the tank circuit. Nor can you get at the cathodes, because they're both solidly grounded at the valve base – and they have to be. Don't even *think* about trying to use bypass capacitors here. I suppose you could hang a 0–500mA meter in series with the output of the HV supply, if you didn't mind its internal metalwork floating about at a couple of kV. On second thoughts we'll forget that bright idea too...

The right way to meter the anode current is to catch it where it's close to ground potential, *ie* in the negative return of the HV supply. This is very easy. Instead of grounding the negative side of the HV rectifier and the reservoir capacitor, you return them to an HV-minus rail (Fig. 11.1). Then connect a low-value resistor between the HV-minus rail and the common chassis ground. 1·0Ω is a convenient value for a legal-limit 4CX pair, because when you connect a 0–0·5V meter across this resistor, you're measuring anode current on a scale of 0–500mA.

By keeping the metering circuitry down at ground potential, you've made everything much safer. You can now move the meter itself into the control unit on the operating desk, leaving the HV PSU on the floor, and you won't need special HV cable trailing around the place. Very convenient, very safe – aren't we clever? Yes, fine if everything works properly but there could be a problem. Specifically, suppose there's a dead short from the HV line to chassis – suppose, for instance, that an HV decoupling capacitor fails, or somebody drops a screwdriver into an open anode compartment (yes, it has happened). In the basic supply we've been describing, a phenomenally high current will promptly flow through the 1Ω resistor – which will probably not take kindly to this onslaught and

ANODE SUPPLIES

may well opt out of existence in a more or less spectacular fashion. If that resistor blows before the HV fuses, or if there are no HV fuses to blow, the HV-minus rail will fly off to −2kV. Somehow I don't think either the meter or a large portion of the associated wiring will survive that, and *you* might not live to fight another day either, if you touch the wrong things in the wrong order.

Come hell or high voltage, the HV-minus rail **must** stay close to ground potential. You can ensure this by several means. One is to connect a couple of big silicon diodes in inverse parallel across the 1Ω metering resistor, as shown in Fig. 11.1. By "big" in this context, I mean capable of carrying the full prospective surge-current of a dead short-circuit. This is why I use a 68Ω surge-limiting resistor in the HV side; it limits the worst-case prospective current to somewhere around 40A, which meaty diodes can easily handle. With these two components across the 1Ω resistor, the voltage across it can never exceed 1V or so. The meter won't worry about this mild overload, since a 0·5V FSD meter will cope with 1V for ever.

However beefy your diodes, the 1Ω metering resistor must still be incredibly reliable and extremely conservatively rated. If you saw the 1Ω resistor in the main GW4FRX HV supply you'd probably laugh your head off. It's made up of five separate chassis-mounting 5Ω 100W Mil-spec Welwyn components in parallel, with one end of each taken individually to chassis ground with a large bolt and solder-tag, extra-nasty shakeproof washers, and locknuts to finish the job. There's a common 14SWG ground busbar with its own separate ground connection as well, just to make sure. One BYX97-1600 and one BYX97-1600R (DO-5 stud-anode and stud-cathode diodes, rated 47A average, 800A surge – 50p each from a rally) are paralleled across the resistor. This probably sounds amazingly Over The Top. But everything still works, despite the supply having been used for several years of tests involving all sorts of abuse and some extravagantly violent flashovers – not to mention generating the HV for working a fair amount of assorted DX and the occasional contest. The supply is still there, and so am I.

By the way, the dangers you're trying to avoid could also occur if the HV-minus lead drops off the resistor and diodes, while still remaining connected to the meter. To prevent this possibility in the 'FRX supplies, the meter and the HV-minus rail are connected to opposite ends of the '1Ω' busbar. And just in case it really isn't your day, locate the series resistor for the 'anode-current' voltmeter inside the HV supply. If all else fails, this will make sure that high currents at HV-minus can't get out into the meter extension wiring.

Bearing in mind Nelson's Rule – "If you can lift the HV transformer, it isn't big enough" – it's as well to make the HV PSU and the amplifier as separate entities. So the next problem is: how are you going to get 2kV safely to the amplifier? The best high-voltage connectors easily available in the UK are the PET 100 series, which are of screw-on coaxial construction and can often be found in scrap nucleonic gear as well as in the component catalogues. For high-voltage cable the easiest solution is ordinary UR43 coax, which has an adequate DC rating for the job and goes straight into a PET 100 plug. This way everything is nice and safe, and any cable or connector fault is most likely to lead to the HV going short-circuit to ground and blowing fuses rather than doing any personal damage. After all, your HV supply can cope with dead shorts to ground now, can't it?

Let's now suppose that you've solved all these problems and built the anode HV supply. You finally managed to obtain a really superb transformer, which is fused in all the right places and which has nice soft-starting, mains filtering, VDRs and a duvet to keep it warm at night. You've put together a tasty rectifier stack and spent some folding stuff on a decent reservoir capacitor and bleeder resistors. You've built it well, it seems to work a treat and you're now basking in the rosy glow that comes from knowing you've done the job properly. Well done – I'm proud of you. Now answer me this –

POWER SUPPLIES & CONTROL UNITS

Where is the ground return from the amplifier to the HV supply?

If you can't come up with an answer in less than half a second, you could be in for a disaster. You **MUST** have a solid and totally reliable connection between the amplifier's chassis and the chassis of the HV supply. And the whole thing **MUST** have a ground (earth) connection you'd stake your life on – *because that's precisely what you are doing*. Suppose you lose the amplifier's negative return while the anode voltage is on. If you're lucky you'll get away with a spectacular fry-up. Alternatively the chassis and metalwork could silently float up to HV potential, just waiting for you to touch it. Or maybe none of the above – the HV-negative return might find some other devious route to ground, and you'd be none the wiser until the day you part some innocent-looking connection with a plug in either hand...

How do you make certain this never happens? By fitting a real, independent safety ground. Don't just rely on the outer of the coaxial HV cable, or the outer braiding of the multiway cable connecting the power supply and control unit to the amplifier. By all means connect the braided shields to ground, but also make one of the wires in the PSCU-to-amplifier multiway a dedicated ground-return line. However, regard *all* of those as mere back-ups for the *real* safety ground: a 'daisy chain' of substantial 30A flat-braid cable between screw terminals on the back of the amplifier, the PSCU and the HV supply, taken to a solid mains 'earth' connection. This will give you a reliable safety ground which you can actually see is present and correct, and which you can (and certainly should) inspect and test from time to time. To do this, disconnect all power, remove all ground connections and then test them one by one for continuity before reassembling the whole system. Solid grounding and screening are vital for safety, and will also contribute quite a lot to the EMC aspects of the station.

That common grounding strap is absolutely vital for your safety – omit it at your extreme peril.

SCREEN SUPPLIES FOR 4CXS

Having sorted out the heavy stuff for the anode volts, let's move on to the screen supply. To recall what I said in Chapter 6, a 4CX250 screen grid needs around 350V for Class-AB1 service. For Class C the data sheet states 250V maximum, mostly because you could exceed the 12W screen-dissipation rating by driving hard in Class C with too many volts on the screen. It doesn't matter if your screen supply isn't exactly 350V or 250V; neither does it matter if the voltage drifts a bit over the course of an hour or two. What does matter for Class-AB1 operation is the stability under dynamic conditions – basically at syllabic rate. That's the tricky bit: the requirement is to source and sink something in the region of 40–50mA peak while keeping the output within better than ±50mV of the design value. And that's where most screen supplies fall down. A simple two-tone test won't show the problem because the average screen voltage sinks or rises to a steady value. Where the poor dynamic screen-voltage regulation really shows is with speech modulation, on the air.

The screen supply must obviously consist of pure DC and be free from oscillations which would create tone sidebands, or noise causing the 'DIY aurora' effect. It's also desirable to prevent the valve's screen dissipation from being exceeded, *eg* by making the screen supply current-limited. The final point is that in a two-valve amplifier the best way to balance the static anode currents is by slightly altering the screen-grid voltage of one valve. This means that each valve needs a separate screen supply, and at least one of them must be adjustable.

Naturally you need a conventional transformer-rectifier-reservoir combination to produce the basic voltage, but then you need some sort of regulator circuit to meet the other requirements. The requirement to sink current rules out the conventional series regulator, so it has to be some form of shunt regulator circuit. Essentially there are three possible ways to implement a shunt-regulator at 250–350V: gas stabilizer (VR) tubes; Zener

SCREEN SUPPLIES

diodes; or an active shunt-regulator with valves or transistors.

Unfortunately, almost every published screen supply for 4CX tetrodes harks straight back to the Days of Old when all amplifiers ran in Class C for AM, FM or CW. In those days, of course, VHF/UHF SSB was a gleam in the eyes of a few pioneering stalwarts. The thing about SSB, of course, is that it requires the power amplifier to run in Class AB – and in the face of the much more demanding requirements of Class-AB service, those old-fashioned screen supplies just don't cut it.

Let's start at the beginning and examine why this is so. Voltage-regulator tubes or Zeners look simple and attractive, but most published circuits are thoroughly inadequate. Neon stabilizer tubes such as the OA2 and OB2 seem to perform quite well, but you need four in series-parallel for each valve in order to sink the necessary 50mA of standing current through them. You'd therefore need eight tubes for Class AB1 and some fancy switching for Class C – and how are you going to vary one screen voltage slightly to balance the two valves?

Zener diodes look more promising, since you can adjust the number of components in each separate Zener chain to set up the right screen voltage and standing anode current for each valve. But you still need to send about 50mA of standing current down each chain, so you need to use 10W Zeners and a big heatsink. Suppose you decide to use four 82V devices per chain together with a few smaller ones to balance the standing currents – messy, but possible. The slope resistance of these diodes will be a few ohms apiece, and with a current variation of about 50mA the dynamic regulation won't be brilliant. With care in selecting your Zeners and a lot of trial and error you can just about manage 0·5V regulation.

Remember that VR tubes and Zener diodes are both extremely noisy and will require copious bypassing (but be careful not to turn a chain of VR tubes into a relaxation oscillator). Remember also that if you have a flashover, an awful lot of kV will appear at the top of the regulator chain. The best way to prevent this from taking out all the devices at once is to use a VDR across the output of the supply, in addition to the VDR between the screen pin of the valve-base and ground.

Another highly desirable feature for any VR-tube or Zener screen supply is a large electrolytic across the output. As well as helping to take the sting out of a flashover, a capacitor of about 50μF will dramatically improve the dynamic screen regulation at higher audio frequencies, and hence the intermodulation performance. Even if you're not prepared to change your old-fashioned screen supply for something better, the rest of us on the band will be grateful if you add this capacitor as a brute-force method of improving its dynamic regulation.

The next step towards state-of-the-art intermod performance from your 4CXs is active screen regulation. One very well-known circuit by Andy Wade G4AJW [2] featured two separate supplies, each consisting of an op-amp driving a valve shunt-stabilizer. It was widely duplicated, and although purists would argue that the EL84 used was somewhat overrun, the G4AJW design will outperform anything using VR tubes or Zeners. Another active supply, this time using transistors instead of valves, was published by myself in *Short Wave Magazine*; however, a later development of this appeared in *Radio Communication* for December 1987 and January 1988 as part of a comprehensive screen and control-grid supply complete with control logic. We'll examine this power supply and control unit (PSCU) in detail at various points in this chapter.

Meanwhile here's a simpler screen-regulator circuit designed specifically for upgrading existing power supplies.

ACTIVE SCREEN REGULATOR

Figure 11.2 shows an active screen-regulator for a single 4CX. It was developed by Melvyn Noakes G4JZQ, initially as a retrofit item for early 'Heatherlite' amplifiers which had no screen-supply regulation at all! This little gem uses power MOSFETs for both the current

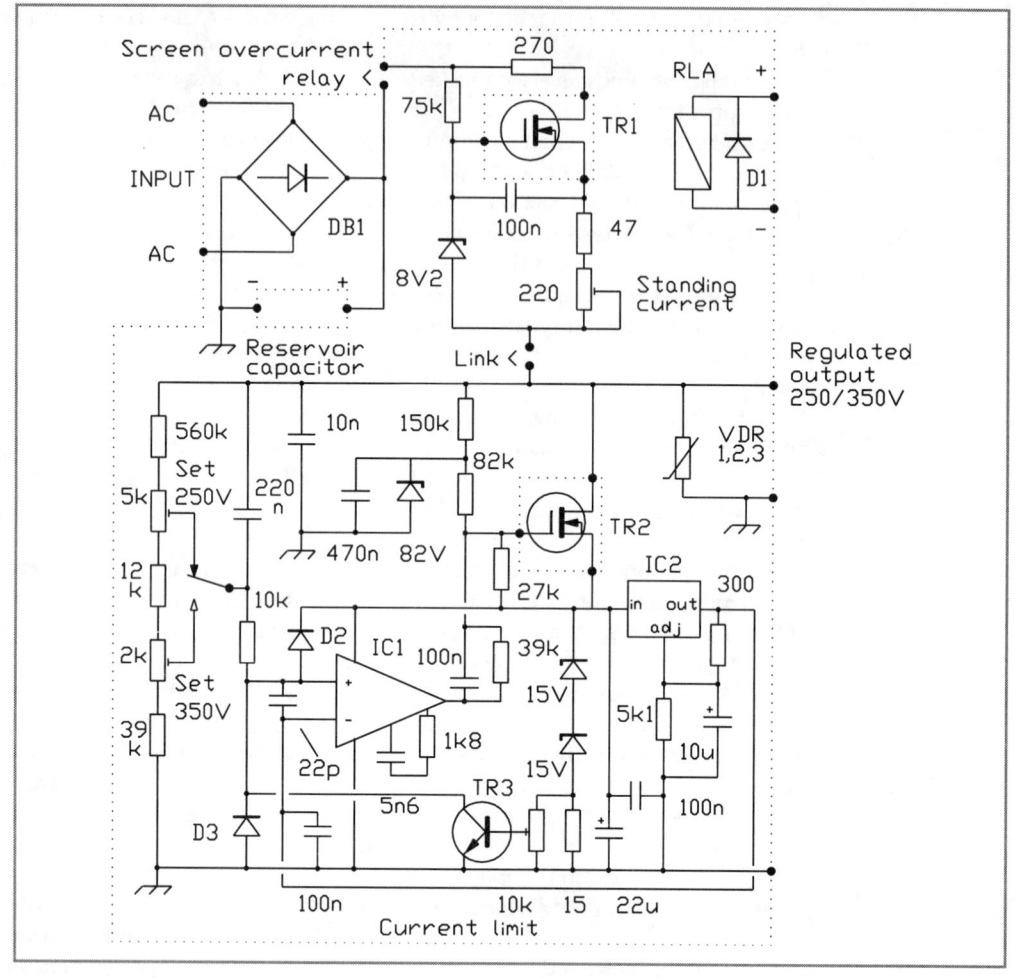

Fig. 11.2. Screen voltage regulator for a single 4CX250/350, by G4JZQ. Relay switching is provided between output voltages suitable for Class C and Class AB1, and each voltage is adjustable. Large dots show interfaces to a PC board

COMPONENTS

D1	1N4003 *etc*
D2, D3	1N914, 1N4148 *etc*
DB1	2KBB100R (1000V PIV 2A bridge)
IC1	LM748
IC2	LM317L
RLA	'BT47'-style 12V DC (energize for Class AB1)
TR1, TR2	IRF840P (500V 8A 125W or greater, with heat sinks
TR3	BC107
VDR1, VDR2, VDR3	V275LA40B (VDR, 275V RMS wkg, 140J capacity), 3 in parallel

source and the shunt stabilizer, and its performance for Class AB1 is as follows:

Output voltage range: 350V ±10V
Max source current: 30mA
Max sink current: 30mA
Output short-circuit current: 52mA approx.
    (with load resistor, 66mA approx)
Ripple rejection: >80dB
Output noise: <0.5mV pk-pk
Regulation at ±30mA: better than ±2mV

The requirements upstream of the regulator are a DC supply of about 420V from a suitable transformer, rectifier and a reservoir capacitor of 16µF or greater. Further smoothing is not required; the regulator itself takes care of that. One option is to connect an overcurrent trip

SCREEN SUPPLIES

relay between the 420V supply and the regulator; for more details, see the description of the comprehensive PSCU later in this chapter.

The upper part of the regulator circuit (Fig. 11.2) is a constant-current source. A constant 45mA flows through TR1, and in the absence of screen current this passes to ground through TR2 and the two 15V Zener diodes. These two diodes also provide a 30V supply for the control circuitry. One input of op-amp IC1 receives a sample of the screen voltage from the voltage divider; the other input of IC1 receives a stabilized reference voltage from IC2.

To analyse the circuit, ask yourself: "What happens if the screen voltage starts to fall?" If that happens, the voltage at the '+' input of IC1 falls, causing the output of IC1 to fall by a much greater amount. This will tend to turn off TR2, which will therefore allow the screen voltage to rise to its correct value. This regulating mechanism will operate until TR2 is fully cut off, allowing the full 45mA from TR1 to flow into the screen grid as positive screen current; beyond that point, regulation is lost and the screen-voltage will fall rapidly.

Negative screen current is sunk through TR2, which would be turned on more and more by IC1 to maintain a constant screen voltage. To prevent excessive power dissipation in TR2, an overcurrent circuit is provided: TR3 looks at a sample of the current being sunk though TR2, and deliberately throws the circuit out of regulation if the current is excessive. This is not only to protect the regulator itself: as you know from Chapter 6, the valve shouldn't be running excessive positive or negative screen current, so the loss of screen-voltage regulation can be used to generate an alarm signal.

VDR protection is provided against flashovers, in addition to the VDR connected directly from screen to ground in the amplifier. A further option is to fit a bleeder resistor directly from screen to ground, to sink a permanent current of 10mA; two 68kΩ 2W resistors will do the job for a 350V supply. As well as preventing the screen grid from ever floating way above ground potential, these bleeder resistors will make a conventional 0–20mA meter display positive and negative screen current on a scale of 10–0–10mA per valve. This is just what you need, because positive or negative screen-current can then be read without the necessity for switching; more on this subject later. Yet another option is to make the supply switchable between 350V for Class AB1 and 250V for Class C, although you'll then need to have two scales on the screen current meter because the bleeder resistors will draw different currents and offset the zero.

The circuit, apart from the power MOSFETs, can be built on a conventional single-sided PCB [3] or on Veroboard, taking care to avoid high voltages between adjacent tracks. It is imperative that the MOSFETs are mounted on an adequate heatsink – failure to do so will destroy them.

Setting-up goes like this. Initially disable TR3 by adjusting the slider of RV3 to ground. Remove the link from the constant-current source to the rest of the circuit and connect a milliammeter across the pins; adjust RV1 for 45mA indicated current. If the two 68kΩ output load resistors are fitted, set the standing current to 45 + 10 = 55mA. (Do not then operate the regulator without the bleeder resistor, because the extra current would then flow through TR2 and would add to its power dissipation.) Next adjust RV2 to set the output voltage to 350V for Class AB1. To adjust TR3 which prevents excessive sink current, briefly increase the current through TR2 to 65mA using RV1, and adjust RV3 until the output voltage just starts to rise above its regulated value; then return RV1 to its proper setting.

In principle this circuit could be used for large single tetrodes if bigger MOSFETs are fitted and a few resistors changed – it could easily do the business for a 4CX1000/1500 or a 7213/DOD006, for instance.

SCREEN SUPPLY FOR A PSCU

Unlike the stand-alone regulator I've just described, the screen supply in Fig. 11.3 forms part of a comprehensive power supply and

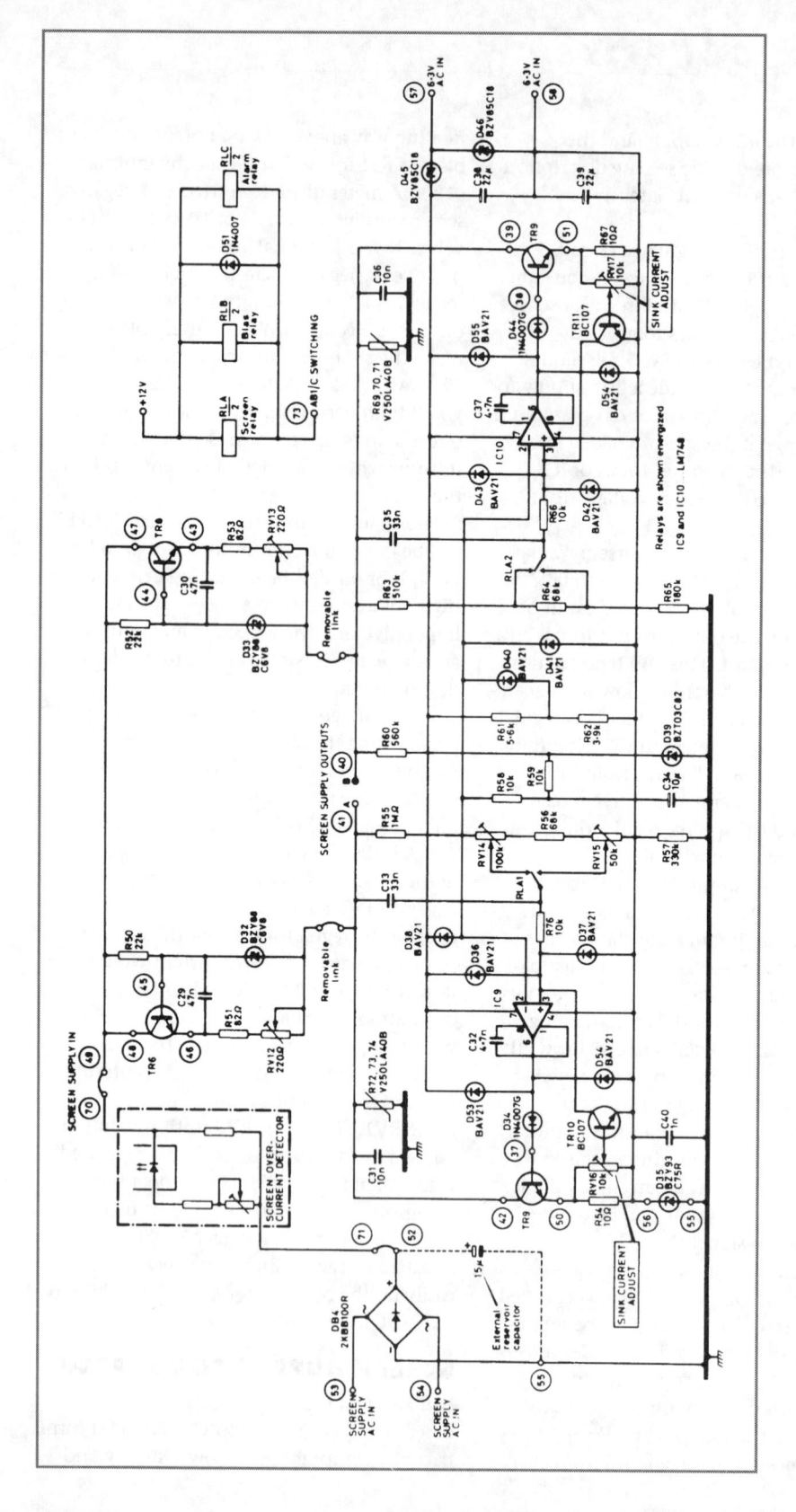

Fig. 11.3. Screen regulator for a comprehensive power supply and control unit (PSCU). See reference [4] for full component details

SCREEN SUPPLIES

control unit [4]. In this case we have the complete circuit for a dual-channel screen supply capable of producing a regulated 350V for Class AB1 or 250V for Class C, which will source and sink some 50mA peak per channel. The description refers to one channel, which is Channel B (the right-hand channel looking at the circuit diagram), and PCB pinouts mentioned in the write-up refer to those on a ready-made board which carries the complete screen, grid bias and logic circuitry [4, 5].

The basic screen-regulator circuit (Fig. 11.3) is an earlier version of Fig. 11.2, using bipolar power transistors and a separate supply rail for the op-amps. The raw DC input – which should be 420V or above – is fed via the opto-isolator IC4b (an overcurrent sensor – see later) to the collector of TR8. With associated components this forms a constant-current source. The 748 op-amp IC10 has its inverting input voltage derived from the 82V reference from D39, which is compared with a sample voltage derived from the output rail in the resistor network R63, R64 and R65. This sample of the output voltage is compared with the reference voltage from D39 in IC10, which drives the shunt stabilizer element TR9. The output voltage is thus kept constant. If you're familiar with the 748, you'll note that in this circuit the compensation capacitor has a considerably greater value than is usual. This 'overcompensation' technique can be a powerful tool for ensuring stability in regulator circuits where the total open-loop gain is extremely high, as it is here. Note that an internally-compensated op-amp such as the 741 cannot be used in this circuit. The supply voltage for the op-amps is produced from a spare 6·3V AC winding in a slightly unorthodox manner, using Zener diodes D45 and D46 as a voltage doubler which also provides protection against transient voltages. Note that this must be a dedicated winding, since it is floating at 75V above ground. TR11, RV17 and R67 are for protection against excessive sink current, which could conceivably damage TR9. The output voltage of the supply is set to either 350V for Class-AB1 working or 250V for Class C by means of RLA2. The two output

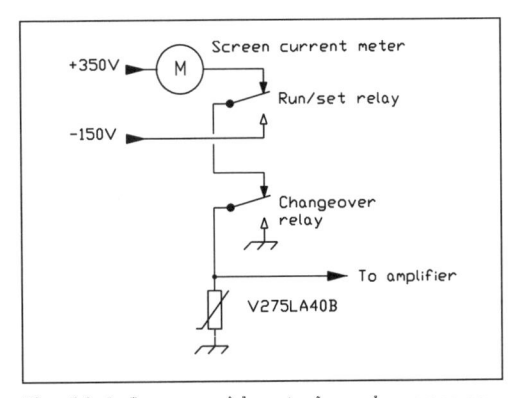

Fig. 11.4. Screen-grid metering, changeover switching and run/set switching within the PSCU (one for each grid). Fig. 11.15 shows an alternative system with SCR 'crowbar' protection for the screen grids

voltages of the B channel are fixed; however, those of the A channel can be varied by RV14 and RV15 to equalize the standing currents in a two-valve amplifier. R69, R70 and R71 are varistors (VDRs) whose function is to protect the screen-supply circuitry in the event of flashover.

The performance of this design is very good. If less than 50mA peak is being sourced or sunk, the consequent variation in output voltage is not measurable on a 4½-digit DVM or visible on an oscilloscope. However, if more than some 40mA is drawn on a sustained basis, the output voltage quickly falls to a safe value so that the screen dissipation of the valve cannot be exceeded even under fault conditions. As you would expect of a shunt stabilizer, it doesn't care if it's short-circuited. The output of the supply is fed to a changeover relay via metering and a run/set relay (Fig. 11.4), which enables each valve to be turned off individually by –150V applied to the screen grid so that the correct standing current can be set.

In normal operation the screen-grid changeover relay grounds the screen on receive and connects it to the output of the screen supply on transmit. The bleeder resistors and VDR prevent the screen grid from floating momentarily as the relay changes over. Grounding the screen grid during receive periods avoids shot

11-13

Power Supplies & Control Units

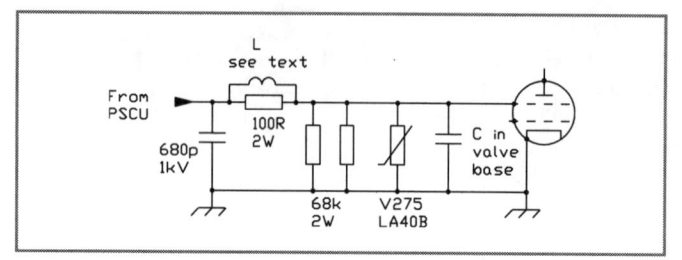

Fig. 11.5. Screen-grid protection circuitry within the amplifier

noise in the receiver, and also positively prevents the screen from floating away to a high voltage. A further advantage of the changeover system is that any alarm in the control unit or failure of the HV can be made to ground the screen grids; the 'ready' relay RLD deals with this. If you can invent a way of electronically commanding the screen voltage between 250–350V and a solid ground without using a relay, let me know all about it for the second edition!

The PSCU circuitry is continued in the amplifier itself (Fig. 11.5). Two 68kΩ 2W resistors and the VDR should be connected directly to pin 1 of each valveholder. Allowance will obviously have to be made in the screen metering for the 10mA drawn by the two resistors, but this can give you a sneaky way of obtaining a centre-zero meter as mentioned earlier. Incidentally, you really need a separate meter for each screen grid if it's a two-valve machine. Don't try and save a meter by switching – it's false economy. You'll curse fluently when it comes to setting the amplifier up and you're trying to remember which switch position does what with which valve.

The series inductor in Fig. 11.5 is formed by winding ten or so turns of 22SWG enamelled wire around the body of a low-value 2W resistor. In conjunction with the built-in capacitor in the valve base and the 680pF capacitor shown on the diagram, this forms an RF lowpass filter. Most 4CX250 amplifier designs include a low-value resistor in series with the screen feed to each valve, but this seems to be one of those hangovers from the Class-C age when the valves had a common supply. Quite what it's supposed to achieve I can't imagine. It can have little or no anti-parasitic effect, and if you have a flashover it'll certainly force the voltage on the screen of the valve to rise to a very high value – which will almost inevitably destroy the capacitor in the valve base. The best thing to do with this resistor is to wind an inductor round it.

Some writers have recommended a fuse in series with the screen-feed lines: this is a dangerous mistake. No known fuse will blow reliably at such a low current as 50mA, and very low-current fuses are prone to early fatigue failures with varying current. With a correctly designed screen supply there is no need for fuses, and they can be positively dangerous if fuse failure leaves the screen grid floating; flashover and destruction of the valve base is likely to result. Screen grids must *never* float with respect to cathode, as we've seen.

Having thought about screen-grid circuits in some detail over the years, I've come to the conclusion that a bit of modification to most published amplifier circuits would be a Good Move. If you've already built an amplifier, look at its screen-grid circuitry in the light of Fig. 11.5. A few alterations along these lines will do a lot to reduce the risk of flashover and the consequent damage – the subject of my dire warnings in Chapter 6.

You may think I'm paranoid about flashovers and destruction of valve bases. Not at all – I'm just trying to make *you* paranoid, so that you'll lie awake at night and worry until you've taken some precautions! If you do incorporate these items, you can be totally relaxed about pops and bangs in the PA. Just for fun, I tested the circuitry and modifications mentioned above by inducing a series of some twenty test flashovers to see whether anything blew up. After each one there was no damage to any part of the amplifier or its power supply apart from blown fuses in the anode HV supply – I could have been back on the air working the DX as soon as the fuses had been changed. For these tests I used a W1SL-type 144MHz amplifier (Chapter 8) with the screen-grid modifications of Fig. 11.5. Deliberate flashovers are easy to induce, by the way. Tune up the amp and drive flat-out into a dummy load. Then switch to receive, remove the load and apply full drive again...

One final thought is that the little plastic-cased meters which most of us use for grid-and screen-current metering are not really designed to be floated at high voltage above a grounded metal front panel. There isn't much insulation between the scale plate and the panel, and the insulation on the internal wires is very thin. Be particularly careful if you're using a meter with internal lighting – you're bringing a ground lead right inside the meter. If you want to use a small plastic-cased meter in any application where the internal workings are at more than about 50V above (or below) ground, either don't use the internal lighting or replace the relevant wires with something more substantial. There isn't much room between the movement and the casing, so use thin PTFE-covered wire and avoid scraping the insulation when you put it all back together.

So that's about it for the screen supply. Obviously the choice you make will depend on what you want and how state-of-the-art you want to be. The main point of discussing it at considerable length has been to convince you that 4CX amplifiers in linear service really do need better screen supplies than the vast majority of published circuits, and then to show you what the options are.

CONTROL-GRID SUPPLY

Happily there's a lot less to the control-grid supply (or "grid bias" as old-timers tend to call it) than there is to the screen supply. Apart from supplying a constant bias voltage at zero current for Class AB1, the control-grid supply only needs to sink current (not source it) when you over-drive into Class AB2, or when you switch to a different voltage for Class C. The voltages should of course be adjustable so that you can set the standing anode current.

The control-grid supply also needs to produce a large negative voltage on receive, to cut the valve off. Now here's another mild quirk of the 4CX250 family: unlike almost every other tetrode ever made, you haven't quite got total control of the anode current via the control grid near cutoff. To turn a 4CX250B right off you not only have to apply a large negative voltage to the control grid, but you also have to take the positive volts off the screen (which means you have to firmly ground it – remember?). Only then will the anode current fall to zero and stay there.

Since the control-grid supply doesn't have to source any current, a simple variable shunt-stabilizer will do nicely. You can reckon that a '250B with 2kV on the anode and 350V on the screen will need about –55V to set a standing current of 100mA for Class-AB1 SSB use. Somewhere around –100V will do both for Class C and to cut off the valves on receive. As long as the circuit will give you about 15V variation either side of those ballpark figures, you should be all right.

What sort of circuitry? Actually, you can get away with nothing more elaborate than a transformer-rectifier-smoothing circuit to provide something like –150V, in conjunction with a switchable potential divider including a variable potentiometer. Hundreds of amplifiers have been built with grid supplies no more complicated than that, and they work just fine. The only snag is that the working point of the amplifier will move about when you're using Class C – which, to be honest, doesn't really matter. It's better to stabilize the supply, though, because otherwise the working point of the amplifier in Class AB1 wanders around a fair bit as the mains wobbles about (or the generator voltage falls when your contest colleagues put the kettle on) so the power and intermod performance vary slightly. Here again, the G4AJW circuit [2] is fine, although it's a bit old-tech now – you could replace the VR tube by a 150V Zener, for example.

A slightly more elaborate Nelson-Noakes circuit (Fig. 11.6) uses an op-amp and shunt stabilizer. The reason wasn't so much a desire for the last word in stability but a wish to eliminate the TX/RX changeover relay and so remove any chance of the control grid ever being allowed to float. In this design the grid voltage is electronically commanded up and down between its 'transmit' and 'receive'

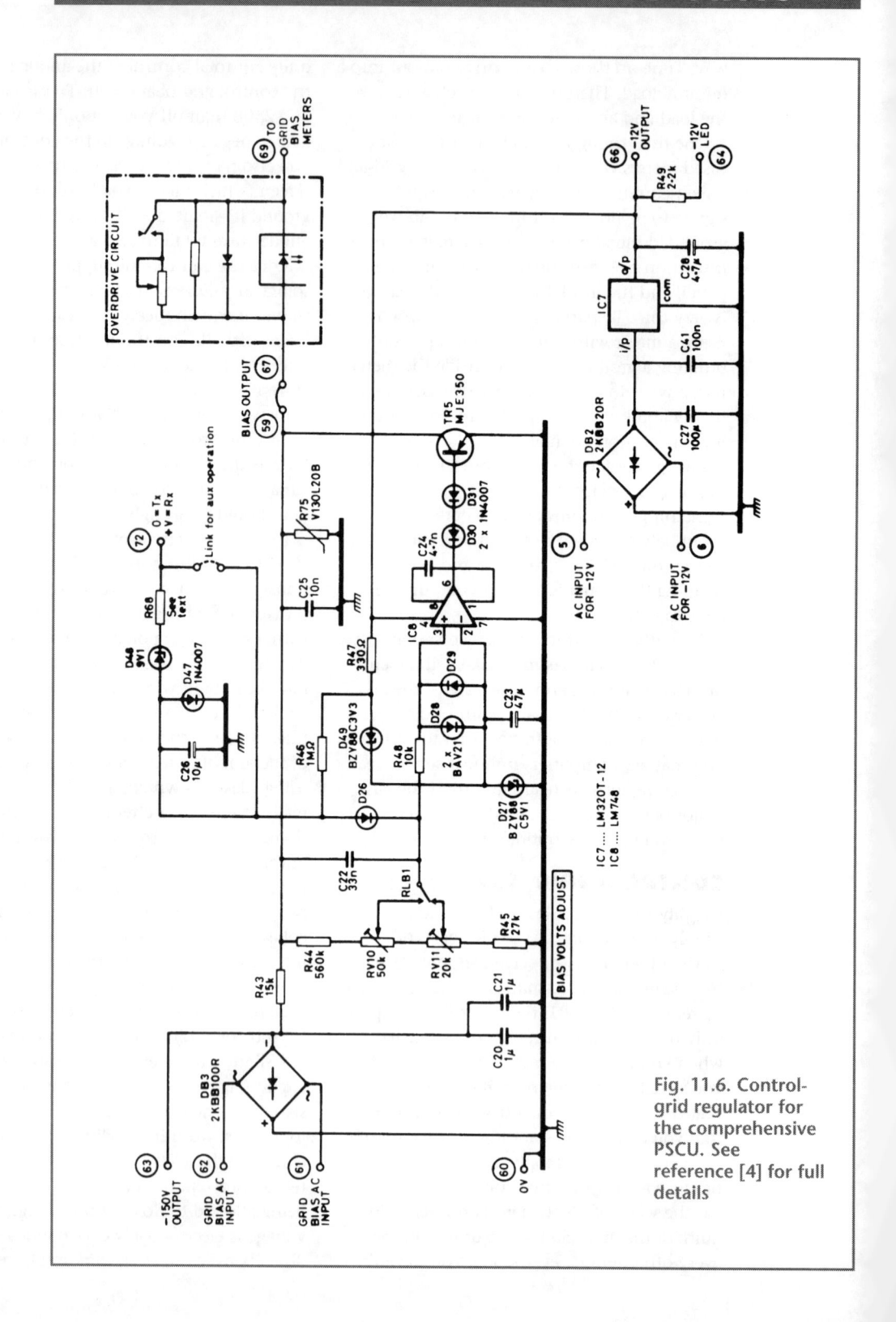

Fig. 11.6. Control-grid regulator for the comprehensive PSCU. See reference [4] for full details

HEATER SUPPLIES

values. A time-constant is included so that the all the changeover relays can settle down and stop bouncing about before the valves turn on and start producing RF to fire at the DX.

Take a look at the details in Fig. 11.6. The heart of this wondrous creation is yet another overcompensated 748 driving a shunt-stabilizer transistor TR5. RLB1 selects the appropriate voltages for Class-AB1 or Class-C use. Operating voltage is derived from a three-terminal regulator IC7. As I just mentioned, a timing element is included in the transmit/receive switching of the supply, so that it takes some 100ms to switch to transmit from receive but only a few milliseconds to return to receive from transmit. Two methods of transmit/receive switching are possible, and are described more fully in reference [4].

The performance of this supply, like that of the screen supply, is very good indeed. Because its reference is a 5V1 Zener diode – a value which has a temperature coefficient of virtually zero – the long-term stability of the output voltage is excellent. Since the supply will sink up to 60mA without difficulty, there are no problems of fluctuation of the working point of the valve as a function of grid current. The output ripple of the bias supply is less than 2mV peak-to-peak.

Each control grid requires its own current meter, of the order of 0–10mA for a 4CX250 (in case you want to go into Class C). A 0–1mA meter will suffice to show that a 4CX350 isn't drawing any grid current at all (except that you can draw up to 1mA per grid when balancing the grid drive – see Chapter 6).

That's all that can be said about the grid-bias supply, so we can heave a sigh of relief and turn to our last area of concern – the heater.

HEATER SUPPLY

Heater supply? Doesn't that just need six volts or something? Surely there isn't much else to say about it? Well, yes and no. Melvyn Noakes and I developed a little circuit to stabilize the heater voltage [6] for six good reasons:

1 The 4CX family data sheets stipulate that the heater voltage should be held to within 5% of the 6V rating to ensure long life and consistency of performance. This specification is in fact impossible to meet if the valve heaters are supplied directly from a transformer connected to the UK mains electricity supply! The reason is that the power companies are obliged by statute to maintain voltage and frequency only to within 6%. In practice, this is probably hair-splitting – but read on.

2 When I moved from London to the middle of nowhere in Wales some years ago, I found the regulation of the incoming mains was more like 10% than the statutory 6%, probably due to the long run of overhead cable. From discussions with amateurs in various parts of the UK, I get the distinct feeling that this situation is not uncommon in rural areas.

3 We made some measurements made on the regulation of apparently adequate generators during portable contest operation. At typical stations the combined line and load regulation seems to be around 15% – and judging by comments made by other contest operators, we rather think that poor voltage-regulation may be common at large portable stations. All the above measurements are, of course, considerably outside the Eimac 5% specification.

4 A power supply then under construction didn't have space for an extra 6V transformer with the appropriate current rating, but there was a spare 6·3V 8A winding on the screen-supply transformer.

5 I had an irrational fit of impatience when faced with the prospect of fiddling about with lengths of wire and resistors in order to reduce 6·3V to 6V at the requisite current: there had to be a better way!

6 The final, deciding factor was a set of life-test data on a large sample of 4CX250B valves, and consequent discussions with two professional users and the valve manufacturers themselves. We learned that the following procedure can give a dramatic increase in the usable life of 4CX tetrodes. The valve is run for the first 100 hours of use with precisely 6V DC on its heater, to allow the emission to stabilize. It is then run under CW conditions at full power whilst the heater voltage is reduced in

0·1V steps. At some point a marked drop in output power (or, if two-tone drive is used, a very pronounced deterioration in intermodulation performance) will be observed. If the heater voltage is then increased by 0·2V from that point, that corresponds to the optimum heater voltage for the valve in that class of service.

With the 4CX250B at frequencies between 120MHz and 300MHz the optimum voltage at maximum ratings would usually be around 5·7V, and between 300MHz and 500MHz it was about 5·5V. Maintaining that voltage accurately and stably would lead to approximately a threefold increase in life. The implication for amateur use is that one valve or pair of valves purchased or obtained from new would probably last longer than the owner.

Another important factor when considering valve heater supplies is the surge current which flows in the heater when you initially switch on. Like the filament of a lamp bulb, the heater of a thermionic valve has considerably less resistance when it is cold than when it has reached normal operating temperature. If the source impedance of the heater supply is very low, the initial current flowing in the heater at switch-on will be several times its normal running value. The resultant thermal cycling and overload places considerable stresses on the heater. To obtain some idea of the magnitude of the problem, the heater of a 4CX250B was fed for test purposes from a suitably metered 6V 30A DC supply. At switch-on the initial surge current took the supply into current limiting; the current being drawn was still 11A five seconds after switch-on. Valves of the 4CX family are built to take this kind of treatment, but it's reasonable to infer that it does nothing to lengthen valve life. Many 6V transformer secondaries are quite able to supply very high peak currents for a short time, and this is probably a good reason not to use a generously-rated heater transformer. For many of its small triodes and tetrodes (although not, as it happens, the 4CX250 family) Varian-Eimac recommends a separate heater transformer which will saturate and limit the switch-on surge. In particular, you shouldn't use a 6V overwind on a multi-purpose transformer with a large core.

So some form of controlled soft-start system is a sensible addition to a heater supply, and this probably applies to other valves as well as the 4CX series. It's fairly simple to incorporate suitable soft-start circuitry into an electronic heater-voltage regulator system such as the one described here.

In other words, there's actually a great deal to be said about valve heaters. It's not so much from the point of view of performance, but more in order to maximize the life of what are – when all's said and done – very expensive devices. When we first published this design [6] for a stabilized heater supply, G4JZQ and I received a good deal of incredulous comment from people who couldn't for the life of them see the point of it. But if by spending about £20 you can triple the life of other components costing anything up to £200, isn't that a bargain?

Anyway, here's the design we came up with – and if you still can't see why we bothered, skip this bit. The heater current of a 4CX250 is specified as 2·6A at 6V, and the corresponding figure for a 4CX350A is 2·9A. For a two-valve amplifier, this implies in round numbers a total heater current of some 6A at 6V. Assuming that a suitably rated 6·3V secondary winding is available for use in the heater supply, the design task comes down to deriving a 6V 6A stabilized DC supply from it. In the final design the value of output voltage was to be finely adjustable, to permit use of the running-in procedure outlined above for extending valve life. Overvoltage and overload protection were to be included, as was a soft-start system and optional remote sensing of the voltage right at the heater pins.

The main problem in getting 6·0V DC from a 6·3V AC supply is that the trough of the output ripple voltage of the smoothed supply will be only about 6·2V at best. This is largely due to the voltage drop in the rectifiers – which will be significant at 6A, even if you use Schottky devices. Conventional IC voltage regulators for high current involve a voltage drop of at least 1·8V, so they're out. Even

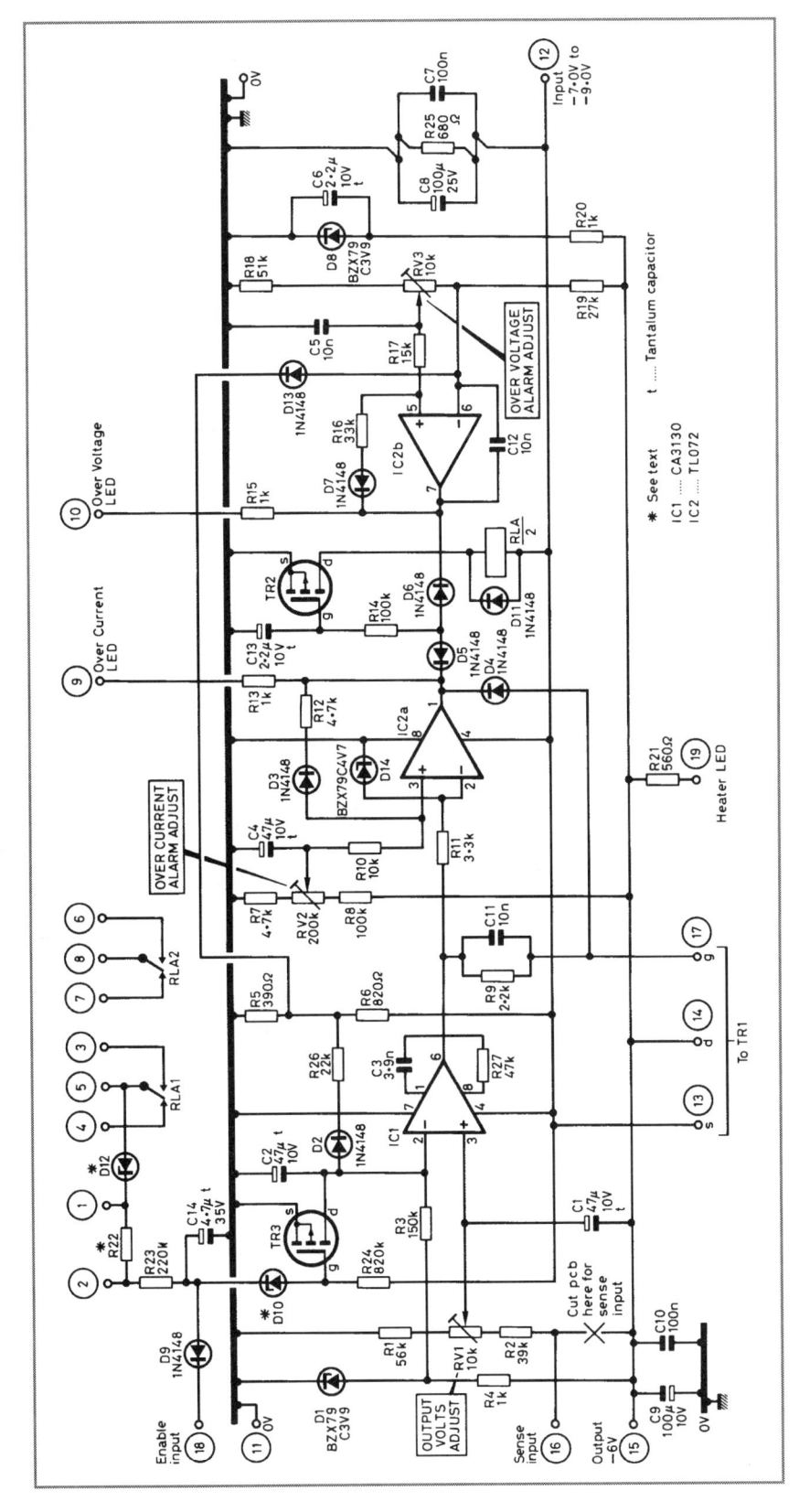

Fig. 11.7. Heater regulator for the comprehensive PSCU. See reference [6] for full details.

POWER SUPPLIES & CONTROL UNITS

discrete regulators using low V_{CE}(sat) bipolar transistors need more than the available 'headroom' of 200mV. A switching regulator is possible and indeed one was used at GW4FRX for some time; it worked very well but there were rather a lot of components in it! Also, some inductors involved were rather difficult to make and did not prove easily repeatable by other constructors, so for that reason it isn't given in this book.

The job became much more simple with the appearance of a new generation of power FETs capable of handling very high currents. The FET equivalent of V_{CE}(sat) is R_{DS}(on), which is simply the minimum resistance of the device when fully turned on. Modern power FETs such as those used in the present design have R_{DS}(on) figures of around 15 milliohms, implying a voltage drop of no more than about 90mV at 6A. The other nice thing about power FETs is that they can also be connected directly in parallel to achieve whatever low level of R_{DS}(on) is necessary, with no need for ballast resistors and no danger of second breakdown. This means that cheap devices found at rallies can simply be paralleled to achieve the required performance. The only constraint is their quite high input capacitance, and any driver circuitry for power FETs must take account of this. Also, the majority of low-R_{DS}(on) devices available at present are N-channel, which means that this particular design has a negative output voltage.

Fig. 11.7 shows the circuit of the final design evolved to meet the requirements outlined above. In essence there's an operational amplifier, IC1, in which a sample of the output voltage from the unit is compared with a reference voltage derived from a 3·9V Zener diode D1. The sample voltage at the non-inverting input is in fact slightly variable by means of RV1, which forms the fine output-voltage control. The output voltage from IC1 is taken to the gate of TR1, which is the normal pass transistor. Note that the output of the unit is –6V, with the positive side of the supply being at 0V (ground). The reason for this is the high turn-on voltage of the FET, which for an N-channel device implies that the gate must be some 4V positive with respect to the drain (output). Since there is only 6V available, the supply has to be turned upside-down so that the FET can be driven from the existing rails. The valve heater doesn't mind which way round its supply voltage comes, of course, so the supply polarity is academic.

Both long- and short-term output voltage stability of this supply are very good, since the reference Zener D1 has only a very small negative temperature coefficient and is fed from the stabilized output. Mounting one of the prototypes in a die-cast box and warming the box to about 40°C produced a measured decrease from 6·00V to 5·98V output.

Because the reference voltage is derived from the output voltage – which obviously isn't there at initial switch-on – some form of start-up circuit is required. This is formed by D2, R5 and R6, which at turn-on time provide some 2V at the inverting input of IC1, enabling the regulator to start. When the reference voltage is fully established, D2 becomes reverse-biased and isolates the unregulated 'start' voltage from the regulator.

IC2 is a dual op-amp which is used for overvoltage and overcurrent detection. IC2b compares the output voltage with an independent reference voltage derived from the 3·9V Zener diode D8. An output overvoltage fault drives a warning LED and also operates the 'fault' relay RL1 via the diode OR-gate D5 and D6 and FET TR2. This relay can be connected to external circuitry as required; typically it should operate the main heater-control relay to remove the heater voltage from the valve. The regulator circuitry itself is directly interfaced with the heater control relay by means of TR3. The intention here is to ensure that the relay never switches high current, either at switch-on or – as a result of a fault condition – when the supply is switched off by TR3 before the relay opens. In the comprehensive PSCU setup, the amplifier airflow switch enables the heater relay via contacts on RL1 but is also tied to the regulator circuitry. Failure of cooling air turns off the heater supply electronically in less than a millisecond, leaving the heater relay to catch

HEATER SUPPLIES

up as described above.

Overload protection is provided in a slightly unusual way. The normal technique would be to sense the voltage developed across a resistor in series with the output of the supply. That implies a voltage drop across the resistor, which we couldn't afford in this design. Instead, the gate voltage of the pass FET is compared in IC2a with a sample of the supply output. If the gate voltage increases past a pre-determined point because of an excessive current demand, the output of the op-amp operates TR2 and RL1 as before and illuminates an 'overcurrent' LED. IC2a's output also switches off the pass transistor via D4 to prevent potential damage. C4 delays the trip function until the correct voltage has been established after switch-on. This method of current sensing is not especially precise, but it can distinguish perfectly well between a pair of 4CX heaters and a dead short-circuit – which is all that's really required. Fig. 11.8 shows how the regulator unit fits into the overall heater supply.

It's important to give some thought to the transformer, rectifier and reservoir capacitor as well as the electronic components in the regulator itself [6, 7]. The current rating of the transformer will obviously have to be adequate for the valves to be used. Supplying a rectifier-reservoir-regulator combination is rather harder on the transformer than supplying a valve heater directly, so a factor of safety is desirable. In round numbers, if you're using this circuit with a pair of valves demanding 6A you'd be wise to choose a transformer capable of delivering perhaps 8 or 9A into an ordinary load. Fortunately, many ex-military or commercial transformers are considerably under-rated (for example, the famous 'Admiralty rating' demands full output from all windings simultaneously, at an external temperature of 80°C, for ever and a day). One of the prototypes was used with a surplus Parmeko 6·3V 5A transformer to supply a pair of 4CX250Rs for a couple of openings and a contest or two, with no ill effects.

The rectifier is the next item to consider. Many packaged silicon bridge rectifiers are available, and one with a rating of at least 25A – such as the obsolete Mullard/Philips BY261 (still commonly available at rallies), International Rectifier 26MB series or the General Instruments KPBC25, for instance – should be used. The chosen device should be bolted to a substantial heatsink (preferably the main chassis of the power supply) using plenty of heatsink compound. Unfortunately, conventional silicon bridge rectifiers will inevitably drop well in excess of 1V at 6A or so; in this circuit one volt is precious. Schottky rectifiers will do rather better. The 30A Philips BYV121 family (which replaces the old BYV20 series) looks a good choice, although you'll need four of them to form a bridge. The heater supply currently in use at GW4FRX uses two Philips PBYR16045 devices; these are Schottky 'duals' which are connected in a bridge configuration. No-one seems to make a packaged bridge Schottky power rectifier, unfortunately, but

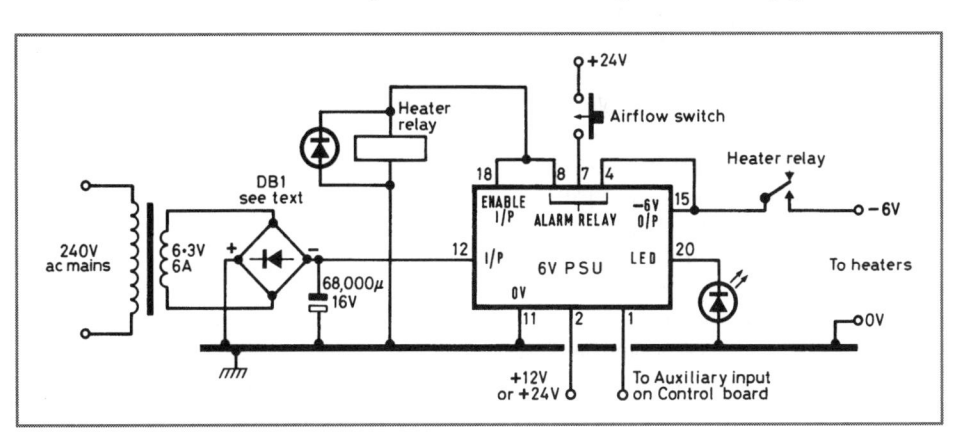

Fig. 11.8. Connections to the heater regulator PC board [6]

using two packaged pairs is the next easiest solution. Having said that, be advised that some of them are configured as common-cathode devices, so you can't use them to make a bridge rectifier.

The reservoir capacitor is an important component in this design. Normally, of course, the use of an active regulator circuit implies that considerably less reservoir and smoothing capacitance can be used to achieve a particular level of output ripple. However, in this application it's essential to provide enough capacitance to keep the trough of the input ripple voltage above the input threshold of the regulator during the worst case of low mains voltage and full load. Calculating on the basis of an imaginary transformer with 5% regulation producing 6·3V at the required current for 240V input and then assuming a worst case of 220V input, full load current and 2V drop in the rectifier and associated wiring, the minimum value of reservoir capacitor which theoretically ought to be used is about 50,000µF. The nearest preferred value above is 68,000µF, and a working voltage of 16V would be appropriate. The ripple-current rating of the capacitor must also be adequate and should be at least 10A for reliability. In practice, large 'computer-grade' components with screw terminals such as those frequently seen on offer at rallies should do the job very well; many of these have their ripple-current ratings marked on them. Remember, however, that surplus electrolytic capacitors should always be re-formed prior to use.

Finally, all wiring carrying the full input and output currents should be substantial; we used 30/0·25mm (1·5mm^2) in the prototypes. Fig. 11.8 shows the basic layout of the completed heater supply and the components involved. Further details of construction, commissioning and integration with the comprehensive PSCU are given in the original reference [6].

When running-in new or nearly new valves, I'd suggest you apply a heater voltage of 6·00V (measured at the socket) for the first two hundred hours' use, and then carry out the tests mentioned in the introduction to this topic. When the optimum heater voltage for your particular application has been established, the overvoltage alarm should be reset so that it trips at some 0·1V higher than the voltage in normal use.

ALARMS & INTERLOCKS

Right – we've ploughed our way through the basics and hopefully got some ideas about what sort of techniques we'll use to generate the appropriate voltages. What safety precautions and interlocks and whatnot need to be provided? And when do we apply them?

I'm sure you're more than capable of operating a few switches. However, for the sake of a bit of time and trouble at the design and construction stage, you can make the PSCU look after all the boring chores of power-up, transmit/receive sequencing, fault and alarm handling and power-down. You only begin to appreciate these features when you stagger out of bed at 0255 for the 0300 sked, throw one switch and know that the PA will look after itself while you try to find the bit of paper with the sked frequency written on one corner. And if you're chief amplifier builder to a contest group, you'll probably sleep better if you know that the amp is being looked after by a PSCU that's a shade more awake than the average night-shift operator...

So how does a 'smart' PSCU start up? First of all, it turns on the air. A good amplifier control unit will say to itself on power-up "Is there any air going through the valves or not? Until there is, I can't allow any voltages on them – sorry and all that". Equally, as mentioned in Chapter 6, we want the control unit to continue to keep an eye on the blower and to do something *pronto* if it fails. When the air is on, the heaters can be powered. The stabilized supply outlined in the previous section starts up electronically, but you can do it very simply with an airflow switch and a relay if you wish. At this stage you can also apply the control-grid bias voltage at its 'receive' level. After the heater voltage has become fully established, you must wait for a minimum of one minute before applying the HV to the anode of a small 4CX – longer than that for larger valves – so the PSCU then starts a timer.

While that minute is ticking away, we ought to note a few things. One is that the screen voltage must *never* be applied to any of the 4CX family if the anode voltage isn't present. If it is, the screen will think it's become the anode and proceed to draw vast amounts of current, which will kill the valve stone-dead. So you must have some interlocking which satisfies two conditions:

1 The screens can't come off ground (*ie* the screen relay isn't allowed to change over) until the anode supply is established.
2 Failure of the anode supply for any reason will immediately remove the voltage from the screen – preferably by forcing the screen changeover relay to release and grounding the screens as before.

Those are the minimum interlock requirements for applying any volts to a 4CX tetrode. If you want, you can do the whole thing with relays and/or switches as in G4AJW's PSCU [2], but in a very simple supply it's up to you to do the right things at the right times. It's better to build in a bit of automation, because then you can also include some fault protection. Here's the way Nelson and Noakes did it [4].

In Fig. 11.9, all the alarm signals are combined into a single 'fault' rail, and each alarm detector drives a front-panel warning LED.

POWER SUPPLIES & CONTROL UNITS

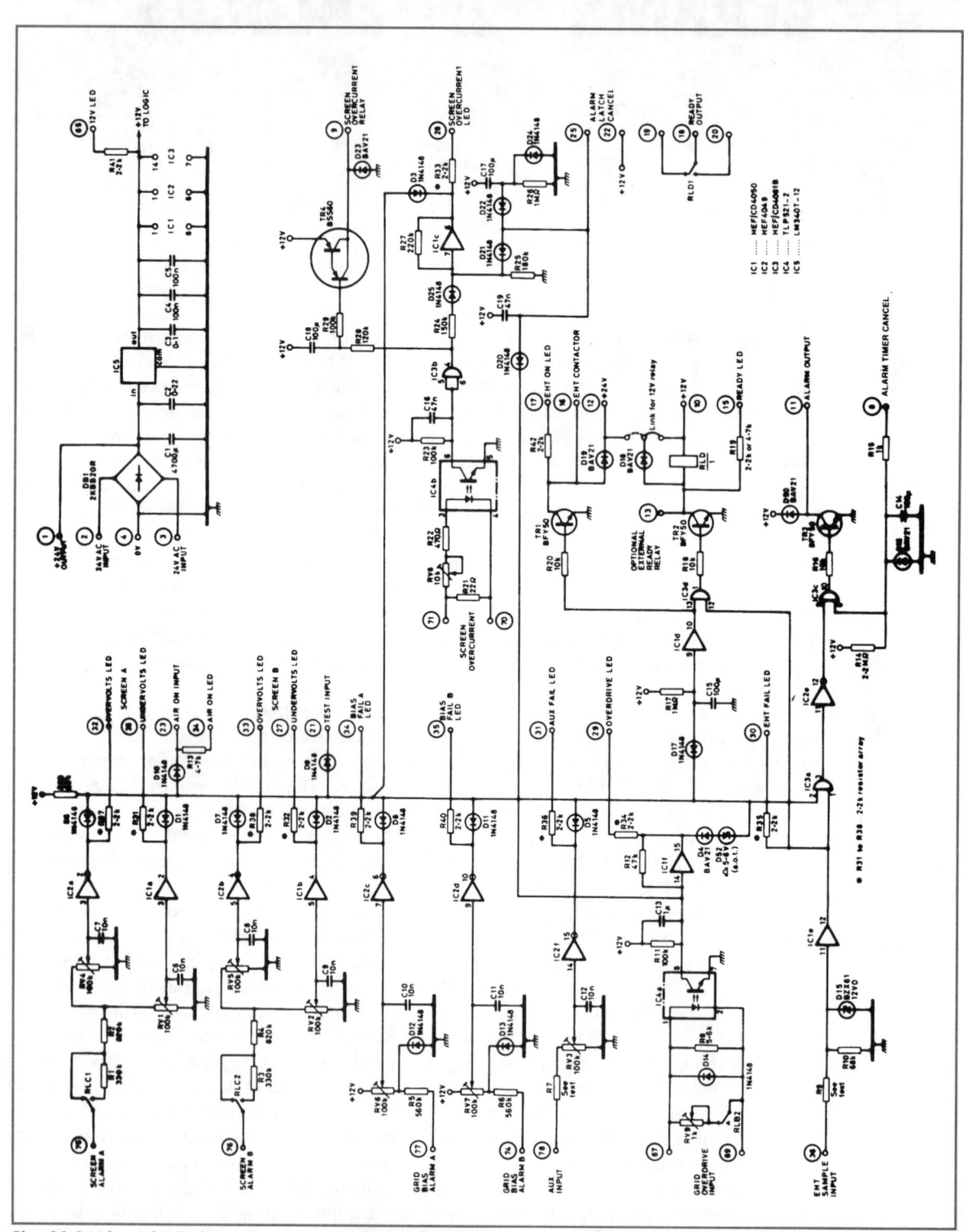

Fig. 11.9. Alarm logic for the comprehensive PSCU [4]

ALARMS & INTERLOCKS

IC1a/b and IC2a/b form over- and under-voltage detectors for the two channels of the screen supply. Typically these will be set to trigger the alarm if the voltage on either screen grid – nominally 350V – strays above 355V or below 340V. RLC1 and RLC2 alter the voltage-divider values to cater for the different voltages produced by the screen supply in Class AB1 and Class C. IC2c and IC2d form under-voltage detectors for the two grid-bias feeds. (Imagine what would happen if the grid-bias voltage disappeared and nothing was done about it – the anode-current meter needle would hit the end stop at about Mach Three, and quite a lot of damage could be done before the HV fuses blew.) IC1f, in conjunction with the opto-isolator IC4a, forms a grid 'overdrive' alarm whose function is discussed later.

IC4b and associated circuitry forms a screen overcurrent detector. The diode part of the opto-isolator is in series with the raw DC feed to the screen-stabilizer circuitry, and any fault which causes excessive current to be drawn will operate a resettable latch (IC1c) and illuminate a LED. Incidentally, since the screen supply is essentially a shunt stabilizer, this part of the circuit cannot be used for overcurrent protection of the valves themselves. If the overcurrent persists for more than about five seconds, a relay disconnects the raw DC feed to the screen regulator.

IC2f is a 'spare' fault input, used as the heater undervoltage/no-voltage alarm in conjunction with the heater regulator described above – and it's easy to 'fan-in' any number of additional active-high alarm voltages to this point using a diode OR-gate. Two further inputs to the common fault rail are derived from the airflow switch and from the 'test' switch, operation of which resets the timer. Other active-low alarms could be fed into the circuitry at this point via diodes if required; in the GW4FRX amplifier the antenna changeover relay supply is monitored in this way.

All the outputs of the alarm detectors are tied to a common rail which is normally pulled up to +12V by R30. A failure condition will force this line low and reset the one-minute timer formed by IC1d, R17 and C15. At initial switch-on C15 is discharged and the input to IC1d is consequently low. After C15 has charged for one minute the output of IC1d switches high, turning on TR1 and energising the HV supply contactor. The output of this timer also goes to one input of the two-input AND gate IC3d. When the HV appears, a sample is fed via IC1e to the other input to IC3d. If there are no other alarm conditions, the output of IC3d switches high, turning on TR2 and energising RLD, the 'ready' relay.

The 'ready' relay is the PSCU's message to the outside world that the machinery is all fired up and ready for action. There are several ways of using this set of relay contacts, but in essence it should be connected to the external circuitry in such a way that the amplifier cannot be placed in circuit unless it has operated. This implies disabling the PTT line, the antenna changeover relays, the screen changeover relay and the bias-supply changeover system unless the PSCU has closed the 'ready' relay. Have a look at Fig. 11.10 [4] for one way of doing it. There is plenty of scope for implementing your own requirements in this area, and another possibility is presented later in Fig. 11.15, in the section on Tx/Rx control.

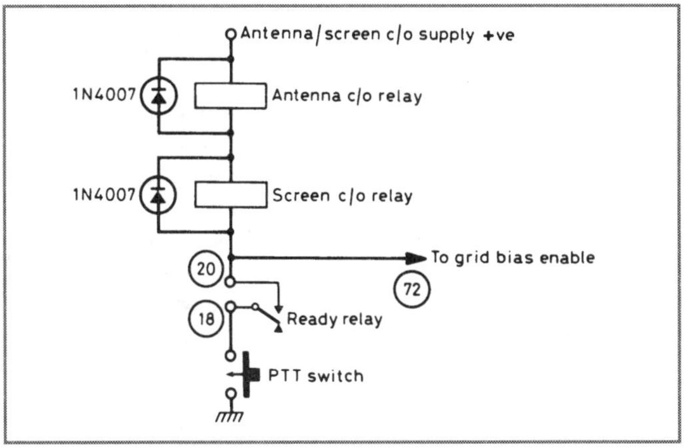

Fig. 11.10. 'Ready' relay logic for the comprehensive PSCU [4]. Fig. 11.15 shows another option

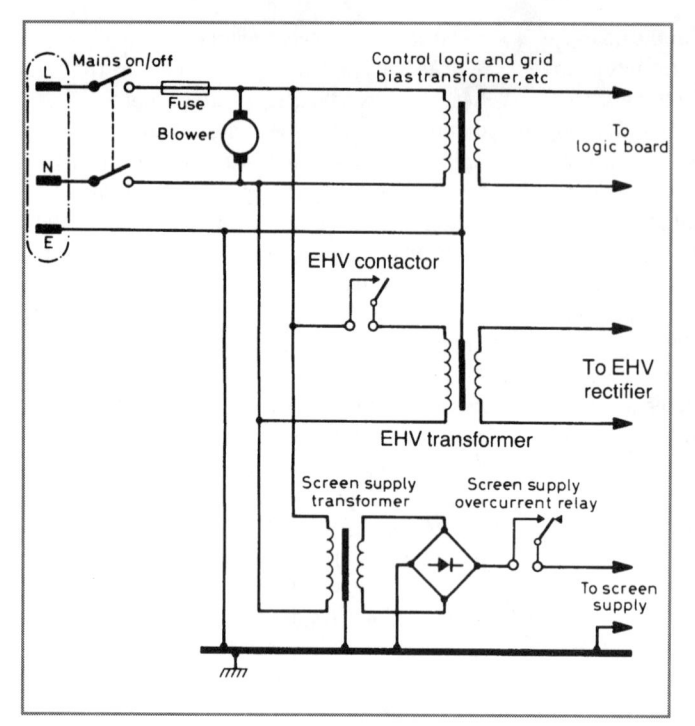

Fig. 11.11. Interconnection to the HV and screen-supply transformers

second pause. This may sound a bit elaborate, not to mention Draconian, but it's just the thing for reminding the contest night-shift that Big Brother is still watching the quality of the signal from the cosy depths of his sleeping bag!

Full construction details are given in reference [4], and PCBs etc are available for the main control board [5]. Interconnection to the HV and screen transformers is shown in Fig. 11.11.

Setting-up is a matter of systematically checking that all alarms operate, that the main timer and audible-alarm timer work and that the 'ready' relay operates when all supplies have been established. With no alarms, the HV contactor output (pin 16 in Fig. 11.9) should go low after about one minute. If a simulated HV sample is then applied to pin 36, the 'ready' relay should operate. If the HV sample is not applied, the master alarm should sound after about 2.5 minutes. Simulating any failure should result in the 'ready' relay opening, pin 16 going high to disable the HV and the master alarm sounding.

So there you are – the all-bells-and-whistles FRX/JZQ control logic, which does everything but make the tea. If all that was too much for you, G4AJW's circuit [2] handled some of the functions but without the over- and under-voltage sensing and some other twiddly bits. The only potential problem with Andy's design is that it depends rather heavily on exact resistor values, precise Zener voltages and a few other bits and pieces; whilst it's easy to get going if you have a passing familiarity with transistor circuitry, it can be a bit of a fiddle if you're not. Even so, it was a classic in its day and a lot of them were built.

In the end it's your choice. You can opt for high-tech or low-tech, and decide how much personal responsibility you want to take for the well-being of the amplifier and PSU. If you can't bear the thought of a PSCU that sometimes seems smarter than you are, that's fine by me. Personally I need all the help I can get to keep skeds at 0300, and it's nice be able to concentrate on working the DX or piling up the contest points without having to worry

If any fault occurs when the amplifier is running, the common fault rail is pulled low and the 'ready' relay drops out. In addition to the specific fault LED, the fault activates a master alarm light and an audible sounder which the operator has to cancel manually. Very handy if you've wandered away to make a cup of coffee while the memory keyer is grinding out a two-minute EME sequence.

The 'overdrive' alarm (IC4a) isn't so much to detect a fault as to stop you winding the drive up beyond a sanitary level. As we've seen in Chapter 6, SSB operation with a pair of 4CXs means that you have to stay in Class AB1 – which implies no grid current. If you overdo the drive, IC4a will detect any grid current and pass the message on to IC1f. This in turn gives the alarm rail a nudge via D4 and D52, forcing it low enough to drop the amplifier off-line, light an LED, and sound the audible alarm until the operator pushes a button to acknowledge the error of his ways. The amplifier comes back on-line after a haughty five-

ALARMS & INTERLOCKS

about the well-being of the PA or whether it might be splattering all over the neighbours. Whatever you decide, it'll probably take you much longer to build the PSCU than it took to build the amplifier!

One final point to ponder is how you're going to convey all the various operating voltages to the amplifier from the PSU and control unit. We've discussed the HV side of it, but what about the grids, screens, antenna changeover, heaters, blower and so on? For preference, go for a multicore screened cable with good reliable connectors. Miniature D-type multiway connectors are fine inside the PSCU, but they're mechanically a bit fragile for the external power cables – so go to the nearest rally and get hold of some nice surplus military-style things like the Cannon 'KPT' or Burndy 'Metalok' series. When you've got them, do spend some time making up the cables and connectors carefully. They're often the things that go wrong in amateur amplifiers, especially if they're taken out on contests and so on – and the results of a cable fault can be more than a bit nasty.

TRANSMIT/RECEIVE CONTROL

You might wonder why we have a separate section devoted to this topic – don't you just press the PTT and let it all happen? Well, no – changing from transmit to receive and back involves a lot more than hooking a bunch of relays together so they all crash over at once. As you might have begun to suspect from the previous section, various things need to be done in sequence. The ideal way to get from receive to transmit and back is as follows:

RECEIVE → TRANSMIT	TRANSMIT → RECEIVE
1: Mute receiver	7: Enable receiver
2: Remove power from preamplifier	6: Apply preamp power
3: Change over coaxial relays	5: Change over coaxial relays
4: Confirm they've changed over	4: Confirm TX is off
5: Apply screen volts to PA	3: Remove screen volts
6: Release cutoff grid bias on PA	2: Apply cutoff grid bias
7: Apply drive	1: Disable driver

In this ideal scheme everything happens in due sequence, and only after the previous event has been checked and verified. The important features are to avoid coaxial relays switching heavy RF power, and to protect the preamplifier device from excessive RF.

COAXIAL RELAYS

Coaxial relays can carry far more RF power than they can switch; the important point is to give the contacts time to stop bouncing and settle into place before you smash the RF through them. For example, the little Magnetic Devices 951-type coaxial relay (RS Components part number 349-686) is usually thought of as an antenna changeover relay for a big PA because its RF rating is only 50W. However, it will reliably carry the UK legal power limit at 144MHz for years if you treat it kindly in a sequenced changeover system. The same applies to the popular CX520 relay, rated at 300W at 1GHz; in a sequenced setup it will carry a kilowatt at 432MHz. If you attempted to 'hot-switch' that kind of power, the contacts wouldn't last for more than a QSO or two.

As far as the preamplifier device is concerned, the most important feature of the coaxial antenna-changeover relay is its isolation or 'crosstalk'. The on-chip dimensions of most preamp devices are extremely small, with very limited insulation and current-carrying capacity. One false move and the device is dead, or damaged so that its gain and/or noise figure deteriorate. Many devices can be damaged by as little as 100mW of RF power, and it's a very good move to design the coaxial changeover system so that the device never sees more than 10mW. In round numbers, if you have a kilowatt flowing between the antenna and TX ports of the relay, the isolation to the RX port must be at least 1000W/10mW = 50dB. Take care when reading the relay specification. Isolation is generally specified with all relay ports terminated in 50Ω, but in the practical setup your RX port is unlikely to look like 50Ω – so the actual isolation may be better or worse than specified.

For EME, where you need the very best in power handling and isolation combined with minimum loss, the HF-400 and HF-2000 relays

TRANSMIT/RECEIVE CONTROL

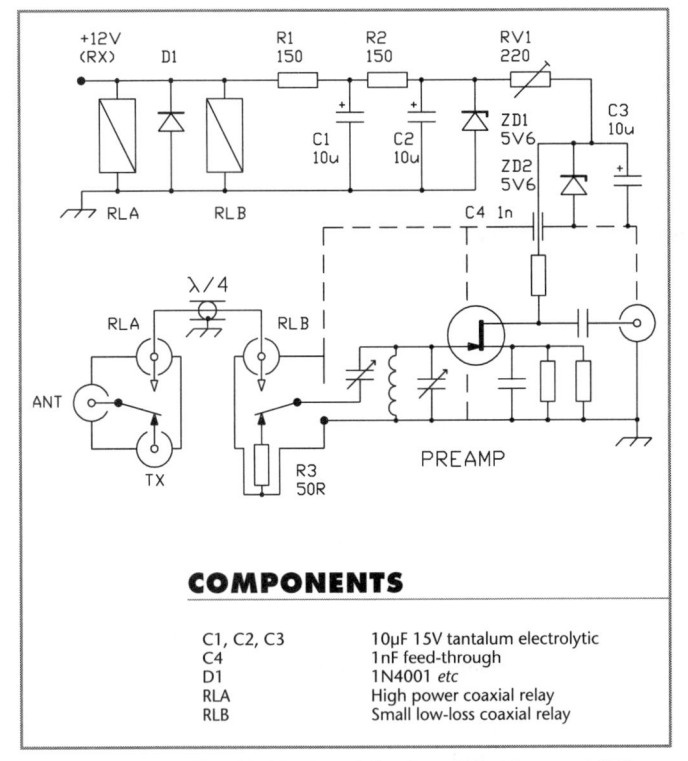

Fig. 11.12. Coaxial-relay switching and DC-supply precautions to protect a GaAsFET preamplifier

COMPONENTS

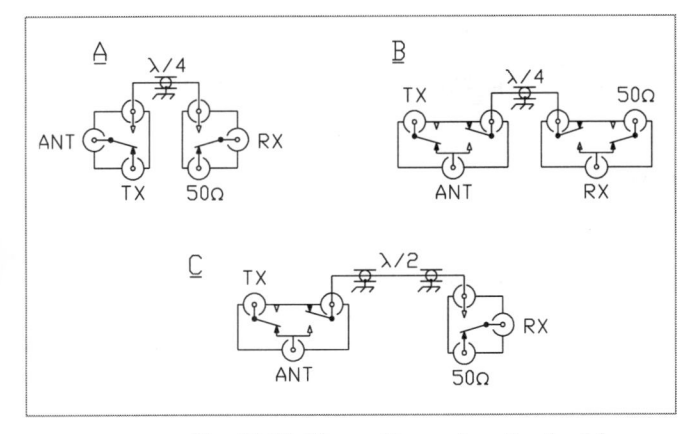

C1, C2, C3	10µF 15V tantalum electrolytic
C4	1nF feed-through
D1	1N4001 *etc*
RLA	High power coaxial relay
RLB	Small low-loss coaxial relay

Fig. 11.13. The optimum length of cable between the two coaxial relays depends on their construction

(available from Piper Communications in the UK) are highly recommended. These relays are beautiful examples of German craftsmanship but they don't come cheap. A more cost-effective solution for general DX-chasing is to use a second small coax relay to protect the preamplifier (Fig. 11.12). By using two relays you separate the two problems: only the main relay RLA carries high power, and RLB only needs to provide good isolation.

What about VSWR and insertion loss? Aren't they worse with two relays? Won't they ruin my receiver noise figure? Happily, no – or at least, not much. Although the measured insertion losses of coaxial relays in a 50Ω system may seem significant, they are due almost entirely to mismatched impedances – meaning that almost no power is actually lost in the relay itself. As long as the relay is truly coaxial and reasonably compact, any impedance 'bump' will be relatively minor except perhaps in a bad case at 432MHz. In a valve PA, the tuning and loading controls will automatically take complete care of any mismatch seen by the transmitter, and even with a fixed-tuned transistor PA you shouldn't have too much trouble. A minor VSWR bump needn't bother the receiver RF stage either. If you connect the noise source (Chapter 12) to the antenna socket and tune for the lowest noise figure with all the coaxial relays and cables in place, you've automatically compensated for the minor impedance mismatches.

There's a right and a wrong way to cascade two coax relays for maximum isolation. You'll be delighted to know that the obvious way, just connecting the two relays directly together, is always the *wrong* way. To achieve maximum isolation you aim for the worst possible impedance mismatch for any RF leakage, and you reach this happy state of affairs by using the impedance-transforming properties of quarter-wave (or sometimes half-wave) coaxial lines. Take a look at Fig. 11.13A. Leakage across the power relay RLA is caused by the open-circuit RX port acting as a short antenna inside the relay, so the RF leakage comes from a high-impedance source. The open-circuit port of RLB is a high-impedance point as well, but you can transform this by adding a quarter-wave of coax line, to present a low impedance to the leakage from RLA. This creates the worst possible impedance mismatch for the RF leakage, which will give the best possible isolation from the two relays together.

What would happen if you connected the two relays directly together or used a half-wave of coax? In both cases the impedance at the RLA end would be high, so the leakage through RLA would be transferred quite efficiently into RLB. In the worst case (with two identical relays and a half-wave line between them) the extra isolation might be only 6dB above that of a single relay!

If the unused contacts of RLA and RLB are shorted to ground (Fig. 11.13B) they both act as low-impedance magnetic pickup loops. Once again you'll need a quarter-wave line for the worst mismatch and the best isolation from the pair of relays. However, if one relay has open contacts and the other one is shorting, a quarter-wave line would provide an impedance match for RF leakage and you'd be in trouble. In that case the correct length would be a half-wave as shown in Fig. 11.13C.

Fortunately the correct length isn't too critical. Most coax relays are only a little short of the desirable isolation, and two give you plenty in hand unless you make a serious error in calculating the length of the connecting line. Even so, it's a Good Move to make some effort to get the length right. Don't forget the velocity factor, and also don't forget to include some estimate of the effective lengths inside RLA and RLB.

DC PROTECTION OF GaAsFETS

Returning to Fig. 11.12, the DC connections to the GaAsFET are protected against voltage spikes on the supply rail. Modern GaAsFETs are quite robust as far as RF leakage is concerned, but a single spike exceeding the voltage ratings will inevitably do them in. Some early published advice has proved rather over-cautious; there's no problem about using the same 12V supply for the GaAsFET preamp and the coax relays, provided you take sensible precautions. The actual details of the preamp don't matter here, although you'll find a full description of a 432MHz design in Chapter 10.

A GaAsFET typically requires 3V at 10mA, or a supply voltage of about 4V if you're using source-resistor biasing. The best way to arrange this from a 12V supply is to regulate the voltage down to about 5V using a 4·7V or 5·6V Zener diode (ZD1 in Fig. 11.12). Why not a 7805 IC regulator? Simply because Zeners provide very fast clipping of voltage spikes, and active regulators don't. You can also turn the dropping resistor for the Zener into a two-pole lowpass filter, which will stretch and further reduce any spikes on the 12V rail. You'll be aiming to send about 10mA through the GaAsFET plus about the same through the Zener, so the 150Ω values shown for R1 and R2 will generally be correct. RV1 and the source bias resistor of the GaAsFET should be adjusted to provide the necessary supply and bias voltages, and can of course be optimized for lowest noise figure. The second Zener diode ZD2 should have the same voltage rating as ZD1; it's normally inoperative but it's there to save the GaAsFET if something else fails. The circuit is liberally sprinkled with 10µF tantalum bypass capacitors and reverse-voltage protection diodes (1N4148s or 1N4001s will do fine). Note that you must use *tantalum* capacitors – ordinary aluminium-foil electrolytics have poor HF performance and won't even notice a microsecond spike. The GaAsFET will – fatally.

SIMPLE TX/RX SEQUENCER

Fig. 11.14 shows a simple TX/RX sequencing circuit suitable for controlling a transverter and a valve PA, which goes with the preamplifier arrangements in Figure 11.12. This circuit doesn't attempt to do anything clever, like verifying that the coax relays have changed over and stopped bouncing – it simply leaves enough time for those things to happen before applying 12V to the transmit stages of the transverter.

The main feature of this circuit is that the coaxial relays are energized on **receive**, which has several advantages. When the entire

TRANSMIT/RECEIVE CONTROL

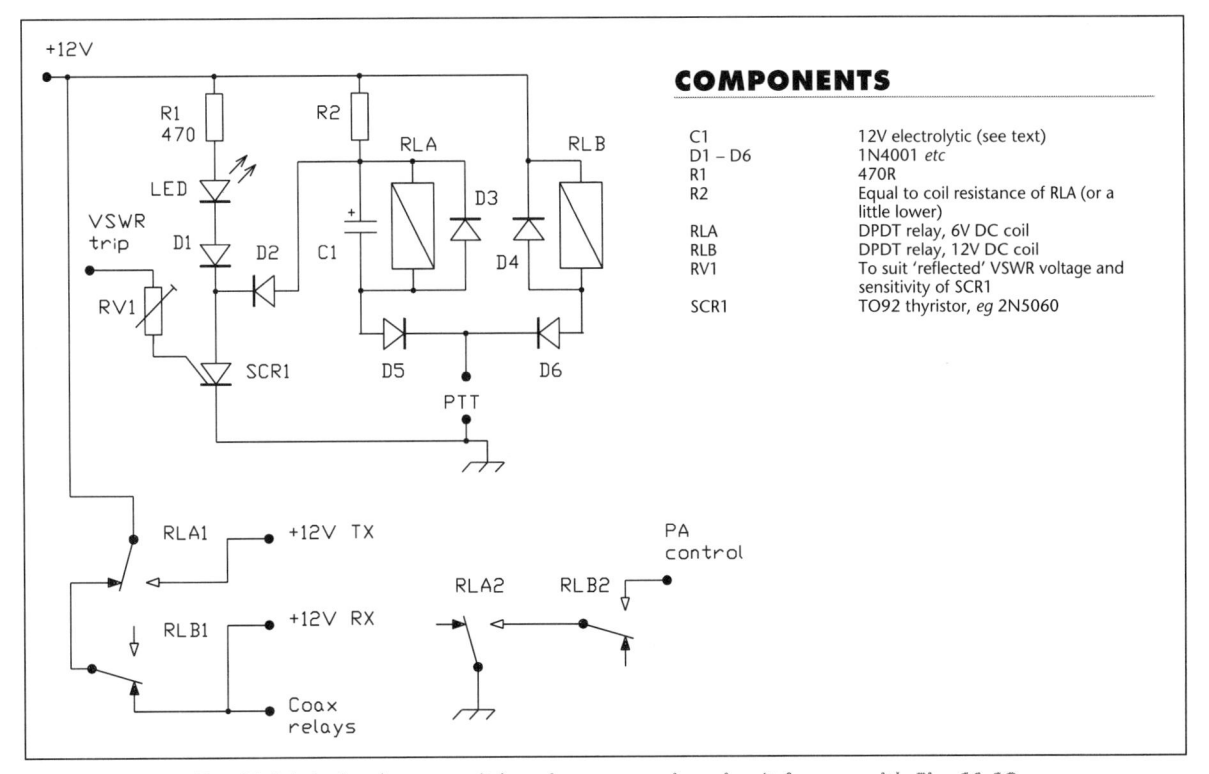

COMPONENTS

C1	12V electrolytic (see text)
D1 – D6	1N4001 *etc*
R1	470R
R2	Equal to coil resistance of RLA (or a little lower)
RLA	DPDT relay, 6V DC coil
RLB	DPDT relay, 12V DC coil
RV1	To suit 'reflected' VSWR voltage and sensitivity of SCR1
SCR1	TO92 thyristor, *eg* 2N5060

Fig. 11.14. A simple transmit/receive sequencing circuit for use with Fig. 11.12

system is switched off, the preamp is protected by two relays against things like static discharges. If the relay power fails, the system fails safe.

All the timing is based on two relays; RLA operates slowly, and RLB quickly. The contacts of RLA and RLB are simply interlocked to connect and remove 12V to the transverter and the coax relays in the right sequence. RLA is a 6V relay, fed from the 12V DC rail by R2 – which should, of course, be the same resistance as the relay coil. When the PTT is switched to ground, RLB switches quickly but the opening and closing of RLA is delayed for a few tenths of a second by C1. The value of C1 depends on the relay in question; experiment with values of a few hundred µF to make the two relays switch in sequence with an audible "ker-lick".

The VSWR protection in Fig. 11.14 is thrown in as a bonus. An excessive 'reflected' DC signal from the VSWR detector triggers the small thyristor, shorting out RLA and disabling the transverter and PA until the PTT is released and the system resets. Note that R2 must have sufficient wattage rating to withstand being connected across the full 12V in this condition.

This simple sequencing scheme can be adapted to many other systems. If you want something a little more sophisticated than two relays, reference [8] describes a 'universal' TX/RX controller which changes over several contacts one after another, and then reverses the sequence.

A potential problem with the circuit in Fig. 11.14 is that the initial changeover signal comes from the main HF transceiver, which means that low-level RF at 28MHz is already present at the transverter input when the transmit stages are energized. In this particular case it doesn't matter, because nothing is being 'hot-switched' with any significant RF power. Even so, it would be nice to disable the prime mover until the transverter and PA are ready to accept a driving signal. The next circuit does just that – and much more.

ADVANCED TX/RX SEQUENCER

If you want to go the whole hog, here's a super-de-luxe bit of circuitry for interfacing practically anything to everything else, from the fertile brain of Melvyn Noakes, G4JZQ. This circuit really looks after you. It won't let you change the PA over to transmit unless everything is in order: the +12V and –12V supplies must be present, the PA run/standby switch must of course be in 'run' mode, and the screen run/set switch(es) must also be set to 'run'. And then the *pièce de résistance*: the circuit not only checks that the coaxial relay supply is on – it also checks that the relays are there. Will it cope with different ways of signalling a transmit condition, by pulling the PTT line low or by driving it to +8V or +12V? But of course...

At first glance Fig. 11.15 looks a bit horrific, but break it down into sections and all becomes vividly clear. On receive, if the run/standby switch is on 'standby', the PTT line has no effect on the PA, and the ALC output to the transceiver is effectively zero. So the transceiver runs barefoot at its full normal power. Switch to 'PA run' and the controller takes over: the ALC output goes to –10V, holding off all RF drive until everything's good and ready.

Receive-to-transmit changeover starts from a PTT input from the transceiver, on whichever PTT line is appropriate. This is the first link in a chain of electronic and relay switches in series. Until all the switches have closed, the screen and grid-bias relays cannot be energized and the ALC output remains firmly negative. If the +12V rail is absent, this will never happen; if the –12V rail has taken the day off, the 82kΩ from the +12V rail makes TR2 clamp the PTT input to ground and so prevents any TX/RX switching.

If the PA run/standby switch is on 'run', and + and –12V are both present, a signal on the appropriate PTT line will turn on TR3. However, nothing will happen unless the main PSCU supervisory circuit votes in favour by closing its 'ready' relay. All being well, this should turn on the coax changeover relays. TR6 is the next switch in the sequence, and it will only turn on if the coax relays have changed over correctly.

Fig. 11.15 shows two alternative arrangements for checking the coax relays. The recommended method uses 'tell-back' DC contacts which are closed by the same mechanism as the RF changeover contacts themselves. The HF-400 and HF-2000 relays provide this option, which is obviously more important on the high-power output relay. If both the input and output coaxial relays have tell-back contacts, the contacts should be wired in series as shown in Fig. 11.15, so that TR6 is only turned on after both sets of tell-back contacts have closed. The relay coils could be connected in parallel if that is more convenient. The alternative method shown in Fig. 11.15 is to sense the current drawn by the coax relay coils, in the hope that the RF contacts will have changed over if their electromagnets are energized. This sensing method is more reliable if the two relay coils

Fig. 11.15. Advanced transmit/receive sequencer by G4JZQ, which can be incorporated into the comprehensive PSCU instead of Figure 11.10. The circuit includes an alternative screen-grid switching arrangement, incorporating SCR 'crowbar' protection which is activated by a rapid rise in anode current

COMPONENTS

D1, D2	40A diodes (see text and Fig. 11.1)
IC1	TIL117
M, M	Screen current meters (see text)
RLA, RLB, RLC 12V DC	Coaxial relays As available
SCR1, SCR2 TIC126N	(800V 12A or greater)
TR1, TR4, TR7	ZVP0106A
TR2	VN10KM
TR3	ZVN2110L
TR5	BCY70
TR6	BFY50
VDR1, VDR2	V250LA40B

Protection diodes across relay coils: 1N4002 etc.
All other unmarked diodes 1N4148 etc.

TRANSMIT/RECEIVE CONTROL

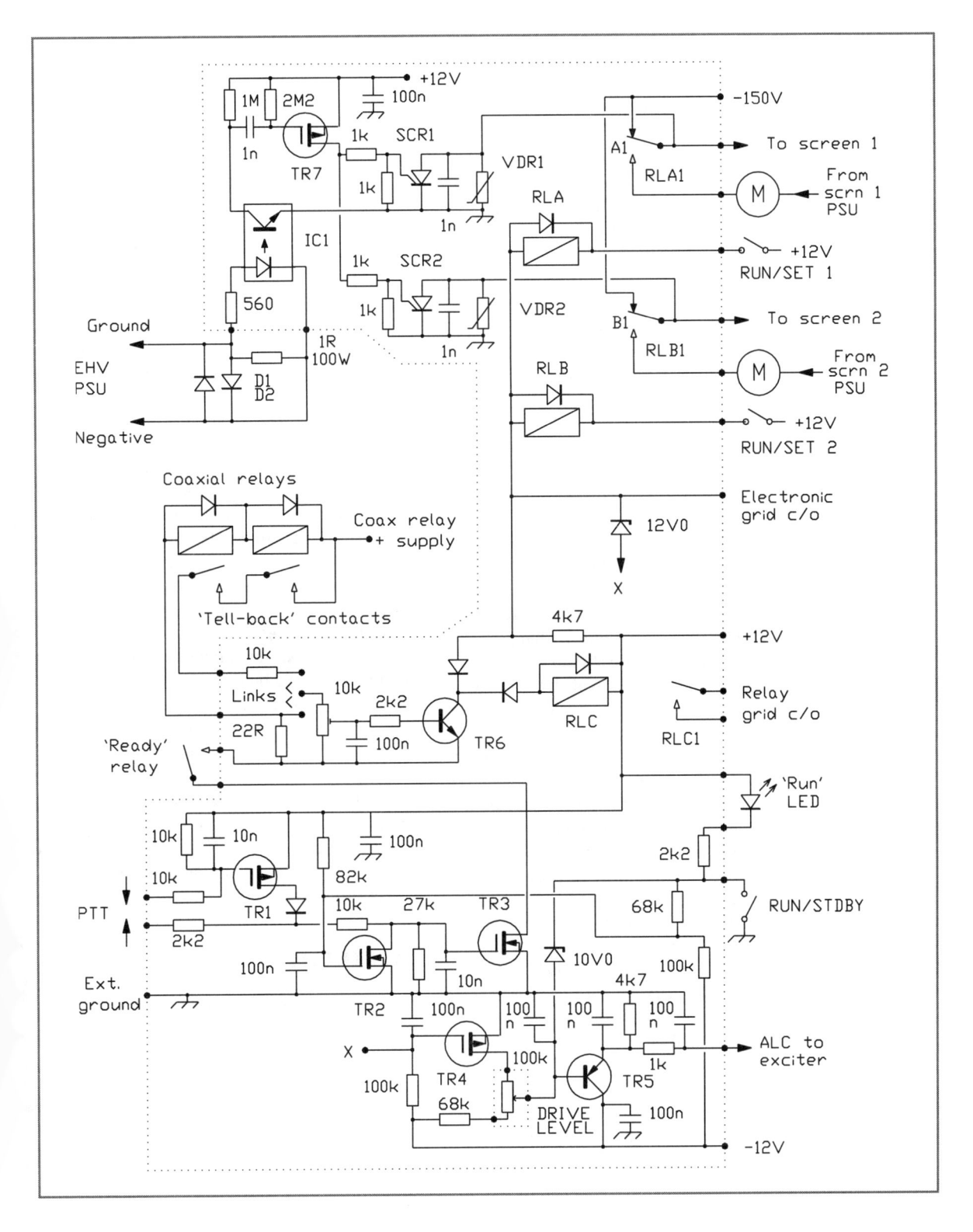

are connected in series as shown, though the voltage-sensing potentiometer could be adjusted to sense if either of two parallel-connected coils had failed.

If all seems well with the coaxial relays, TR6 turns on the screen relays (which cunningly double in this circuit as run/set relays), and enables the control-grid bias to bring up the PA. Then and only then will the circuit release the ALC and permit RF drive to flow. The final delight is that the ALC level on transmit is adjustable, to provide the variable-drive feature that I recommended in Chapter 6. If the PSCU finds anything wrong in the PA and opens the 'ready' relay, the antenna and screen changeover relays drop out just after the grid-bias supply returns to the receive condition, and the ALC voltage goes negative to turn off the drive from the transceiver. In other words, even Walter can't blow anything up with this little lot!

Fig. 11.15 also shows an improved way of switching the screen-grid voltages and protecting the screens against flashover. This circuitry improves upon the switching arrangements in Fig. 11.4 by making the screen changeover relay double as the run/set relay; in other words, you save two relays at the expense of some modifications to the earlier PSCU circuit board [5].

The other feature is the optoisolator-coupled trip circuit which monitors the anode current *via* the 1Ω resistor on the HV-minus rail. If the circuit detects a fast sharp rise in current, characteristic of a flashover, it fires a thyristor on each screen rail to provide positive 'crowbar' protection to the valves and bases, leaving the rest of the circuit to sort itself out as before. This system is inspired by the pulse-transformer circuit in the G4AJW PSCU [2]; the RC-coupled arrangement adopted here is rather more reliable. Ideally the crowbar thyristors should be in the amplifier itself, though the HV-minus rail is more accessible in the PSCU. An alternative therefore would be to move the current sensing into the amplifier, by monitoring the voltage drop across a 1Ω resistor in the HV+ supply. A suitable optoisolator will withstand the voltage, but care would be required with the PC board layout to isolate the HV+, screen and logic voltage levels.

FINAL

By this stage of the book, you might think you're beginning to detect an obsession with reliability and doing the job as well as you possibly can. Guilty, m'Lud. The reason is quite simple – it's the unashamed pursuit of what that Zen Motor Cyclist called "quality". Quite apart from any aesthetic or creative pleasures you may experience from building amplifiers and power supplies – which for me are considerable and lasting – the chief reason for doing so is to work the DX more effectively. In hard practical terms, the more you fret and worry about the amplifier and its power supplies when you're building it, the less you'll have to think about it when you really need to use it for its intended purpose.

Electronic components nowadays are extremely reliable if they're chosen with care and run within their ratings, so the odds against something critical going bang just as the DX is starting to appear are quite remote. I don't mind too much if a random component failure once in a blue moon makes me miss one prime contact. But I mind very much indeed if I miss some DX because I didn't do my job properly when I built the PSCU.

When the band opens, I want to switch on, work the DX, have a wash-up with the locals and switch off to write out the QSL cards. I *don't* want to spend half the opening trying to make the amplifier work properly, and half a week afterwards sorting out why it wouldn't.

As the saying goes, "train hard, fight easy".

REFERENCES

[1] Printed-circuit boards for bridge and biphase rectifier stacks are available from Melvyn Noakes, G4JZQ*.

[2] Andy Wade G4AJW, Power supply and control circuits for a 4CX250B amplifier. *Radio Communication*, October 1977.

[3] PCBs for the screen regulator are available from Melvyn Noakes, G4JZQ*.

TRANSMIT/RECEIVE CONTROL

[4] John Nelson GW4FRX and Melvyn Noakes G4JZQ, A power supply and control system for tetrode amplifiers. *Radio Communication*, December 1987 and January 1988.

[5] Printed-circuit boards for the power supply and control unit are available from Melvyn Noakes, G4JZQ*.

[6] John Nelson GW4FRX and Melvyn Noakes G4JZQ, Recipe for longer life – keep the heaters under proper control. *Radio Communication*, July 1988.

[7] A printed-circuit board for the regulated heater supply is available from Melvyn Noakes, G4JZQ*.

[8] TR time-delay generator. *ARRL Handbook* (*eg* 1986 edn, Chapter 31)

Figs. 11.3 and 11.6-11.11 are reproduced by kind permission of the Radio Society of Great Britain.

* G4JZQ can also supply PCBs completely built, tested and guaranteed.
QTH: 55 West Drive, Highfields, Caldecote, Cambridge CB3 7NY.

It's an evening in early November, and 144MHz has returned to normal after three days and nights of a monster tropo opening to Eastern Europe. You're having a natter with your friendly rival down the road, exchanging congratulations and commiserations:

"G3XBY from GW4FRX – fine, Dale, all copied and very good indeed. Congratulations on that OK in JN98. I heard you working him and I listened for all I was worth but couldn't hear a thing – I nearly fell off my chair when you gave him 539. Isn't it funny how tropo can vary like that? We're not that far apart, our antennas are the same and our take-offs that way are quite similar. I wish I understood tropo a bit better...

"I wonder whether Fred, over your way – what's his call, G0*** – heard the OK too? He's got a super site to the east; I bet he worked a lot of good stuff. Have you spoken to him, over?"

"From G3XBY. Well, it's funny you should say that. I heard Fred on the first night, the Tuesday, but I didn't hear him working anyone after that. I guess he must have gone off to 432, over."

"Break, from G0***"

"Hello Fred, you must be clairvoyant – we were just talking about you!"

"Yes, I know, I've been listening to you for quite a while. Actually, I didn't do too well in the opening because I was down to 10 watts from the prime mover. You probably won't believe this – well, you probably will, knowing me – the amplifier blew up again. That's three times this year and to be honest I'm kicking myself!

"I heard all the SP stations same as you did, and that OK was quite strong too, but I just couldn't get through the pile-ups. There was a very weak HG at one time too, but I couldn't get through to him either. I really only called in to let you know that I won't be on the air much for the next couple of months – I've decided to rebuild the whole thing, and I'll do it properly this time!"

Poor old Fred – a dozen tasty squares to the bad and his street cred completely blown. You resist the temptation to quote the parable of the Wise and Foolish Virgins at him, and just offer to lend him your *VHF/UHF DX Book*. Maybe something in there might help...

CHAPTER 12

TEST EQUIPMENT & STATION ACCESSORIES

TEST EQUIPMENT & STATION ACCESSORIES

by Roger Blackwell G4PMK

This chapter describes some of the test equipment, techniques and accessories which will help you get the most from your station. While a lab-full of expensive and sophisticated RF testgear might be very nice, it is possible to make the measurements that we amateurs want – those directed at maintaining or improving our station's performance – without taking out a second mortgage. In true amateur tradition, by making our own equipment, by taking more time than an engineer in a busy development lab could afford, and by setting our sights just a little lower, we can do very nicely. For example, while professional gear might have impressive absolute accuracy and wideband frequency coverage, we often find that relative measurements within a narrow amateur band are quite sufficient for our needs.

By taking advantage of recent developments in technology and design, it is possible to start out with a very good idea of what your equipment **should** be doing, perhaps by using some analysis and design programs running on the shack microcomputer. This allows you to substitute a few simple instruments (together with a modicum of common sense) for a whole array of expensive test gear.

This chapter isn't about the re-invention of the multi-meter, grid-dip oscillator or other standard items of testgear. You can find plenty about these in the general-purpose amateur radio handbooks. Instead, I'll concentrate on those techniques and items of equipment which either haven't been covered in the standard books or deserve wider appreciation than they've had up to now.

RF POWER AND VOLTAGE MEASUREMENT

If you want to achieve good performance from your station, and keep it that way, RF power measurements are fundamental. After all, everything that we do is about radiating and receiving RF power.

Power levels can be expressed in watts or milliwatts – or more conveniently for many purposes, in decibels relative to 1W or 1mW (see the panel opposite).

RF DETECTORS

Amateur RF power measurements rely almost exclusively on diode detectors. These produce a DC voltage in response to the applied RF voltage. You then calculate the RF power using a suitable formula, or calibrate a meter scale accordingly.

Most detector diodes have a minimum turn-on or 'barrier' voltage below which they respond non-linearly to the applied RF signal. If the applied RF voltage is much higher than the turn-on voltage of the diode, the resulting DC voltage is directly proportional to the applied RF voltage.

At relatively high power levels (1W or more) the DC output of a diode detector is proportional to the applied RF voltage. So the power can be calculated from the standard relationship:

$$P = \frac{V^2}{R}$$

where P is the RF power and R is the RF impedance across which the detected RF voltage is developed. V in the above formula should be the RMS RF voltage, though a diode detector normally gives a reading closer to the *peak* value of the RF voltage. Applying the

---

POWER LEVELS AND DECIBELS

Power levels are often measured in decibels relative to 1 watt or 1 milliwatt.

Powers of 1W or more are usually expressed in dBW, decibels relative to 1 watt. The conversion formulae are:

$dBW = 10 \log_{10}(\text{power in watts})$

Power in watts = $10^{(dBW/10)}$

You should already be familiar with the fact that the 400W UK legal limit on RF power output represents 26dBW (if you're uncertain about that, find a calculator and check it out). Power levels less than 1W are *negative* when expressed as dBW.

The low levels of receiver input signals or outputs of oscillator chains are often expressed in dBm, which are decibels relative to 1 milliwatt. Once again, powers above 1mW will be positive when expressed in dBm; power levels below 1mW will be negative in dBm.

$dBm = 10 \log_{10}(\text{power in milliwatts})$

Power in milliwatts = $10^{(dBm/10)}$

dBm = dBW + 30dB

dBW = dBm + 30dB

The virtue of expressing powers in dBW or dBm is that any calculations involving attenuators, amplifiers and signal levels become simple additions and subtractions. For example, a signal level of –3dBm (3dB below 1mW) fed through an amplifier of 22dB gain and then a 6dB attenuator results in an output signal of +13dBm.

–3dBm input + 22dB gain + –6dB attenuation = +13dBm output

TEST GEAR & STATION ACCESSORIES

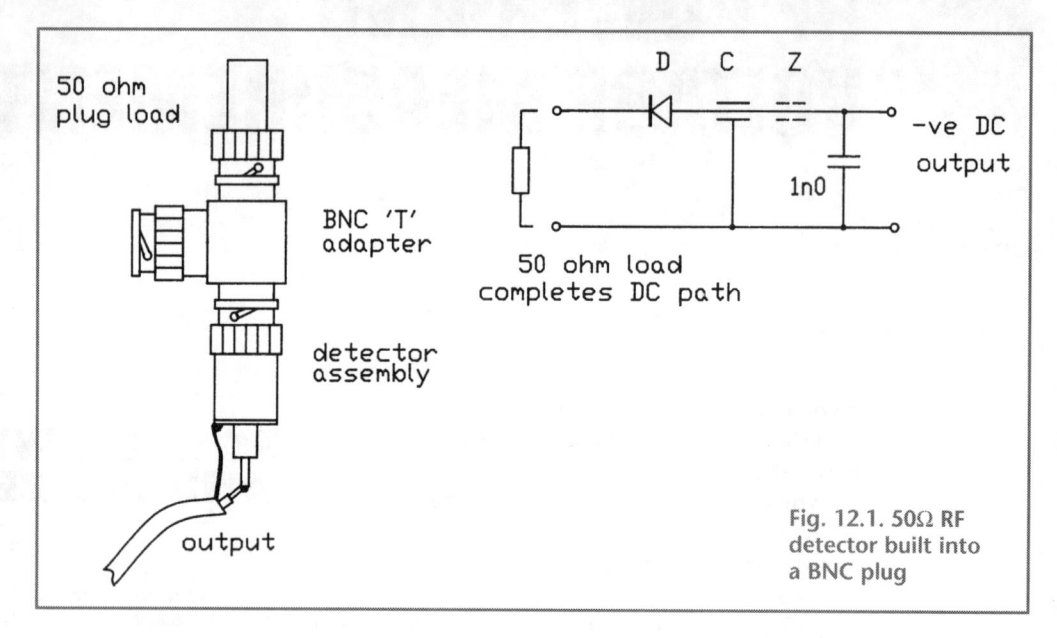

Fig. 12.1. 50Ω RF detector built into a BNC plug

usual correction factor of $\sqrt{2}$ to V gives a modified formula:

$$P = \frac{V^2}{2R}$$

Unfortunately, inefficiencies in RF detector diodes and stray reactances in their associated circuitry may prevent them from indicating the full peak RF voltage. For example a major source of error arises from loading the output of the diode with a low-resistance DC voltmeter. When the DC output is only a volt or two, another problem is the turn-on voltage of the diode. Yet another formula for diode detectors is therefore:

$$P = \frac{(V - V_o)^2}{2R}$$

where V_o is the turn-on voltage of the diode; about 0·6V for an ordinary silicon RF diode such as the 1N914/916/4148 or about 0·3V for a Schottky-barrier type such as the HP5082-2800. 'Low-barrier' and 'zero-barrier' Schottky diodes have correspondingly lower turn-on voltages.

LOW-LEVEL SQUARE-LAW DETECTORS

At low RF voltage levels, of the order of a few hundred millivolts, a detector diode is operating within its turn-on region. The DC voltage from a diode RF detector is no longer directly proportional to the RF voltage: it is instead proportional to the *square* of the input voltage. The diode is then said to be operating in its 'square-law' region. If the DC output of a square-law detector is directly proportional to the square of the RF voltage, it must also be directly proportional to the applied RF power. Such a detector is valuable because its output meter can have a linear scale calibrated directly in RF power.

The main advantage of a square-law detector, however, is its sensitivity – it operates in the turn-on region below the useful range of the RF voltage detectors described above. Almost any RF diode can be used as either a square-law or a voltage detector, subject to voltage breakdown limitations at higher levels of RF. The transition between the square-law and voltage-detector modes depends more upon the RF signal level than on the diode itself. Some diodes are optimized for operation as square-law detectors and possess low turn-on voltages.

SIMPLE PRACTICAL RF DETECTORS

There are many situations where a simple RF detector is just about indispensable. All you

POWER & VOLTAGE MEASUREMENT

need is a Schottky diode and a 50Ω load fitted into a type N or BNC plug as shown in Fig. 12.1. This can serve as a simple milliwattmeter, or as an RF sensing element in a VSWR meter for example. What deserves a wider application is the increased sensitivity and power linearity which results from the application of a little forward bias. You can accomplish this by using a small constant-current source – nothing more than a large resistor fed from a voltage of a few hundred millivolts – into the diode from the load side. It's worth making the polarity of the diode such that it can easily be biased from the associated piece of test equipment, as shown in the tuned AF level meter described later.

As a basic RF 'sniffer' for probing around in transmitters, oscillators and frequency multipliers, don't despise the simple pickup loop connected to a diode detector. Use a sensitive meter with as large a scale as you can find. For years I used a 5µA meter with a 4-inch scale, mounted in a hole in a cardboard box – just 'temporary', you understand...

RF MILLIVOLTMETER

An electronic RF millivoltmeter can be a useful aid when setting LO drive levels or testing the low-power stages of a transmitter. The circuit shown in Fig. 12.2 offers excellent performance with a flat frequency response of 1MHz to at least 200MHz, and is usable up to 500MHz. Unlike many simple instruments it also boasts a linear scale. The range is 10mV to 1V RMS, with a 50Ω input impedance. With suitable calibration (or a squaring circuit as in [1]) it could also be used as a low-level power meter.

The circuit uses a pair of Schottky diodes (matched if possible), one of which is used in a DC feedback circuit to compensate for low-level non-linearity, a technique described for a slightly different application in [1]. Construction of the RF detector circuitry should follow good VHF/UHF practice; the prototype units had a 50Ω microstripline input circuit with chip capacitors and 100Ω chip resistors. The rest of the circuit operates at DC and its layout is not critical. I built several prototypes, and when checked against precision commercial RF millivoltmeters their accuracy was better than ±2% across the entire range. This is good enough to justify the use of a 3½-digit LCD panel meter module as an alternative to the

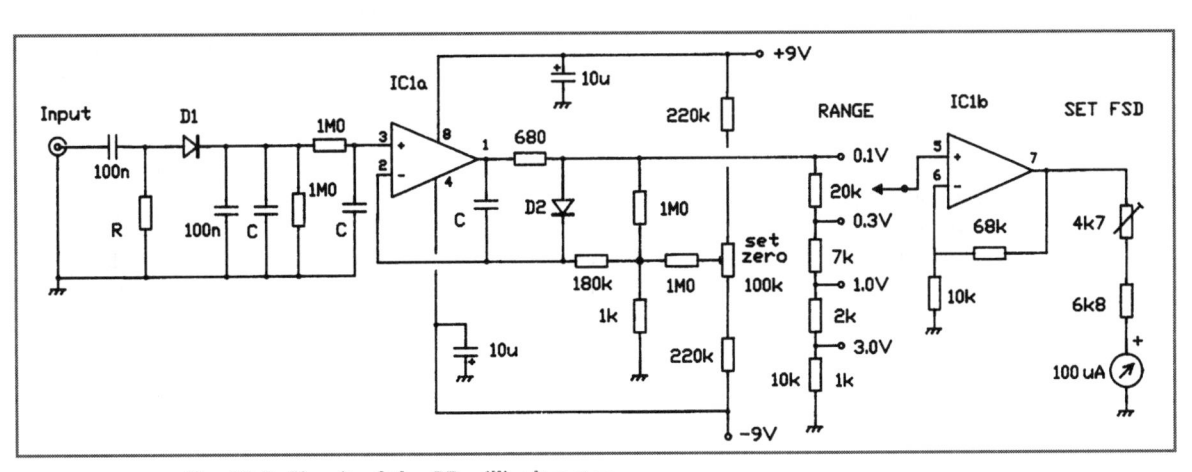

Fig. 12.2. Circuit of the RF millivoltmeter

COMPONENTS FOR THE RF MILLIVOLTMETER
IC1   TL072 or TL082 dual op-amp
D1, D2   Matched pair of Schottky diodes. If no matched pair is readily available, select a pair which show similar forward resistances from a batch of HP2800 or similar.
C   1n0 chip capacitor
R   50Ω load formed from two 100R chip resistors

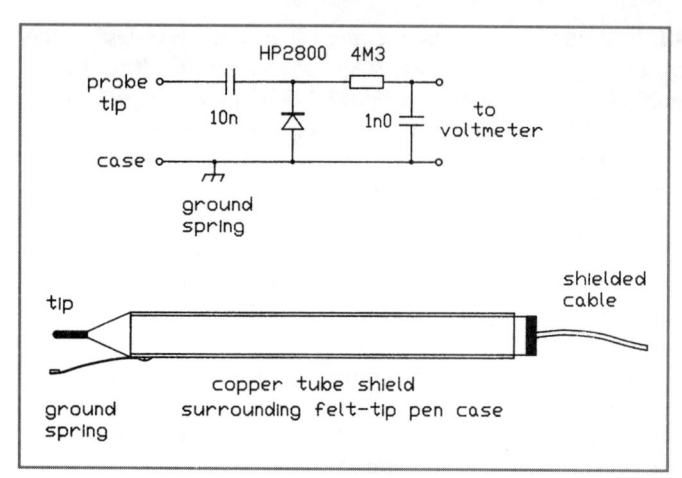

Fig. 12.3. A high-impedance RF probe, suitable for a 10MΩ voltmeter

analogue meter shown.

An alternative high-impedance probe for use with an external solid-state 10MΩ input-impedance millivoltmeter is shown in Fig. 12.3. Since the diode acts as a peak detector, you would normally need to divide the meter readings by $\sqrt{2}$ (1·414) to calculate the RMS voltage. In this circuit the 4·3MΩ series resistor forms a 1·43:1 potential divider with the 10MΩ input resistance of the external meter, so the meter indicates near-enough the RMS value of a sine-wave RF signal. The probe is usable between 100mV and about 10V RMS, the maximum depending on the peak inverse voltage-breakdown rating of the diode. I built my version in a plastic felt-tip pen case, screened with copper tube. Use zero lead lengths for the best UHF performance. A springy contact from a broken relay provides a short ground connection, useful for measurements on circuits built on a copper ground-plane. Although the lead length is too great to give accurate results at UHF, the readings are reproducible for relative measurements.

RF MILLIWATTMETER

Although not often seen in amateur stations, an RF milliwattmeter is an indispensable item in a professional RF lab – once you have one, you'll wonder how you ever managed without it! For example it is almost essential equipment for aligning the transverters described in Chapters 8, 9 and 10. This version is a development of the design by DC9RK described in [2]. Two diodes *of different types* are the innovation in the design, their differing sensitivities with frequency combining to broaden the frequency response.

The 'business end' should be made in a small screened box – mine is soldered together from double-sided PCB – which holds the socket and the board. The 50Ω load was made from two 100Ω chip resistors soldered across the 50Ω microstrip joining the socket inner to the diodes. High-frequency performance is critically dependent on layout and short lead lengths, and Fig. 12.4 should make this clear. Although the instrument is intended only for the range 10–600MHz, I have found that by adding separate calibration pots for 1·3 and 2·3GHz, reasonable results can be obtained on these bands as well, albeit at slightly reduced accuracy. Useful comparative power indications are even obtainable at 10GHz.

The meter is housed in a plastic box, together with the calibration pots and range switches. If the meter is calibrated on the 10–600MHz range for 10mW and 50mW scales, simply adjusting the 1·3 and 2·3GHz range presets for full-scale deflection gives very good results with the same two scales. With careful construction and calibration (take it along to a measurement stand at a rally) this milliwattmeter will repay your efforts many times over.

If you have a hankering for nice scales calibrated in dB, and also have a pocket deep enough for a zero-bias Schottky diode or two, you will find an interesting microwattmeter with a 30dB dynamic range described in [3].

FREQUENCY COUNTERS

There are several important points to consider about the measurement of frequency at VHF and beyond. The first is the ability to count the frequency, which means often using a prescaler to divide the input frequency down to something within the counter's range. This in turn affects the resolution of the counter, because the prescaling means that a longer gate time is needed to achieve the full resolution of the counter. Finally, the requirements for the reference frequency standard of the counter become more severe at higher frequencies. I'll consider each point in turn.

PRESCALERS FOR VHF/UHF

Most basic digital frequency counters have an upper limit of a few tens of MHz, owing to counting-speed limitations of the logic. If you want to count higher frequencies, you must use faster (and more expensive) circuitry to divide the input frequency down to something within your counter's range. Usually this means one or more specialized divider chips, possibly with input amplification.

Modern prescalers divide by 10 or 100, simplifying the alterations (mental or electrical) to the counter display. In general, avoid the prescalers meant for use with frequency synthesizers, which divide by powers of 2 or other non-decimal ratios. If your prescaler divides by 10, the counting time must be increased tenfold in order to retain the same

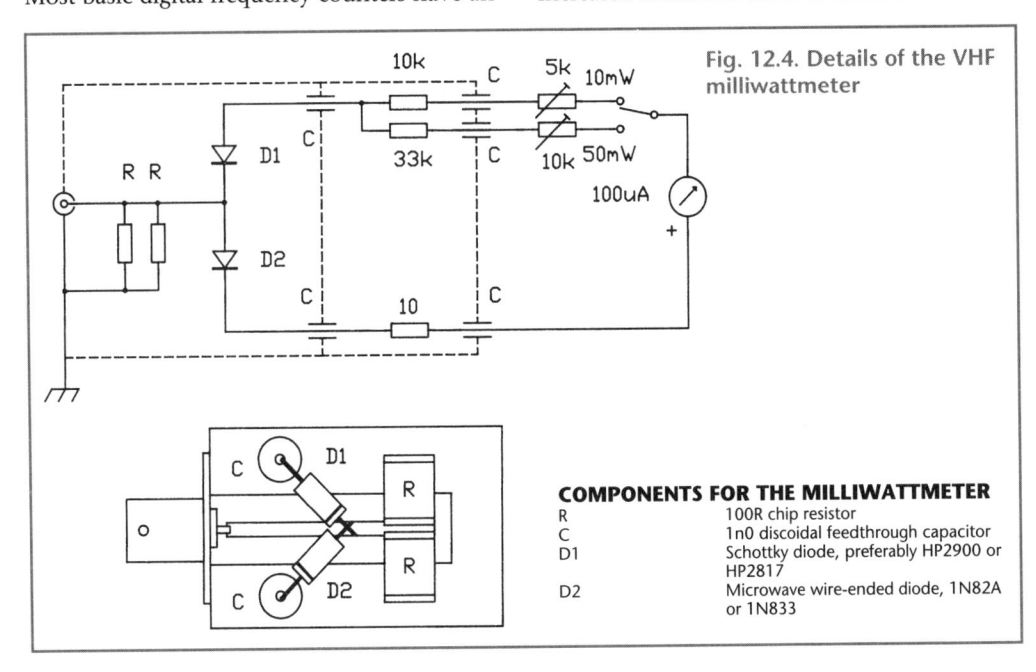

Fig. 12.4. Details of the VHF milliwattmeter

COMPONENTS FOR THE MILLIWATTMETER
R   100R chip resistor
C   1n0 discoidal feedthrough capacitor
D1   Schottky diode, preferably HP2900 or HP2817
D2   Microwave wire-ended diode, 1N82A or 1N833

Fig. 12.5. UHF prescaler for frequency counters. The separate VHF unit begins at the point marked 'VHF'

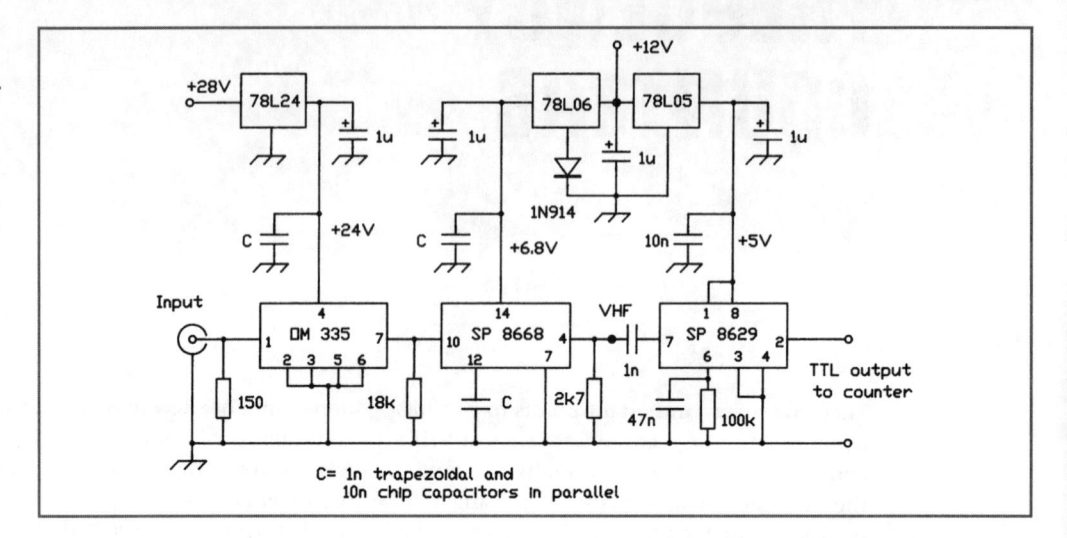

resolution as you had with the basic counter; the decimal point must also be moved one place to the left. If you had not increased the counting (or 'gate') time, the resolution of your counter would have deteriorated 10-fold.

For example, a counter with 100Hz resolution up to 150MHz would be more than adequate for setting up your transceiver to locate the 144·3000MHz SSB calling frequency. Provided that the frequency standard is also accurate – see later – you could hit the spot within ±100Hz ±1 count, *ie* within ±200Hz. But if you add a 10:1 prescaler to extend the frequency range up to 1·5GHz without also increasing the gate time by a factor of 10, the accuracy would deteriorate to ±2kHz at *any* frequency – which is probably not the kind of improvement you were looking for. So there's more to adding a prescaler than just bolting the chip on to the front-end of the counter.

With modern LSI frequency-counter chips, it's usually quite easy to make the necessary changes when adding a VHF/UHF prescaler. Often more than one prescaler is necessary to achieve unbroken coverage from HF on up; my own counter uses a single-chip divide-by-100 prescaler to cover up to about 200MHz with good sensitivity, and a completely separate divide-by-1000 unit plus a hybrid-module input-amplifier block to continue the coverage up to about 1·5GHz. These prescalers are shown in Fig. 12.5 and can be built either as add-on units or as an integral part of the counter.

Both prescalers have a 50Ω input impedance and a 5V (TTL-compatible) output. The VHF prescaler uses the same circuit as the final stage of the 1·5GHz unit, *ie* everything to the right of the point marked '*VHF*'. If the 24V supply for the OM335 is a problem, you can use an OM361 which requires a 12V supply; in that case, the output (pin 8) should have a 1µH choke to the 12V rail, with a 1nF chip capacitor providing a DC block before the input of the SP8668 divider. UHF prescaler chips are invariably low-impedance devices, so don't hit them with the direct output from the PA and expect them to survive! One of the advantages of using a hybrid module preamplifier rather than the more obvious and cheaper MMIC is that the hybrid module will stand a few hundred milliwatts of RF input without overloading the prescaler chip.

FREQUENCY ACCURACY, STABILITY AND ADJUSTMENT

The accuracy of a frequency counter depends greatly on the accuracy of its internal frequency standard. This determines the 'gate time' during which the cycles of the test signal are counted, so any error in the frequency standard contributes directly to the errors in the displayed frequency count. Many simple counters which use an inexpensive crystal and

FREQUENCY COUNTERS

an 'on-chip' oscillator have accuracies of only ±5 parts per million, at best. This corresponds to an error at 144MHz of something like ±720Hz (plus the counting errors already mentioned). Use such a counter at 101MHz to set up the 404MHz LO of your 432MHz transverter and you could miss the SSB calling channel completely. Even if you can temporarily set the crystal accurately on frequency, it won't stay there for long unless you are very lucky.

You can improve things considerably by upgrading the crystal standard in your counter. Consider using a separate high-quality crystal oscillator, with either a zero-temperature-coefficient crystal or one in an oven, or use a TCXO module (Temperature-Compensated Crystal Oscillator) occasionally seen as a rally bargain. As a minimum, you could fit a lower-value trimmer in parallel with the existing one (*eg* 0–3pF in parallel with the typical 20pF trimmer) to make fine adjustment a little easier.

Once you've achieved an adequately stable timebase, how do you check the accuracy? In the old days it could be very simple to zero-beat a harmonic of your oscillator against a standard-frequency transmission. But that's no longer easy in the UK, following the QSY of BBC Radio 4 to 198kHz and the demise of the HF standard-frequency transmissions from MSF at Rugby. Personally I've never had much success in using HF standard-frequency transmissions from the other side of the world. The GB3BUX beacon on precisely 50·000MHz may be usable for zero-beating if you can copy it strongly enough, but this too has to compete with QRM from computers and other digital devices.

There is a *much* better way, though, which uses a VLF standard-frequency transmitter. This is MSF on 60kHz from Rugby, whose frequency is accurate to 1 part in 10^{12}. This method avoids all the problems of phase-locking reference oscillators or whatever. It simply provides a visual comparison of the

Coil details:	370 turns of 34swg enamelled solid copper or litz wire, wound in four layers 1" long, in the centre of an 8" x $3/8$" dia ferrite rod.

Fig. 12.6. MSF phase-comparison unit

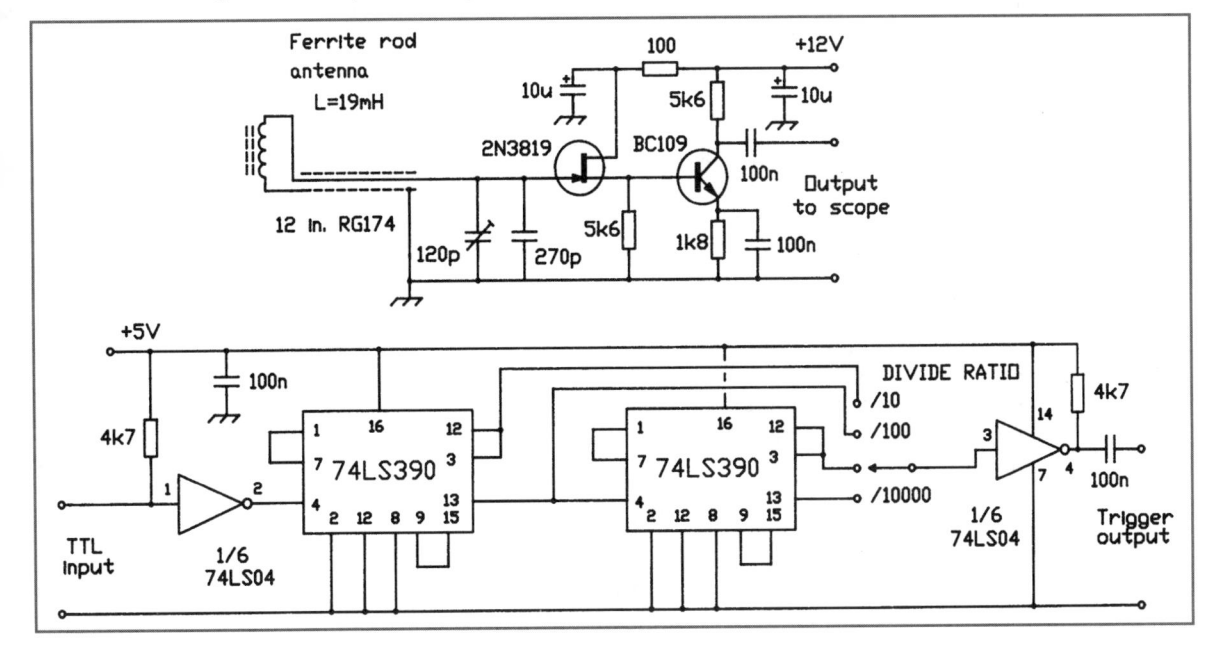

phase of your own standard (calibrator, counter clock or whatever) with MSF, using an oscilloscope which has a triggered timebase. The technique is based on an article about using the American WWVB 60kHz service for this purpose [4].

Firstly, you must be able to display the MSF signal on an oscilloscope using a simple antenna and tuned amplifier. You **must not** use any form of frequency conversion, of course, since the frequency accuracy of the original signal will be lost. The modulation on MSF is irrelevant since the phase of the signal is preserved through the carrier breaks. Second, you need to derive a signal from the oscillator-divider chain of your counter which is a sub-multiple of 60kHz. Usually this means no more than a couple of IC dividers to produce 10kHz, for example. This signal is used to trigger the sweep of the oscilloscope when it is displaying two or three cycles of the MSF waveform applied to the Y input.

The difference in frequency between your reference signal and MSF shows up as a drift of the sine-wave display across the screen. Drift to the right indicates your local reference is high in frequency, and drift to the left means that it's low. You can find the exact error by noting the time it takes for a complete sine-wave cycle to drift past a fixed point on the screen. The error of your reference is then:

Error in ppm = 16·6667 / time in seconds

I use the front-end of an MSF receiver intended for a time-code clock [5], and at a range of about 200km it can receive about 500mV pk-pk using an indoor ferrite-rod antenna. The entire setup is shown in Fig. 12.6, including the divider chain used with the auxiliary output of my frequency counter's external oscillator. If the oscillator in your counter has adequate short-term stability and some means of adjusting the frequency to fine limits, it's easy to obtain ±0·1 ppm accuracy, which will for instance give you better than 15Hz measurement accuracy at 144MHz. This ought to be good enough for most purposes! One warning – MSF is off the air for maintenance on the first Tuesday of every month.

Even if your counter's internal frequency-standard is not a multiple of 10kHz, it may still be possible to adapt the same technique. In much of western Europe and several other parts of the world, another easily receivable frequency source of high stability is the PAL TV line timebase at 15·625kHz. In most countries this frequency is maintained to a very high accuracy, and of course your TV timebase is locked to the sync pulses. A coil of a few hundred turns, wound on a ferrite rod and draped over the back of the TV set, should provide enough signal for the Y input of your 'scope. To extract a suitable signal from your counter for comparison, play around with a calculator until you find a division ratio that will give the required 15·625kHz for triggering the 'scope sweep. For example with a counter using a 3·90625MHz crystal the division ratio would be 250 (two divide-by-5s followed by one divide-by-10). Naturally this frequency standard can only be used when the TV is receiving a station and the line timebase oscillator is locked to the transmitted sync pulses.

IMPEDANCE AND VSWR MEASUREMENT

50Ω LOADS

The universal standard impedance in RF communications is 50Ω, so the first essential in impedance measurement is an accurate 50Ω load to form the local reference against which your workaday '50Ω' loads are judged. It's worth obtaining one good standard load, *eg* one in an N or SMA plug body, which can sometimes be found at rallies for a few pounds. Keep this load in good condition by *not* using it as a dummy load for your transmitter! A precision load will – almost by definition – be equipped with an N or SMA connector; devices with BNC plugs or sockets aren't going to give the same performance, as the connector itself will introduce errors with prolonged use.

For many day-to-day applications, such as terminating the branches of directional couplers or providing low-power loads for the 50Ω detector I described earlier, you can manage with a load consisting of an ordinary resistor in a BNC plug. Although these are relatively cheap to buy, they're even cheaper to make. The BNC plug should be a captive-contact type. Prepare it by soldering a small disk of thin copper or brass sheet to the top (hexagonal end) of the clamp nut. When cool, drill a 1mm hole in the centre of the disk. Now take a 51Ω 0·25W metal film (MF) resistor, which is exactly what the cheaper commercial loads use, and crop the lead at one end to about 7mm. Slip the upper insulator (the one that normally goes next to the cable) over the lead, then seat and solder the pin. Fit the bottom insulator, then carefully push the assembly into the plug body. Now slip the modified nut over the long lead of the resistor and tighten the nut fully. Finally trim the projecting wire back and solder. While this isn't a precision load, it is perfectly adequate for a lot of applications.

ATTENUATORS

Attenuators have many uses, and a good attenuator – which means one whose performance is accurately known and can be trusted – is almost worth its weight in gold. If you can, try to acquire at least one commercially manufactured unit of good performance, preferably with N-type connectors. These are available at rallies from time to time, and a few pounds spent on one or two will amply repay the modest investment. The most useful values will be 10 or 20dB, but anything in this general range will be usable as a known value of attenuation. Having obtained your attenuation 'standard', try to keep its accuracy by not using it as a power attenuator beyond its dissipation, which for most semi-precision types will be a watt or less.

You'll need to build up a range of good-quality attenuators designed for use in a 50Ω system. These should include a switched set giving any combination of 1, 2, 3, 4, 10 and 10dB so that you can select 1 to 30dB in 1dB steps, together with a selection of fixed attenuators of handy values such as 3dB, 6dB, 10dB and 20dB. You can often pick these up as surplus, although if your budget for test gear is tight (and whose isn't?) by all means extend your selection of attenuators with a few home-made devices. You'll find some information on designing your own attenuators a little

IMPEDANCE, RETURN LOSS & VSWR

The design impedance of almost everything in our RF systems – transmitters, receivers, preamplifiers, cables, connectors and antennas – is 50Ω. We often need to express the impedance of a particular component in terms of how well it is matched to that system impedance. There are three common ways of doing this: by expressing the impedance in terms of its return loss, its VSWR or its reflection coefficient. These are ways of comparing an impedance with the prevailing system impedance.

Because they are so-called scalar quantities (magnitude but no phase), return loss, VSWR and reflection coefficient don't show the value of the mismatched impedance unambiguously. A VSWR of 2 in a 50Ω system could be caused by a resistive impedance of either 50 x 2 = 100Ω or 50/2 = 25Ω. Equally, it could be caused by many possible combinations of resistance and reactance, but a simple VSWR measurement provides no information about this.

VSWR stands for 'Voltage Standing Wave Ratio'. Standing waves are formed by the addition and cancellation of the forward-travelling and reflected waves on the transmission line. If V_{FWD} is the voltage of the forward-travelling wave on a transmission line, and V_{REF} is the voltage of the reflected wave, the maximum voltage of the standing wave will be ($V_{FWD} + V_{REF}$) and the minimum voltage will be ($V_{FWD} - V_{REF}$).

VSWR is simply the ratio of maximum to minimum voltage on the line, and is thus given by:

$$VSWR = \frac{(V_{FWD} + V_{REF})}{(V_{FWD} - V_{REF})}$$

Without changing the basic definition, a small twist of algebra rearranges this into a more useful form:

$$VSWR = \frac{\left(1 + \frac{V_{REF}}{V_{FWD}}\right)}{\left(1 - \frac{V_{REF}}{V_{FWD}}\right)}$$

Since V_{REF}/V_{FWD} is a voltage ratio, and power is proportional to the square of voltage, it's just as true to say:

$$VSWR = \frac{\left(1 + \sqrt{\frac{P_{REF}}{P_{FWD}}}\right)}{\left(1 - \sqrt{\frac{P_{REF}}{P_{FWD}}}\right)}$$

Reflection coefficient (ρ, the Greek letter rho) is defined simply as (V_{REF}/V_{FWD}), so another way of expressing the VSWR equation becomes:

$$VSWR = \frac{(1 + \rho)}{(1 - \rho)}$$

or

$$\rho = \frac{(VSWR - 1)}{(VSWR + 1)}$$

Return loss is simply the ratio (in dB) between the forward and reflected power. In other words, it's an inverse measure of VSWR.

$$L(dB) = 10 \log_{10}\left(\frac{P_{REF}}{P_{FWD}}\right)$$

or

$$L(dB) = 20 \log_{10}\left(\frac{V_{REF}}{V_{FWD}}\right)$$

or

$$VSWR = \frac{1 + \text{antilog}_{10}\left(-\frac{L}{20}\right)}{1 - \text{antilog}_{10}\left(-\frac{L}{20}\right)}$$

The higher the return loss, the better the match. It's a very useful way of expressing impedance match, particularly at low VSWRs. Later in this chapter I'll give details of a return-loss bridge which can be used for a variety of impedance measurements.

To sum up these different ways of expressing the same thing, here's a small table of equivalent values. In any practical measurement, VSWR values smaller than 1·02 or return losses better than 40dB should be taken with a large pinch of salt!

Return Loss dB	Reflection coefficient	VSWR
0	1	infinite
3	0·707	5·85
10	0·316	1·92
20	0·100	1·22
30	0·032	1·07
40	0·010	1·02
infinite	0	1·00

IMPEDANCE & VSWR MEASUREMENT

further on. Home-made attenuators can be reasonably accurate at VHF, and if built with care can even work well into the UHF region. Excellent information for building your own attenuators can also be found in [6] and [7]. It isn't too hard to measure their performance accurately with simple testgear, provided, of course, that you have one or more units of reasonable accuracy to calibrate them against.

USING ATTENUATORS TO STABILIZE IMPEDANCES

Almost all your measurements are going to be made in a 50Ω system, *ie* feeding the device under test from a 50Ω source impedance and terminating it in a 50Ω load. For low-level measurements, such as those described in the section on receiver measurements, you can ensure this by placing the device under test between two 50Ω attenuators of 10–20dB. Any power reflected from a mismatch at the far side of an attenuator will itself be attenuated on the way back, so the impedance 'seen' by the device under test can't stray far away from 50Ω. For example, the return loss looking into a good 20dB attenuator cannot in theory be less than 40dB – which is the same as saying the VSWR cannot exceed 1·02 – no matter what is connected at the other end. In practice the VSWR would probably be rather higher, owing to impedance errors in the attenuator itself.

If you make all your gain and loss measurements using impedance-stabilizing attenuators at the input and output of the device under test, you'll pretty much know where you are. If you don't, you'll get confusing and misleading results. For example, even your so-called 'standard' attenuators won't give the correct attenuation unless they themselves 'see' source and load impedances of 50Ω.

CUSTOM-MADE POWER ATTENUATORS

There are a number of occasions when you need a particular attenuator, perhaps permanently built into equipment, of a value which isn't available off-the-shelf. For example, you might wish to attenuate the output of your multi-mode transceiver to a sensible level for driving an external PA (Chapter 6). It's quite a simple job to make such attenuators using standard carbon- or metal-film (not wire-wound!) resistors. Rather than present a table of values as most of the handbooks do, here are the necessary formulas to design your own 50Ω Pi and T attenuators as shown in Fig. 12.7.

Fig. 12.7. Pi and T attenuators

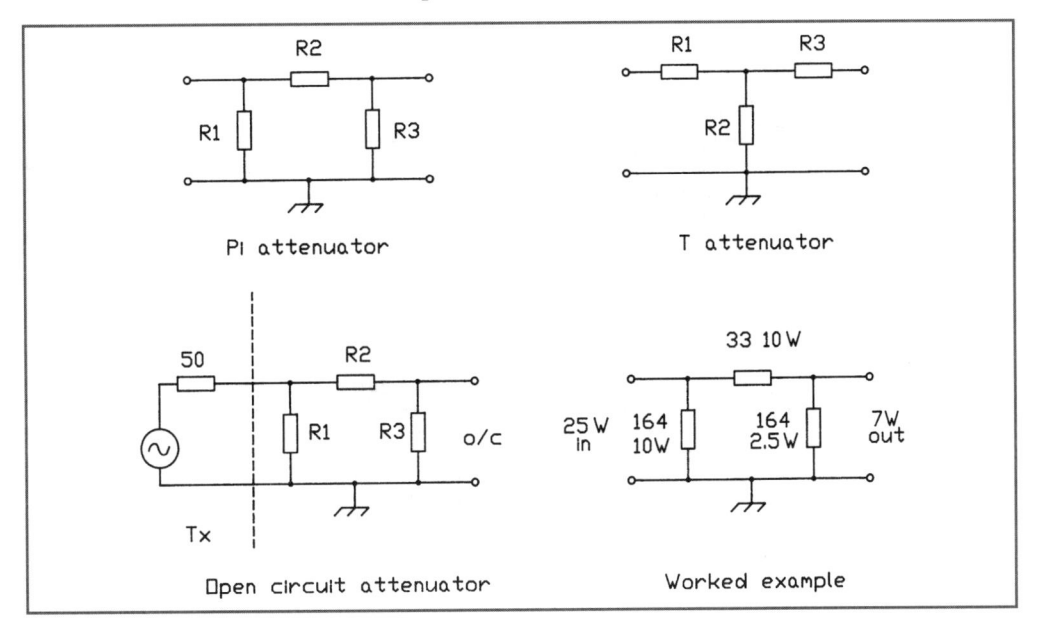

If A is your desired attenuation in dB, first calculate the factor n, from:

$$n = 10^{\left(\frac{A}{20}\right)}$$

For a Pi attenuator:

$$R1 = R3 = \frac{(n+1)}{(n-1)} \times 50 \ \Omega$$

and $R2 = \frac{n^2 - 1}{2n} \times 50 \ \Omega$

For a T attenuator:

$$R1 = R3 = \frac{(n-1)}{(n+1)} \times 50 \ \Omega$$

and $R2 = \frac{2n}{(n^2 - 1)} \times 50 \ \Omega$

If you want the attenuator for anything other than low-level work, the power dissipation of the individual resistors is important. If the attenuator is normally terminated in a 50Ω load, then:

For the Pi attenuator –

$$P_{R1} = P_{IN} \times \frac{50}{R1} \ W$$

$$P_{R2} = P_{IN} \times \frac{50}{R2} \times \left(1 - \frac{1}{n}\right)^2 \ W$$

$$P_{R3} = \frac{P_{IN}}{n^2} \times \frac{50}{R3} \ W$$

For the T attenuator –

$$P_{R1} = P_{IN} \times \frac{R1}{50} \ W$$

$$P_{R2} = P_{IN} \times R2 \times \left(1 - \frac{1}{n}\right)^2 \ W$$

$$P_{R3} = \frac{P_{IN}}{n^2} \times \frac{R3}{50} \ W$$

If there is a chance that the attenuator will not be terminated, the power ratings of some of the resistors may have to be increased. You can find the power dissipated in each resistor by considering the power source as a 50Ω resistor in series with a voltage source giving twice the voltage that the rig would give across a normal 50Ω load.

Here's a example of a home-brew power attenuator: A multi-mode rig has a maximum output of 25W, and is required to drive a linear amplifier which must not receive more than 3·5W of drive. You decide to run the transceiver at 12W RF output to improve the linearity of the driving signal (Chapter 6) so the required attenuation is 5·5dB. Plugging

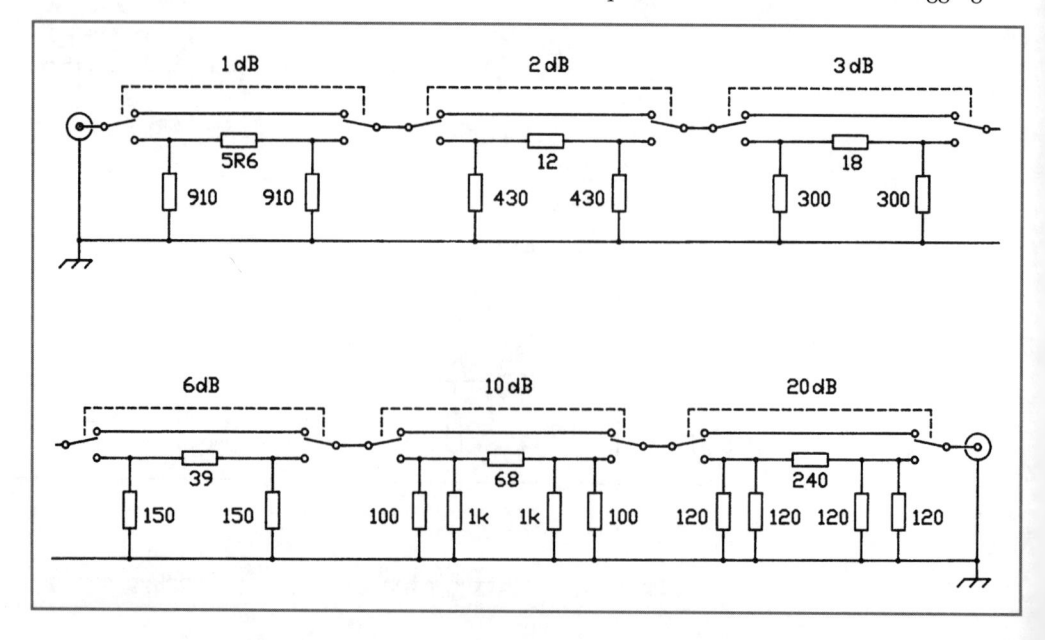

Fig. 12.8. A switched RF step attenuator

IMPEDANCE & VSWR MEASUREMENT

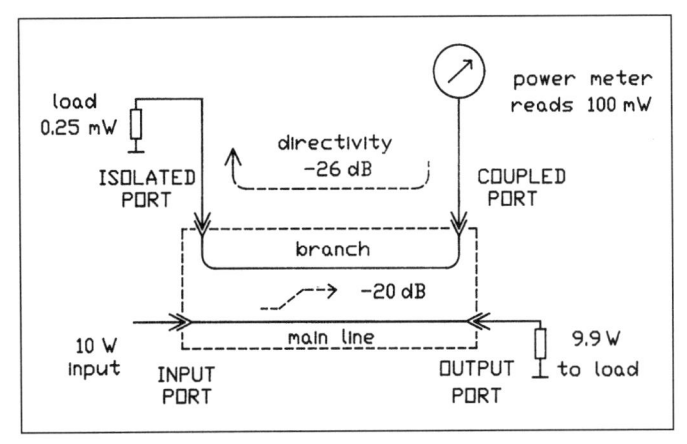

Fig. 12.9. Power measurement using a directional coupler

this value into the Pi-attenuator formulas above, we find that R1 = R3 = 163Ω and R2 = 34Ω. For safety, design the attenuator to accept the full 25W of available drive, in which case the powers dissipated in the resistors R1, R2 and R3 are 7·7W, 8·1W and 2·1W respectively. Since neither you or I have junk-boxes stuffed with 163Ω high-power non-inductive resistors, they have to be made from paralleled higher-value components. Five 820Ω 2W carbon resistors will give us a 164Ω 10W combination for R1; similarly R2 can be made from ten 330Ω 1W components and R3 from five 820Ω 0·5W. Since we're not interested in the last fraction of a dB in accuracy for this application, you can play around endlessly with different combinations of stock resistor values. The attenuator could be built by soldering the resistors to wide PCB tracks, adding a box of soldered-together PC board for shielding.

A SWITCHED ATTENUATOR SET

References [6] and [7] mentioned earlier include descriptions of switched attenuator sets, capable of working into the VHF range if carefully made. Fig. 12.8 shows details of my own set. It is made with nothing more than common slide switches in a box made from soldered together PCB strips with internal screens between stages, and uses ordinary 1% metal-film resistors with short leads. The table below shows its theoretical and measured performance. The 'theory' attenuation is what you would expect if the resistors used had their correct nominal values. As you can see, the measured performance is quite respectable up to at least 200MHz, amply repaying the small effort of making the unit. The measurements also dispel the popular myth that it's the higher values of attenuation that suffer at high frequencies.

MEASURED PERFORMANCE OF SWITCHED ATTENUATOR SET

Nominal	Attenuation, dB					
	1	2	3	6	10	20
Theory	0·98	2·07	3·02	6·16	10·08	19·9
Measured						
1MHz	0·98	2·04	2·88	6·10	10·06	20·0
10MHz	0·97	2·04	2·87	6·09	10·06	20·0
30MHz	0·97	2·04	2·81	6·11	10·10	20·0
50MHz	1·00	2·05	2·90	6·07	10·03	19·9
70MHz	1·02	2·03	2·89	6·00	9·93	19·8
100MHz	1·00	2·06	2·92	6·19	10·28	20·2
145MHz	1·00	2·07	2·93	6·09	10·09	19·9
200MHz	1·01	2·10	2·90	6·15	10·12	19·8
300MHz	1·06	2·19	3·13	6·43	10·45	19·9
400MHz	1·34	2·25	3·30	6·73	10·73	19·8
432MHz	1·27	2·32	3·45	6·70	10·56	19·7

DIRECTIONAL COUPLERS AND VSWR INDICATORS

A directional coupler couples a known fraction of power flowing in one direction in the main line to a second transmission line, as illustrated in Fig. 12.9. For the applications in which we're interested, both the lines are of 50Ω impedance and both are terminated in 50Ω loads. If the coupler has two secondary lines, the power in both forward and reverse directions can be measured simultaneously. Hence the VSWR can be calculated from the relationships given on page 12-12.

The coupling factor is defined as the ratio of the incident power to that coupled from the main line to the auxiliary branch. Coupling is usually expressed in dB, so a '20dB coupler' is one where the coupled power is 20dB smaller than the power in the main line. Directivity is a measure of the 'goodness' of the coupler, and is the ratio of the power appearing at the

TEST GEAR & STATION ACCESSORIES

isolated port to that at the coupled port. Fig. 12.9 clearly shows all these relationships. The directivity sets limits on the accuracy of the coupler; for example, the directivity of 26dB means that an indicated VSWR of 1·20 could in fact be anywhere between 1·08 and 1·33.

This is also a good moment to remind you of the error introduced by the feeder loss when measuring VSWR at the transmitter end of the feeder. For example, an indicated VSWR of about 1·2:1 (400W forward, 3W reflected) at the end of a feeder with 3dB loss is actually a VSWR at the antenna of a little more than 1·4:1, since the forward power at the antenna is only 200W, and the 3W of reflected power arriving back at the transmitter end was 6W at the antenna.

There are two common types of 'VSWR bridge' – those based on directional couplers formed from parallel transmission lines, and those such as the Bruene bridge based on current transformers. Both give outputs proportional (in some fashion) to the forward and reflected power in the main line. Simple parallel-line couplers have a frequency-dependent sensitivity, so using them on more than one band may not be practical. However, they will work at UHF and above, whereas the practical limit for current-transformer types is the 144MHz band.

SIMPLE VSWR INDICATORS

A simple VSWR indicator has several uses apart from the time-honoured one of checking whether your antenna's still there on a dark and windy night! For example, you can insert a VSWR indicator in the line between your transverter and a valve linear amplifier to ensure that the latter's input circuit always presents something reasonably like a 50Ω load to the transceiver (Chapter 6). For that type of monitoring job the VSWR indicator doesn't need to be particularly accurate. For more precise measurements, and for looking at the VSWR of devices not intended for carrying power, you'll have to use something better – for example the return-loss bridge which I'll describe later.

UHF IN-LINE POWER AND VSWR METER

The simple instrument shown in Fig. 12.10 is based on a directional coupler using 50Ω microstrip lines etched on a double-sided epoxy-glassfibre PC board. It will work up to 1·3GHz (although its sensitivity increases with frequency) and it can be used for powers up to about 100W at 432MHz.

VSWR BRIDGE AND POWER METER FOR 28-144MHZ

This unit is suitable for measuring VSWR and power with reasonable accuracy up to the 144MHz band. Designed by W1FB [8], the circuit diagram is shown in Fig. 12.11. It is a conventional toroidal-transformer Bruene bridge, apart from the details of the current transformer and the balancing capacitor.

Fig. 12.10. Details of the stripline directional coupler and detectors. The large layout is reproduced at full scale

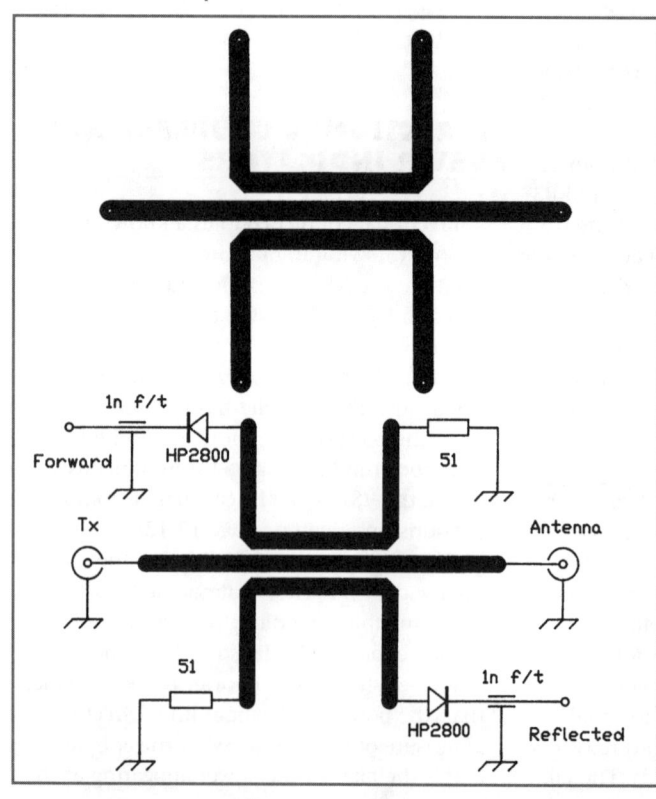

IMPEDANCE & VSWR MEASUREMENT

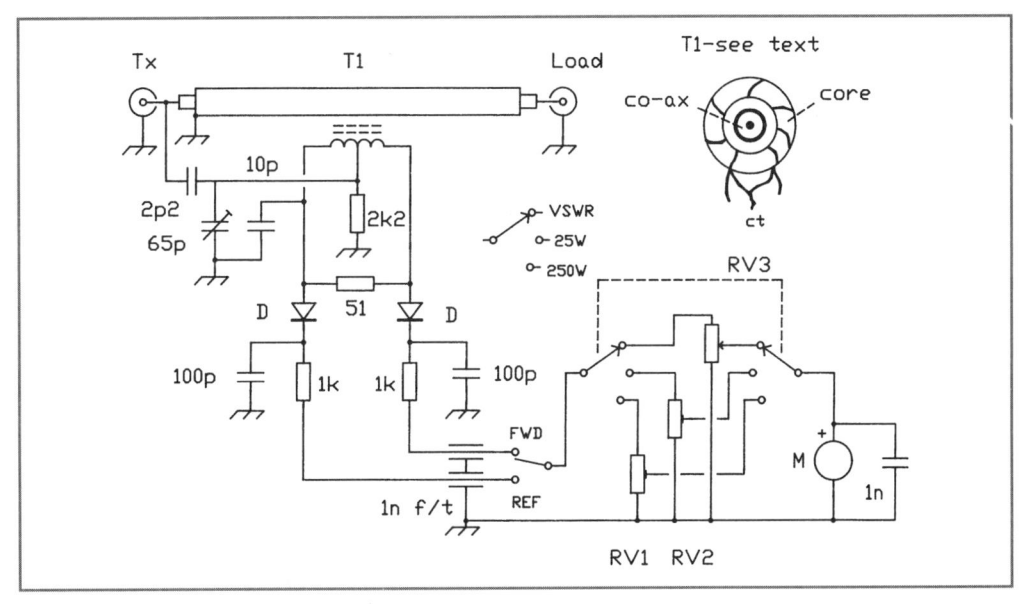

Fig. 12.11. VSWR/power meter suitable for 28–144MHz

COMPONENTS	
D	1N4148 or 1N914
M	50 or 100µA meter
RV1, RV2	25k preset
RV3	25k panel-mounted potentiometer
T1	Core: Amidon T50-61 or Fair-rite 59-61000401 toroid Primary: 2in length of URM43 coaxial cable, braid grounded at input socket only Secondary: 8 turns of bifilar 28swg enamelled wire (twisted 3 to 4 twists per inch). The start of one wire is connected to the end of the other to form the centre-tap.

Transformer T1 is wound with a twisted-pair transmission line formed from two lengths of 28swg enamelled wire, twisted together (using a hand drill) so that there are about 3 twists per inch. The beginning of one wire is joined to the end of the other to form the centre-tap. This winding technique offers a much better balance at VHF than the usual method of using one continuous winding centre-tapped half-way. The 50mm length of coaxial cable (URM43 or similar) forming the primary of the transformer is grounded at one socket only, so that the shield forms a Faraday (electrostatic) screen. Before use, the bridge must be nulled by adjusting the trimmer capacitor for minimum reflected power reading when using an accurate 50Ω load connected to the antenna port. Do this at a power level of about 10W to ensure that the null you see is real, and not just due to the lack of sensitivity of the diode detector at low levels. You can either calibrate the scale to read power directly on the two ranges, or use a calibration graph.

PCB-MOUNTING VSWR METER FOR 50MHz AND 70MHz

The toroidal current-transformer conventionally used for a Bruene bridge is not readily mounted on a PC board. By using a two-hole core instead, the transformer will fit vertically on a PCB. The original unit was designed by G4DGU for incorporation on to the PCB of a transverter, providing outputs for both ALC control and high VSWR detection. Details of the transformer and associated circuit are shown in Fig. 12.12. The RG405 (UT085) semi-rigid cable needs to be carefully bent in a 'U'-shape around a suitable mandrel. Slip a couple of pieces of thin-walled sleeving over the arms of the 'U' to insulate the ferrite from the coax outer.

The rectified DC output is about 4V for 20W throughput, although with a larger ferrite core and coax the unit is suitable for powers of up to about 100W.

TEST GEAR & STATION ACCESSORIES

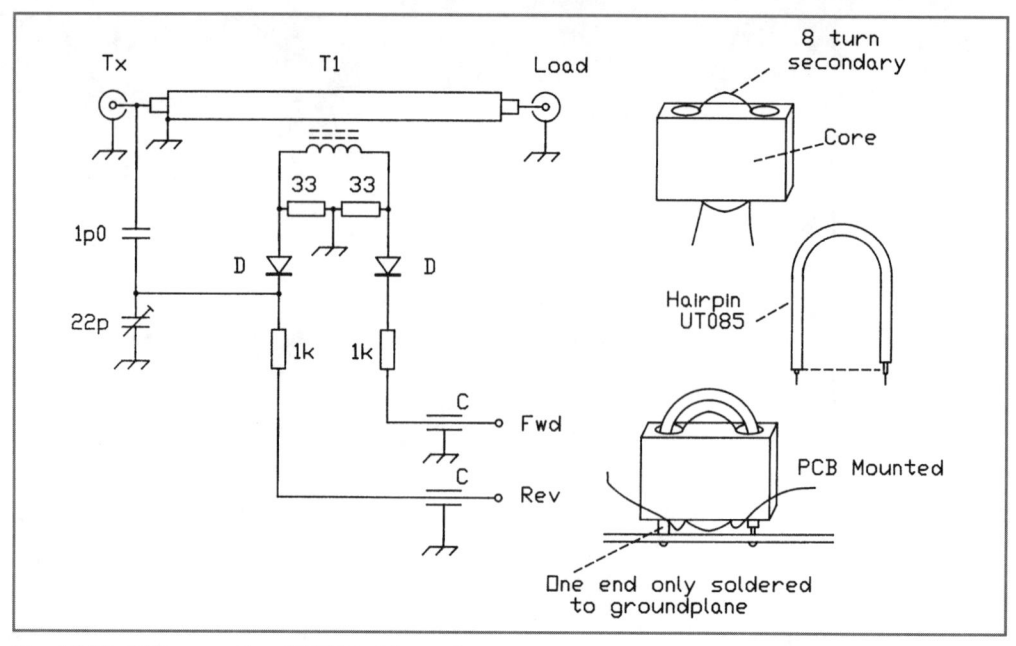

Fig. 12.12. PCB-mounting VSWR bridge using a two-hole ferrite core

COMPONENTS
D   OA90 or OA91
C   1n0 feedthrough or disk ceramic
T1   Core: Siemens two-hole ferrite core B62152-A4-X30
Primary: Hairpin loop of RG405 (UT085) semi-rigid coax. Outer sheath grounded at one end only
Secondary: 8 turns of 28swg enamelled wire through holes. See sketch.

RETURN-LOSS BRIDGE

A return-loss bridge (RLB) allows the accurate measurement of impedance matching at very low power levels. For example you can use the RLB to look at the input impedance at the antenna socket of a VHF/UHF low-noise amplifier without fear of damaging an expensive front-end device. It also enables you to make antenna measurements and adjustments safely since the bridge only requires a few mW for operation.

The circuit and construction of a VHF–SHF return-loss bridge are shown in Fig. 12.13. If two of the ports are terminated in 50Ω, the impedance looking into the third port will also be 50Ω. If the REFerence and TEST ports have 50Ω loads, and you feed RF into the INPUT port, the voltages at either side of the diode are the same, so no output results. But if the impedance at the test port is not 50Ω, the bridge will not be balanced and an output proportional to the return loss will appear across the diode. This bridge is really intended for use with amplitude-modulated RF sources and the tuned audio-level meter described later; by keeping the RF levels within the square-law region of the detector and using the level meter, the return loss can be read directly in dB.

The bridge is constructed using two N sockets joined by four tapped-brass spacers, with a 'U'-shaped copper or brass strip wrapped and soldered around them. The exact spacing will depend on the size of the 51Ω resistors you choose; use 1% metal-film resistors and arrange lead lengths as close to zero as you can manage. The use of 51Ω resistors rather than 50Ω does not affect the accuracy of the bridge, all other things being equal – the important thing is that the resistors are as nearly equal in value as possible. Detected audio is taken from a screened lead connected to the feed-through capacitor.

To set up the bridge, fit an accurate 50Ω

IMPEDANCE & VSWR MEASUREMENT

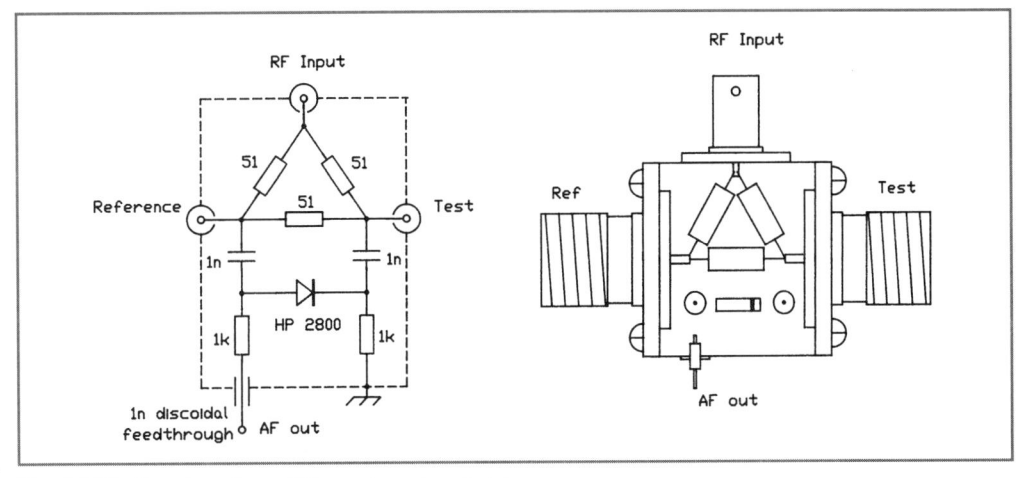

Fig. 12.13. Construction of the return-loss bridge

reference load to the REF port and a type N short-circuit to the TEST port. Apply a few mW from a modulated source at the BNC RF port, preferably at the highest frequency you intend to use. Adjust the controls on the level meter so the meter reads 0dB. Now replace the short-circuit with a second identical reference 50Ω termination and adjust the level meter for an on-scale reading, which will typically be around −30dB. Since the two loads are supposedly the same, any deviation from the ideal (infinite) return loss is due to minor imbalance in the RLB. You can improve the balance by either bending the bodies of the two capacitors or experimenting with a small copper tab attached to one side of the diode. It should be possible to achieve a residual balance of −40dB or so. To check the overall symmetry of the bridge – or possibly to expose differences between your two supposedly-identical 50Ω loads – exchange the two loads and see whether the bridge remains well balanced.

In normal use, keep your best 50Ω load permanently connected to the REF port as the station impedance standard. Having set up the bridge, try connecting two 50Ω loads in parallel to the TEST port by means of a T-adapter. You should now see a return loss of 6dB, which corresponds to a VSWR of 2 (reflection coefficient of 0·5) due to the 25Ω impedance you've presented to the bridge. Don't forget to set zero return-loss using the type N short-circuit before each measurement.

Another application of the RLB is to prepare cables which are an accurate electrical half-wavelength long (or any multiple thereof). To do this, connect a 50Ω load to the TEST port via a T-adaptor, check the bridge balance, then connect the cable to the other side of the T-piece. When the open-circuited cable is a multiple of a half-wavelength long at the frequency at which the bridge is energized, it will present a very high shunt impedance so the bridge will again balance. This method automatically takes into account variations in velocity factor along the cable due to manufacturing tolerances. Accuracies of ±1mm can be achieved at 432MHz regardless of the overall length of the cable.

THE SIMPLEST TDR

How would you like to examine a suspect joint at the top of your feeder, even though it's well after midnight, and blowing Force 8 with hailstones the size of golfballs? Simple – use a Time-Domain Reflectometer (TDR) and do it all without leaving your cosy shack. A TDR is a complex and expensive piece of testgear found usually only in the more fortunate professional labs. It allows the display on an oscilloscope of variations in impedance along the length of a transmission line. Commercial units allow you to inspect, for example, the minute changes in impedance at the joint between two connectors at the far end of a long cable. If you're prepared to lower your sights somewhat, and have a 'scope with a reasonably fast timebase, the simple circuit shown in Fig. 12.14 will do quite a lot. You can measure cable impedance and length, and discover open- or short-circuits in feeders before tearing the tower down. It's based on an earlier design [9], but with an improved 50Ω line-driver following the pulse generator.

The operation of a TDR is not to difficult to follow; it's rather like an echo-sounder. Imagine what happens when a pulse from a 50Ω source is applied to a cable. As the pulse begins its journey down the line, it 'sees' the impedance of the line, so the instantaneous voltage on the line is that due to a potential divider formed from the generator impedance – carefully arranged to be 50Ω – and the cable impedance. Now imagine what happens when there's a change of impedance, say an open-circuit. When the pulse arrives, the voltage at that end of the line rises to something near the open-circuit output of the generator. This higher-voltage reflected wave now propagates back down the cable, and eventually it reaches the generator end where the rise in voltage is indicated on the 'scope. What you see in this case is a short pulse of half-screen size followed by a rise to full amplitude. The half-size pulse shows the length of time taken for the pulse to reach the end of the cable and return. The final rise to full screen height shows that there's an open circuit at the far end. If the amplitude had instead dropped to zero it would have meant a short-circuit. So the 'scope gives you a picture of impedance along the line.

The circuit of the 'Simplest TDR' (Fig. 12.14) consists of a 555 multivibrator which generates square-wave pulses of about 6µs duration at a frequency of around 10kHz. These are fed to the 2N4401 50Ω line driver, which is connected to the line under test and the 'scope via an attenuator. To use the Simplest TDR, connect it to one branch of a BNC T-piece via a short length of 50Ω cable, and fit the T-piece directly to the front panel input of the 'scope Y amplifier. You need a 'scope with a DC amplifier usable up to about 30MHz and a triggered timebase of 1µs per division. A sweep multiplier also helps. Set the 'scope for DC input and a sensitivity of around 100mV/cm. If your 'scope has an external trigger input, connect it to the trigger output from the TDR. You should now see the pulse from the TDR on the screen. Adjust the controls so that the timebase triggers on the rising edge of the pulse, and the pulse occupies the full height of the screen.

Now to begin testing. Check the operation first by putting a 50Ω load on the other

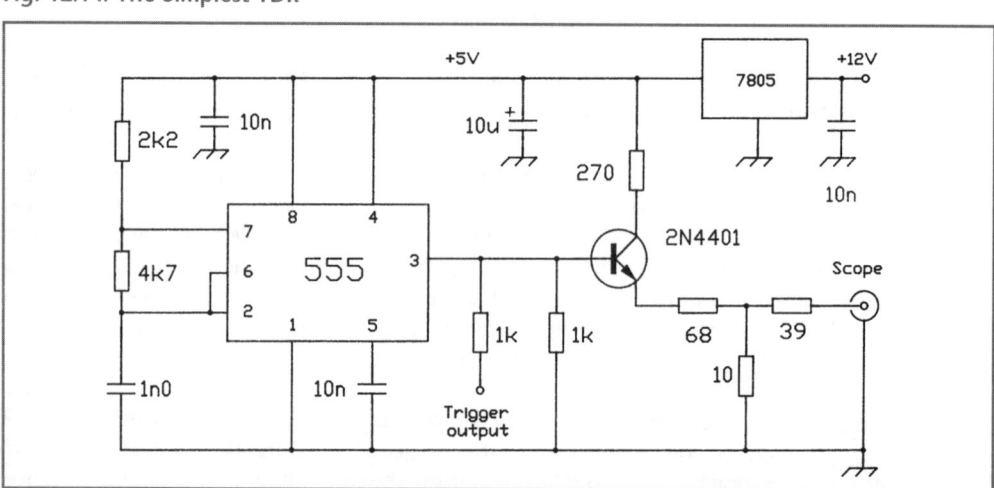

Fig. 12.14. The Simplest TDR

IMPEDANCE & VSWR MEASUREMENT

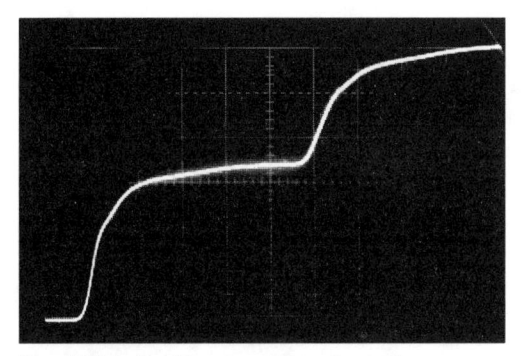

Fig. 12.15. Oscilloscope trace from the Simplest TDR connected to 20m of cable with the far end open-circuited

length, to move the reflection away from the initial rising edge. While the Simplest TDR doesn't offer the performance of a commercial instrument, it doesn't cost an arm and a leg either!

branch of the T-piece. The pulse height should drop to around half amplitude if all is well. If you like, you can roughly calibrate the screen for different impedances by connecting a range of different loads (resistors with short leads).

I tested the prototype by connecting it to a part drum of RG174, open-circuited at its far end. The 'scope timebase was set to 40ns/cm (200ns/cm with x5 multiplier). Fig. 12.15 shows the result on the screen. Zero ohms is at the bottom of the screen, 50Ω is half-way up and infinite impedance is at the top. The start of the pulse and the start of the cable can be seen at the left, and the step upwards at the open-circuit end is just to the right of centre. The length of the cable thus corresponds to about 5cm on-screen, *ie* 200ns. The length of the cable in metres is then given by:

$$\frac{(\text{time in nanoseconds}) \times 299 \cdot 8 \times (\text{velocity factor})}{2000}$$

ie about 20m. When measured the hard way, the length on the drum was found to be... just over 20m! It was possible to see a 5m length of 75Ω cable at the far end quite clearly.

You can use the Simplest TDR on your feeder cables – but remember that since it employs a pulse technique, frequency-dependent loads such as antennas will display their DC characteristics, *ie* short- or open-circuits. Because this simple TDR has only limited time (and hence length) resolution, it often helps to insert a few metres of 50Ω cable before the test

RECEIVER MEASUREMENTS

GAIN AND LOSS MEASUREMENTS USING MODULATED SOURCES

Low-level measurements, especially those relevant to receiver performance, largely concern the measurement of power gains and losses of system components such as preamps, filters and mixers. Since almost all such measurements are made in 50Ω systems, we're concerned here with the gains and losses you see when the object under test is inserted between a 50Ω source and a 50Ω load. If you're building a new system, or adding to one, stage-by-stage gain and loss measurement, combined perhaps with an performance-analysis program such as TCALC (see Chapter 5) is going to be a tremendous help in checking out the actual performance.

Although losses and gains are concerned with *relative* power levels, measurements do need to be made at about the right absolute level. For example you can't make meaningful measurements on a preamplifier at the one-watt level – you need to use powers of tens to hundreds of microwatts to make sure the device is operating well below its compression point. Making accurate direct power measurements at these low levels isn't easy, but a simple technique – one that's been used for decades in professional microwave measurements – can be used to shift the burden of accuracy down to AF where precision is a lot easier and cheaper to come by. This technique

Fig. 12.16. Using a tuned audio-level meter to measure the gain of an amplifier

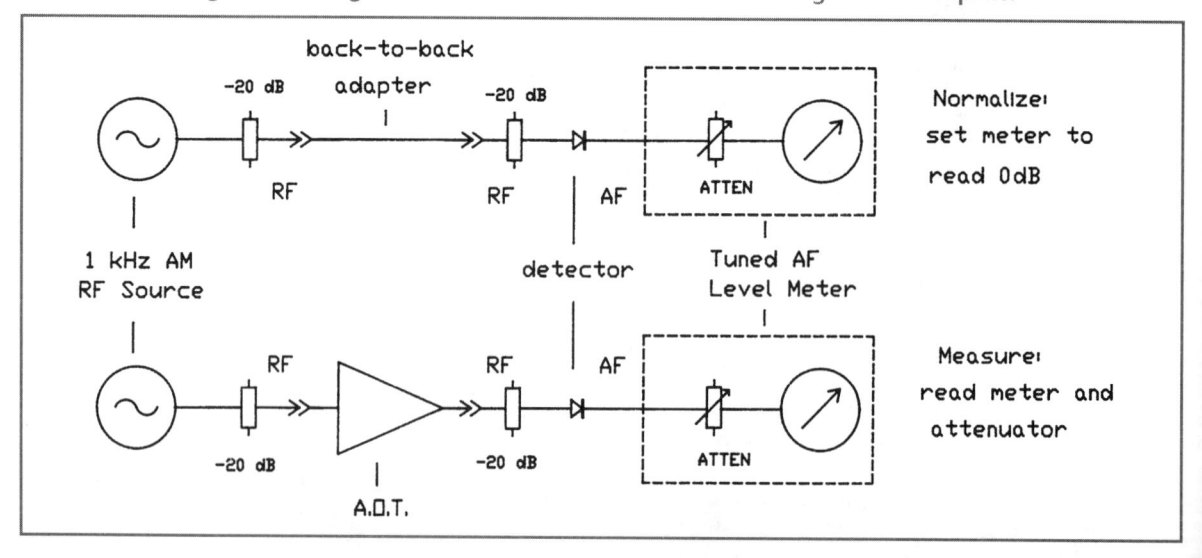

RECEIVER MEASUREMENTS

uses an amplitude-modulated RF source, a simple square-law diode detector and a tuned AF level meter.

The RF equipment used with this technique is fairly conventional, *eg* a VSWR bridge with only minor modifications to the diode detectors. But instead of amplifying the DC component of the rectified RF, the level meter works with the AF modulation. This side-steps all the problems of amplifying and measuring low-level DC signals. Since only a single audio frequency is of interest, broadband AF noise in the amplifiers can be kept in check by using sharp audio filters. Attenuators can also be included in the AF amplifier chain, where accuracy isn't difficult to achieve. I'll show you how to make the level meter a little later.

Have a look at Fig. 12.16, which shows a simple setup for measuring the gain of an RF amplifier. A modulated RF source is connected via an impedance-normalizing attenuator of 20dB or so to the input of the amplifier. The output of the amplifier is connected via another 50Ω attenuator to the diode detector, and the output from the detector goes to the tuned AF level meter. First you connect a coaxial back-to-back adapter in place of the amplifier under test and set the level meter to read 0dB. Then insert the amplifier, switch in some AF attenuation to bring the signal on-scale and read the gain within a fraction of a dB. The technique can equally well be used to measure any kind of RF power gain or loss, *eg* to measure the loss in a length of coaxial cable or to compare the gains of antennas.

As well as measuring the gain of an RF amplifier you can turn it around and measure its reverse isolation; for assured stability the reverse isolation should be several dB more than forward gain. This home-made test setup doesn't provide the same amount of detailed information as a proper network analyser, because it measures only amplitude and not phase – but it costs about a thousand times less. As already mentioned, the same equipment can also be used to measure VSWR with a return-loss bridge, and it reads return loss directly in dB. You'll find more information on the modulated-signal technique in [10], and the use of the technique to measure antenna gains is described in [11].

TUNED AUDIO LEVEL METER

The circuit of my tuned AF-level meter in shown in Fig. 12.17. The input transformer provides a suitable AC load for the diode detector, gives a moderate voltage step-up, and allows the injection of a small amount of bias (of suitable polarity) for optimum square-law detector performance. Three 'gain blocks', each followed by a active filter, are used to provide the necessary gain and selectivity. An active rectifier drives a large panel meter which provides a usable scale of 10dB. A switched 0–50dB attenuator covers any larger gain variations in 10dB steps. To ensure freedom from overload and good signal-to-noise ratio, the calibrated attenuation takes place in two stages along the signal chain.

Construction isn't too critical – just build the instrument using standard hi-fi preamplifier practice, put it all in a metal box and connect the circuit ground to the case at the input socket only. Power should come from two internal 9V batteries, to avoid problems with ground loops. My prototype was built using Veroboard, laying the stages out in line along the board. Use a two-wafer switch for the attenuator to avoid leakage from the higher-level stages back to the input (with consequent loss of calibration accuracy) and make all connections to the attenuator and SET LEVEL control with screened cable. Each AF filter has a small frequency adjustment to allow for component tolerances. The active rectifier is followed by a buffer amplifier to drive the meter. Don't skimp on the meter – use the biggest and best-quality meter you can get, since you will be reading from the scale.

Mark the meter face from full-scale deflection (FSD) downwards with a dB scale according to the following table:

% FSD	dB	% FSD	dB
100	0·0	25·1	6·0
79·4	1·0	20·0	7·0
63·1	2·0	15·9	8·0
50·1	3·0	12·6	9·0
39·8	4·0	10·0	10·0
31·6	5·0	1·0	20·0

Fig. 12.17. Circuit of a tuned audio-level meter

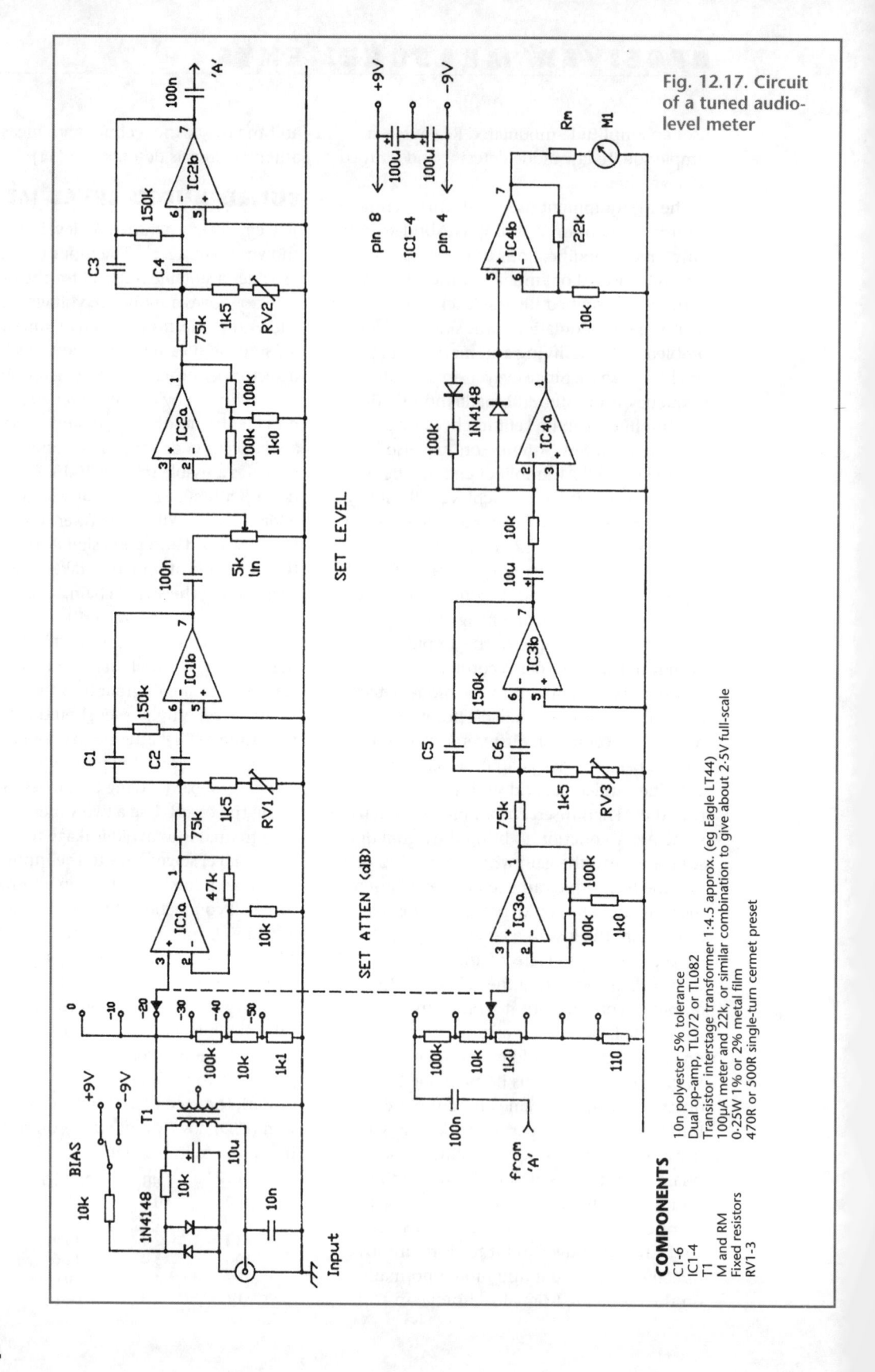

COMPONENTS

C1-6　　　　　10n polyester 5% tolerance
IC1-4　　　　　Dual op-amp, TL072 or TL082
T1　　　　　　Transistor interstage transformer 1:4.5 approx. (eg Eagle LT44)
M and Rm　　 100µA meter and 22k, or similar combination to give about 2·5V full-scale
Fixed resistors　0·25W 1% or 2% metal film
RV1-3　　　　 470R or 500R single-turn Cermet preset

RECEIVER MEASUREMENTS

This comes from the decibel formula, modified for a square-law detector:

10 antilog$_{10}$(dB below full scale) = % FSD

Note that the meter is scaled in terms of power, not voltage. If you're a perfectionist, you can use an accurate digital multimeter with an external adjustable current source to mark the meter scale directly. This compensates for any non-linearity in the meter movement, which can be quite detectable.

Setting up the instrument is simplicity itself. First feed a modulated RF signal into the diode detector and peak each filter response. Then short-circuit the input, switch the attenuator to its most sensitive range (0dB) and adjust the SET LEVEL control to about 50%. If all is well you should see about 20% deflection on the meter due to circuit noise.

MODULATED RF SOURCES

To use the AF level-measuring technique just described, you'll need some amplitude-modulated RF sources. You may have a suitable signal source already – the modulating tone of most RF signal generators is 1kHz, which is also the standard modulating frequency for these measurements. The local oscillators already described in the earlier chapters of this book can be adapted as in-band RF signal sources with appropriate substitutions of crystals and minor adjustments to the tuned circuits. To modulate these or any other low-level RF sources, the simple circuit of Fig. 12.18 can be used. A 1kHz oscillator using a 555 timer IC provides a square-wave drive to a simple RF chopper using a PIN diode, generating nearly 100% AF modulation. The output impedance isn't going to be 50Ω so the modulator must be followed by a normalizing attenuator; this might as well be built-in permanently as shown.

SENSITIVITY AND NOISE FIGURE MEASUREMENTS

For modern VHF/UHF receivers, the best methods of sensitivity measurement are those based on noise figure – or better still, the fundamental concept of noise temperature.

The traditional method using a signal generator cannot sensibly be used with a modern VHF/UHF front-end which can probably detect signals as small as 20nV, as explained in Chapter 4 and reference [12]. Most signal generators simply can't produce such low levels with the required accuracy, so it's extremely difficult to make measurements based on the signal level required to give a 3dB increase in receiver output. If you have a reasonably screened generator and good attenuators, you can use it to measure the sensitivity of an HF transceiver used with a

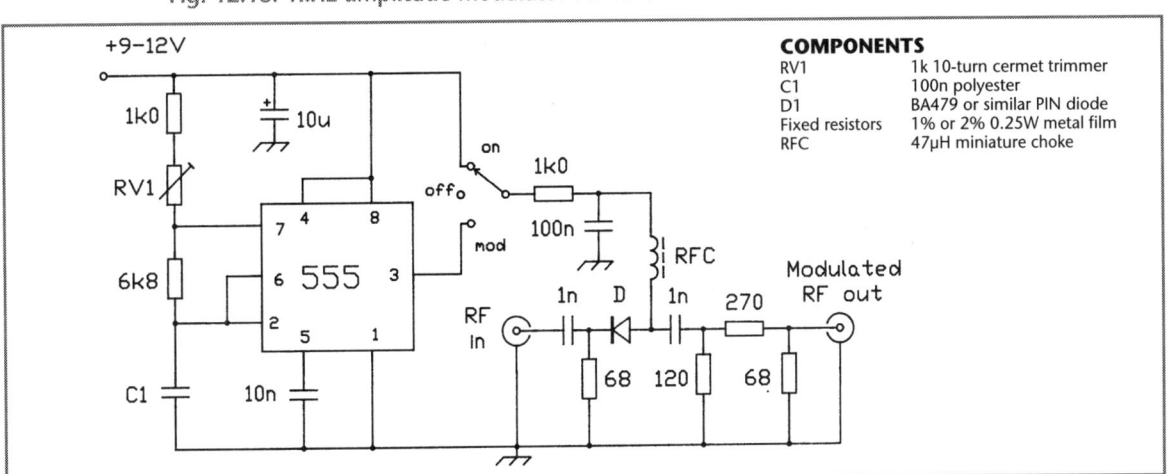

Fig. 12.18. 1kHz amplitude modulator for RF sources

COMPONENTS
RV1   1k 10-turn cermet trimmer
C1   100n polyester
D1   BA479 or similar PIN diode
Fixed resistors   1% or 2% 0.25W metal film
RFC   47μH miniature choke

TEST GEAR & STATION ACCESSORIES

transverter, but that's about all.

Absolute noise-figure or noise-temperature measurements are not an easy task for the amateur. In the absence of a calibrated noise source, the only viable solution is to measure noise temperature using the 'hot/cold' method. This involves measuring the change in thermal noise from a 50Ω resistor at the receiver input, first when the resistor is hot and then when it's cold. Typical combinations of temperatures for amateur measurements are boiling water and melting ice, or room temperature and melting dry ice or liquid nitrogen. The details of this method are described in [12]. The thermionic noise diode can no longer be considered a reliable absolute noise-source – it was good enough in the days of valve front-ends (which those of us old enough to remember are still trying desperately to forget) but it's no good in the GaAs-FET era.

So where does that leave us? If you're not an adept with cryogenics, relative measurements are about all that's left. Quite good enough too, for most purposes. If your front-end has been designed to meet a noise-figure target as described in Chapter 4, tweaking for maximum signal-to-noise ratio is probably all you need to do. This kind of relative measurement is fairly simple if you watch what you're doing. If absolute numbers still worry you, take your pet preamp to the measurement stand at a rally, or find someone who boasts of a professional noise-figure meter at work.

But even relative NF measurements have their pitfalls. Unless you're prepared to spend hours on repeated measurements of signal+noise and noise levels, you'd better build an automatic relative NF indicator. The almost universal choice is the G4COM alignment aid [13], which uses the difference between the receiver outputs when a noise source is switched on and off to generate a signal proportional to NF. A number of modifications to the basic design have been published [14], though the main cause of erratic performance seems to be the behaviour of the SSB IF filter through which measurements are usually made; in any noise-based measurement, it's best to use a bandwidth of at least several kHz. One genuinely worthwhile improvement to the original G4COM design is to increase the square-wave voltage driving the noise source to about 12V.

The traditional noise-source using a microwave mixer diode is not satisfactory for modern FET front-ends. The noise source must present an accurate 50Ω impedance to the front-end in both its 'on' and 'off' states. Otherwise the change in the match can considerably alter the gain of a FET stage whose own input impedance is not close to 50Ω, and this will render any NF 'measurement' completely invalid. Worse than that, you can't even tweak the front-end for lowest indicated NF because the main effect of your adjustments will be not be upon the front-end performance but upon the measurement error itself! A well-matched noise source avoids these problems, so a front-end optimized by amateur methods using a well-matched source will actually be in better shape than one that's been 'measured' using the older generation of professional noise sources.

A good noise source [15] which is eminently suitable for use with the G4COM alignment aid is shown in Fig. 12.19. It uses the reverse base-emitter breakdown characteristics of a BFW92 transistor to generate a generous amount of noise well up into the microwave region, allowing sufficient attenuation to be placed on the output to swamp the on-off changes in source impedance. I've shown a 16dB attenuator built-in, though you may want to include some extra external attenuation. Build the noise head in a shielded box, with the shortest possible lead lengths for the output circuit. For those of you who can read the original reference in Dutch [15], or are brave enough to build from the circuit diagram alone, that article also described an improved NF meter which can have its scale directly calibrated in NF. That might be worth trying, assuming you can check the calibration against a reliable professional instrument, although the long-term stability of any noise source is problematical.

A noise-figure measuring instrument of

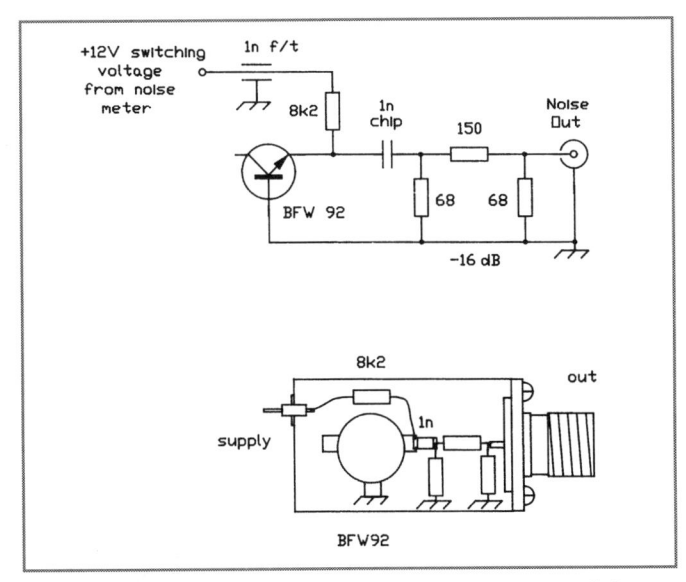

Fig. 12.19. Circuit and arrangement of the noise head using a BFW92 transistor

clean, stable power. The local-oscillator designs in Chapters 8–10 would make a good starting point and other ideas appear in [12]. The diode doubler [18] is particularly recommended as good spectral purity is far easier to attain than with conventional multipliers – an important design consideration if you don't have ready access to a spectrum analyser.

A level of several mW is needed for gain compression measurements, while –30dBm is sufficient for reciprocal mixing tests. Intermodulation measurements require two separate sources. Since the necessary isolation between the sources is achieved by means of 30dB attenuators as well as a hybrid combiner, an initial output level of at least +10dBm is needed from each source in order to arrive at a combined output level of the order of –20dBm.

almost professional quality has been described in *DUBUS* magazine [16], together with a precision noise source capable of operating up to 10GHz. A follow-up article explains in detail how to use a noise-figure meter for accurate measurements, and the importance of impedance stabilization in the noise source [17].

STRONG-SIGNAL SOURCES AND PERFORMANCE

Good sensitivity isn't the only characteristic that the VHF/UHF DXer requires from a receiver. Strong-signal handling is vitally important too. If you're not a convert yet, go back and re-read Chapter 5. To measure the strong-signal performance of your system you will need some strong-signal sources. The requirements for these are pretty stringent, since the shortcomings of the receiver are often not easily distinguishable from those of a poor signal source. I'll simply outline the bits of specialized testgear you'll need for making the measurements; for the theory behind it all, refer back to Chapter 5.

Measurements of the strong-signal handling performance of a receiver require sources capable of delivering up to a few milliwatts of

TRANSMITTER MEASUREMENTS

DUMMY LOADS

A dummy load is a must for transmitter testing – and I don't mean that Big Dummy Load in the Sky at the other end of your feeder cable! For low or medium power, commercially available loads are not too expensive. Make sure that the one you buy is adequately rated in terms of power. Also make sure that the VSWR is good at VHF and if possible at UHF as well. A load whose VSWR is already deteriorating at 30MHz isn't going to be particularly useful at 432MHz. High-power loads are a bit more of a problem, and good UHF loads rated at more than 50W are hard to find at reasonable prices. You can sometimes pick up good surplus bargains at rallies; if this appeals to you, do take a pocket digital multimeter along to check at least that the DC resistance is correct before you buy.

Home-made loads using paralleled resistors, as described in most of the handbooks, are likely to be OK for a few watts up to 144MHz or so, especially if you're careful with the construction. Coaxial loads based on 50Ω carbon cylindrical resistors are another possibility. Although the resistors themselves are hard to find, one relatively inexpensive source of this and other components for a HF/VHF dummy load is the Heath 'Cantenna' kit. If you wish, you can use the components for a load of improved VHF/UHF performance by following the directions in [19], but even then the VSWR at 432MHz isn't brilliant.

If the VSWR of your dummy load is unacceptably high you can always 'flatten' the VSWR using a matching device. Yes, I do mean a sort of antenna tuning unit for 144MHz or 432MHz! Such devices have been described by Joe Reisert W1JR [20] and the idea is shown in Fig. 12.20 – both units bear a close resemblance to the three-screw tuners used for impedance matching in waveguide. The first option requires a straight length of square-section coaxial airline, $3/8$ of a wavelength long, so it is physically suitable only for 432MHz. The nominal impedance of the line should be 50Ω although the precise impedance doesn't matter because any errors will be tuned out along with the rest of the VSWR. The physical dimensions of the inner and outer conductor depend on the materials available, *eg* square-section aluminium chassis boxes and copper water-pipe. To make your trial calculations, use the impedance formula given in Chapter 7 for RF power combiners.

The coaxial-cable option proposed by W1JR has the same overall electrical length as the air-line matching unit but is (literally) more flexible. It uses two sections of 50Ω cable, each $3/16$ of an electrical wavelength long, so you need to allow for the velocity factor of the cable. These lengths of cable can be folded up inside an outer screening box, so even a 50MHz unit is feasible. By the way, there's no reason why you shouldn't make the coaxial-cable design in air-line if you wish, or *vice-versa* – the two designs merely have slightly different impedance-matching capabilities.

The variable capacitors for either of the designs shown in Fig. 12.20 should each have a maximum value of about 180pF for 50MHz, 120pF for 70MHz, 60pF for 144MHz, or 20pF for 432MHz. Minimum capacitance should be

TRANSMITTER MEASUREMENTS

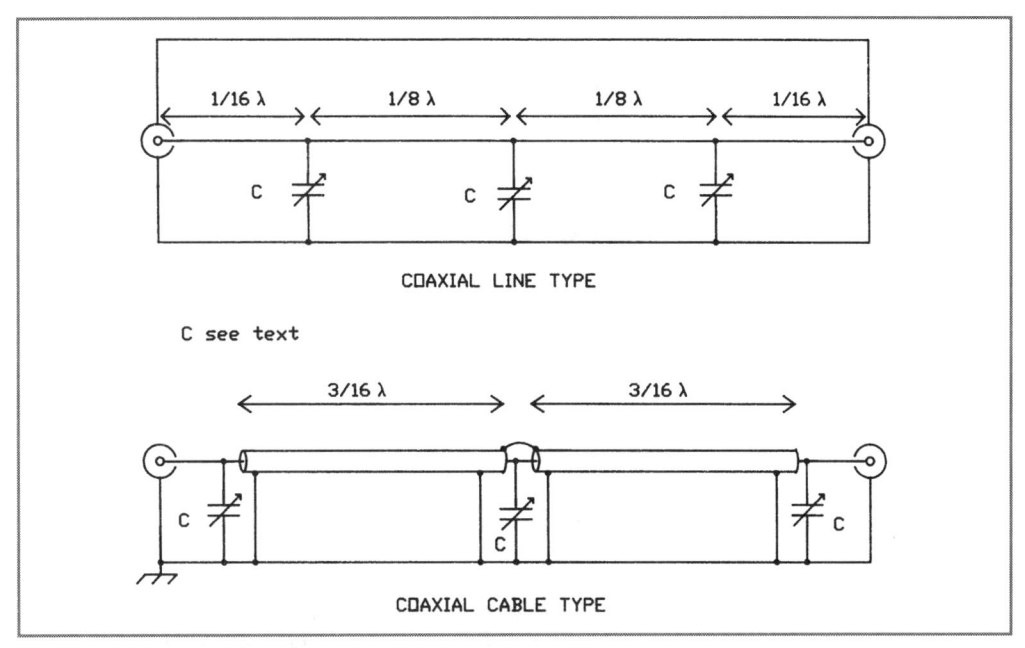

Fig. 12.20. Impedance matching units for VHF/UHF

no more than 10% of the maximum. Good choices would be mica trimmers or small air variables for the larger values, or small plastic-film or piston trimmers for 432MHz. The kinds of trimmers used in the output network of your PA for the same band are the obvious choice.

To use such a matching device with your dummy load, you must first adjust it at low power using a good VSWR indicator. Set all capacitors to minimum, then adjust each in turn for lowest VSWR, starting with that closest to the load. In most cases there will be several possible combinations of capacitor settings; you're aiming for the lowest-capacitance, lowest-Q solution because that also involves the lowest circulating currents and losses in the capacitors.

You'll probably have guessed that the impedance-matching devices could also be used for 'flattening' the VSWR of a poorly matched antenna of the real variety. But I wouldn't recommend it: if the VSWR at the transmitter end of your feedline is unacceptable, you ought to fix the problem at the antenna itself.

MEASUREMENT OF HIGH POWER

Two distinct sorts of measurement come under this heading, with different techniques to achieve them. The first is the measurement of the output power of a transmitter or power amplifier in a fairly accurate way, for example to verify that you're operating within the legal RF power limit or when calculating the efficiency of a new PA. Second are the everyday checks to verify that the usual amount of power is being produced. For the latter, one of the simple VSWR/power indicators I described earlier will usually be adequate, given the occasional check on its calibration. The more accurate RF power measurements call for a little more time and trouble, though they can still be done with relatively simple home-made equipment. Of course it's possible to buy equipment to do this job (for example the Bird 43 meter and elements) but I'd urge caution with some of the cheaper imported devices. One Japanese VSWR/power meter which I came across underestimated the power on its 500W range by a factor of two, so it pays to

TEST GEAR & STATION ACCESSORIES

check rather than trust the calibration! Within an amateur band your home-produced power measurements can be at least as accurate as a Bird 43, and decidedly more accurate than many power meters made for the amateur market.

DIRECTIONAL COUPLER AND MILLIWATTMETER

The customary way to measure high power (watts rather than milliwatts) in professional circles is to use a directional coupler to obtain a known fraction of the power in the main line. A low-power instrument, such as the milliwattmeter I described earlier, can then be used to measure the power in the branch line; hence the power in the main line can be easily calculated. A typical setup was shown earlier in Fig. 12.9. For accurate measurements it is vital that the main line is correctly terminated in 50Ω, or else the coupler will not work as it should. So that's yet another reason for needing a good dummy load capable of dissipating the full legal power limit and then some.

For most applications I'd suggest a coupling factor of about –20 to –30dB. With a –30dB coupler, 500W in the main line results in 500mW appearing at the coupled port on the branch line (Fig. 12.9). You can of course extend the range of a –20dB coupler by adding a suitably-rated attenuator at the input of the milliwattmeter. For a good treatment of directional couplers, with some practical designs, see [6] and [21]. Directional couplers are usually rather frequency-conscious so you'll need to calibrate them for each band. Generally a coupler which uses parallel lines has an upper frequency limit where the length of the branch line approaches a quarter-wavelength. The lowest useful frequency is set by the falling coupling factor.

Although making a good high-power directional coupler is a rather difficult task without some machining facilities, you can measure its performance quite accurately using the return-loss bridge, modulated signal source and tuned AF level meter which were described earlier. Before the days of network analysers, that's exactly what the professional RF labs did use.

A directional coupler using semi-rigid coax is shown in Fig. 12.21. Based on a design described in [22], it offers coupling of about –55 to –45dB over the range 100–500MHz, is usable up to at least 2·3GHz and can be simply made with hand tools. This coupler can form the basis of a power monitor for a high-power 432MHz transmitter for example. You first need to find some ¼-inch semi-rigid cable, since the more common UT141 is too small. To make the coupler, first file identical slots in two pieces of semi-rigid cable as shown in the diagram. Then slightly bend one of the pieces so that there will be enough room for all the connectors when the two slots are offered up to each other. When everything is positioned correctly, bind the cables together with wire. Then solder the joint with a big iron or hot-air gun, or very carefully with a blowtorch. Finally fit the connectors and calibrate the coupler.

Another alternative to a directional coupler is to use a power attenuator (*eg* 20–30dB) with a milliwattmeter. If you can come across a large power attenuator at a good price, it forms the basis of an excellent power meter and dummy load.

PEP CONVERTER FOR WATTMETERS

All the power-measuring devices I've described in this chapter are suitable for measuring

Fig. 12.21. Directional coupler made from semi-rigid coaxial cable

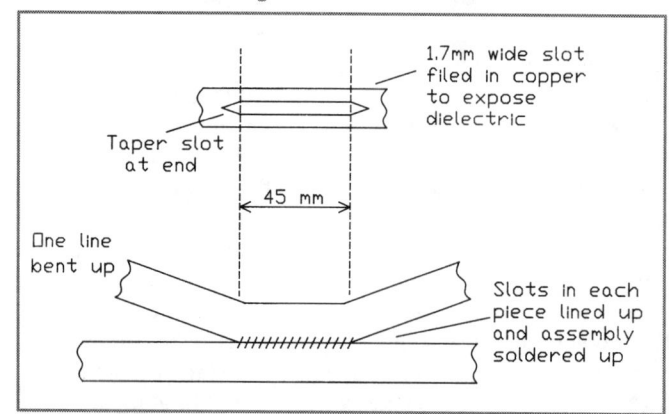

TRANSMITTER MEASUREMENTS

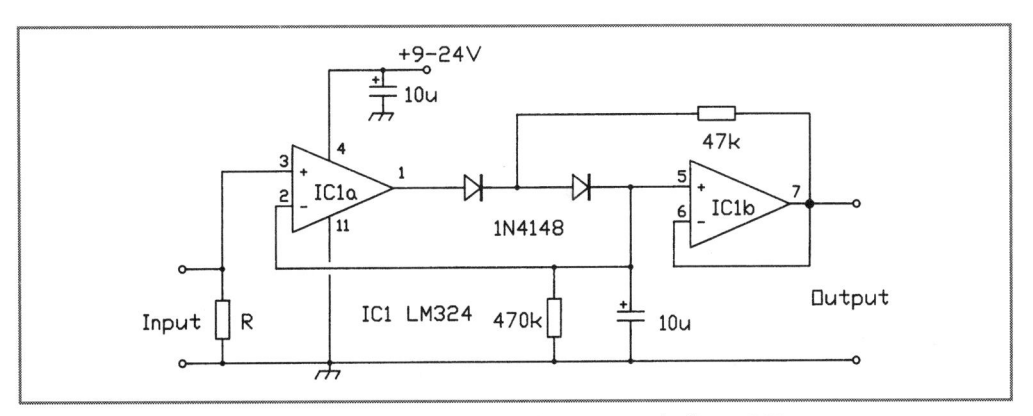

Fig. 12.22. Circuit to convert a wattmeter or VSWR meter to indicate PEP

average power (sometimes incorrectly termed RMS power). That's fine if you only want to measure the 'key-down' power on CW – place brick on key and take reading. But that method is not suitable for SSB where you want to read peak envelope power (PEP – for a definition see Chapter 6). Because of the peaky nature of the transmitted signal and the uncertain ballistics of meter movements, the PEP of an SSB signal cannot be read directly from the meter.

The circuit shown in Fig. 12.22 overcomes this problem. Simply place it between the forward-power detector of the wattmeter and the DC voltmeter part. Resistor R should be a reasonable approximation to the resistance normally seen by the forward-power detector diode, including the meter multiplier resistor. The circuit in Fig. 12.22 is a simple detector to register the peak value of a voltage which is fluctuating at an audio-frequency rate. The circuit has unity gain, so the existing meter calibration will still hold true. The decay time-constant of the detector is about 5 seconds, long enough to allow the meter needle to respond fully.

If the DC output voltage of the power sensor in your VSWR meter is less than about 7V, you can conveniently use a 9V battery; for higher voltage outputs, increase the supply voltage up to a maximum of 24V. Since the cheap and readily available LM324 op-amp is a quad package, you could build two of these PEP adapters on one board.

LINEARITY AND TWO-TONE TESTING

The reasons and methods for using RF wattmeters check the linearity of your transmitter or power amplifier have been covered in depth in Chapter 6. Making the assumption that I'm already preaching to the converted, I'll confine my remarks to the test equipment required. Graphs of output versus input power – what you might call static tests – can be done very simply using attenuators, a power meter and of course a dummy load, all of which I've already described.

Driving the transmitter with a two-frequency (or two-tone) signal source is the accepted method of testing a linear amplifier for intermodulation distortion under more dynamic conditions. If you really wish to measure only the performance of your linear amplifier, unsullied by the contribution of your exciter, the testgear requirements are quite ambitious. You'll need a pair of CW signal sources with a frequency separation of a few kHz, an attenuator after each CW source to help isolate it from the other one, and a hybrid power combiner. Isolation between the two signal sources is imperative – without it, intermodulation will occur in the signal sources themselves, ruining your measurement of intermodulation in the amplifier under test. Because of the attenuator that follows it, each CW source needs to generate quite a sub-stantial power output in order that the attenuated and combined two-frequency signal emerges at the correct drive level for the power amplifier. If you're prepared to contemplate all that, I reckon you won't need any more guidance from me!

For most of us, a two-tone intermodulation test means injecting a two-tone audio signal into the microphone input of the SSB exciter. This has the virtue of testing the entire transmitting setup, though individual sources of intermodulation can become very difficult to unravel. For amateurs, however, the main consideration is that a two-tone audio source is far easier to build than one based on two CW transmitters. The audio two-tone source I use (Fig. 12.23) is based on the design in the 1986 ARRL Handbook [23]; it works just fine, so I make no apology for recommending it here. As you see, it consists of two separate Wien-bridge oscillators and active low-pass filters, with a combining amplifier. Particularly good features are the inclusion of a step attenuator (calibrated in dB of course) and the ability to generate a single tone which allows static linearity testing as well.

The particular bugbear of Wien-bridge oscillators is the amplitude-stabilizing element – frequently either an expensive thermistor or an incredibly obscure low-current bulb. In this design, the necessary bulbs are readily available in many countries from Radio Shack or Tandy stores. It's useful to make the output socket a common DIN type, and to have a number of short adapter leads with microphone plugs to suit your various SSB rigs. Putting a PTT switch in the box is a good idea too.

COMPONENTS FOR TWO-TONE SOURCE (FIG. 12.23)

	700 Hz oscillator	1900 Hz oscillator
R1, R2	11k	470R
Adjust on test for exact frequency		
C1, C2	1n0 + 1n0	1n0 + 1n0
C3	2n7 + 2n7	1n0 + 1n0
C4	22n + 12n	12n
C5	100p + 39p	47p
Where two values are shown these should be paralleled. All the above capacitors should be polystyrene, 5% tolerance or better.		
B1, B2	Tandy (Radio Shack) 272-1141 12V 25mA bulb	
IC1-3	TL072 or TL082 dual op-amps	
RV1, RV2	470R or 500R 10-turn cermet trimmer	
RV3	470R or 500R panel-mounting potentiometer	
RV4	1k panel-mounting potentiometer	

Fig. 12.23. Two-tone source based on ARRL design

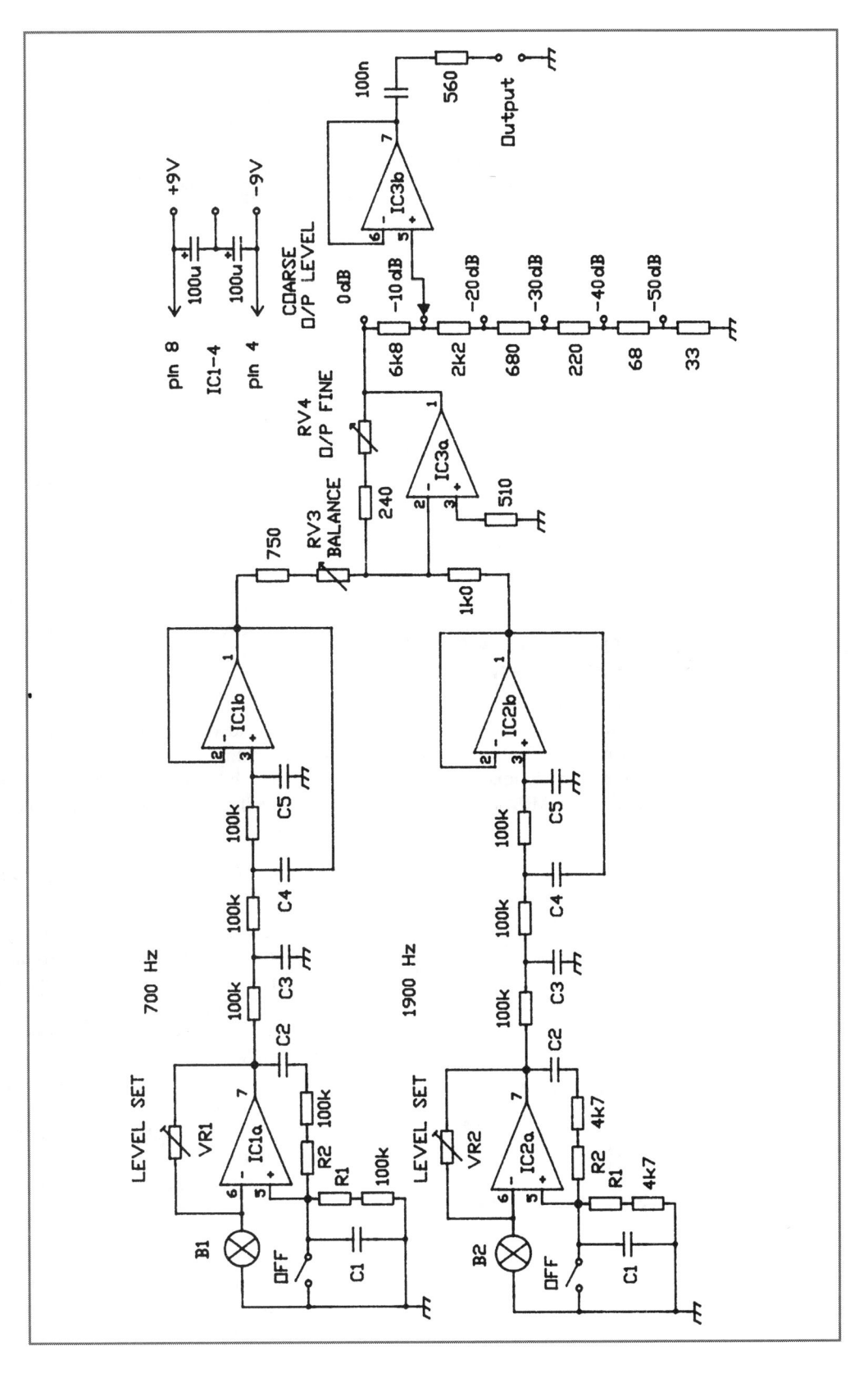

FILTERS FOR THE VHF AND UHF BANDS

In this section I'm going to describe a number of different applications of filters, with examples for each band. The accent will be on modern, reproducible designs. Most have been designed or analysed using computer programs; all have been built and tested to make sure that their performance is up to expectations.

BANDPASS FILTERS FOR LOW-POWER STAGES

The filters in this section are intended either for use in the receiver chain – for example, to provide front-end selectivity after a low-noise preamp – or after the mixer in a transmit converter. For 144MHz and 432MHz there's little to be gained by building your own filters since off-the-shelf Toko helical filters are available from several UK component suppliers, so I've concentrated here on bandpass filters (BPFs) for 50MHz and 70MHz. I've also included a design for 28MHz since equipment reviews show that the transverter outputs of many HF rigs will benefit from the extra selectivity to clean up spurious signals. All the BPFs shown in Fig. 12.24 are two-resonator Butterworth designs and have 50Ω input and output impedances so that they can be cascaded if necessary. Reference [24] describes a technique for tuning multi-resonator filters, which is particularly easy using the modulated-source testgear described earlier in this chapter.

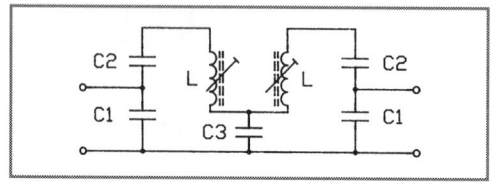

Fig. 12.24. Bandpass filters for 28, 50 and 70MHz

COMPONENTS FOR BANDPASS FILTERS

Filter	C1	C2	C3	L	
28MHz	220p	22p	470p	1·4µH	Toko TKXNS 22250N
50MHz	270p	27p	910p	0·39µH	Toko S18 violet
70MHz	220p	18p	680p	0·30µH	Toko S18 blue

PERFORMANCE DATA FOR THE FILTERS OF FIG. 12.24

Centre frequency (MHz)	Bandwidths, MHz			Insertion loss (dB)
	3dB	20dB	30dB	
28·5	2	7	10·5	2·5
50·5	2	5·5	9·5	2·2
70·2	2·5	6·5	12·0	2·5

LOWPASS FILTERS FOR TRANSMITTERS

I presume that you've already read Chapter 6, so you know why LPFs are so much better for harmonic suppression than the old-fashioned single high-Q circuit. Here is a family of seven-pole lowpass filters for the 50, 70, 144 and 432MHz bands, shown in Fig. 12.25. All should offer a second-harmonic rejection of at least 35dB, and a third-harmonic rejection of better than 60dB, together with very low insertion loss.

The accompanying components list shows coil and capacitor details. For power levels up to about 100W the metal-cased mica capacitors found in transistor PAs are a good choice. For high power on 50–144MHz use good-quality 1kV capacitors, possibly two in series or in parallel to share the heat losses due to

FILTERS FOR VHF & UHF

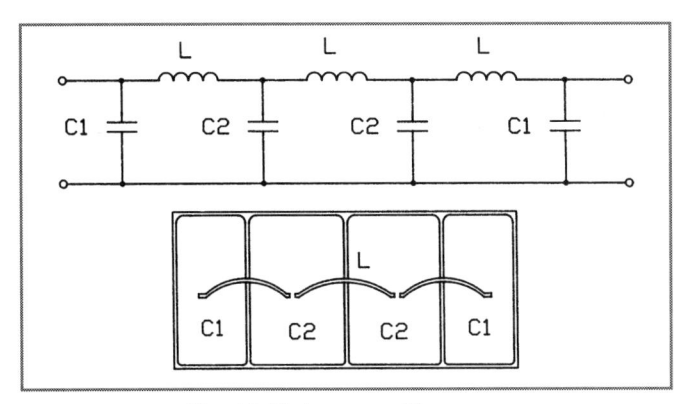

Fig. 12.25. Lowpass filters

COMPONENTS

50MHz:
- C1: 68p
- C2: 120p
- L: 160nH – 10 turns 0·5mm enamelled wire on T37-12 toroid; or 4 turns 1·5mm enamelled wire on 9·5mm diam. former, close-wound

50MHz with 100MHz notch:
As above, but replace one C2 with a series-tuned circuit of a 110nH inductor (see 70MHz details) and a 5-35pF trimmer. Tune for best suppression at 100MHz.

70 MHz:
- C1: 47p
- C2: 82p
- L: 110nH – 8 turns 0·5mm enamelled wire on T37-12 toroid; or 3 turns 1·5mm enameled wire on 9·5mm diam. former, close-wound

144 MHz:
- C1: 22p
- C2: 39p
- L: 55nH – 3½ turns 1mm wire on 5mm diam former, spaced wire diameter

432 MHz:
- C1: 5p6
- C2: 8p2
- L: 12nH As Fig. 12.25

Printed-circuit capacitors can be made from double-sided board as shown. C1 and C2 require areas of about 300 and 550mm² respectively for G10 (glass-epoxy) board; for Duroid 5870 (glass-PTFE) board these areas would be 630 and 1150mm²

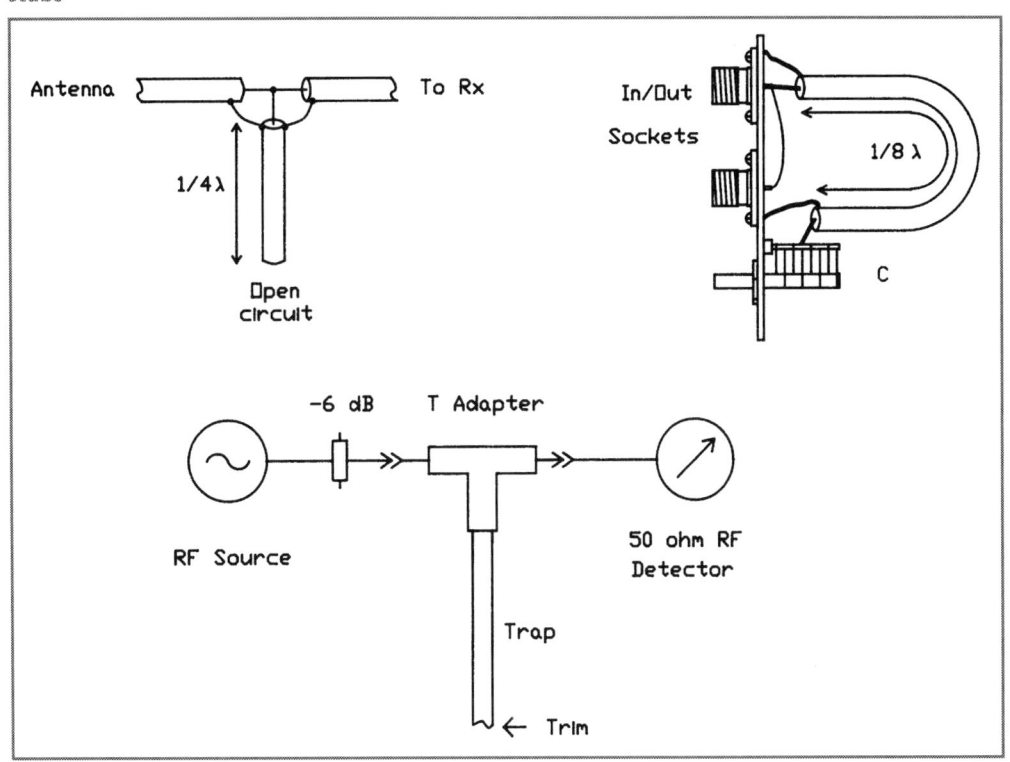

COAXIAL TRAP DETAILS
All dimensions assume the use of URM43 cable

Band	λ/4	λ/8	C(max)
50MHz	1·0m	0·5m	45p
70MHz	0·71m	0·35m	35p
144MHz	0·35m	0·17m	12p

Fig. 12.26. Notch filters using coaxial-cable stubs

the large circulating currents – and if they don't survive, they weren't "good quality"! For the 432MHz LPF, Fig. 12.25 gives an alternative layout using PCB capacitors.

Lowpass filters for the 50MHz band require extra suppression of the second harmonic, which falls within the FM broadcast bands. The component list in Fig. 12.25 includes a variation of the standard 7-pole LPF in which a 100MHz series tuned trap is substituted for one of the capacitors, giving an extra 20dB of rejection.

STUB NOTCH FILTERS FOR RECEIVER PROTECTION

These are equally applicable to domestic equipment and multi-band contesting. You might find them useful if you are unfortunate to live close to a commercial VHF transmitter site as well. They are made of coaxial cable and can be simply fitted with a T-adapter at the receiver input. Ordinary cable will give a rejection notch of 25–30dB; lower-loss cable will produce a resonator with a higher Q and will deepen the notch by a further 10dB or so. For increased rejection, it is possible to cascade these stubs by fitting them a quarter-wavelength apart along the main line. If you want to notch out a band of frequencies, several stubs – each of different lengths – can be connected so that their individual responses overlap. A 100MHz stub might be rather a good idea at the output of a 50MHz transmitter, in addition to the existing LPF. Fig. 12.26 shows the general idea – a simple quarter-wave line, open-circuited at its far end, connected across the receiver antenna input. Of course, you must allow for the velocity factor of the cable when calculating the length. In fact it's probably better to start with the cable a little on the long side and trim it *in situ*. Such stubs, by the way, will have a notch at every frequency where the stub is an odd multiple of a quarter-wavelength long.

Cutting cables to length can be a tedious business. One improvement to this technique is to make the stub electrically short, say $1/_8$th of a wavelength, and tune its far end with a trimmer capacitor [25]. Fig. 12.26 and the components list also give details for this alternative, together with suggested values for 50, 70 and 144MHz. To adjust any of these stubs, a test setup using equipment I've already described is also shown.

STATION ACCESSORIES

Here, to finish with, are a few ideas to make operating your thoroughly tweaked and tested station more pleasant and successful.

KEYERS AND ACCESSORIES FOR METEOR SCATTER

There are plenty of good memory keyers described in recent books and magazines, so I won't add to that list here. All I ask is that you spend some time analysing your operating habits – in other words, what you really do send time after time – before buying or building the memory keyer of your dreams. Unless you're a meteor-scatter or moonbounce addict, you'll probably find that the most common use for a keyer is merely to grind out your own version of "CQ CQ CQ DE G4PMK G4PMK G4PMK", or "CQ CQ CQ TEST" for contests. If so, a very simple keyer [26] with an EPROM memory will satisfy over 90% of your operating requirements. Most keyers will adapt quite easily to the addition of an EPROM message facility. It only remains to find your local friendly computer hacker with an EPROM burner.

If you are keen on meteor scatter, an EPROM memory is still useful for the unchanging messages such as "CQ MS G4PMK" and "RRRRRRRR G4PMK", in addition to the main read/write memory.

To read ultra-high-speed CW for MS operation, you naturally have to slow it down first (Chapter 3). The accepted way of doing this is to use a variable-speed tape or cassette recorder. MS signals are recorded at a faster tape speed than normal, and the motor is run more slowly than normal when the captured burst is replayed. Modifying a standard cassette recorder for this purpose is a usually a simple job, provided the motor has an electronically-regulated power supply. You merely have to doctor the potential divider which sets the speed, usually identified by a trimmer potentiometer. If the motor has a fixed mechanical governor, by the way, forget it and buy another tape recorder.

Since the necessary change in playback speed may be a factor of ten or more, a meteor burst recorded at normal pitch is replayed at a burbling 100–200Hz which doesn't make for easy copy. Unless you permanently shift the BFO for MS reception, or your receiver has an unusually wide IF shift, you can't pitch the incoming signal much higher than 2kHz without risk of losing it over the top edge of the IF filter. The most popular way around this problem is to use an audio up-converter. The audio signal fed to the tape recorder is shifted up in frequency by mixing it with a tone of about 7kHz, so the MS bursts are recorded at an audio frequency of 8–9kHz and played back at a comfortable 800–900Hz.

The popularity of the idea is due in no small measure to LA8AK, whose elegant implementation of this idea using a readily available consumer IC is shown in Fig. 12.27; this and earlier versions of his idea have appeared in several publications [27]. The TBA120 IC contains both the double-balanced mixer and the active elements for the 7kHz oscillator, which is pretty good going because this IC was originally meant to be an FM IF amplifier and discriminator! Setting-up merely consists of confirming oscillator operation by means of a

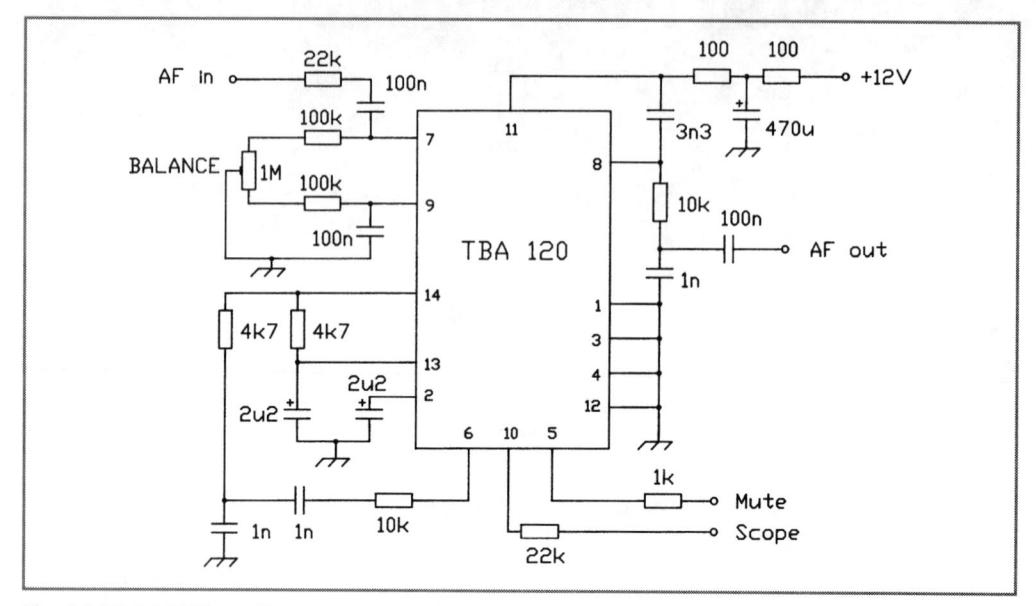

Fig. 12.27. LA8AK's audio up-converter for CW meteor-scatter

'scope connected to the test point and adjusting the balance preset for minimum 7kHz feedthrough with no AF input. Although the circuit specifies a TBA120, other suitable devices include the TBA120A, TBA120S, SO41P and SN76660N.

One unusual idea for MS recording is to turn the memory keyer idea on its head, so to speak. G4ASR converts the output from the receiver into a logic signal, which is clocked into the memory at a high rate. When a burst is heard, the loading is stopped, the clock-rate slowed down, and the output from the memory switched to a tone generator.

Still on the subject of keyers, it's now possible to digitize speech and store it in a memory for repeated replay with good audio fidelity [28]. Soon the 'SSB MS throat' will be a thing of the past, and the digital voice recorder can also take the strain of those endlessly-repeated "CQ contest" calls.

PIP-TONE UNIT

Used with common sense, a pip-tone at the end of a transmission can be a very positive aid to weak-signal SSB communication. It is most useful in conditions of deep fading or in quick-break meteor-scatter operation. The circuit shown in Fig. 12.28 is suitable for the majority of rigs, which have a push-to-talk (PTT) line sitting at about +12V on receive and needing to be grounded to transmit. The unit monitors the state of the PTT line and detects the rising edge when the PTT is released. It then pulls the PTT line back down to ground, holding the rig on transmit, injects a 1kHz tone into the microphone input line for about 30ms and finally releases the PTT. In many rigs the necessary connections to the PTT line, microphone input line and +12V power supply are all available at an accessory socket, so the pip-tone unit can simply be plugged into the back of the rig.

The unit is based on two CMOS logic ICs; a 4011 quad 2-input NAND gate (IC1a-d) and four elements of a 4049 hex inverter (IC2a-d). IC1a, IC1b and IC2a form a detector which monitors the state of the PTT line. On receiving a trigger pulse from IC1b, the monostable IC1c-IC2b delivers a 30ms pulse via IC2c, which pulls down the PTT line through TR1 and also enables the tone generator formed by IC1d and IC2d. The PTT sensing circuit is a little complex because it has to respond only to a deliberate release of the PTT switch. This is achieved by gating the rising-edge detector

STATION ACCESSORIES

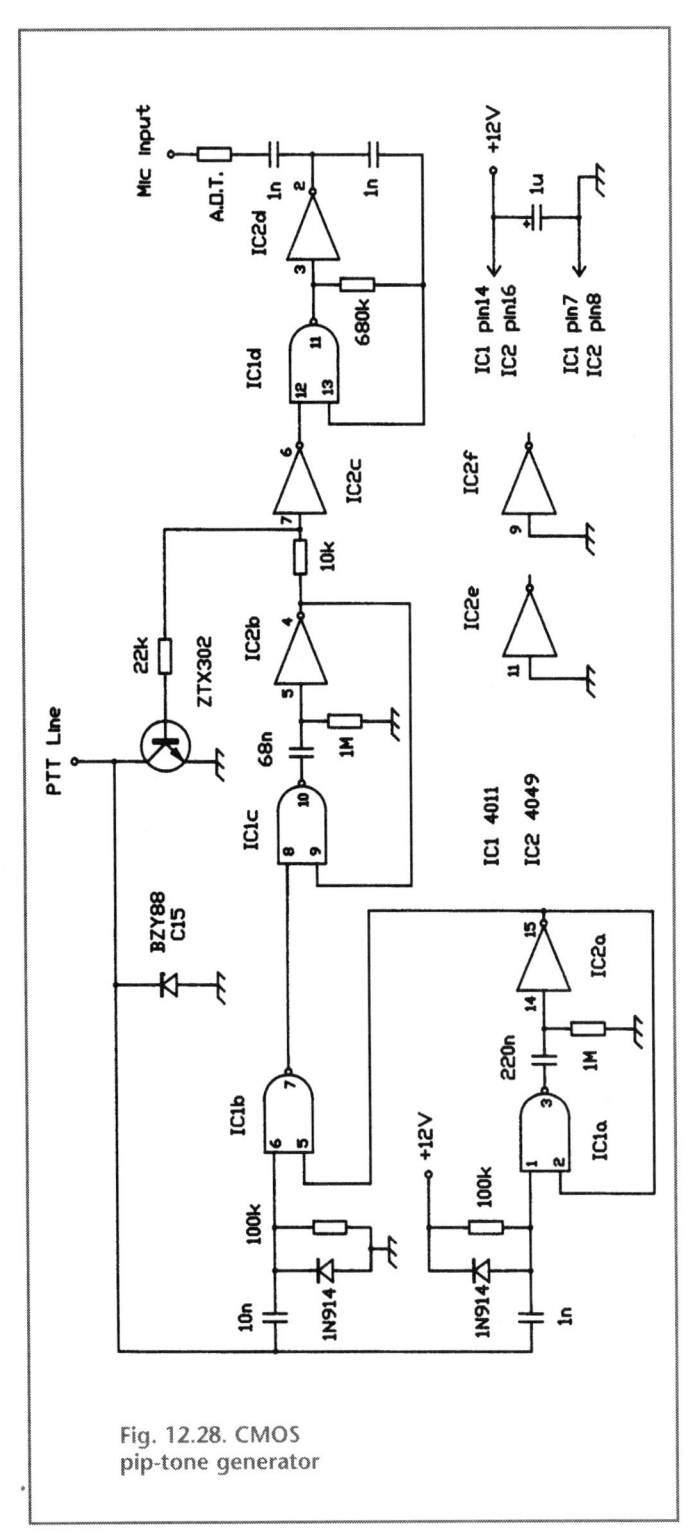

IC1b so that it will not respond for about 100ms after IC1a has detected a falling edge, the delay being set by monostable IC1a-IC2a. The net effect is that the detector ignores contact bounce after the PTT switch is closed, and it also ignores the rising edge when the unit itself releases the PTT line after generating the pip-tone.

Adjustment involves only setting the level of the tone to correspond to full modulation without overdriving the microphone input. The output of the unit is a square wave and you are relying on external filtering (ultimately the SSB IF filter) to clean up the tone. This works well enough in practice, especially if the entire audio signal then passes through some form of speech processor which includes further filtering. Grounding the input of IC2c will generate a continuous tone, and the value of the output resistor can then be adjusted with the aid of a separate receiver to make sure that the signal does not spread.

SIMPLE SPEECH CLIPPER

The virtues of speech clipping have already been loudly proclaimed in Chapter 6. Speech clipping has two functions: to increase the intelligibility or 'talk power' of your signal, and also as a positive method of preventing overdriving. Both functions require careful adjustment of the speech processor and the other audio and RF drive levels, or else the results will be worse than before! Many modern rigs include IF speech processing, which is generally the best kind because it permits a considerable degree of clipping without undue distortion. Add-on speech processors such as the Datong range convert the audio signal into SSB, clip and filter it, and then convert it back to audio; these too are highly recommended. However, even a very simple audio clipper will positively prevent overdrive, and will give a modest increase in talk power as a bonus.

The circuit of Fig. 12.29 is based on an old design by DJ4BG [29], but unlike the original is intended to be built permanently into the rig. An audio signal of a few hundred millivolts peak-to-peak is typically available at the

Fig. 12.28. CMOS pip-tone generator

TEST GEAR & STATION ACCESSORIES

microphone gain control, and this level is suitable for direct clipping by the long-tailed pair TR1 and TR2 which are biased as a symmetrical peak limiter. Ideally the transistors should be a matched pair or a dual transistor to provide symmetrical clipping of positive and negative peaks. Clipping is a form of distortion and produces copious intermodulation products and harmonics. Products which fall outside the normal audio passband are removed by the lowpass filter (TR3 and TR4) but those which fall within the audio passband cannot be removed in this manner; too much audio clipping will thus produce a heavily distorted signal with reduced intelligibility. If you want more talk power, you'll have to use a more advanced and complex speech processor such as the Datong – this simple AF unit is more for *our* benefit than for yours.

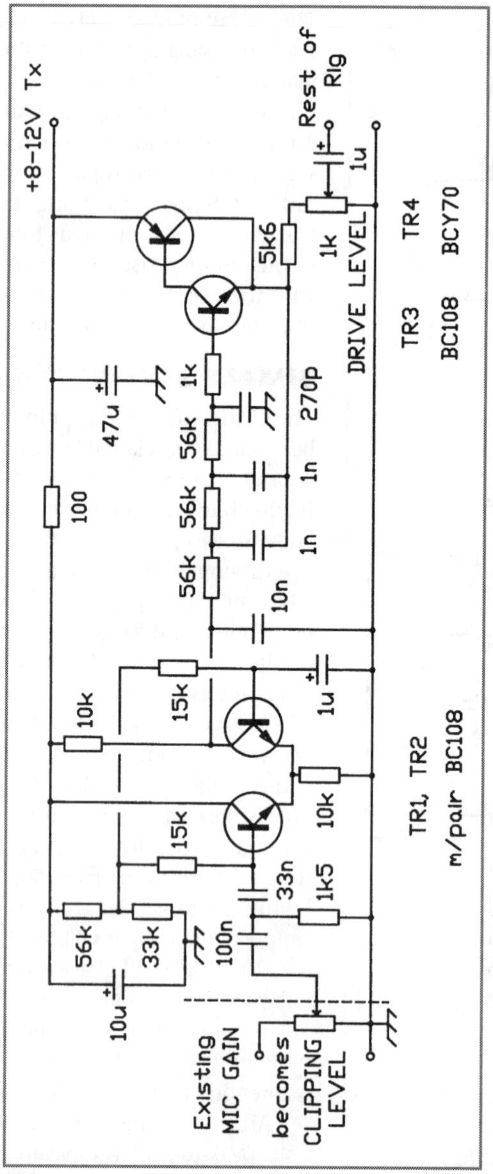

Speech clippers always need to be set up with great care and Chapter 6 gives full instructions on how to do this. The only preliminary alignment required for this module is to is to adjust its overall gain to unity, so that the following stages of the rig are not over- or under-driven. Feed a 1kHz tone into the module, at a level too low to suffer any clipping, and with the aid of a 'scope adjust the preset control to give equal signal levels at the input and output of the module. Then follow the general setting-up instructions in Chapter 6. In this particular arrangement the CLIPPING LEVEL control referred to in Chapter 6 is the rig's own microphone gain control which comes before the clipper module; the DRIVE LEVEL control is the rig's own drive control.

Fig. 12.29. A simple audio speech-clipper

ROTATOR SPEED CONTROL

Large VHF/UHF antenna arrays place great demands on the rotator. Even if your rotator has a substantial mechanism, the momentum which builds up while the array is in full swing can be quite alarming and the torque when it suddenly stops or reverses rotation may in time cause damage to the mechanism. Large arrays also bring reductions in beamwidth, especially at the higher frequencies, so aligning the array accurately on a weak signal can be very frustrating. If your rotator makes a 360° turn in 60 seconds, it takes a little under 2 seconds to sweep right though the 10° beamwidth of a high-gain antenna. The result is a lot of full-speed starting and stopping, and that's what damages the rotator.

One way of modifying your existing rotator to cope with both problems is to slow it down

STATION ACCESSORIES

using a pulse-width controller. All that's required is an add-on board with a couple of 555 timer ICs and a triac. The circuit of the controller is shown in Fig. 12.30, together with details of how it is connected to a typical rotator. A NORMAL/SLOW switch allows full-speed operation if required. The board can be small enough to fit inside the control box of most rotators, and you don't need to touch the tower end of things at all. The prototype, based on ideas in [30], has completely tamed the aiming of an eight-Yagi EME array for 432MHz.

FINIS

I hope you've found some ideas in this chapter to help you obtain the best performance from your station, and that you get as much satisfaction in making the measurements as you do from the results. Just keep a sense of proportion, and don't lose sight of your objectives. The ultimate aim is not to achieve perfection in the art of measurement, but to make your station as effective as you can – and then to use it for chasing DX!

Fig. 12.30. Rotator-speed controller. The original rotator components are shown inside the dashed line

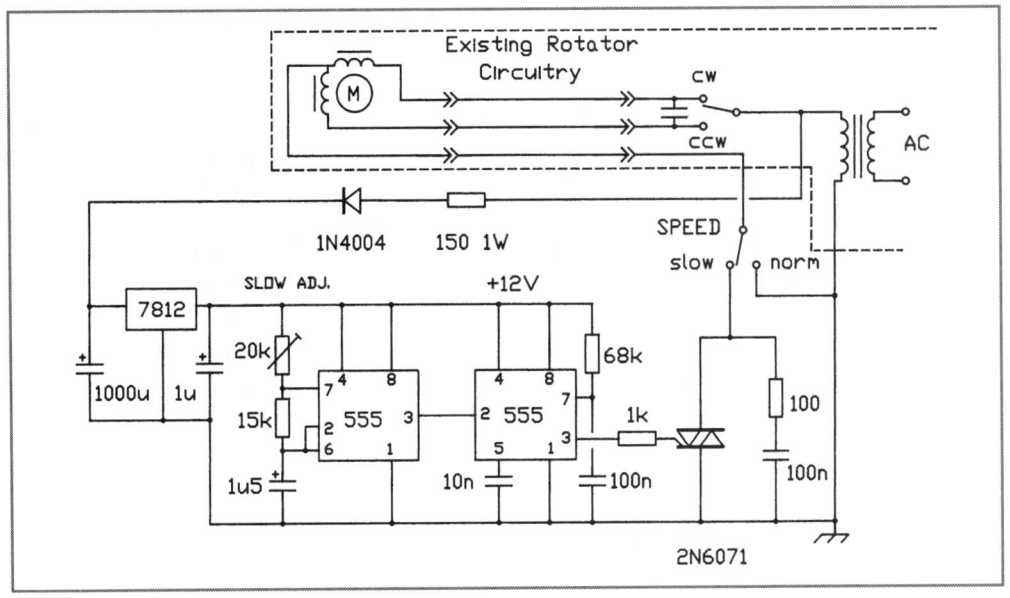

REFERENCES

[1] J. Grebenkemper KA3BLO, The Tandem Match – An Accurate Directional Wattmeter. *QST*, January 1987.

[2] A Milliwattmeter with Direct Power Calibration. *The UHF Compendium* Volume 3, Editor K. Weiner DJ9HO.

[3] R.E. Six KA8OBL, Wide-range RF Power Meter. *Ham Radio*, April 1986.

[4] J. Cowan W4ZPS, Frequency calibration using 60kHz WWVB. *Ham Radio*, March 1988.

[5] N.S. Hoult G4CIK, A Rugby MSF Time-Code Clock. *Radio Communication*, February 1979.

[6] *The UHF Compendium* Volumes 1 and 2, Editor K. Weiner DJ9HO.

[7] J.N. Gannaway G3YGF, The Effect of Preamplifiers on Receiver Performance. *Radio Communication*, November 1981.

[8] D. DeMaw W1FB, How to Build and Use a VHF Wattmeter. *QST*, December 1987.

[9] W. Unger VE3EFC, A Time Domain Reflectometer. *Ham Radio*, November 1983.

[10] Ian White G3SEK, Measurements on Modern VHF/UHF Front-Ends. *Radio Communication*, October and November 1986.

[11] F. Brown W6HPH, Antenna Gain Measurements. *QST*, November and December 1982.

[12] Ian White G3SEK, Modern VHF/UHF Front-End Design. *Radio Communication*, April-July 1985.

[13] J.R. Compton G4COM, An Alignment Aid for VHF Receivers. *Radio Communication*, June 1976. Re-published in *Microwave Handbook* Volume 2. RSGB, 1991.

[14] *Microwave Newsletter Technical Collection*, RSGB.

[15] H. van Leeuwen PA0DBQ and K. Joosse PE1BCQ, Automatische Ruisgetalmeter (Automatic noise-figure meter). *Electron*, July 1980 (In Dutch).

[16] R. Bertelsmeier DJ9BV, Novel Approach to Automatic Noise Figure Measurement. *DUBUS* 2/90.

R. Bertelsmeier DJ9BV and H. Fischer DF7VX, Construction of a Precision Noise-Figure Measurement System. *DUBUS* 2/90 and 3/90.

[17] R. Bertelsmeier DJ9BV, How to use a Noise-Figure Meter. *DUBUS* 4/90.

[18] Wes Hayward W7ZOI and Doug DeMaw W1FB, *Solid State Design for the Radio Amateur*. ARRL, 1977.

[19] Super Dummy and UHF Gallon Dummy. Chapter 34 of *The ARRL Handbook*, 1986-.

[20] Joe Reisert W1JR, VHF-UHF World – Impedance-Matching Techniques. *Ham Radio*, October 1987.

[21] R.E. Fisher W2CQH and R.H. Turrin W2IMU, UHF Directional Couplers. *QST*, September 1970.

[22] A Directional Coupler. *RSGB Microwave Newletter*, 04/82.

[23] A Two-Tone Audio Generator. Chapter 25 of *The ARRL Handbook*, 1986-.

[24] I.F. White G3SEK, A Simple Way to Design Narrowband Interdigital Filters. *Radio Communication*, February 1984.

[25] J.E. Fleagle W0FY, Traps for VHF Interference. *QST*, March 1988.

[26] D.J. Robinson G4FRE, An EPROM Keyer for Beacon Usage. *Radio Communication*, November 1986.

[27] Contributions by J-M Nöding, LA8AK to *Technical Topics. Radio Communication*, October 1981, September 1982 and November 1985.

[28] Martin Dearman, Digital Record and Playback Module. *Maplin Magazine*, February 1989.

[29] D.E. Schmitzer DJ4BG, Speech Processing. *Radio Communication*, May 1972.

[30] W.R. Gabriel WB4EXW, Pulse-Position Control of the CDE Tailtwister rotor. *Ham Radio*, January 1981.

INDEX

INDEX

A

A, Ap index	2-38
AB1, class	see Amplifiers
Absorption, radio waves	2-7, 2-10
ionospheric	2-50, 2-54
polar cap (PCA)	2-50
AF	see Audio
ALC	see Automatic level control
Alligators	4-13
Amateur radio	1
Amplitude, EM wave	2-4
AMSAT-UK	3-30
Amplifiers	
class AB1/C	6-18, 6-25
GaAsFET	4-3, 10-8, 10-25
MMIC	8-5, 8-7, 9-5, 9-23, 10-13, 10-15
Norton	10-7
power	**6, 11**
power, solid-state	6-14, 9-34
power, 50/70MHz	9-34
power, 144MHz W1SL	8-31
power, 432MHz K2RIW	10-27
see also Preamplifiers	
Analyser, spectrum	5-7, 6-3
Analysis	
intermodulation	5-14, 5-19
noise/gain	4-4, 5-3, 5-19
Antennas	**7**
array	7-9
backfire	7-12
bayed	7-8, 7-17, 7-36
broadside	7-9
capture area	2-19, 7-6, 7-8, 7-17
commercial	7-30
construction	7-24, 8-37, 9-39, 10-34
directivity	7-2
feed methods	7-5, 7-32
feedlines	7-32
gain	2-19, 7-2
height	7-14
helical	7-12
impedance	7-5, 7-36, 7-37, 7-39
isotropic	2-18, 7-2
log-periodic	7-12
making work	7-28, 9-39
noise	7-6
optimization	7-13, 7-16, 7-30
optimum choice	7-19, 7-30
pattern	7-3
parabolic	7-11
polarization	2-4, 2-10, 7-2, 7-4, 7-8, 7-9, 7-12, 7-20, 7-21, 7-39
radiation angle	2-16, 7-14, 7-15
rhombic	7-12
rotator controller	12-40
stacked	7-8, 7-17, 7-36
tuners	12-28
wave angle	see Antennas, radiation angle
Yagi	**7**, 8-37, 9-39, 10-34
Anticyclone	2-27
Aperture, antenna	see Capture area
Arcing	see Flashovers
Arrays	see Antennas
Atmosphere	
humidity	2-13, 2-27
Standard	2-14, 2-23
refractive index	2-13
Atoms	2-5
Attenuators	
50Ω	12-11
impedance-stabilizing	12-13
power	12-13

INDEX

Bold numerals indicate a reference to a whole chapter

preamplifier	5-24
standard	12-11
switched step	12-15
values	12-13
Audio	
level-meter, tuned	12-23
speech processor	12-39
two-tone source	12-31
Aurora	2-31, 3-13, 7-21
Auroral Es	2-47
Automatic level control	6-10, 6-12, 6-37, 6-39, 8-7, 8-10, 8-23, 9-8, 9-16, 9-24, 9-30, 10-16
Awards, operating	1-2, 1-3, 3-2

B

Backfire antenna	7-12
Backscatter	see Scattering
Balance	
antenna feed	7-37
signal handling	4-8, 5-2
transmit and receive	4-13
Baluns	7-37
Bandwidth	
antenna	7-5
aural	4-11
front-end	5-3, 5-6, 5-13
noise	4-3, 4-9, 4-11, 4-12, 5-17
receiver IF/AF	4-11
transmitted	6-3
Bases, valve	6-19
Bazooka balun	7-38
Beam antennas	see Antennas
Bias	
control-grid	6-28, 6-32, 11-15
gain compression	5-26
screen-grid	6-26, 6-32, 11-8
transistor PA	9-35
Bifilar windings	9-10
Blaster, good	6-2
Bleeder resistors	11-4
Block-diagram analysis	**4, 5**
Blocking	see Gain compression
Blowers	6-21
Boltzmann's constant	4-3
Boundary fence, auroral	2-37
Bridge	
return loss	12-18

VSWR	12-15
Broadside arrays	7-9
Burst	see Meteor-scatter
Butterworth filters	12-34

C

C, class	see Amplifiers
Cables, coaxial	7-32
Calibration	
attenuators	12-11
frequency counters	12-8
power meters	12-29
Calling frequencies	3-8, 3-10, 3-17
Callsigns, recognising	3-3, 3-4, 3-29
Capacitors	
smoothing	11-3
PA tuning	8-31, 9-34, 10-28
RF ratings	9-8, 9-31, 9-34, 12-28, 12-34
Capture area	2-19, 7-6, 7-8, 7-17
Choke-input	11-3
Chordal hops	2-51
Circular polarization	2-4, 7-12, 7-21
Clippers, speech	6-38, 12-39
Coaxial cables	7-32
Coherence	2-5, 2-8, 2-29, 2-30, 2-48, 2-53, 2-57
Colinear	see Antennas
Combiners, power	see Dividers, power
Compression, gain	5-6, 5-8, 5-26, 6-7
Computer programs	2-36, 2-50, 2-58, 2-61, 5-18, 7-13, 7-26, 7-28, 7-30, 8-2, 8-24, 10-6, 10-34, 12-2, 12-34
Conductivity	7-33
Connectors, coaxial	7-33, 12-11
Contact, definition	3-5
Contests	1-3, 3-12, 3-27
Control grid	
dissipation	11-17
protection	11-15
power supplies	11-15, 11-25
Convection cells	2-30
Conversion, frequency	see Mixers
Cooling	
transistor PAs	9-37
valve PAs	6-16, 6-21
Coronal holes, solar	2-33, 2-38
Correction, Yagi element length	7-26

ix

INDEX

Bold numerals indicate a reference to a whole chapter

Counters, frequency	12-7
Couplers, directional	12-15, 12-30
Coupling, PA output	see Loading
CQ calls	
answering	3-3
making	3-4
Cray 70MHz transverter	9-20
Critical frequency f0	2-16, 2-40
Crystal oscillators	see Oscillators
CW	
advantages	3-6, 3-11, 3-13, 3-15, 4-11, 4-12
keyers	12-37
operating	3-6
slowing down M-S	3-21, 12-37

D

D region, ionosphere	2-14, 2-38, 2-54
dB, dBm, dBW	see Decibels
dBc	5-9
dBd, dBi	7-2
Decibels	12-3
Delta match	7-39
Design	
antenna	**7**
for DX	1-4
power supply	**11**
receiver system	**5**
RF system	**4**
test equipment	**12**
transmitter	**6**
50 and 70MHz	**9**
144MHz	**8**
432MHz	**10**
Detectors	
50Ω RF	12-3
square-law	12-4
Diffraction	2-7, 2-12
Diodes	
mixer	see Mixers
Schottky	12-4
varicap	5-26
Zener	11-8, 11-30
Directional couplers	
calibration	12-30
semi-rigid coax	12-30
UHF PCB	12-15
Directivity	
beam antenna	7-2
couplers	12-15
Dissipation, anode	6-16, 6-24
Distortion	see Compression, Intermodulation
Dividers, power	
transmission-line	7-34
Wilkinson	10-13
Doppler shift	2-21, 2-32, 2-62
Driven elements	7-5, 7-36, 7-37, 8-38, 9-43, 10-35
Driver stages	6-10
Ducting, tropo	2-25, 2-27
Dummy loads	6-10, 12-28
DX, VHF/UHF	**1**
Dynamic ranges	5-12, 5-17, 5-19, 8-3

E

E region, ionosphere	2-15, 2-31, 2-39, 2-48, 2-56
Earth currents	2-39
Echos	
moon	1-2, 2-59, 4-12
long-delayed	2-63
Effective radiated power	4-10, 6-5, 6-12, 7-19
Effective receiver sensitivity	4-10
Efficiency, antenna	7-7, 7-30
Electric field	2-4
Electromagnetic compatibility	6-42
Electromagnetic waves	2-2
Electrons	2-5, 2-31
Electron oscillators	2-5
Elephants	4-13
EM waves	see Electromagnetic waves
EMC	see Electromagnetic compatibility
EME	see Moonbounce
Emission, secondary	6-26
Endfire arrays	7-11
Equipment	
Test	**12**
50 and 70MHz	**9**
144MHz	**8**
432MHz	**10**
Equivalent noise temperature	4-4
ERP, EIRP	see Effective radiated power
ERS, EIRS	see Effective receiver sensitivity
Es	See Sporadic-E

F

F region, ionosphere	2-15, 2-49

INDEX

Bold numerals indicate a reference to a whole chapter

F1, F2 layers	2-15
Fading	2-20
Faraday	2-62
libration	2-63
FAI	see Field-aligned irregularities
Faraday rotation	2-62
Feedback, RF	6-40
Feedline	
antenna	4-6, 4-10, 4-13, 7-32
coaxial	7-32
open-wire	7-36
Field	
alignment	2-48
electromagnetic	2-3
Field-aligned irregularities	2-48, 3-14, 7-20
Filters	
AC mains	11-5
IF	5-2, 5-13
power supply	see Smoothing
RF bandpass	8-5, 12-34
RF lowpass	6-8, 8-11, 9-8, 9-24, 12-34
RF notch	12-35
Flashovers	6-26, 6-40, 11-14
Flux, solar	2-50
FM signal, weak	5-25
Folded dipole	7-26, 7-28, 7-36, 7-39
Forward scatter	see Scattering
Free space	2-4
characteristic impedance	2-4
Frequency	
accuracy	12-8
counters	12-7
critical, f0	2-16, 2-40
Maximum Usable	2-16, 2-50
random MS	3-17
setting	3-7
standards	12-9
upconverter, AF	12-38
Frontal systems	2-27
Front-end, receiver	**5**
Fuses	11-4

G

Gain	
antenna	4-9, 7-2
antenna stacking	7-8, 7-17, 7-36
compression	5-6, 5-8, 5-26, 6-7
cumulative	4-5
ground	2-61, 7-14
measurement	12-22
power	4-9, 12-3
Gamma match	7-38, 9-43
Grid	see Screen grid, Control grid
Ground	
gain	2-61, 7-12, 7-14
noise	4-7, 4-15, 7-7
Grounding, safety	11-8
Group velocity	2-6

H

Harang discontinuity	2-35
Harmonics	
distortion	5-6
filters	12-34
transmitters	6-6
Hearing	4-11
Heat sinks	8-14, 9-14, 9-25, 9-37, 10-20
Heater supplies	11-17
Height, virtual	2-15
High-voltage (HV) supplies	11-2
transformer selection	11-3
transformer protection	11-4
Horizon, radio	2-23
Humidity	2-13, 2-27
HV	see High voltage

I

IF filter	5-2, 5-13
Incidence, angle of	2-8
Incoherent	see Coherence
Intermodulation	
analysis	5-14, 5-19
receivers	5-6
transmitters	6-3
Intermodulation intercepts	5-14
estimating	5-26
Impedance	
antennas	7-5, 7-36, 7-37, 7-39
matching	7-26, 7-32, 12-28
measurement	12-11
noise sources	12-26
return loss / VSWR	12-12
stabilization	12-13
transmission line	7-32

INDEX

Bold numerals indicate a reference to a whole chapter

Improving station	4-13
Injection, local-oscillator	8-5, 9-5, 9-21, 10-10
Interfacing	
power amplifier control	11-28
transverter to HF transceiver	8-26, 9-2, 9-30, 10-3
Interference	
amateur-band	5-6, 6-2
to broadcasts	2-28, 2-65, 6-6, 6-42
Inverse-square law	2-18
Inversion, temperature	2-25
radiation	2-25, 2-27
subsidence	2-27
Ions, ionization	2-5
Ionosonde	2-15, 2-40
Ionosphere	2-14
Ionospheric scattering	2-54
Isoflectional map	2-36, 2-48

K

K, Kp index	2-38
Key clicks	6-8
Keyers	12-37

L

Legal power limit	6-12, 6-38
Level-meter, tuned AF	12-22, 12-23
Libration fading	2-63
Light, velocity	2-4
Light waves	2-4
Lightning-scatter	2-64
Linear amplifiers	see Power amplifiers
Linearity	
two-tone tests	12-31
see Intermodulation, Gain compression	
Link coupling	8-34, 8-36
Loading, PA output	6-35, 8-34, 10-28
Loads	
50Ω standards	12-11
dummy	12-28
Local oscillator (LO)	5-7
Locator, IARU QTH	3-2, 3-3, 3-29
Loops, phase-locked	see Oscillators
Losses	
attenuators	12-11
cables	7-32
dielectric	7-32
measurement	12-22
noise from	4-6
path	2-18, 4-10
skin-effect	7-33
Lowpass filters	6-8, 8-11, 9-8, 9-24, 12-34

M

Magnetic field	
EM waves	2-4
indices	2-38
interplanetary	2-32
Magnetic poles	
earth and sun	2-33
Magnetic storm	2-31
Magnetometers	2-38
Magnetosphere	2-33
Magneto-tail	2-33
Matching	see Impedance
Maximum Usable Frequency	2-16, 2-50
Measurements	4-13, 5-20, **12**
Meteors	2-55
Meteor-scatter	2-55, 3-15, 7-20, 12-37
Meteor showers	2-56, 3-25
Meteor trails	2-55
Millivoltmeter	
RF 50Ω	12-5
RF probe	12-6
Milliwattmeter, RF	12-6
Mixers	
diode-ring	5-26, 8-5, 8-7, 9-5, 9-6, 9-22, 9-24, 10-10, 10-15
overload	5-15, 5-26, 8-7
Mixing, reciprocal	5-6, 5-11, 5-17
MMICs	see Amplifiers
MSF	12-9
Modulator, 1kHz	12-25
Modular amplifiers	see Power amplifiers
Moon, motion	2-59
Moonbounce	
antennas	7-21
HF net	3-29
propagation	2-59
operating	3-22
single-Yagi	4-12
Morse code	see CW
MS	see Meteor-scatter
MUF	see Maximum Usable Frequency

INDEX

Bold numerals indicate a reference to a whole chapter

N

Neighbours, psychology	6-42
Nets	
HF liaison	3-28
telephone warning	3-26
Neutralization	6-29
Newsletters	3-30
Noise	
digital	5-10
factor	4-6
figure (NF)	**4**, **5**, 8-3, 8-24, 9-2, 9-20, 10-3, 10-25, 12-25
floor	4-9, 5-12, 5-14, 5-19
galactic	2-60, 4-3, 4-7, 4-15
generator	12-25
ground	4-7, 4-15
man-made	4-7
oscillator	5-8, 5-17
pedestal	5-9, 5-11
phase	5-9
power	4-3
sidebands	5-9, 5-11, 5-17
thermal	4-2
sky	2-60, 4-7, 4-15
sun	4-7, 4-15
Noise figure (NF)	**4**, **5**, 8-3, 8-24, 9-2, 9-20, 10-3, 10-25
measurement	12-25
Noise/gain analysis	4-4, 5-3, 5-18
Noise temperature	**4**, **5**
antenna	4-7, 4-8, 7-6
measurement	12-25
sky	2-60, 4-7, 4-15
system	4-4
target	4-8, 5-3, 8-3, 8-24, 9-2, 9-20, 10-6
Norton amplifiers	10-7
Notch filters, RF	12-36

O

Obstacles, reflection	2-16
Obstacle gain	2-16
Open-wire feedline	7-36
Operating	**3**
Oscillators	
AF two-tone	12-31
crystal	5-2, 5-9, 8-6, 9-5, 9-21, 10-11
electron	2-5
noise	5-8
synthesized	5-9
VCO	5-10
VFO	5-2, 5-9
Oval, auroral	2-31, 2-33
Overdense	2-8, 2-57
Overdriving, transmitter	6-4, 6-7
Overload, receiver	**5**

P

Path loss	2-18, 4-9, 4-10
capability	2-18, 4-9
free-space	2-18
moonbounce	2-59, 4-12
with reflection	2-19
troposcatter	2-29
Pattern, antenna	7-3
Peak envelope power	6-5
meter converter	12-30
PEP	see Peak envelope power
Phase, EM waves	2-4
coherence	see Coherence
Phase-locked loops	5-10
Phase velocity	2-4, 2-6
Phasing, antennas	7-34, 7-36
Phonetics	3-7
Pile-ups	3-3, 3-5, 3-9, 3-11, 3-14, 6-12
PIN diodes	8-7, 9-5, 9-23, 12-25
Ping	see Meteor-scatter
Pip-tone	3-16
accessory	12-38
Plasma	2-14, 2-32
PLL	see Phase-locked loops
Polarization	2-4, 2-10
circular	2-4, 7-12, 7-21, 7-39
horizontal	2-10, 7-2, 7-8, 7-9, 7-12
linear	2-2, 2-4
rotation	2-62
Power, RF	
EM waves	2-4
gains and losses	4-9, 12-3
low (QRP)	4-13, 6-12
high	4-13, 6-12
legal-limit	6-12, 6-38
measurement	12-29
peak envelope	6-5, 12-30
Power amplifiers	**6**

INDEX

Bold numerals indicate a reference to a whole chapter

modular	8-10, 9-6, 10-19
transistor	6-6, 9-24, 9-34
valve	**6**, 8-31, 10-27, **11**
Power dividers	7-34
Power supplies	**11**
Preamplifiers	5-21, 12-22
setting gain	5-25
Prescalers	12-7
Procedures	see Operating
meteor-scatter	3-18
moonbounce	3-23
Probe, RF	12-6
Processing	see Speech processing
Programs	see Computer programs
Propagation	**2**
auroral	2-31
F2-layer	2-49
ionospheric scatter	2-54
media	2-13
meteor-scatter	2-55
modes	2-22, 2-65
moonbounce	2-59
sporadic-E	2-39
trans-equatorial	2-52
tropospheric	2-23
Protection	
GaAsFET	10-25, 11-29
power amplifier	6-30, 11-23
HV transformer	11-4
personal	see Safety warnings
Protons	2-32
PSU (power supply unit)	see Power supplies
PSCU (power supply/control unit)	see Power supplies

Q

Q, circuit	6-40, 12-28
Quality	6-2, 11-34
QRP	see Power, low
QSL cards	3-6, 3-9
QSO	see Contact

R

Radiation	
electromagnetic	2-3
solar	2-31
Radiation inversion	2-25, 2-27
Radio waves	2-2
Random meteors	2-56
Range, maximum	2-15
aurora	2-37
geometric	2-15
meteor-scatter	2-59
sporadic-E	2-43
tropo ducting	2-28
Ranges, dynamic	See Dynamic ranges
Receivers	**5**
measurements	12-22
Reciprocal mixing	5-6, 5-11, 5-17
Rectifiers, HV	
biphase or bridge	11-3
circuit	11-5
diodes	11-3
Reflection	
radio waves	2-7, 2-9, 2-19, 2-64
standing waves	12-12
Reflectometer	
Simplest TDR	12-19
Reflectors	
antenna	7-11, 7-12
natural	2-64
Yagi	7-13, 7-24
Refraction	2-7, 2-8
angle of	2-8
Refractive index	2-8
atmosphere	2-13
radio	2-13
Regulation, voltage	6-27, **11**
Relays, coaxial	11-28
Return loss	12-12
bridge	12-18
Rhombic antenna	7-12
RF interference	6-42
Ring mixers	see Mixers
Rotator speed-controller	12-40

S

S-meter	2-19, 5-17, 5-22
Safety warnings	2-38, 6-32, 9-37, 11-3, 11-8
Satellites, amateur	1-5, 7-21, 10-3
Saturation	see Gain compression
Scattering	2-7
auroral	2-31
ionospheric	2-54
tropospheric	2-29

INDEX

Bold numerals indicate a reference to a whole chapter

Screen grid	
dissipation	11-13
protection	11-14
power supplies	6-26, 11-2, 11-11
secondary emission	6-26
Secondary electrons	2-3
Sector boundaries, solar	2-33, 2-38
Sequencing, transmit/receive	11-28
SID	see Sudden Ionospheric Disturbance
Signals	
weak	1-3, 3-3, **4**, 5-2
strong	**5**
Signal/noise ratio	4-8, 5-25
Signal sources	
AF two-tone	12-31
noise	12-25
RF modulated	12-25
RF strong-signal	12-27
Signal-strength meter	see S-meter
Skin effect	7-33, 7-38
Skip, ionospheric	2-16
Sky noise	2-60, 4-7, 4-8, 4-15, 7-6
Sleeve baluns	7-38
Smoothing	11-3
Snell's Law	2-8
Software	see Computer programs
Solar flux	2-50
Solar wind	2-32
Solo, Han	6-2, 6-28
Spectrum analyser	5-7, 6-3
Spectrum, electromagnetic	2-3
Speech processing	6-38, 12-39
Splatter	6-4, 12-40
Sporadic-E	2-39, 3-14, 7-20
auroral	2-39, 2-47
formation	2-45
maximum distances	2-43
multi-hop	2-44
temperate-zone	2-39
times of occurrence	2-40
Sporadic meteors	2-56
Spurious signals	see Harmonics, Intermodulation
Stacking, antennas	7-8, 7-17, 7-36
Standard Atmosphere	2-14, 2-23
Standing wave ratio	see VSWR
Station	
assembling	**4**

improvements	4-13
Step attenuator	12-15
Sub-refraction	2-23
Subsidence inversion	2-27
Sudden Ionospheric Disturbance	2-50, 2-54
Suffolk 144MHz transverter	5-3, 5-16, 5-19, 8-2
Sunspots	
cycle	2-49
number	2-50
Super-refraction	2-23
Switching, transmit/receive	11-28
SWR	see VSWR
Synthesizers	see Oscillators

T

T-match	7-39
TCALC program	5-18, 8-2, 8-24, 9-4, 10-6
TDR	see Time-delay reflectometer
TEP	see Trans-equatorial propagation
Telephone nets	3-26
Temperature	
noise	see Noise temperature
ceramic seals	6-21, 6-24
Test equipment	**12**
Tetrodes	6-16
Thermal noise	see Noise
Thunderstorms	2-45, 2-64
Tilts, ionospheric	2-16, 2-51
Time, Universal Coordinated (UTC)	3-16
Time-delay reflectometer (TDR)	12-19
Toroids, winding	9-10
TR	see Transmit/receive
Transceivers	
VHF/UHF	5-2, 5-22
HF	5-2, 5-8, 5-23, 8-3, 8-26, 9-2, 9-30, 10-3
Transequatorial propagation (TEP)	2-52
Transformers	see HV, Toroids
Transformation	see Impedance
Transmit/receive control	11-28
Transmitters	**6**
Transverters	1-4, 5-2, 5-23
50MHz	9-2
70 MHz *Cray*	9-20
144MHz *Suffolk*	5-3, 5-16, 5-19, 8-2
432MHz	10-2
Trifilar windings	9-10
Triodes	6-17

INDEX

Bold numerals indicate a reference to a whole chapter

Troposcatter	2-29, 7-19
Troposphere	2-13
Tropospheric propagation	2-23, 3-10, 7-19
backscatter	2-30
ducting	2-24
forward scatter	2-30
side-scatter	2-29
Tubes	see Valves
TV	
interference	6-42
interference warnings	2-28, 2-65
PAL timebase	12-10
Two-tone testing	6-3, 12-31
TX/RX	see Transmit/receive

U

Universal Coordinated Time (UTC)	3-16
Unbalance, antenna	7-37
Underdense	2-8, 2-57

V

Valves, transmitting	6-14
Varistors	see VDRs
VDRs	6-30, 8-32, 10-32, 11-5, 11-9
VCO	see Oscillators
Velocity	
radio waves	2-4
group and phase	2-6
VFO	see Oscillators
VHF nets (on HF)	3-28
Virtual height	2-15
Voltage-controlled oscillator (VCO)	see Oscillators
Voltage-dependent resistors	see VDRs
Voltmeters, RF	12-3, 12-5
VSWR	
definition	12-12
bridge, 28-144MHz	12-16
bridge, 50/70MHz	12-17
bridge, UHF	12-16

W

Water vapour	2-13, 2-27
Wattmeters, RF	12-3, 12-6, 12-16, 12-29
Waves	
radio, electromagnetic	2-3
gravity	2-45
standing	12-12
Wavelength	2-4
Weather	2-27
Wind shear	2-46
WWVB	12-10

Y

Yagi antennas	**7**, 8-37, 9-39, 10-34
DL6WU designs	7-24, 7-30, 7-39, 8-37, 10-34
Year planner, DXer's	3-25

Z

Zen	6-2, 11-34